Geography of Deprivation

By the Same Author

- Geography of Himalayan Kingdom: Bhutan
  with Neil Fraser, Anima Bhattacharya
- Urban Development in India: Supce Pre- Historic Times
- Urbanization, Urban Sustainability and the Future of Cities

About the author

B. Bhattacharya, educated at Calcutta University and at the London School of Economics and Political Science (as a Commonwealth Scholar). B. Bhattacharya has consistently pursued a career of teaching and research, having taught in many institutions, including Gauhati University (Assam), Indian Institute of Technology (Kharagpur IIT, W. Bengal), Presidency College (Kolkata), and the University of North Bengal (W. Bengal) at the last of which he had been a Senior Professor, Head of the Department as well as the Dean of the Faculty of Science. His research in Geography spans a large area with publications in academic Journals in India and abroad, and a number of books including 'Samaj Bignaniya Bhugol' (Social Geography in Bengali) jointly with A. Bhattachara published by the West Bengal Government, 'Urban Development in India since Pre-historic Times' (New Delhi), 'Sikkian-Land and People' (New Delhi), 'geography of a Himalayan Kingdom : Bhutan, in joint authorship with A. Bhattacharya and Neil Fraser (New Delhi), and 'Urbanization, Urban Sustainability and the Future of Cities' (New Delhi).

Geography of Deprivation

An Unfair World

Bimalendu Bhattacharya

Formerly, Professor and Head of the Department of Geography
and Applied Geography, and Dean of the Faculty of Science,
University of North Bengal

CONCEPT PUBLISHING COMPANY PVT. LTD.
NEW DELHI-110059

ISBN-13: 978-81-8069-993-1 (HB)
ISBN-13: 978-93-5125-031-9 (PB)

First Published 2014

Published and Printed by

Concept Publishing Company Pvt. Ltd.
A/15-16, Commercial Block, Mohan Garden
New Delhi-110059 (India)
T: 25351460, 25351794, *F*: 091-11-25357109
E: publishing@conceptpub.com, W: www.conceptpub.com
Editorial Office: H-13, Bali Nagar, New Delhi-110 015, India.

Cataloging in Publication Data--*Courtesy:* D.K. Agencies (P) Ltd. <docinfo@dkagencies.com>

Bhattacharya, Bimalendu, 1931-
Geography of deprivation : an unfair world / Bimalendu Bhattacharya.
p. cm.
Includes bibliographical references (p.) and index.
ISBN 9788180699931 (hb.)
ISBN 9789351250319 (pbk.)

1. Poverty. 2. Hunger. 3. Malnutrition. I. Title.

DDC 339.46 23

Dedicated to

Monai (Rupsha), Bebu (Bashabi) and Neil
for their unfailing love and encouragement

importantly, lack or inadequacy of such facilities like clean drinking water, healthcare, education, etc, has a severe effect on the life of these people and thus perpetuating their poverty as fragile health and lack of education/training together greatly enhance their incapability in improving their condition, reducing further their entitlement to immediate needs, including food, making "them chronically poor" , as defined by the United Nations (2007), amounting to 420 million living below the poverty line for more than five consecutive years. A situation like this is redeemable only by taking necessary measures for providing the services without discrimination since poverty has a tendency to bounce back if not checked or restrained by improving 'human capital' with education/training, healthcare along with clean drinking water, sanitation as sustaining measures, all of them bearing serious implications in fighting poverty as the ultimate goal.

Undoubtedly, poverty as the worst form of human sufferance, a complete denial of human dignity is still flourishing in an unequal world in the absence of an effective approach to its final abolition.There are promises, strong commitments, goals defined, targets taken and methods discussed at the international levels for reducing and finally doing away with poverty from the face of the earth. Progress achieved so far in this regard is, in a way, quite impressive in some areas, the United Nations sustaining the hope of reaching its target of halving poverty and hunger by 2015, but in reality doubt still persists about its feasibility when almost 925 million (or 923 million people in 2007 worldwide according to FAO 2008) remained hungry in 2009, and there has been no progress in the reduction of hungry people in 28 countries till date, in mid-2010. One of the major reasons behind the increase in the number of hungry people in recent years is cited as the rise of food prices across the world triggered by factors both beyond human control, such as unprecedented droughts, wild fires as in Russia causing large damage to wheat crops, catastrophic floods, etc. and those factors within human control, namely the rising prices of petroleum leading to the conversion of more land to the production of crops used for increasing the production of ethanol causing reduction in food production and consequent rise in food prices denying access of many people to food available in the market.

The worsening situation can be improved by enhancing the capability of the poor through education/training and healthcare making them better citizens offering better human capital for widening the scope of higher income as a gateway to greater access to basic needs. In a highly competitive market where working opportunities as the only way to earn a living are not expanding equal to the rate of demands always rising, and more in youthful societies of the developing world – India, for example – education/training and a good health act as the deciding factors of higher eligibility. People lacking in them are left behind in this struggle of existence with their requirements unfulfilled and remain steeped in unending poverty with hunger as an unavoidable condition. Offering financial help is no effective solution to the situation as it is an extremely temporary measure without making any fundamental change in the people's condition providing them with proper/greater entitlement or access to food or the basics of life on a permanent basis which are available only through people's greater marketability by way of offering labour having saleability in any economic system. Alarmingly, though 90 per cent of countries under the stress of poverty have made progress in making primary education universal, the pace of progress, so far, is not adequate to make the target universal, more so because of higher rates of dropouts caused by various factors, more economic than social. Similarly, achievement in upgrading health is far from satisfactory when the target of reducing both infant and child under five years old mortality rates and maternal mortality ratio by two-thirds looks like unachievable by 2015, keeping a large section of people, particularly in developing countries, very ill-equipped for facing the challenge of poverty.

Poverty and hunger as the prime deterrents to human progress do not find an achievable solution as long as factors hindering their progress cannot be rooted out and prioritised to be dealt with with unflinching determination and complete honesty. The responsibility lies largely with respective national governments, public institutes, the international bodies providing funds and expertise, and the rich nations, all fully aware of the forebodings of a disastrous future unless the remorseless plunder of the natural resources of the earth, worth 1.5 times the planet's capacity (according to the data compiled by the Zoological Society of London brought out in the latest biennial 'Living Planet Report, 2010) getting drained away for meeting the unbridled lust for consumption by man - having doubled in the last 40 years or so—is drastically checked without further preamble. This seems to be the only way to avoid the inevitable clash of interests if inequality between one and another world, containing extremely dissimilar proportions of world population, continues.

To grasp the real implications of the growing challenge of remaining alive for a considerable section of the world population located in many of the developing countries (when even the developed world cannot be fully exempted from this bleak prospect) – one of them being India—one needs to visualize the situation in which the sheer insecurity makes life highly uncertain for multitudes of them living even above or at the margin of the poverty line. A striking example in this regard is provided by domestic helps found in all the big cities of India (and also most of the urban and some rural centres), where comprising mostly women, they work with a family or families (household/households) generally as part-time helpers or in other capacities paid as a common practice by monthly salary on the basis of oral contract (between the employer and the employed) without any written validity. Interestingly, as a general rule, there is no fixed rate for salary, not usually based on a scale with any prospect of annual increment and retirement benefits (or fringe benefits).The recruit works usually seven days a week with no fixed leave during the entire tenure of service, and works for the hours fixed at the time of appointment running the risk of getting sacked without prior notice, or for prolonged illness without any terminal benefit. An extremely uncertain situation like this faced by millions of women (as well as men) engaged in domestic services as the only means of livelihood for the entire family (when in many cases the husband does not work on various pleas in a highly patriarchal society) makes life highly vulnerable in the complete absence of any social security to tide over the period of unemployment, death of the earning member, serious illness in the family, or at times of getting dislodged due to natural calamities, or even through eviction in the name of 'development' or 'cultural issues using the space by the government for holding Olympic games or such exhibitionism'.

Even the job they do as domestic help is pathetically inadequate considering the fact that they are invariably very poorly paid. To take up a case as an example, I know a young woman , let us call her Mandira, about thirty years old, working in three households as a part-timer in each, who earns Rs. 3000 (about £40) altogether a month of which nearly a quarter is spent on house rent and electricity (in an inexplicably poor living condition sharing bath and latrine as well as drinking water, the latter if at all exists with the rented house) while the rest of her earnings go into feeding the family of four, education of two school-going boys, non-working husband (though physically fine, but following the example of many other non-working males of his class), and paying for her railway monthly ticket for travelling daily from the suburb (besides financially helping her mother and married sister at times of distress). Interestingly, Mandira has a little education (her husband is illiterate), is highly intelligent and could do much better in life if she had the opportunity but contrastingly leads an extremely vulnerable life which breaks down any moment if she loses her ultra-temporary jobs in a highly competitive market, a highly likely prospect for a woman like her working to exhaustion as getting up very early from bed she has to cook for the family and leave without breakfast to catch the early train for avoiding being late for three successive jobs, returning home quite late taking a very much crowded train with little energy left to care for the family though that is another compulsion she cannot avoid. Unquestionably, the existing net of social security does not offer any protection to people like Mandira and others, exposing the million-faces to a highly unpredictable future verging on the brink of imminent poverty and destruction of family-fabric.

A desperate situation like this, prevailing across the world for millions cannot be improved and poverty avoided without taking the challenge with an ungrudging aim of accomplishing a drastic improvement in the entitlement as well as eligibility of the people left aside by the society in its onward march making life more comfortable for others and this can only be done when education and healthcare are based on a rightful accessibility to basic services, such as improved sanitation, enhanced facility of safe drinking water and the like, having cumulative effect on factors like infant and children under five mortality rates, fertility rates and maternal mortality ratios unhinging human confidence in life when families are already in the risk of untimely death under the constant threat of starvation and poor health.

Arguably, educating children up to primary and secondary levels along with necessary training for acquiring proper skill is itself a time-consuming process and demands substantial investment. The necessary gestation period required to accomplish it needs to be covered by a means of livelihood for the families involved. For this purpose one can emulate the example of Bangladesh where provision of a small credit to people (without a proper livelihood) as made available by the Grameen Bank has very richly been rewarding in providing creditors with a source of earning a steady income—thus fulfilling a long-lost dream for them- which went into rescuing the families from dire poverty. Besides, making food available from income they derived on investment of the credit offered by the Grameen Bank, it enabled them further in taking care of education and health of the family members. As a result, the small credit made available to the poor not only enhanced their entitlement to food and other basics of life on the one hand, but also ultimately added to the efforts of other organizations (primarily NGOs) engaged in making education and healthcare universal on the other. Having a firm foothold in the national economy make them better citizens, narrowing the gap of inequality. An economic process like this initiated with low credit is likely to expand the horizon of new opportunities for many people in need of jobs which should be the ultimate aim to be achieved as the only means to fight poverty at its root.

Bimalendu Bhattacharya

Acknowledgements

I am deeply indebted to Bashabi Fraser and Neil Fraser — my daughter and son-in-law – of Edinburgh Napier and Edinburgh University, respectively, for their constant support, critical appreciation and encouragement, without which this book could not have been completed. Neil had taken great care in editing the work with valuable suggestions which, I have no doubt, came to be of great help. My grand-daughter's faith in my work has been sustaining throughout. A very big 'thank you' to all of them.

I am very much thankful to the library of the Edinburgh University for giving me access to its collections which proved invaluable for writing this book.

It is, indeed, of no less importance to mention my debt to various scholars for strengthening my views and ideas through their work, acting as a great stimulus at all stages, duly acknowledged at appropriate places.

At the same time, both Dr. Ballari Bagchi and Mrs. Jyotsna Baidya highly deserve my thanks for giving their valuable time in helping with the cartographic content of the book, making a very important part of this work. Lastly, I will be failing in my duty if I do not mention how I had been helped by Rajat Shubhra Chakraborty in using the computer at every stage.

And above all, my efforts would have come to nothing without the unstinting help and cooperation from my publishers for which I remain immensely grateful to them.

Kolkata
28th May, 2012

Bimalendu Bhattacharya

Contents

List of Tables

List of Figures

1
Poverty, Hunger and Malnutrition

A. POVERTY: A UNIVERSAL PHENOMENON DEFINED

In the long indistinct past when man first set out in search of food following the movement of animals, there were differences in them irrespective of geographical locations in the manner how men did it and such differences were not prominent enough even when man had formed into groups, making joint endeavours in acquiring food wrested from nature. The growing resistance between different groups did not initially give rise to any noticeable difference in social behaviour for a long time to categorise any of these human groups into particular classes apart from the fact that some sort of hierarchical order at its nascent level might have emerged with the formation of a number of people into a group in the geographical settings where they made their living directly dependent on nature's bounty. Obviously, no precise economy had emerged at this stage defining the status of these people based on functions which, confined to hunting (and gathering to some extent), did not have a productive character making a base for the emergence of an economy in the true sense of the term to call it productive. In all likelihood this did not happen till the production of crops – a cultured practice – substituting for hunting and gathering to an effective scale came to dominate the activities of man bringing in one of the earliest, most significant landmark changes in human history, enabling the appearance of permanent settlements.

It was undoubtedly a very long passage through time, thousands of years, to achieve this stage in a new era of human activities, marking dynamic changes in the pattern of human life and activities and behaviour followed by the emergence of cities based on a surplus production of food resulting from what was then an advanced agricultural technology. For the first time in human history man did not have to run after vanishing food in the wake of migrating animals seeking newer pastures as domesticated animals gradually came to be a part of life along with the domestication of crops. This was, indeed, the most important cultural revolution happening till then in the history of mankind making very strongly a new direction in the efforts of human survival perhaps ever unmatched in its profound impact on human society, changing it from a mobile existence to a permanent one highly characterized by a great diversity in human activities for the first time. Since, he could settle down permanently at a place producing food to sustain himself and store it for future use, and also for the first time in human history each person was not required to produce his own food, a large number of people in each society – which had taken shape by then – were released to utilize their talents in other fields of occupation, multiplying slowly, such as manufacturing, engineering works related to keeping irrigation of crops going, construction and maintenance of irrigation channels, organizational work required for keeping accounts, storage of food and its distribution at times of need, administration, construction in the cities quite often devastated by floods and above all, trade in manufactured products, raw materials and people visiting distant lands for making commercial contacts, in consequence giving rise to a great exchange of ideas.

To all appearance, this is the stage of human success achieved in various fields that provided the scope for stronger economic foundations making better use of resources more in tune with greater response to human talents. The changing character of the growing economy with further diversification into non-primary activities naturally was a greater incentive for human contacts over space, initiating primary steps towards global commerce.

Poverty as a concept to define a social condition has come to be recognized with growing consciousness from about the time of the Industrial Revolution when employment opportunities enhanced by industrial development made the social distinction sharper between the wage-earners, or the workers with steady and assured income, and the workers in other fields, such as agriculture or rural pursuits. Prior to that the divide between the poor and non-poor (not the rich) was not sharp enough to make the distinction categorical. Although there were rich and poor with separate existences in society since its formation, consciousness was lacking in those societies to bring the difference to a point of conflict. As a result, poverty as the most significant manifestation of deprivation was not a contesting issue till awareness grew among those deprived. It is, rather, from the nineteenth century that writers and social reformers started exposing the horrendous situation, writing about the sub-human treatment of the poorer classes of people who constituted the bulk of the labour force, the society ignoring their claim to all opportunities accessible for the affluent class alone.

The inequalities on an international scale stemming from lack of employment opportunities, characterizing the non-industrial countries as a prevailing phenomenon, were at the root of burgeoning poverty prompted further by growing population, till effective changes started taking place in many of them with the end of colonial rule. Conscious efforts considered as a "great awakening", as Gunnar Myrdal calls it, for the poorer countries "demanding higher standards of income, a bigger share in the good things in life and greater equality of opportunity" (Myrdal,1963:8). There is a graphic account of the nature of poverty prevalent in the countries called the 'Third World', in the Brandt Report, where hundreds of millions of people are "preoccupied solely with survival and elementary needs. For them work is frequently not available or, when it is, pay is very low and conditions often barely tolerable. Homes are constructed of impermanent materials and have neither piped water nor sanitation. Electricity is a luxury. Health services are thinly spread and in the rural areas only rarely within walking distance. Primary schools, where they exist, may be free and not too far away, but children are needed to work and cannot easily be spared for schooling. Permanent insecurity is the condition of the poor. There are no public systems of social security in the event of unemployment, sickness or death of the wage-earner in the family. Flood, drought or disease affecting people or livestock can destroy livelihoods without hope of compensation. In the North, ordinary men and women face genuine economic problems – uncertainty, inflation, the fear if not the reality of unemployment. But they rarely face anything resembling the total deprivation found in the South" (Brandt Report, 1980: 49).

Economic development, revival in other words, in most of these countries since the end of the colonial rule has brought in a visible change in the overall situation, making life more bearable for people in general, although it has not removed gross inequality for a major section of the population of the Third World countries. The abject poverty sustained by lack of employment opportunities, lack of education, health services, other basic requirements such as potable safe drinking water, sanitation, housing, electricity and the absence of any social security at times of utter need make life a mere day-to-day stressful existence for them bereft of any future hopes for themselves as well as for their children. The deprivation in this sense is appalling in both dimension and depth, turning a major section of the society almost sterile in many of these countries, causing unaccountable loss for each country when untapped human potential with all its capabilities and talents goes completely to waste which, otherwise, properly utilized could have made a great contribution to build up a strong economic foundation for these countries.

The prevailing inequality brings out the difference between any two situations comprising the major two sections of the society in terms of basic human rights to live referring to various aspects of social and economic conditions affecting the life of people. As such there is no difficulty in assuming that in some cases it may even apply to the difference or inequality existing between two affluent countries in some areas of socio-economic manifestations, though in actual fact inequality primarily connotes deprivation in a proper sense afflicting a major part of the world, keeping it far from a justifiable level of human dignity. In short, the nature of deprivation is wide enough to include most of the aspects shaping human life. Beginning from opportunities offering a steady source of living or employment, it includes the very basic human needs of adequate opportunities for proper housing rights with facilities for water, sanitation, electricity, health, education, shopping, recreation and the freedom to exercise power in the formulation and execution of policies for a sustainable development process. Undoubtedly, here employment means it is capable of providing access to all the basic needs without which inequality remains. Ensuring social justice and complete negation of social exclusion impeding social mobility happen to be essential major steps in the removal of inequality arresting the progress of these societies, because of the creation of an intolerant section of the community, a small section which has enjoyed all the wealth and privileges without sharing them with others.

An awakening to the overwhelming presence of inequality of this nature, slowing human progress, has happened with society in many parts of the world, shaking the earlier belief that people not affected by inequality have no responsibility in making right efforts for the removal of inequality by way of restoring human rights to live with equal dignity as others. Nothing short of this absolute necessity can be justified as every human being has an equal right to benefit, accruing from the use of natural resources. Arguably, people suffering from economic backwardness cannot be held responsible for the prevailing situation as this is the outcome of exploitation perpetrated by another section of the society under the encouragement of colonial rule which derived its benefit from such a situation on a massive scale. As a result, poverty among the masses perpetuated, remaining as a characteristic feature of the society in the developing countries even long after the cessation of colonial domination. The entire economic backbone of most of these countries seemed to have weakened enough during colonial rule, losing direction and courage to make determined efforts to get a strong hold on actions leading to a healthy start for a robust economic course. Apart from the fact that the economy in these countries was in a shambles when the colonial power left them to their own fate, they were already drained of many of their resources through the long period of exploitation for use in the interest of the 'home' country and as such the entire economy replacing the 'native' one was developed primarily oriented to the needs of the colonizing country itself. Needless to say, whatever infrastructure came to be developed during the period of colonial rule was mostly tailored to fulfil the requirements of the ruling country without the least concern for the country they ruled. In consequence, when they left the country it was lacking infrastructure and other basic services required for building up the economy. Prolonged foreign rule put the country with all its lacuna in an uncertain situation and it took a long time for most of them to emerge from this disadvantageous position to start on a course of development beneficial to itself. It has, therefore, not so far been an overall success story in gearing up the economy to a stage enabling all the people of the working group to have a livelihood and hence poverty has remained as the most challenging issue for a major section of the population till now in most of the developing countries.

To define poverty in this perspective becomes an important issue confronting many for assessing its grip on countries of the developing world. According to Amartya Sen, there are two distinct problems in measuring poverty, (a) 'identifying the poor among the total population', and (b) 'constructing an index of poverty using the available information on the poor'. The first of them involves the selection of a 'poverty line' in terms of real income per head and then finding out who falls below that line. However, Sen is far from satisfied with this method and suggests a new approach to measurement (Sen, 1976:219). Elsewhere, he, making a very penetrating observation, says all forms of deprivation severely restrict the "capabilities that a person has, that is, the substantative freedoms he or she enjoys to lead the kind of life he or she has reason to value in this perspective, poverty must be seen as the deprivation of basic capabilities rather than merely as lowness of incomes, which is the standard criterion of identification of poverty" (Sen, 1999:87). He observes, "poverty must be seen as an exercise of description assessing the predicament of people in terms of the prevailing standards of necessities... facts relate to what is regarded as deprivation" (Sen, 1982:21). Sen considers 'poverty' as 'unfreedom.' He appears far more convincing in asserting that, "There are good reasons for seeing poverty as a deprivation of basic capabilities, rather than merely as low income. Deprivation of elementary capabilities can be reflected in premature morbidity, significant undernourishment (especially of children), persistent morbidity, widespread illiteracy and other failures" (Sen, 1999:20). It requires further to be made clear that starvation from which people in poverty generally suffer should not be equated with famine, 'a virulent manifestation' causing widespread death. The violent outbursts of famines, as such, present altogether a different situation than persistent poverty, with effects having an element of almost perpetuity compared with the case of the former, that is, famine with situation improving much faster.

Besides, poverty is a prolonged process of deprivation owing to lack of entitlement, limiting people's capabilities to avoid hunger and access to other basic benefits making life viable. In contrast, famines deprive people of entitlement for food but keep their status in regard to other basic requirements (or services) unaffected and eventually it does not take very long for the society (or the government) to bring back the situation to normal. Furthermore, one can anticipate the appearance of famine and prevent it from happening with adequate measures. Moreover, in some cases the famines are man-made caused by wrong policies of the government as happened in Bengal (India) in 1943, known as 'the Great Bengal Famine', having no equal in its virulence, taking away the lives of almost 3 million people over a short period entirely based on a completely faulty policy of the government in power which, corroborating Sen (1982), is succinctly explained by Fraser (2006) as follows:

> "The fear of Japanese attack" (during the Second World War, 1939-1945) "was palpable in Bengal and to ensure the 'denial policy' (which involved the 'scorched earth policy') in anticipation of an advancing army of Japanese occupation, 'surplus stock' of paddy and items like boats and bicycles were destroyed, creating an underclass with denuded livelihoods, and the means of sustenance of alluvial farmers, fishermen, boatmen and other dependent occupations, were taken away from them overnight. The 'denial policy' once put into effect in Bengal, further impoverished the economy, destroying arable land and

crops, a factor that contributed to the man-made famine of 1943 in Bengal" (Fraser, 2006:15).

Such artificial scarcities through political upheaval taking place in many parts of Africa today, may create similar situations of man-made famine affecting a large part of the displaced population deprived of entitlements to food.

Watts (1968), on the other hand, giving a definition of poverty, made two different approaches, the economic concept and the cultural concept, of which the former is defined "in terms of the external circumstances that condition a person's behaviour, especially, the behaviour he displays in economic transactions: buying consumption items, selling productive services, securing professional advice, etc.", while the cultural concept "focuses on the internal attitudes and behaviour patterns that a person brings to any particular set of circumstances... (Hence) eliminating economic poverty will measure its success by the increase in command over goods and services... (and) eliminating the culture of poverty will measure its success by changes in the complex of attitudes and behaviour patterns characteristic of that culture" (Watts, 1968:131). Orshansky (1965), talking about counting the poor, thought it is not possible to state unequivocally "how much is enough" but it should be "possible to assert with confidence how much, on an average, is too little" and in her opinion "the measure of income used should reflect at least roughly an equivalent level of living for individuals and families of different size and composition". Thus, according to her it may be possible to "sketch a profile of poverty based on a particular income standard that makes allowance for different needs of families with varying numbers of adults and children to support" (Orshansky, 1965:4).

However, Ravallion finds "interlinkage between income poverty and undernutrition" inconclusive since "chronically undernourished" people "may be so unproductive that they do not get hired at any wage". Contrastingly, in his opinion "income poverty is a cause of inequalities in education and health, which in turn perpetuate income poverty...Public action to improve access to non-market goods and services – clean drinking water, sanitation, healthcare, epidemiological protection, elementary education, and so on – has often paid off, and the benefits should clearly not be assessed solely in terms of incomes...Low income, for example, is likely to be both a cause effect of poor health and schooling" (Ravallion, 1996"121).

The 'Human Poverty Index' introduced in 1997 by the United Nations in its Human Development Report, viewed poverty in terms of a lack of basic human capabilities which consisted of literacy, life expectancy, access to safe water, availability of health services and the proportion of underweight children aged five and below. The World Bank in its debate on the meaning of poverty defined in 2000 the key elements of poverty as – the inability to satisfy basic needs, lack of control over resources, lack of education and skills, poor health, malnutrition, lack of shelter and lack of access to water supply and sanitation, vulnerability to shocks and lack of political freedom and voice. Griffin (1978), in making a study of seven Asiatic countries for determining 'the trends in the absolute and relative incomes of the rural people' used two separate types of measurements and a 'poverty line' was derived from an estimate of the level of income necessary to ensure a minimum diet, though the issue of what forms a minimum diet was contentious in view of local situations. The interesting part of his study shows that the proportion of the population below the 'poverty line' either has remained constant over the period of study or has tended to increase in the later years (Griffin, 1978:122).

As a matter of definition, Sachs distinguishes between three degrees of poverty, such as extreme or absolute poverty, moderate poverty and relative poverty when "extreme poverty means that households cannot meet basic needs for survival. They are chronically hungry, unable to access to healthcare, lack the amenities of safe drinking water and sanitation, cannot afford education for some or all of the children, and perhaps lack rudimentary shelter…Unlike moderate and relative poverty, extreme poverty occurs only in developing countries. Moderate poverty generally refers to conditions of life in which basic needs are met, but just barely (Fig. 1.1). Relative poverty is generally construed as a household income level below a given proportion of average national income. The relatively poor, in high-income countries, lack access to cultural goods, entertainment, recreation, and to quality healthcare, education, and other prerequisites for upward social mobility" (Sachs, 2005:20). However, in simple terms, "poverty can be defined as the condition where individuals or households are unable to afford what might be perceived to be the normal necessities of life" (Daniels *et al.* 2005:247).

According to Eastwood and Lipton the best way to draw an absolute poverty line is to use the 'Food Energy Method' (FEM) which "identifies the level of consumption per person at which a household just achieves an expected value

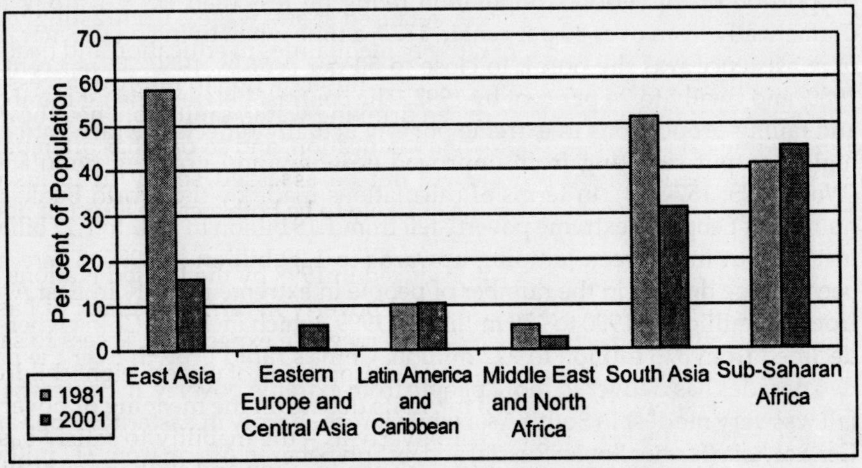

Fig. 1.1: People Living in Extreme Poverty

Source: Sachs Jeffery, 2005.

of dietary calorie intake, per equivalent adult, equal to requirements" (Eastwood and Lipton, 1999:4). Peter Townsend offers a definition for poverty and according to him people are in poverty, "when they lack the resources to obtain the types of diet, participate in the activities, and have the living conditions and amenities which are customary, or are at least widely encouraged and approved, in the societies they belong" (CPAG, 2004:1).Poverty, in that sense, may be measured directly by people's income or material circumstances including material deprivation, people lacking essential services they cannot afford; or, indirectly using benefit or tax credit receipt and socio-economic classification and on the basis of these criteria there were 12.7 million people in the UK (22%) classified as 'income poor' or living in poverty (CAPG, 2004:2). Mandle (2003), on the other hand, uses a direct measure of poverty, such as the percentage of population living with incomes less than US $ 2 a day, and he divides countries into three groupings: (a) group more advanced among them, with per capita GNP levels between $ 6,240 and $ 8,120: (b) middle group with a range between $ 2,940 and $ 3,670; and (c) the lowest ranking group, with levels of $ 1,650 and below. In his opinion, while more than 80 per cent of the population of the poorest nations lives on US $ 2 or less per day, that percentage is less than 40 per cent in the more advanced countries. The poverty level employed by Mandle is the "desperately low level of US $ 2 per day and almost two in five people are recorded as living below that threshold in countries, such as Mexico, Thailand, Turkey and Brazil" (Mandle, 2003: 21-23).

Importantly, the World Bank has defined extreme poverty on the basis of an income of US $ 1 a day per person at 1985 international prices. Interestingly, Bourguinon and Morrison also used that figure to define extreme poverty since 1820, leading to three conclusions: first, the number of desperately poor people increased from about 900 million in 1820 to a peak of from 1.3 to 1.4 billion between 1960 and 1980, falling to just under 1.3 billion in 1992; second, the proportion of the world's population living on less than US $ 1 a day fell dramatically from over 80 per cent in 1820, a time when living on the margins of subsistence was the norm, to close to 50 per cent by 1950, 32 per cent in 1980, and finally to 24 per cent by 1992. The contrast between rising numbers and falling proportions in extreme poverty actually reflects the race between higher output resulting from improved economy and growing population (Wolf, 2005: 157-158). In terms of calculations made by the World Bank, the number of people in extreme poverty fell from 1.18 billion in 1987 to 1.17 billion in 1999, but in between jumping upwards to 1.29 billion in 1990. There has been a large decline in the number of people in extreme poverty in East Asia, from 486 million in 1990 to 279 million in 1999, which includes China where it declined from 376 million to 222 million. China's rapid growth over the past two decades has delivered more people from extreme poverty. In contrast, the fall was very modest in South Asia while it rose sharply in eastern Europe and Central Asia (former Soviet Russia) and, in Sub-Saharan Africa, from 217 million in 1987 to 241 million in 1990, and then to 315 million in 1999, from 47.4 per cent to 49.0 per cent. At the same time, the incidence of poverty fell sharply in

East Asia, from 30.5 per cent of the population to 15.6 per cent in 1999. Excluding China, it fell from 24.12 to 10.6 per cent while in China it went down from 33 per cent to less than 18 per cent between 1990 and 1999. Similarly, the incidence of poverty declined sharply in South Asia, dominated by India, from 45.0 per cent in 1990 to 36.6 per cent in 1999 (Wolf, 2005: 160).

Historical estimates suggest that the number of people living in extreme poverty has consistently been above 1 billion since the late nineteenth century since which time the global population has quadrupled. The fast growth of the world's population not commensurate with economic growth on a uniform keel in many countries is at the root of prevailing persistent poverty. Notably, it was not until the beginning of the nineteenth century that the world population reached its first billion around 1804, and it took 123 years to add a second billion in 1927. But it took only 33 years to have the third billion, and 15 years for the fourth billion, while in another 12 years both the fifth and sixth billion marks were achieved. Perhaps, the most dramatic achievement in this regard is that while the first 1900 years from Christ's birth added 1.3 billion people, the last 100 years have made an addition of 4.4 billion (Daniel's *et al.*, 2005: 89).

In another estimate made by the World Bank, using an income of US $ 1 a day per person to determine the numbers of extremely poor around the world, there were about 1.1 billion people living in extreme poverty in 2001, the number, however, coming down from 1.5 billion in 1981.The major share, 93 per cent in 2001, of the world's extremely poor live in three regions: East Asia, South Asia and Sub-Saharan Africa. Nearly, half of Africa's population live in extreme poverty, the proportion having risen slightly over the period while the proportion in East Asia comprised 58 per cent in 1981 and dropped to 15 per cent in 2001; and in South Asia the proportion reduced from 52 per cent to 31 per cent during the same period. The rate of extreme poverty is around 10 per cent and has remained unchanged. In Eastern Europe, it rose from a 'negligible' level in 1981 to 4 per cent in 2001 (Sachs, 2005: 21.). The moderate poor, living between US $1 and $ 2 per day, as calculated by the World Bank, are similarly concentrated mostly in East Asia, South Asia and Sub-Saharan Africa, containing 87 per cent of the world's 1.6 billion moderately poor. The number of them, in fact, has increased in East Asia and South Asia due to social mobility whereas the number of them has remained constant at 15 per cent since 1981 in Latin America.

According to Chen and Ravallion (2007), the number of extremely poor people decreased by more than 300 million between 1990 and 2004 (United Nations, 2007: 97). In other words, the proportion of people living in extreme poverty fell from nearly a third to less than one-fifth between 1990 and 2004 which, however, does not conceal the fact that more than a billion, comprising 21 per cent of the world's population, are still in extreme poverty. And 400 million are "chronically poor", defined as living below the extreme poverty line for more than five consecutive years. Global trends mask the differences occurring in the prevalence of poverty. In Sub-Saharan Africa it has increased by 85 million in contrast to East Asia where China alone reduced poverty by 160 million people, but it increased in Eastern Europe and Central Asia by some 15 million people (United Nations, 2007: 108) (*See Note A*).

It is further estimated that the number of people in developing countries living on less than US $1 a day fell below 1 billion in 2004 from 1.25 billion in 1990, reducing the proportion of these people in those countries from 31.6 per cent to 18.4 per cent. The major contribution to this decline, as mentioned earlier, came from China and East Asia but elsewhere the numbers of extremely poor people have mostly increased or remained roughly unchanged (Table 1.1). As the table shows, the total number of people living in extreme poverty in the developing world (including Eastern Europe) has a declining trend, falling from a total of 1,120 million in 1999 to 986 million between 1999 and 2004. But in South Asia there is a contrasting picture of the share of poverty steadily rising from 42.41 per cent in 1999 to 45.45 per cent in 2002 and 46.86 per cent in 2004. A similar rising trend is found in Sub-Saharan Africa, holding the next place in the total number of people living in extreme poverty, its share rising from 26.43 per cent in 1999 to 27.74 per cent in 2002 and 30.22 per cent in 2004. In comparison, East Asia taking third position in terms of the numbers of people living in extreme poverty shows a steady declining trend, going down from 24.73 per cent in 1999 to 17.14 per cent in 2004. The only other region of consequence in terms of extreme poverty, Latin America and the Caribbean, has less than 5 per cent of the total in all the three periods of count, and maintains almost a static position, its share remaining nearly unaltered. However, the fact revealed by these figures in Table 1.1 indicating a general declining trend in the proportion of people living in extreme poverty in all the regions is of high significance in spite of the fact that Sub-Saharan Africa tops

Table 1.1: Extreme Poverty in Developing Countries by Region, 1999-2004

Region	*Number (millions)*			*Proportion of Population (percentage)*		
	1999	*2002*	*2004*	*1999*	*2002*	*2004*
Eastern Asia	277 (24.33)	227 (21.27)	169 (17.14)	15.5	12.3	9.1
Eastern Europe and Central Asia	18	6	4	3.8	1.3	0.9
Latin America and the Caribbean	49 (4.38)	48 (4.50)	47 (4.77)	9.7	9.1	8.6
Middle East and North Africa	6	5	4	2.1	1.7	1.5
South Asia	475 (42.41)	485 (45.45)	462 (46.86)	35.8	34.7	32.0
Sub-Saharan Africa	296 (26.43)	296 (27.74)	298 (30.22)	45.8	42.6	41.1
TOTAL	1,120	1,067	986	22.3	20.4	18.4

Note: Figures within brackets indicate percentage of value.

Source: United Nations, 2007 (Taken from World Bank, Global Monitoring Report, 2007).

the list in having the highest proportion of its people in that category replacing South Asia although the latter has a much larger size of such people all through the period. The reason lies in the larger population of South Asia than that of Sub-Saharan Africa (Fig. 1.2).

B. THE CAUSES OF POVERTY

Poverty as an incidence arises out of certain situations which may be treated as the causes of poverty. They are –

(a) 'Lack of income and assets to attain basic necessities', such as 'food, shelter, clothing, and acceptable levels of health and education',
(b) 'Sense of voicelessness and powerlessness in the institutions of state and society', and
(c) 'Vulnerability to adverse shocks, linked to an inability to cope with them'.

Apparently, economic growth is intimately connected with reduction in poverty and "the rate at which growth translates into lower poverty depends on the initial level of inequality in the distribution of income and how that distribution changes over time. Growth – and its effectiveness in reducing poverty – also depends on sound, stable governance" (The World Bank, 2000/2001: 35).It becomes imperative that governments should focus on the removal of socio-economic inequalities and build 'sound institutions' to take the responsibility of finding ways and means for providing a 'socially sustainable basis for overall growth ensuring that poor people gain substantially from that growth'.

Voicelessness and powerlessness deprive poor people of their legitimate rights to express their views on any development programme taken up by the government, as a result of which their needs often get bypassed, keeping their position economically unchanged. Besides, they are often humiliated, subjected to violence, shame, extortion and unnecessary harassment on trifling pleas, getting no protection from the administration. This is intolerable, and becomes the greatest hindrance to any development programme not taking up their cause as a primary issue. Further, there is inequality between men and women pervasive in all societies, and poor women are discriminated against in the household and in land, labour and credit markets, though the latter has largely been removed in Bangladesh under the efforts of the Grameen Bank programme giving preference to poor women. The existing inequality between men and women in other countries causes poverty and reinforces poverty as women's role is a powerful source of gain for the society—especially for the children. Discrimination based on other issues, such as race, religion, ethnicity and social status has similar effects contributing to poverty.

Living in constant uncertainty in life, the poor feel extremely vulnerable to any major shock upsetting their support system, in the absence of any protection from the society or the administration. This is a major deprivation

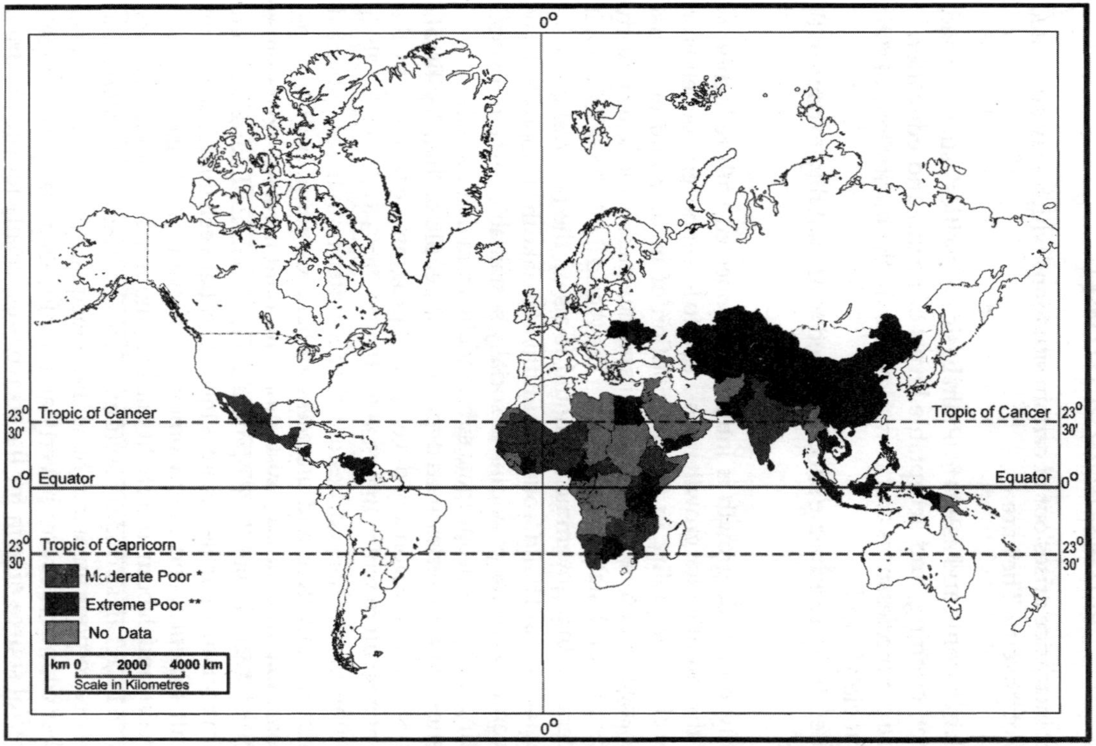

Fig. 1.2: People in Moderate and Extreme Poverty Worldwide

Note: * more than 25 per cent of population lives below $ 2 per day

** more than 25 per cent of population lives below $ 1 per day

Source: Jeffrey Sachs, 2005.

in the case of farming on marginal lands dependent on rainfall, they suffer from crop failures due to uncertain rainfall. They become easy victims in employment in the formal or informal sector and remain at higher risk of diseases, such as malaria and tuberculosis in the absence of healthcare and nutrition. Social exclusion, and women becoming victims of violence and crime, enhance their vulnerability. Forming the lowest stratum of the society, the poor people become constrained to give up whatever little they possess, land or livestock, to face a challenging economic situation in the absence of any mitigating process offered by the state. Needless to say that easy and necessary access to irrigation, infrastructure, public health services, 'honest policy and fair legal system, public work schemes in times of stress, micro-credit to help people through the aftermath of an adverse shock, famine relief in extreme circumstances', and above all a sympathetic unbiased helpful attitude on the part of the society inclusive of the people in power will reduce vulnerability for the poor (The World Bank, 2000-2001:36-37).

C. GEOGRAPHICAL DISTRIBUTION OF PEOPLE LIVING IN EXTREME POVERTY

Extreme poverty is particularly a phenomenon of the developing countries but all of them do not suffer equally (Table 1.1). An analysis exploring its prevalence and the proportion of people living in extreme poverty in these countries as well as in countries beyond the corresponding world will provide a much clearer perception of the whole issue. The data in this regard provided by the United Nations show that of the 20 Arab States, 4 (Algeria, Jordon, Morocco and the United Arab Emirates) have less than 2 per cent of their population living in extreme poverty and this exceeds 3 per cent in Egypt and 15 per cent in Yemen (Table 1.2). In the region of East Asia and the Pacific, consisting of 28 states, in 4 of them (Indonesia, Korea-Republic of, Malaysia and Thailand) the percentage falls below 2 per cent, but rises above 10 per cent in China (16.1%), Mongolia (13.9%), Philippines (14.6%) and Vietnam (17.7%), peaking to 26.3 per cent in Lao, Peoples Democratic Republic. In the third region of Latin America and the Caribbean, consisting of 33 developing countries the percentage of people living in extreme poverty falls below 2 per cent in 5 countries (Chile, Dominican Republic, Guyana, Jamaica and Uruguay) and rises above that level in Brazil (9.9%), Costa Rica (6.9%), Mexico (8.0%), and Panama (7.6%). It is at a higher level in another seven states of this region, rising above 10 per cent in Trinidad and Tobago (12.4%), Bolivia (14.4%), Colombia (14.4%), Venezuela (15.0%), Peru (15.5%), Guatemala (15.5%) and Paraguay (19.5%). Further, there are 3 states recording a higher percentage, El Salvador (21.4%), Honduras (23.8%) and an astounding high of 82.3 per cent in Nicaragua. The fourth region, South Asia, with 9 countries, has only one country, Iran-Islamic Republic having less than 2 per cent of its people living in extreme poverty, while the percentage rises to 6.6 per cent in Sri Lanka, 13.4 per cent in Pakistan, and far exceeds 10 per cent in Bangladesh (36.0%), India (34.7%), and Nepal (37.0%).

Table 1.2: Level of Poverty by Region and Country, 1990-2001. Population below an income of US $ 1 a day considered to be living in extreme poverty (Poverty Line is equivalent to US $ 1.08)

Region	*Country*	*Population in %*					
		<10 and below	*11-20*	*21-30*	*31-50*	*51-70*	*<70*
Arab States	Algeria	<2					
(20 countries)	Egypt	3.1					
14:nil	Jordan	<2					
	Morocco	<2					
	United Arab Emirates	<2					
	Yemen		15.7				
East Asia and the Pacific	Indonesia	<2					
(28 countries)	Korea, Rep. of	<2					
19:nil	Malaysia	<2					
	Thailand	<2					
	China		16.1				
	Mongolia		13.9				
	Philippines		14.6				
	Vietnam		17.7				
	Laos. People's Dem. Rep.			26.3			
S. America and Caribbean	Chile	<2					
(33 countries)	Brazil	9.9					
13:nil	Costa Rica	6.9					
	Dominican Republic	<2					
	Guyana	<2					
	Mexico	8					
	Panama	7.6					
	Uruguay	<2					
	Bolivia		14.4				
	Colombia		14.4				
	Guatemala		16.0				
	Paraguay		19.5				
	Peru		15.5				
	Trinidad & Tobago		12.4				
	Venezuela		15.0				
	El Salvador			21.4			
	Honduras			23.8			
	Nicaragua						82.3
South Asia	Sri Lanka	6.6					
(9 countries)	Iran, Islamic Rep.	<2					
3:nil	Pakistan		13.4				
	Bangladesh				36.0		
	India				34.7		
	Nepal				37.7		

(*Contd.*)

Table 1.2 (*Contd.*)

Region	*Country*	*Population in %*					
		<10 and below	*11-20*	*21-30*	*31-50*	*51-70*	*<70*
Southern Europe (2 countries) 1:nil	Turkey	<2					
Sub-Saharan Africa (45 countries) 15:nil	Swaziland	<2					
	South Africa	<2					
	Côte d'Ivoire		12.3				
	Tanzania		19.9				
	Botswana			23.5			
	Kenya			23.0			
	Mauritania			28.6			
	Senegal			26.3			
	Cameroon			33.4			
	Ghana			44.8			
	Lesotho			43.3			
	Madagascar			49.1			
	Malawi			41.7			
	Mozambique			37.9			
	Namibia			34.9			
	Rwanda			35.7			
	Zimbabwe			36.0			
	Burkina Faso					61.2	
	Burundi					58.4	
	Central African Rep.					66.6	
	Gambia					59.3	
	Niger					61.4	
	Sierra Leone					57.0	
	Zambia					63.7	
	Ethiopia						81.9
	Mali						72.8
	Nigeria						70.2
	Uganda						82.5
Central and Eastern Europe (27 countries) 3:nil	Armenia		12.8				
	Azerbaijan	3.7					
	Belarus	<2					
	Bulgaria	4.7					
	Croatia	<2					
	Czech Republic	<2					
	Estonia	<2					
	Georgia	<2					
	Hungary	<2					
	Kazakhstan	1.5					
	Kyrgyzstan	2					
	Latvia	<2					

(*Contd.*)

Table 1.2 (*Contd.*)

Region	*Country*	*Population in %*					
		<10 and below	*11-20*	*21-30*	*31-50*	*51-70*	*<70*
	Lithuania	<2					
	Moldovia, TYFR	<2					
	Moldova, Rep. of			22.0			
	Poland	<2					
	Romania	2.1					
	Russian Federation	6.1					
	Slovakia	<2					
	Slovenia	<2					
	Tajikistan		10.3				
	Turkmenistan		12.1				
	Ukraine	2.9					
	Uzbekistan		19.1				
High-Income OECD (23 countries) 2:nil	Portugal	<2					
Other UN Member countries (6 countries) +11:nil							

Source: UNDP, Human Development Report 2003, (Compiled from Table 1.3).
"Data refers to the most recent year available during the period specified.

Of the two countries of Southern Europe, Turkey has less than 2 per cent of its people living below the poverty line. In sharp contrast, Sub-Saharan Africa with its 45 countries has only 2 countries – Swaziland and South Africa – having less than 2 per cent people suffering from extreme poverty, but the percentage rises sharply in other countries of this region. It is between 10 and 20 per cent in 2 of them- Cote d' Ivoire (12.3%) and Tanzania (19.9%); 20 to 30 per cent in 4 states – Botswana (23.5%), Kenya (23.0%), Mauritania (28.6%), and Senegal (26.3%).There are 9 states recording percentages of poverty-stricken people, ranging between 33 and nearly 50 per cent of the total population. They are Cameroon (33.4%), Ghana (44.8%), Lesotho (43.1%), Madagascar (49.1%), Malawi (41.7%), Mozambique (37.4%), Namibia (34.9%), Rwanda (35.7%), and Zimbabwe (36.0%). And 50 -70 per cent extremely poor is recorded in 7 states where 57 to almost 67 per cent of the total population of the country remain steeped in extreme poverty. The countries are Sierra Leone (57.0%), Burundi (58.4%), Gambia (59.3%), Burkina Faso (61.2%), Niger (61.4%), Zambia (63.7%), and Central African Republic (66.6%). Finally, the proportion of people living below the extreme poverty line increases further, varying between 70 and above 82 per cent in 4 other states of Sub-Saharan Africa: Nigeria (70.2%), Mali (72.8%), Ethiopia (81.9%), and lastly Uganda (82.2%).

The picture, however, is not so dismal in another region, Central and Eastern Europe and CIS, which comprising 27 countries has at least 14 countries where less than 2 per cent people live below the poverty line. They are Belarus, Croatia, Czech Republic, Estonia, Georgia, Hungary, Kazakhstan, Kyrgyzstan,

Latvia, Lithuania, Macedonia TYFR, Poland, Slovakia and Slovenia. There are 5 countries in which the percentage of people suffering from extreme poverty is between 2 and 10 per cent of the total population: Romania (2.1%), Ukraine (2.9%), Azerbaijan (3.7%), Bulgaria (4.7%) and the Russian Federation (6.1%). However, in countries of this region the percentage is more than 10 per cent in the following: Tajikistan (10.3%); Turkmenistan (12.1%); Armenia (12.8%); Uzbekistan (19.1%), and exceeds 20 per cent in only one country, Moldova, Republic of, rising to 22 per cent. On the other hand, among the high-income OECD countries, consisting of 23 states, only Portugal has a record of people with extreme poverty, though it affects less than 2 per cent of the total population of the country.

In summarizing all these details some very interesting as well as important facts emerge. Specifically, deprivation as demonstrated by extreme poverty has a very wide range the world over although it happens to be almost an identifying feature of the developing world. But, as the data provided by the United Nations 2003, the problem is not entirely confined to the developing countries alone; it goes beyond them, but in a much lower profile though, because of much better economy.

According to the data made available by the United Nations 2003, there are altogether 94 countries having records of extreme poverty, with people living below the poverty line (as defined by the World Bank, an income of US $1.08 a day), and in 32 of these countries less than 2 per cent people are affected. In another 11 countries, it is more than 2 but less than 10 per cent; in 19 countries the proportion of total population remaining below the poverty line varies between 10 and 20 per cent; in 8 countries the proportion is between 21 and 30 per cent; in 12 countries the percentage of such people ranges between 31 and 50 per cent; but it exceeds 50 per cent in another 12 countries of which 5 countries have more than 70 per cent of their population living in extreme poverty, peaking the highest at 82.3 per cent in Nicaragua.

Looking from another angle, region-wise Sub-Saharan Africa in particular stands out much more prominently than any other region in the world in respect of people afflicted by extreme poverty where one finds 11 out of a total of 12 countries of the world having 50 per cent of their total population living below the poverty line, exceeding 60 per cent in 4 of them, while between 40 and 50 per cent in another 4 countries of this region.

In spite of improvement in various fields and higher GDP growth in the mid-1990s, India's household surveys indicate poverty reduction has been sluggish, the growth becoming 'less potent' in reducing poverty even in the 1990s. The country faces significant challenges and needs to take firm decisions to realize its potential. For this purpose, policy action is required to lift the 'more than 300 million poor, 34 per cent of the population and increasingly concentrated in the poorer states, out of poverty' which demands more spending on education and health, providing better access for the poor, females and other disadvantaged groups with improvement in basic services. The World Bank thinks that "major challenges in reducing poverty and getting India's population ready for the twenty-first century are raising literacy rate from the current 62 per cent (50 per cent for females); enrolling the over 30 million children, mostly poor, who are out of school; and increasing the overall average years of quality schooling. In addition, inequalities faced by women in participating fully in political, legal and economic systems need to be addressed" (The World Bank, 2000: 3). Notably, East Asian countries, despite economic crisis, have a much lower rate of poverty and better social indicators than India. Indonesia, for example, having similar per capita income to India in the mid-1960s and which was the hardest hit by the East Asian crisis, has a literacy rate of 80 per cent and had less than 2 per cent of its population living below the poverty line in 1999-2001.

In estimating the incidence of poverty in India the Planning Commission (Government of India) divides the total population of the country into six groups based on their consumption expenditure of which the first four groups are taken together to represent the 'common people'. According to this classification, the first group, comprising the "extremely poor" are those having a monthly per capita consumer expenditure of up to the three-fourths of the official poverty line, that is "an average of Rs.8.9 per capita per day in 2004-05; the second group "poor" consists of those between extremely poor and up to the official poverty line (average expenditure of Rs.11.6 per capita per day); the third represents the "marginally poor with per capita consumer expenditure of only 1.25 times the poverty line, that is, Rs. 14.6 per capita per day while the fourth called "vulnerable" have per capita consumer expenditure of only two times the poverty line, that is Rs. 20.3 per capita per day. According to this classification, the extremely poor constituted 6.4 per cent, the poor 15.4 per cent, the marginally poor 19.0 per cent, the three together constituting 41 per cent of the population, and including the vulnerable group, the total accounts for 77 per cent of the population which represents a total of 836 million people, with an income roughly below US $ 2 in PPP terms (Table 1.3) (Government of India, 2008:6. *Note*: that India's official poverty line is not the same as that used by the United Nations in the table quoted above for incidence of extreme poverty throughout the world).

Table 1.3 shows that there has been a significant decline in extreme poverty in India, climbing down from nearly 12 per cent of the total population in 1993-94 to a little over 6 per cent in 2003-04. This also applies to the decline in the share of poor among the total population in the country, from over 19 to slightly over 15 per cent during the same period. In contrast, however, the share of the marginally poor remained almost unchanged when the share of those vulnerable increased from 32 to 36 per cent. Nonetheless, the overall share recorded a significant gain, sliding down from a high of about 82 to 77 per cent over a period of about 11 years, though in effect, it represented a larger number of people, increasing from 732 to 836 million in tune with a growing population of the country.

However, according to a recent classification made by the Planning Commission, on behalf of the Government of India, the percentage of people

Table 1.3: Population in Different Expenditure Class in India (in millions)

	Expenditure Class	*1993-94*	*1990-00*	*2004-05*
1.	Extremely Poor (up to 0.75 PL*)	103	87	70
2.	Poor (up to 0.75 PL to PL)	172	174	167
3.	Marginally Poor (PL to 1.25 PL)	168	200	207
4.	Vulnerable (1.25 PL to 2 PL)	290	349	392
5.	Middle Income (2 PL to 4 PL)	139	167	210
6.	High Income (>4 PL)	24	26	44
7.	Extremely Poor & Poor (1+2)	274	262	237
8.	Marginal and Vulnerable (3+4)	458	459	599
9.	Poor and Vulnerable (7+8)	732	811	836
10.	Middle and High Income (5+6)	162	193	253
11.	TOTAL	894	1004	1090

	Expenditure	*Percentage Share & Average Consumer Expenditure per capita per day in Rupees*		
1.	Extremely Poor (up to 0.75 PL)	11.5	8.7	6.4
2.	Poor (up to 0.75 PL to PL)	19.2	17.3	15.4
3.	Marginally Poor (PL to 1.25 PL)	18.8	19.9	19.0
4.	Vulnerable (1.25 PL to 2 PL)	32.4	34.8	36.0
5.	Middle Income (2 PL to 4 PL)	15.5	16.7	19.3
6.	High Income (>4 PL)	2.7	2.6	4.0
7.	Extremely Poor & Poor (1+2)	30.7	26.1	21.8
8.	Marginal and Vulnerable (3+4)	51.2	54.7	55.0
9.	Poor and Vulnerable (7+8)	81.8	80.7	76.7
10.	Middle and High Income (5+6)	18.2	19.3	23.3
11.	TOTAL	100.0	100.0	100.

Note: *PL refers to Poverty Line

Figures in brackets denote average consumer expenditure per capita per day in Rupees

Source: Government of India 2008, Report on Conditions of Work and Promotion of Livelihoods in the Unorganised Sector

remaining in poverty in 2004-05 in urban areas was 25.7 per cent at the all-India level, and 41.8 per cent in the rural areas. Based on the consumption poverty line, the estimates emerged from a calculation of 'household goods and services consumed by households at the borderline separating the poor from the non-poor'. By this exercise, poverty at all-India level in 1993-94 was 50.1 per cent in rural areas, 31.8 per cent in urban areas and 45.3 per cent in the country as a whole as compared to the 1993-94 official estimates of 37.2 per cent rural, 32.6 per cent urban and 36.0 per cent combined (Government of India, Planning Commission, Nov. 2009).

China, on the other hand, having the largest population in the world, presents quite a different picture regarding poverty as compared with that of India, with the second largest population. In 1995 there were 70-170 million people living in poverty in China. As mentioned by Shujie Yao, rural per capita income in China in 1978 was only 285 yuan in 1990-prices, which was significantly lower than the official poverty line of 318 yuan, and much lower than the poverty line of 454 yuan defined by the World Bank, implying that majority of the rural population must have lived in poverty in 1978. Effective change to this situation came with Deng Xiaoping coming to power and resolving to reduce and eliminate poverty from China. Deng resorted to reforming agriculture first before reforming urban and industrial sectors. "Initially, the household production responsibility system was introduced without breaking up the communes (Lin, 1988: 199-224; Johnson, 1998: 225-245). This system allowed farmers to retain a certain proportion of outputs after fulfilling a production quota set by the production team. Such a simple reform method, reinforced by a few farm product price increases, released enormous energy from the peasantry. Grain output increased from 305 to 407 million tonnes between 1978 and 1984. Real per capita income more than doubled, rising by 14.9 per cent per year" (Yao, 1999: 106) In the rural areas, non-farming activities, particularly the Township and Village Enterprises (TVEs) became a dynamic and powerful force, more important than agriculture in terms of output value in the early 1990s, employing more than a quarter of the total rural labour force, contributing about 40 per cent of per capita rural income. The rapid development of TVEs, increasing from 1.6 to 24.5 million in number between 1978 and 1993 is probably the greatest achievement of China's economic reforms with their gross output value increasing by 25.6 per cent per year and total employment increasing from 9.2 to 28 per cent of the total rural labour force (Yao, 1999: 119). This resulted in an admirable record of poverty alleviation in China where since 1978, more than 200 million people have been lifted out of poverty. By 1995, the number of absolute poor was 70 million when in official statistics the absolute poverty line was an annual per capita income of 318 yuan in 1990-prices and "in material context, absolute poverty line in China means that a household does not have enough staple food to eat, cannot afford to buy enough warm clothing, and are unable to have all their children educated up to primary school level. These households usually live in very poor housing conditions, with many people crowded in the same house without clean water and electricity" (Yao, 1999: 123-124). It may be noted that a higher line of poverty was 454 yuan in 1990-prices and there were 170 million people living below the higher poverty line in 1995 which was less than 19 per cent of the rural population. However, importantly, poverty in China is almost exclusively a rural phenomenon and according to official statistics, only 0.3 per cent of urban population, amounting to less than one million people residing in urban areas, lived below the low poverty line in 1981.

Talking of urban poverty, in general, it is a real problem in the urban areas of the developing world. The majority of people migrating to cities are joining the ranks of urban poor, contributing to the proliferation and expansion of slums. The number of people living in slums and informal settlements is projected to reach 1.4 billion by 2020 (United Nations, 2007: 247). In many developing countries, health conditions and malnutrition in urban areas, are often comparable to, or worse than, those in rural areas, while drug abuse, alcoholism, crime and environmental degradation are often consequences of the massive movements of people to urban areas. In recent years, there have been sustained efforts in many developing countries in the form of improvement of slums and prevention of their formation, making a visible progress towards the goal of reducing the global number of slum-dwellers by 100 million. Migration has turned out to be a major factor in all the developing countries in enlarging the urban slums in a situation when urbanization has become a major driver of internal migration primarily based on the growing lack of livelihood opportunities in the rural areas and prevailing rural-urban wage differentials. For example, an increase in the demand for labour in the wake of improving economy can push up urban wages and increase migration. Rural-urban differences in average incomes increased in many South and East Asian countries during the 1990s, particularly in China resulting in movement of job-seekers from impoverished rural areas into nearby or distant cities. It is reported that "whether or not migration is poverty reducing, migrants travel and live under very different conditions. Poor immigrants usually stay in slums or even less secure accommodation. The estimates for Indian metropolises are sobering; roughly half of the population in Mumbai and 40 per cent in Delhi live in slums. Roughly 500,000 people migrate to Delhi every year mainly to work in the 95,000 factories there, or as workers in teashops, vendors, and drivers. Of these, 400,000 move into the 1,500 illegal colonies and 1,000 slums in the city, with very poor access to clean water, sanitation or electricity. "Even those who earn reasonable amounts face constant threats of eviction, disease, sexual abuse, underpayment and police harassment" (International Organization for Migration, 2008: 33-34). A study in Madhya Pradesh in India found migrants work long hours in harsh condition without proper medical care or compensation, lacking water, fuel, sanitation or security. Another study in Gujarat in India found that over half of the migrants slept in the open, and the rest had very uncertain accommodation, constantly facing harassment, abuse, theft and forcible eviction. The enormity of this problem involving people living in urban slums of India may be gauged from the fact that 607 towns and cities reporting on slums in India, had more than 22 per cent of their total population living in slums (Census of India 2001, Statement I)

D. HUNGER AND MALNUTRITION

Hunger and malnutrition as the two most visible characteristics or manifestations of poverty represent deprivation in its crudest form. Josue De Castro, writing his '*Geography of Hunger*' in 1952, defined hunger in a new light. In his opinion, hunger does not mean lack of food to satisfy appetite but "lack of any of the forty or so food constituents needed to maintain health" (Castro, 1955: 5). As such, the numbers of people having died from famine is much less than the numbers of people dying from a 'diet inadequate to maintain health', causing 'nutritional diseases'. Hunger used in this sense makes, according to pre-war (Second World War, 1939-45) estimates, 'two-thirds of the population of the world hungry'. Hunger, as the most potent source of social misfortunes, has largely been the product of the colonial system for that colonial economy, 'raising European civilization towards its zenith of splendour,' was at the root of leaving millions hungry in the colonies, though the imperial Europe was never prepared to accept the stark reality of starvation of people they exploited. "Hunger has been chiefly created by the inhuman exploitation of colonial riches by the latifundia and one-crop culture which lay waste the colony, so that the exploiting country can take too cheaply the raw materials its prosperous industrial economy requires. For economic reasons misery was hidden from the world's eyes; tragedies like that of China, where in the nineteenth century some 100 million individuals starved to death, or like that of India, where 20 million people died of hunger in the last thirty years of the century (nineteenth) were glossed over" (Castro, 1955: 16).

To ward off hunger, the primary need is food and as it mainly comes from land it becomes imperative to assess how man is endowed with this resource and whether it is enough to sustain human need under proper use. Famines, often man-made, do not occur from a kind of natural law. Analysis of some fundamental statistics will justify this notion. The solid part of the earth represents about 29 per cent of the surface, leaving 71 per cent under oceans, and covers an area of about 56 million sq. miles, 30 per cent of which is forested, 20 per cent in grassy plains, 18 per cent in mountains and 32 per cent desert, either hot or cold. Thus, as estimated by agricultural scientists, there is only 25 million sq. miles – half the land surface of the planet – that can be used for agriculture. Even by conservative calculations there is some 16 billion acres of land to cultivate, and according to estimates made by these authorities, about 2 acres per person will supply the 'indispensable elements of a rational diet'. As Castro pointed out, cultivation on that basis would require one-fourth of the world's arable land when only 2 billion, one-eighth of earth's natural possibilities, were being used. This amply justified the assumption that famine was not the result of natural laws (Castro, 1955: 22). Besides, he also found more possibilities in extending the horizon of cultivation beyond conceivable limits at his time, noticing instances of admirable success achieved in hundreds of thousands of acres of tropical desert made fertile by modern irrigation and a wide strip of polar desert in Russia was made productive with advanced agricultural techniques. Castro cites another example, drawing instances of Britain, the latter demonstrating how necessity can inspire an agricultural expansion far beyond the limits of normal expectation. He mentions that before the Second World War the U.K. produced only one-third of her food-needs, depending on imports for the remaining two-thirds. But coming under the stress of maritime blockade during the war period, Britain stepped up her efforts in increasing home production to reach 45 per cent of the national requirements "through an increase of about 60 per cent in the area under cultivation, which rose from 8.8 million to 14.5 million acres between 1939and 1944. And all through the war, surprisingly enough, Britain's nutritional situation, far from deteriorating, improved greatly" (Castro, 1955:25).

However, production of food cannot be increased indefinitely though one might agree to the notion that 'we are still a very great distance from maximum limits'. Opinions may widely differ on this issue intimately connected with wide scale hunger in the world. As the primary resource necessary for producing food is land, a study indicates that there are at most, about 3.2 billion hectares of land, or 7.86 billion acres potentially suitable for cultivation, about half of which is under cultivation while the conversion of the remaining land to cultivation proves to be too expensive, and according to an FAO report, 'opening more land to cultivation isn't economically possible', while the former is much lower in amount than estimated by agricultural scientists earlier, as mentioned by Castro in 1955. According to the authors of the "Report of the Club of Rome (1972), about 0.4 hectares of arable land per person is needed for meeting human needs at the level of productivity existent at that time though the requirement would rise to 0.9 hectares per person to be fed at United States' standard, cautioning at the same time that land available for cultivation is on the decrease as it is converted to urban-industrial use.

Land-Grab for Food Security

In recent years development in another direction is becoming of growing concern as land as the source of food is fast changing hands in the face of stiff competition. Rich countries and large corporations are buying up the rights to millions of hectares of agricultural land in developing countries in an effort to secure their own long-term supplies which is quite alarming for poor people (*see* Note B). The head of the United Nations FAO, Jacques Diouf has warned that such controversial deals are going to create a type of "neo-colonialism" in the agricultural systems of the world making poor states produce for the rich people while keeping their own people hungry. The land-grab, in this manner, is taking a serious dimension creating a "scramble for Africa", in particular, where the South Korean firm, Daewoo Logistics is negotiating a deal to buy a 99-year lease on as much as 1.3 million hectares of arable land in Madagascar, half the size of Belgium, aiming to produce 5 million tonnes of corn a year by 2023, and produce palm oil from a further lease of 120,000 hectares (296,000 acres) using South African labour so that South Korea's dependence on imports is lessened. With a likely expenditure of $6 billion over 25 years, the Daewoo project is the biggest of its kind in the world. The Governments and Corporations in some 'cash-rich but land-poor' countries, mostly in the Middle East, are making similar efforts to own long-term access to food buying up land in poorer countries. Saudi Arabia, for example, is planning an investment on 500,000 hectares of land in Indonesia to grow 'basmati rice', while tens of thousands of hectares of arable land in Pakistan have been sold to Abu Dhabi investors. Arab investors including the 'Abu Dhabi Fund for Development' have bought direct stakes in Sudanese agriculture. United Arab Emirates was considering large-scale agricultural projects in Kazakhstan to ensure a stable food supply and further plans to acquire 400,000 hectares of land in a dozen countries, including Australia, Pakistan, the Philippines and Ukraine. Qatar has finished a deal with Vietnam to invest in food production for export to Qatar. Having only 1 per cent of its land suitable for farming, Qatar has asked Kenya to lease it 40,000 hectares of land to grow crops and in exchange of that it will fund a £ 2.4 billion port for popular tourism off the north coast of the country. Being rich in oil and gas reserves Qatar has already struck deals with Cambodia to grow rice, maize and meat in Sudan and vegetables in Vietnam. Even China with plenty of land started exploring land deals in South-East Asia. Recently, China sought to lease 1 million hectares of land for rice in Vietnam. The Philippines has given Chinese companies access to 1.24 million hectares of land to grow crops while China has made a deal to farm 80,000 hectares of Russian land. In Laos, between 2 and 3 million hectares of land – as much as 15 per cent of the country's territory – has been signed away, some of these leases stretching for 70 years. Thai, Vietnamese and Malaysian companies dominate the southern lowlands of Laos, where rubber, sugar and cassava plantations use large areas. Japanese, Indian and Scandinavian farms cultivate fast-growing eucalyptus and acacia trees for paper in the centre, and Chinese operators lease most of the land in the north mainly for the production of rice and rubber trees. The desperation to find land means that rural farmers have been stripped of their holdings for minimal compensation. They are mostly subsistence farmers living on less than US $2 a day, and some can no longer feed themselves. Sudan is trying to attract investors for almost 900,000 hectares of its land and Ethiopia is trying to get into a similar deal with Saudi Arabian investors (Table 1.4).

The facts remain incomplete without referring to contenders eyeing Brazilian farmland, the total area amounting to more than 16 times the size of the U.K., making it one of the most attractive options for countries seeking agricultural acreage overseas. On its 400 million hectares of agricultural land, Brazil produces more sugar, coffee, beef and poultry than anywhere else, and is the world's leading soybean exporter. Saudi Arabia was looking in 'search of possibilities'. Iran is also exploring the possibilities to ensure the supply of beef, soybeans, sugar and corn. China contemplates making long-term investments to buy land in Brazil promising healthy financial returns. In 2007, the Japanese Mitsui Conglomerate bought 100,000 hectares of farmland in Brazil.

Interestingly, all the principal land-grabbing countries till now are in Asia, numbering altogether five. India however has only a very small share in the total land-grabbing race, amounting to over 7.6 million hectares of arable land (involving land already taken on lease or under negotiation) spread across the world unevenly. Asia having been involved in the largest share (52%) of the total, is followed by Africa with little over 31 per cent and South America still trails far behind with about 18 per cent of the deals. Australia and New Zealand, the former having leased land out to China and the latter to Japan, together

Table 1.4: Land Purchased for Agriculture by the Government and Private Companies (Corporate Houses) in Different Countries

Purchasing Country	*Country in which Purchased*	*Amount of Land Purchased (in hectares)*	*Total (in hectares)*
South Korea	Madagascar	1,300,000	2,306,000
	Sudan	690,000	
	Indonesia	25,000	
	Mongolia	270,000	
	Argentina	21,000	
China	Australia	43,000	2,090,796
	Tanzania	300	
	Uganda	4,046	
	Cameroon	1,000	
	Mexico	1,050	
	Cuba	5,000	
	Russia	80,400	
	Kazakhstan	7,000	
	Philippines	1,240,000	
	Laos	700,000	
Japan	United States	216,862	324,262
	Brazil	100,000	
	Egypt	1,600	
	China	100	
	New Zealand	5,700	
Saudi Arabia	Indonesia	1,600,000	1,610,117
	Sudan	10,117	
United Arab Emirates	Pakistan	900,000	1,282,500
	Sudan	378,000	
	Philippines	3,000	
	Algeria	1,500	
India	Paraguay	10,000	10,000
TOTAL			7,623,675

Source: *The Guardian*, U.K., 22 Nov. 2008.

have a share of less than 1 per cent of the total land thus handed out or in the process of the same, while the United States' share is much higher than that – over 2.8 per cent of the total – though leased out to only one country so far which is Japan. On the other side, of the five major countries taking land on lease from outside, confined mostly to the developing countries, South Korea has, so far, taken the lead with over 30 per cent of the total amount of land grabbed, the next place going to China with more than 27 per cent, followed by Saudi Arabia taking over more than 21 per cent, the United Arab Emirates with nearly 17 per cent while this comes down to about 4.3 per cent for Japan. As such, there are four major players in the world in competition between them for land-grabbing beyond the margins of their own countries and doing it at the cost of local farmers' own food where the land is being grabbed and thus becoming directly responsible for increasing food scarcity and consequent poverty. Among the individual countries sacrificing land in this manner, Indonesia giving out 1,600,000 hectares on lease heads the list followed by Madagascar with 1,300,000 hectares, the Philippines with 1,240,000 hectares and Pakistan coming next leasing out 900,000 hectares.

Food and Hunger

It is highly alarming to note that man is already overusing the natural resources of the planet, using more than 30 per cent of the resources than the Earth can replenish each year (according to a recent report by the WWF, the earth's population is using the equivalent of 1.5 planet's worth of natural resources at the moment), is leading to 'deforestation, degraded soils, polluted air and water, and dramatic declines in the number of fish and other species'. In the 1960s most countries lived within their ecological resources but today three-quarters of the world's population live in countries which consume more than they can replenish, thus nearly doubling the ecological footprint of man by this time. The problem is getting worse with time as populations increase and consumption keeps growing faster than available technology can find new ways of expanding what can be produced from the natural world. It is predicted that, if the situation does not change by 2030, mankind could need two planets to sustain its lifestyle. As warned by James Leape, WWF International Director General, "we are living beyond our means", when the United States and China account for more than two-fifths of the ecological footprint, with 21 per cent each. A person's footprint ranges vastly across the globe 'from 8 hectares or more "global hectares (20 acres or more) for the biggest consumers in the United Arab Emirates (9.5 hectares), the United States (9.4 hectares), Kuwait and Denmark (8 hectares), the U.K. (5.5 hectares) to half a hectare in the Democratic Republic of Congo, Haiti, Afghanistan and Malawi against a global average consumption of 2.7 hectares when, in 2005, the sustainable capacity was 2.1 hectares per person. In China it is 2.1hectares and in India 0.9 hectare" (*The Guardian*, the U.K., 2008).

World food production plays a fundamental role in this regard in fighting hunger but the growth of population being not particularly commensurate with growth in food production complicates the situation. Interestingly, growth in grain harvest exceeded that of population from 1954 to 1984, raising the production per person from 247 kgs. to 342 kgs. resulting in a 38 per cent increase. In the following 14 years, growth in grain production fell behind that of population, declining per person from its historic high in 1984 to about 312 kgs. in 1998, a decline of 9 per cent. In other words, grain productivity rising at 2 per cent a year between 1950 and 1970, fell nearly 1 per cent from 1990 to 1997. The actual reason for this decline in grain production since 1984 may be traced back to slower growth in irrigation and fertilizer use. For example, irrigated area per person increasing 29 per cent from 1950 until 1978, but declined by 4 per cent afterwards, thus falling behind the growth of population. Similarly, increase in the use of fertilizer in the world has slowed down since 1990, slowing down the annual growth in world fertilizer use from 6 per cent between 1950 and 1990 to scarcely 2 per cent in subsequent years (Brown *et al.* 2000: 33-35). As a result, the availability of grain varies widely making the developing countries very hard hit in spite of their gain in food production over the last half century or so. India, in this regard, provides an important example where grain consumption per person is less than 200 kgs. per year, and the diet is dominated by a single starchy staple – generally rice or wheat – consumed directly, leaving little for conversion for animal protein. In sharp contrast, for an average American, the per head consumption per year is 900 kgs. taken in indirectly in the form of beef, pork, poultry, eggs, milk, cheese, ice cream and yogurt. For the moot question how many people the Earth can support, the answer is if the world grain production of 1.85 billion tonnes were increased to 2 billion tonnes in the coming years, it would support 10 billion Indians but only 2.2 billion Americans (Brown *et al.* 2000: 35-36).

However, as Smil (1994) observes, "current global food production is sufficient to provide adequate food for the world's population. Even if all of the poor world's populations would raise their health status and activity level to those prevailing in rich countries, the average 1990 global food energy consumption could have to increase by only 10 per cent, and it would still remain well below the available mean supply. Quoting Bender (1993), Smil maintains that this conclusion "starkly confirms the current pattern of malnutrition and hunger is unrelated to food availability, but instead a function of global entitlements to food" (Smil, 1994: 264). Smil further states that in 1990 'world agriculture had recoverable inefficiencies equivalent to about 60 per cent of actual food consumption which would be' "enough to feed ... an additional 3.1 billion people, or a total global population of about 8.4 billion ... without cultivating more land and without increasing fertiliser and water inputs. If everybody were to receive 2,200 kcal/day" (the standard set by the World Bank) "the total would be reduced to 7.7 billion people, 2.4 billion above the 1990 total" (Smil, 1994: 272) which is much higher than the number of world population at present. In this connection, a look at efficiency gains in global food production achievable by the year 2050 may shed light on future global food prospects (Table 1.5) (see Note C).

Similarly, as Dyson (1996) observes, demographic growth having been the increasingly predominant cause of the expansion of world food production, and the trend is likely to continue and assuming that the "levels of consumption in the two developed world regions remains constant, then nearly 3 billion tons of cereal will need to be harvested in 2020 in order to satisfy global demand" ... and this demand barring any unforeseen calamity, "the world's farmers will certainly be able to meet" (Dyson, 1996:204) based on reaching a global cereal yield rate of about 4 tons per hectare which seems feasible.

Nevertheless, despite major technological advances in food production, its availability is still far from adequate to meet the requirements of the people in the developing countries. In spite of growing increase in food production in many parts of the world since 1950, many people suffered from hunger. Defining hunger in terms of malnutrition, FAO has shown that the number of

Table 1.5: Conservative Estimates of Efficiency Gains in Global Food Production Achievable by 2050

Changes Compared to 1990 practices	*Gains Equivalent to Global Food Energy Consumption (per cent)*
Improved field efficiencies	
Better agronomic practices (raise average yields 20 per cent)	22
Higher fertilizer uptake (raise nutrient use efficiency 30 per cent)	7
Reduced irrigation waste (raise water use efficiency 30 per cent)	7
Reduced waste	
Post harvest losses (lower by 20 per cent)	6
End-use waste (lower by 20 per cent)	8
Healthier diets (limit fat intakes to 30 per cent of total energy)	10
TOTAL GAIN	60

Source: Smil, 1994.

Table 1.6: Growth of Population and Undernourishment

World/Region	*Annual Population Growth Rate 1975-2001 (in %)*	*Total Fertility Rate (per woman) 2000-2005*	*Number Undernourished (millions) 2001*	*Per cent Population Undernourished 1998-2000*
Sub-Saharan Africa	2.8	5.4	206.1	33
Arab States	2.7	3.8	37.7	13
South Asia	2.1	3.3	349.2	24
East Asia and the Pacific	1.4	2.0	–	–
Latin America and the Caribbean	1.9	2.5	62.7	12
Central/Eastern Europe and CIS*	0.5	1.4	36.9	9
OECD	0.8	1.8	–	–
World	1.6	2.7	693.2	11.3

Note: *CIS: Commonwealth of Independent States

Source: Daniels *et al.* 2005: Table 7.1

hungry people in the world fell from 786 million in 1990 to 693 million in 2001, and as projected, the number is likely to fall to around 440 million by 2030. However, as indicated in Table 1.6, the number of the world's worst-fed people is increasing the fastest, with Sub-Saharan Africa having one-third of the world's total number of undernourished people. Though the hungry people are mostly concentrated in the developing countries, certain groups within the societies in the developed countries as well go hungry. In California, for example, up to 5 million people are short of food.

Looking at the prevailing situation of hunger, it is necessary to note that growth in food production has outpaced population growth, mentioned earlier, as early as the early sixties of the twentieth century. For example, between 1961 and 1999, the average daily supply of food per person increased 24 per cent globally, and in developing countries it rose by 39 per cent to 2,684 calories which had gone up by 82 per cent to 3,044 calories in China by 1999, from a barely subsistence level of 1,636 calories in 1961. In India, it increased by 48 per cent to 2,417 calories, from 1,635 calories in 1950. As per the estimates made by the FAO, the average need of an adult is between 2,000 and 2,310. Thus, the supply of food in the developing countries has gone, on average, "from inadequate to adequate". Nevertheless, hunger persists, but according to the FAO the number of people suffering from chronic undernourishment has declined from 920 million in 1969-71 to 790 million in 1997-99, or from 35 to 17 per cent of the population of the developing countries (Wolf, 2005:165). But the trend was not the same everywhere, as may be found in Sub-Saharan Africa where between 1979-81 and 1997-99, the share of undernourished people declined from 38 to 34 per cent, but the absolute number, in a rapidly growing population, increased from 168 million to 194 million.

Admittedly, progress in food security has, rather, been a slow process though the number of undernourished people in the developing world declined, as estimated by the FAO in 2006, by 3 million between 1990-92 and 2001-06, from 823 million to 820 million. But, according to the same source if China's remarkable progress in reducing hunger is taken into account, the actual number of undernourished people in the rest of the developing world would increase by 50 million during this period. South Asia has the largest number of people in hunger, amounting to about 300 million while the highest increase of 37 million of them took place in Sub-Saharan Africa. However, according to the FAO (2006), the proportion of undernourished people declined by 3 per cent – from 20 to 17 per cent – between 1990-92 and 2001-03. In this respect the highest gains were recorded by East Asia in contrast to Sub-Saharan Africa, which is facing prevalence of severe undernourishment of people. Needless to say the problem will remain critical for years to come in the developing world where millions of people will live in hunger exceeding the World Food Summit's target of halving the number of them by 170 million (FAO, 2006) by 2015 (United Nations, 2007: 97).

However, considered globally, reduction in the number of people in hunger is not very impressive. It has declined since 1990, dropping from 823 million to 814 million as estimated by the FAO in 2005 which includes 50 million in China while the number of such people increased by 35 million in Sub-Saharan Africa. Similarly, the number of undernourished people grew by 5 million during this period in Eastern Europe and Central Asia. A report jointly prepared by the FAO, IFAD (International Fund for Agricultural Development) and the WFP (World Food Programme) places 854 million people in the world as "chronically hungry", the great majority of which – 96 per cent – live in the developing countries, 75 per cent living in rural areas, 60 per cent of the total number are women, and almost 18 per cent or one-fifths are children under the age of five. The highest number – 64 per cent – of the chronically hungry people are concentrated in Asia, and the highest rate of under nutrition – 33 per cent – is in Africa. According to this report, reduction in the number of them amounting to only 9 million represents a mere 1 per cent of the total number of undernourished people in the developing world. "With the number of chronically hungry people worldwide actually growing by an average of 4 million per year, hunger needs to receive a greater share of resources and attention than it is currently receiving" (United Nations, 2007:214). The major reasons cited for the persistent increase in their number are ascribed to population growth, the spread of HIV/AIDS and conflict and natural disasters– mainly droughts and floods. It is estimated that more than 2 billion people in the world, one in three, suffer from micronutrient deficiencies. Nearly 3 million pre-school children in developing countries are blind due to deficiency of vitamin A. Anemia arising from iron deficiency results in one out of four maternal deaths in the developing world. Iodine deficiency is the world's leading cause of mental retardation affecting over 2 million children, who suffer from lowered IQ. It is considered to be a great shame to watch 'the daily massacre of 30,000 people dying of famine-related causes at a time when the world is capable of feeding twice as many people' (United Nations, 2007).

UNICEF estimated that as early as in 1978 alone more than 12 million children under the age of five died of hunger. Strangely enough food production in all the developing countries increased by over two-and-a-half times between 1950 and 1975, but demand for food during the same period rose by 3 per cent due to the increase in population and income. This resulted in huge imports of cereals, gradually increasing to nearly 80 million tons a year by 1978-79 needed mostly by the poorer countries of Africa and Asia, many of them lacking exportable items or additional aid to finance such massive imports (Brandt *et al.*, 1980:90) The contention that growing poverty and hunger are the offshoots of food shortage or famine is untenable. The study made by Griffin (1978) for seven Asiatic countries – Bangladesh, India, Malaysia, Pakistan, Sri Lanka, the Philippines and Sumatra – reveals that in only one of them did population expand faster than domestic food production. This happened in Bangladesh where even GNP per capita was falling. In India, population and the food growth rate have grown at the same rate with cereal production expanding even faster than the population; in Pakistan, cereal production has been significantly faster than population growth. This has been the case, as found by Griffin, in all other countries, 'notably Sri Lanka and Malaysia, here the difference in growth rates has been large' (Griffin, 1978: 127).

Malnourishment Around the World

The presence of malnutrition, as one of the leading indicators of deprivation, gives a clear idea about its geographical distribution across the world with its significant concentration in the developing countries. Interestingly, its presence, amounting to less than 2.5 per cent of the total population in 31 countries, 26 of which belong to the developed world, may be considered quite negligible which, in all likelihood, is more due to physical reasons, that is, a manifestation of congenital weakness than resulting from loss of nutrition caused by financial factors, or lack of adequate income giving access to nutritious food. (see Note D).

However, the data on malnutrition for the 31 countries with less than 2.5

per cent of their total population missing on natural health are not shown in the adjoining Table (1.7), while these are given for 108 selected countries (out of the total of 177 countries listed by the Report from which they are taken) on malnutrition, as shown in Table 1.7, representing six different regions of the world. The data are given for each of the 108 countries for two periods – 1990-92 and 2002-04 – as average for each of the periods.

Apparently, the range of malnutrition varies widely between regions, and becomes more contrasting when compared between different countries. In general, the range varies from less than 2.5 per cent of the total population, as the minimum for a period, to the highest of 75 per cent though there are only two cases where it goes above 70 per cent for a period in each of the countries (in Eritrea it was 75% in 2002-04 and in Congo Democratic Republic it was 74% in 2002-04) which was noticeably much lower in the preceding period in Eritrea affecting 70 per cent of the total population in 1993-95 and 31 per cent in Congo Democratic Republic in 1990-92.

However, the wide disparities in the prevalence of malnutrition among the people of different countries becomes more palpable in comparing the situation between the regions on the one hand, and between specific countries on the other. For example, as a region Sub-Saharan Africa presents the most dismal picture compared to all other regions since not only the highest level of malnutrition occurs in two of its countries, mentioned above, having almost three-quarters of their total population remaining undernourished but what is much more alarming is that the level of undernourishment has aggravated during the recent period, increasing fast from 31 per cent to 74 per cent over a period of hardly eight years between 1990-92 and 2002-04 in Congo Democratic Republic though having similar trend the rate of increase has not been that fast in Eritrea, increasing from 70 per cent in 1993-95 to 75 per cent in 2002-04.

A careful analysis of the data, however, reveals that the proportion of undernourished population remains below 20 per cent of the total population in, at least, 7 out of a total 36 countries included in the list, but exceeds in the remaining 29 countries, ranging between 21 and 50 per cent (at least in one of the periods) in 22 countries while it rises further in another 7 countries, in 5 of them the level exceeding 60 per cent of the total population in either of the periods. These countries having the highest proportion of undernourished population are Eritrea (2002-04), Congo Democratic Republic (2002-04), Ethiopia (1990-92), Burundi (2002-04) and Mozambique (1990-92). Interestingly, the rate of decline in the proportion of undernourished population has been widely varied, ranging from 1 per cent to the highest of 26 per cent (Ghana), 11 countries recording 10 per cent or more decline in contrast to the record increase of 43 per cent (Congo Democratic Republic) in one country alone while in the remaining 7 countries with increase in proportion, it remains below 10 per cent in all of them during the same period.

In comparison to the situation prevailing in Sub-Saharan Africa, Latin America and the Caribbean region fares comparatively better with 14 out of a total of 23 countries listed in Table (1.7) having the proportion of

Table 1.7: Proportion of Population Suffering from Malnutrition in the World (Among Selected Countries), 1990-1992 and 2002-04

Region	Country	Undernourished Population as % of the Total Population											
		10 and below a		11-20		21-30		31-50		51-70		Above 70	
		1990-92	2002-04	1990-92	2002-04[a]	1990-92	2002-04[a]	1990-92	2002-04[a]	1990-92	2002-04[a]	1990-92	2002-04[a]
Arab States	Algeria	5	4										
	Egypt	4	4										
	Jordon	4	6										
	Kuwait		5			24							
	Lebanon	L 2.5	3										
	Libiya	L 2.5	L 2.5										
	Morocco	6	6										
	Saudi Arabia	4	4					31	26				
	United Arab Emirates	4	L2.5										
	Yemen							34	38				
East Asia & the Pacific	Cambodia							43	33				
	China			16	12								
	Indonesia	9	6										
	Korea Rep. of	L 2.5	L 2.5										
	Lao, Rep. of				19	29							
	Malaysia	3	3										
	Mongolia						27	34					
	Myanmar	10	5										
	Philippines				18	26							
	Thailand					30	22						
	Vietnam						16	31					
Latin America and the Caribbean	Bolivia					28	23						
	Brazil		7	12									
	Chile	8	4										
	Colombia			17	13								
	Costa Rica	6	5										
	Cuba	7	L 2.5										

Region	Country	Undernourished Population as % of the Total Population											
		10 and below a		11-20		21-30		31-50		51-70		Above 70	
		1990-92	2002-04	1990-92	2002-04[a]	1990-92	2002-04[a]	1990-92	2002-04[a]	1990-92	2002-04[a]	1990-92	2002-04[a]
	Dominican Republic					27	29						
	Ecuqador	8	6										
	El Salvador			12	11								
	Guatmala			16			22						
	Guyana		8			21							
	Haiti								46	65			
	Honduras					23	23						
	Jamaica		9	14									
	Mexico	5	5										
	Nicaragua					30	27						
	Panama					21	23						
	Paraguay			18	15								
	Peru	6	5		12			42					
	Surinam		8	13									
	Trinidad and Tobago		10	13									
	Uruguay	7	L 2.5										
	Venezuela			11	18								
4. South Asia	Bengladesh							35	30				
	India					25	20						
	Iran, Islamic Rep.	4	4										
	Nepal			20	17								
	Pakistan					24	24						
	Sri Lanka					28	22						
5. Sub-Saharan Africa	Algeria	5	4										
	Benin			20	12								
	Botswana					23			32				
	Burkina Faso				15	21							

Region	Country	Undernourished Population as % of the Total Population											
		10 and below a		11-20		21-30		31-50		51-70		Above 70	
		1990-92	2002-04	1990-92	2002-04[a]	1990-92	2002-04[a]	1990-92	2002-04[a]	1990-92	2002-04[a]	1990-92	2002-04[a]
	Burundi								48		66		
	Cameroon						26	33					
	Central African Rep.								44	50			
	Chad								35	58			
	Congo								33	54			
	Congo, Dem. Rep.							31					74
	Cote d' Ivoire				18	13							
	Eritrea										70[b]		75
	Ethiopia								46	63[b]			
	Gabon	10	5										
	Gambia					22	29						
	Ghana				11			37					
	Guinea						24	39					
	Kenya							39	31				
	Madagascar							35	38				
	Malawi							50	35				
	Mali					29	29						
	Mauritania			15	10								
	Mauritius	6	5										
	Mozambique								44	46			
	Namibia						24	34					
	Niger							41	32				
	Nigeria	9		13									
	Rwanda							43	33				
	Senegal				20	23							
	Sierra Leone							46		51			
	Swaziland			14			22						
	Tanzania							37	44				
	Togo				19	24							

Region	Country	Undernourished Population as % of the Total Population											
		10 and below a		11-20		21-30		31-50		51-70		Above 70	
		1990-92	2002-04	1990-92	2002-04[a]	1990-92	2002-04[a]	1990-92	2002-04[a]	1990-92	2002-04[a]	1990-92	2002-04[a]
	Zambia							48	46				
	Zimbabwe							45	47				
6. Central and Eastern Europe and the CIS	Albania	5[b]	6										
	Armenia						24			52[b]			
	Azerbaijan		7					34b					
	Belarus	–	4										
	Bosnia & Herzegovina	9b	9										
	Bulgaria	8[b]	8										
	Croatia		7	16[b]									
	Georgia		9					44[b]					
	Kazakhstan	–	6										
	Kyrgyzstan		4			21[b]							
	Latvia	3[b]	3										
	Lithuania	4[b]	L 2.5										
	Macedonia TYER		5	15[b]									
	Moldova	5[b]			11								
	Russian Federation	4[b]	3										
	Tajikistan					22[b]			56				
	Turkmenistan		7	12[b]									
	Ukraine	–	L 2.5										
	Uzbekistan	8[b]					25						

Sources: Compiled from data available in UNDP: Human Development Report 2007/2008.

Note: L = Less than a : data refer to the average for the years specified. b: data refer to the period 1993/1995

undernourished population below 20 per cent of the total population. It does not exceed 30 per cent in 8 other countries, exceeding that in only one country (Haiti), climbing to 65 per cent during one period but decreases to 46 per cent during the subsequent period.

Similarly, both East Asia and the Pacific and South Asia as regions appear to be better off in this respect with 5 out of 12 countries in the former having less that 20 per cent undernourished among the total population, which goes down to below 2.5 per cent in the Republic of Korea during both the periods, and climbs to the highest of 43 per cent in Cambodia in one period. Besides, except in two countries – the Republic of Korea and Malaysia – where the proportion of the undernourished has remained unchanged between the two periods, it has come down in all the other countries which seems to be encouraging. On the other hand, South Asia, with six countries included in the list, presents almost a similar situation, where the highest proportion of the undernourished population reaches 35 per cent in one of the periods in Bangladesh, but it does not go above 30 per cent in any of the countries.

A more promising picture is presented by the Arab States, twelve of them included in the list, where the proportion is visibly much less than in any of the regions discussed above, lying far below 10 per cent in 10 out of 12 countries, falling below 2.5 per cent in three of them. This is further evident from the absence of a very high proportion of undernourished people, maximum of which rises to 31 per cent during one of the periods in only one country (Sudan). Moreover, the region has not witnessed any increase in the proportion in any of the countries though in 4 (Egypt, Saudi Arabia, Libya and Morocco) out of 12 of them it has remained constant.

Lastly, the region comprising Central and Eastern Europe and the CIS, in comparison, presents a contrasting picture unlike some of the other regions mentioned above where though in 13 out of 19 countries listed, the proportion of undernourished population does not exceed 20 per cent, there are, at least, two countries (Armenia and Tajikistan) where the proportion goes beyond 50 per cent of the total population. Second, there has been fast decline in the proportion in three countries (Georgia: 9 per cent; Armenia: 24 per cent and Azerbaijan: 7 per cent) in the succeeding period (2002-04), but very strikingly the proportion increased by 34 per cent in Tajikistan while it remained unchanged in three countries. On the whole, the situation in this regard looks promising in the European part of the region than elsewhere.

To sum up, for any consideration of poverty, hunger and malnutrition representing the three important facets of human deprivation one cannot obviate the fact that it becomes highly complicated to define poverty in the absence of a clear agreement among the experts working in the field. While one can, however, explain hunger and malnutrition related to poverty, one cannot avoid using World Bank/United Nations definition of poverty (based on income, in particular, US $ 1 a day to identify extremely poor) as a measure for comparing the situation between different countries without which fighting poverty becomes difficult. This is very much pertinent in upholding human rights to live on equal terms with dignity, enjoying adequate security and justice as well as inclusion.

2
Towards Ending Poverty, Hunger and Malnutrition

There have been dramatic improvements in the developing countries during the last three decades or so as indicated by life expectancy increasing by eight years, illiteracy cut by nearly a half to 25 per cent, and the number of people living on less than US $ 1 a day nearly halved in East Asia in the 1990s alone. Nevertheless, overall progress in human development has been rather slow since the 1990s, which was a 'decade of despair' as 54 countries were poorer in 2000 than they were in 1990 when a larger proportion of people were going hungry in 21 countries, more children were dying before the age of five in 14 countries, primary school enrolments were shrinking in 12 countries and life expectancy had fallen in 34 countries. Further, the Human Development Index (HDI) – based on the measure of living a long and healthy life, being educated and having a decent standard of living – declined in 21 countries. As reported by the United Nations, 2003, poverty rates, already high, increased in 37 of 67 countries having data. In the 1990s, in 19 countries more than one in four was going hungry, with further worsening of the situation with time. The mortality rates of children under five increased in the 1990s and in 7 countries nearly one in four children will not reach the fifth year of their life. The per capita income growth was less than 3 per cent in 125 developing and transition countries in the 1990s and the average per capita income fell in 54 of them. Among the 54 countries with declining income, 20 were in Sub-Saharan Africa, 17 in Eastern Europe and the CIS, 6 were in Latin America and the Caribbean, 6 in East Asia and the Pacific, and 5 were in Arab States. The spectacular achievement of China in taking 150 million people out of poverty in the 1990s was primarily confined to the coastal regions while deep pockets of poverty persisted in some inland regions (UNDP 2003: 2-3)

As mentioned earlier, more than 1.2 billion people – one in every five - make do or survive somehow on an income of less than US $ 1 a day which is described as extreme poverty. During the 1990s, the share of people suffering from extreme poverty fell from 30 per cent to 23 per cent, meaning a reduction of 123 million people, but excluding China, the number of people in extreme poverty, in fact, increased by 28 million considering the growth of world population. While China made remarkable progress in eliminating poverty for 150 million or 12 per cent of its population in the 1990s, the number of people surviving on less than US $ 1 a day increased during the same period in Latin America and the Caribbean, the Arab States, Central and Eastern Europe and in Sub-Saharan Africa.

A. MILLENNIUM DEVELOPMENT GOALS

As a pious as well as highly daunting wish of getting rid of poverty in the world, the United Nations made a declaration in 2000, known as UN Millennium Declaration, adopted at the largest ever presence of heads of states, which included '8 major Goals with 17 concrete targets' for "advancing development and reducing poverty by 2015". Interestingly, the first goal and the first two concrete targets were on 'eradicating extreme poverty and hunger', and the first of them was to halve poverty between 1990 and 2015 and halve the proportion of people in hunger during the same period. Though achieving the remaining 'goals' and 'targets' was very much integral to eradicating poverty and hunger, it becomes imperative to look first at the policy and planning made so far in making actual progress in the process of elimination of these two principal forms of deprivation.

There need not be any ambiguity about the fact that poverty or associated hunger is caused much more by economic (and social) than by geographical factors, though the latter has a serious role so far as natural resources are concerned. On the whole, it is a 'social phenomenon which results, as a rule, from failure to make proper use of natural resources, or from improper distribution of consumer goods and basic services, largely controlled by the socio-economic status of the people, giving rise to extreme economic inequality'. As has been observed quite some decades ago by Castro, with the forces of nature having been harnessed for mass production, technological advance has made the eradication of poverty and hunger a great possibility. It is no longer beyond the capability of the society to achieve this objective on the basis of "a better adjustment of men to the lands they occupy, and a better distribution of the gifts of the land among men. That this has not come to pass is because there has been no effort to enlarge mass consumption until it corresponds to mass production, and so to balance the economy of humanity" (Castro, 1955:235)

This, however, depends largely on the provisions of social security provided for every individual as a guarantee of minimal protection and survival. This leads us on to the existing relevance of entitlement and deprivation, as observed by Dreze and Sen (1989), who state that the 'only substantial asset that a person owns is his or her ability to work, i.e. labour power' and if he or she cannot sell or offer it for securing an employment, he or she loses the chance of acquiring food (and other basic requirements) resulting in 'serious deprivation – possibly even starvation and death'. Further, in the words of Dreze and Sen (1989), "A person can be reduced to starvation if some economic change makes it no longer possible for her to acquire any commodity bundle with enough food to survive. This 'entitlement failure' can happen because of a fall in her own endowment (e.g. alienation of land or loss of labour power due to ill health) or because of an unfavourable shift in her exchange entitlements (e.g. loss of employment, fall in wages, rise in food prices, drop in the price of goods or services she sells, decline in self-employed production)" (Dreze and Sen 1989: 23). (*See* Note A)

It had already been ascertained by the 1970s that economic growth is one of the most effective weapons dealing directly with joblessness which remains at the back of all forms of deprivation. Ensuring jobs becomes the first priority for people in poverty and hunger which can only be assured by a development policy in which people involved must have a role in the framing and execution of them. As pointed out in the report of the United Nations 2003, to achieve sustainable growth countries must attain basic thresholds in several key areas, such as, governance, health care, education, infrastructure and access to markets without which growth will flounder and fail in achieving the desired results. Certain structural constraints linked with a country's soil fertility, climatic shocks, natural disasters and rampant diseases like malaria are pointed out as obstacles to sustainable growth. But as "geography is not destiny", these challenges can be overcome by careful handling to help ultimately the country from falling into a 'poverty trap'. Accordingly, six major bunches of policies have been recommended to avoid the 'poverty gap'.

1. Generous investment from the beginning in basic education and health, encouraging gender equality which constitute the primary preconditions to sustained economic growth. In turn growth generates employment and raising income acts as feedback for further improvement in education and health.
2. Productivity of the small farmers has to be increased in unfavourable environments as they constitute the majority of the hungry people in the world – when 70 per cent of the world's poorest people live in rural areas. This needs structural changes as, in the majority of cases, a high proportion of rural land is held on a tenant or sharecropping basis with the landlord appropriating large shares of the total crop. Ending rural poverty and increasing food production depend on agrarian reforms, which means entitlement of farming land to farmers who till the soil and are primarily engaged in the production of crops.
3. Improvement in basic infrastructure, such as roads, power, communications and ports will lower costs in business and overcome geographical barriers.
4. An industrial development policy encouraging entrepreneurial enterprise is needed to help in diversifying the economy so that it does not remain dependent on primary commodity exports, but

takes an active role in fostering small and medium-size plants utilizing available resources for providing value-added exportables.

5. Conscious efforts should be there with respect to human rights, to the removal of discrimination or social exclusion, in securing justice with a whole-hearted approach to the well-being of all people and not governed by prejudice.
6. Lastly, environmental sustainability needs to be guaranteed and has to be supported by sound urban management ensuring that development improvements are long-term projects." (UNDP, 2003: 4).

"Growth has to be the 'principal' strategy for raising the incomes and hence consumption and living standards of the poor" (Bhagwati, 2004: 54). Here, growth, as explained by Bhagwati, is not a passive, trickle-down strategy for helping the poor but an active, 'pull-up strategy', as he calls it. According to him, improvement in production, or growth is not always welcome as it may lead to market saturation, a situation called by him "immiserizing growth" where the export- earning may fall. This can be avoided through diversification, giving exports larger acceptability. If growth is outward- oriented consisting of labour-intensive goods and light manufactures exported in far larger quantities, it is likely to increase the demand for labour and help the poor much more than any other type of growth, becoming 'more effective in reducing poverty through salutary effects of increased demand for unskilled labour' (Bhagwati, 2004: 57). At the same time, to improve access of the poor to growing incomes by way of self-help enterprise, micro-credit programmes have been very effective in which very small sums are lent out to poor individuals for investment that improves their ability to earn a livelihood without risking high interest, as has been proved to be a great success in Bangladesh, conducted through Grameen (Rural), Banks, primarily a gender-oriented programme. In the words of the founder of Grameen Bank in Bangladesh, Nobel laureate Md. Yunus, "if the goal of economic development include improved standards of living, removal of poverty, access to dignified employment and reduction in inequality, then it is quite natural to start with women. They constitute the majority of the poor, the under-employed and the economically and socially disadvantaged. And since they were closer to children, women were our key to the future of Bangladesh" (Yunus, 1998: 89). Starting the operation in 1996, the movement swelled into 112 branches all over the country by 1998, employing 12,000 people, dealing with 2,300,000 small borrowers 'face to face each week on their doorstep', lending out more than $35 million worth of Bangladeshi currency in tiny loans and, most interestingly, almost a similar amount comes back to the banks in repayments (Yunus, 1998: 116). This happens to be the way individual entitlements in Bangladesh have undergone a dramatic rise in the recent years for a large section of the population, hitting hard at the root of perpetual poverty, associated hunger and malnutrition leaving out millions in the cold without any charted-out path for survival. Interestingly, as pointed out in UNICEF report, 'The State of the World's Children 2007", helping women, giving them greater priority in the family, as done in Bangladesh Grameen Bank programme, is, perhaps, the most pragmatic approach to ensure social security since "a world fit for children is also a world fit for women. They are inseparable and indivisible - one cannot exist without the other... Women's rights are good for children and ultimately good for the world " (UNICEF, 2007: 15).

B. AN APPROACH TO THE ERADICATION OF POVERTY AND HUNGER

Efforts to eradicate poverty and hunger necessitate progress in a range of interlinked areas, such as economic growth creating work opportunities in decent conditions, social integration, demographic issues and environmental sustainability.

Economic growth in recent years has been quite impressive in many developing countries, with claims for each of the regions, bringing 135 million people out of poverty, with biggest drop in the number of people living on a dollar a day in East Asia. But there cannot be any complacency in this regard as progress has not been equally impressive and fast in many regions, particularly in Sub-Saharan Africa and the Least Developed countries. Sustained economic growth is a major factor in reducing poverty but it alone cannot attain the desired objective unless accompanied by strengthened institutional help, equitable delivery of public services, active social inclusion, bridging the gap between urban and rural development, and empowerment of the poor allowing their full cooperation and involvement in the developmental process. At the same time, emphasis needs to be put on climate-friendly development and "green growth".

World food security seems to be a major hurdle in the path to meeting the challenges to poverty and hunger reduction. Significantly, economic poverty reduction is not translated into improvements in food security. The majority of the world's poor living in rural areas are small farmers, marginalized in the context of global markets and hence unable to come out of poverty. Added to this is the monopoly of energy supply, and rising prices encouraging the expansion of biofuel production, conversion of agricultural land from the production of food crops into biofuel crops, naturally leading to further rise in the prices of food crops, such as the price of wheat pushed to a level never experienced before, with enhanced food insecurity. Glaring instances of land grab on a massive scale in some countries in recent years, as discussed earlier, is posing a serious threat to the global food situation, especially affecting the poor most, already deprived of access to enough food. It is reported by an authentic source that many millions of acres of maize, wheat and other edible crops planted in the United States and elsewhere in 2007 were used to provide biofuel for cars rather than food for the people, and the Food and Agriculture

Organization reported that 2008 had seen the biggest increase in undernourished people in decades (*See* Note B).

As reported by the FAO, the increasing food prices have triggered an increase in hunger worldwide, the number of chronically hungry people increasing by 75 million over and above the FAO's estimate of 848 million in 2003-05, which was 80 million higher than in 1990-92, bringing it to 923 million in 2007 and likely to increase further with continued rise in the prices of staple cereals and oil crops. As observed by IFPRI (International Food Policy Research Institute:1), 2002, each day 800 million people going hungry in the world include 170 million children under five years old suffering from malnutrition (IFPRI, 2002) (*See* Note C).

The rising food prices the world over has nearly reversed the progress achieved in reducing the share of hungry people, reduced from 20 per cent in 1990-92 to less than 18 per cent in 1995-97 and to a little over 16 per cent in 2003-05, moving back to 17 per cent in 2007 (Fig. 2.1). The impact was most severe in the region of Asia and the Pacific and in Sub-Saharan Africa, together accounting for 750 million or 89 per cent of the undernourished people in 2007, the rising food price adding a further 41 million and 24 million respectively to the total number of them in the two regions. As a result, 15 countries in Africa become the home to over 35 per cent of their population living in hunger. Latin America and the Caribbean, the Near East and North Africa as well have experienced increases in hungry people due to food prices. To be more specific, according to the FAO index, after 'four decades of

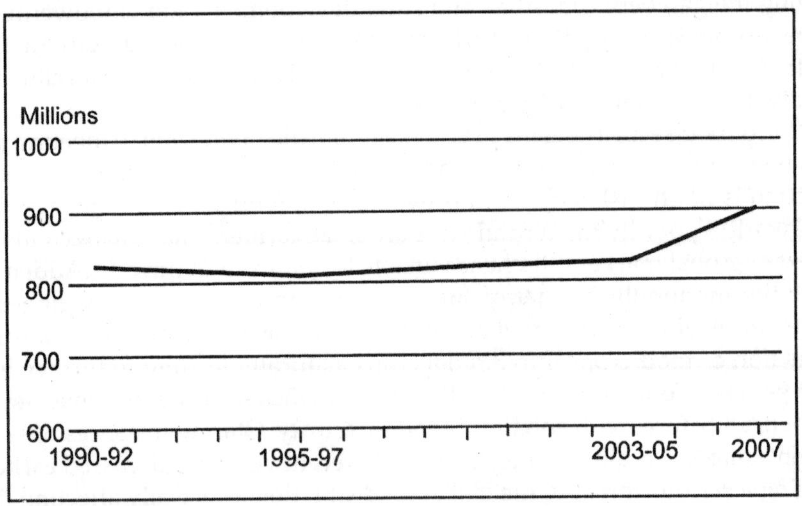

Source: FAO, 2008

Fig. 2.1: Number of undernourished people in the developing World, 1990-92 to 2007

predominantly declining trends', food prices doubled between 2002 and 2008, becoming 64 per cent higher in mid-2008, very much concurrent with the rise in petroleum prices which more than tripled since 2003. As observed in the report of the FAO 2008, "Petroleum and food prices are highly correlative. The rapid rise in petroleum prices exerted upward pressure on food prices as fertilizer prices nearly tripled and transport costs doubled in 2006-08" (FAO 2008: 10), hitting very hard especially the small farmers. This was also preceded by an unfavourable situation of extreme weather conditions in 2005-07, associated with droughts and floods affecting major cereal –producing regions resulting in a fall in world cereal production by 3.6 per cent in 2005 and 6.9 per cent in 2006 before recovering in 2007.

The entire scenario appears to be very gloomy for the food situation, which though production recovered by 2007, increasing cereals by 4.7 per cent in 2007, with a projected increase of 2.8 per cent in 2008, is not likely to be followed by a slump in food prices in the next decade for three reasons. First, "the economic growth in the developing countries", particularly in large emerging economies, such as China and India, "is expected to continue at least at 6 per cent per year, further raising the purchasing power and changing the dietary preferences of hundreds of millions of customers" (according to the pre-budget Economic Survey 2008-09, India could grow up to 7.75 per cent in 2009-10, up from 6.7 per cent in 2008-09 while the World Bank predicted India to grow 8.1 per cent in 2010, ahead of China with 7.5 per cent: *India Digest*, London, July 15, 2009). Second, "biofuel demand is likely to continue its rapid growth and according to the International Energy Agency, the share of the world's arable land devoted to growing biomass for liquid biofuels could triple in the next 20 years. Third, in addition to land and water constraints, including higher fertilizer prices, rising transportation costs resulting from higher petroleum prices, are likely to affect food production adversely, compounding the challenge of meeting the global food demand" (FAO 2008: 11).

However, economic growth based on agricultural development takes a very important role in assuring food security and thus lessening the burden of undernourishment for any country. But small farmers, ones who are most affected by any change in the scenario, face many challenges in their efforts to increase agricultural productivity (Fig. 2.3). 'Almost 85 per cent of the world's 460 million farms are small having less than 2 hectares (Table 2.1) and 90 per cent of the small farms are located in the developing countries. In productive efficiency, however, the small farms do better, giving higher yields per hectare of land compared to large farms. Nonetheless, the small farms face a number of problems dwarfing their efforts in adopting a more useful role in the lives of people dependent on them. Among the problems, (i) 'the high transportation costs; (ii) low capacities to implement food safety and quality measures; (iii) low access to finance, technology and information have an important role. It is suggested that policy actions in some areas, such as (a) improving access to markets and reduction in transaction costs substantially; (b) strengthening land

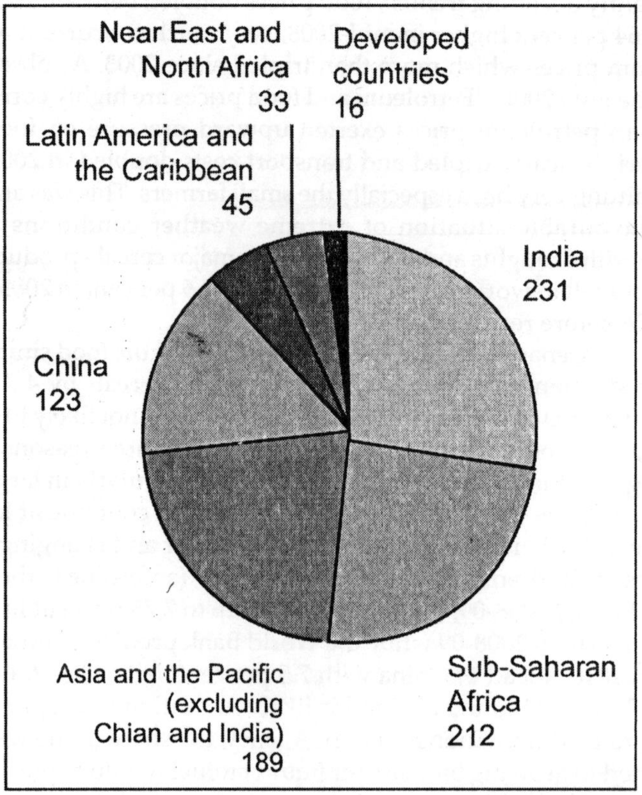

Fig. 2.2: Number of undernourished people in the World, 2003-05 (millions)
Source: FAO, 2008.

Table 2.1: World Farm Size and Distribution

Farm Size (hectare)	*% of all farms*	*Number of Farms (millions)*
<2	85	387
2-10	12	54
10-100	3	13
>100	1	2
TOTAL	100	456

Source: United Nations, 2007 (Strengthening Efforts to Eradicate Poverty and Hunger) (*Sourced* from Von Braun, 2004)

policies to secure rights of the farmers to land they cultivate; (c) creating a structure of incentives providing benefits to all the farmers; (d) improving

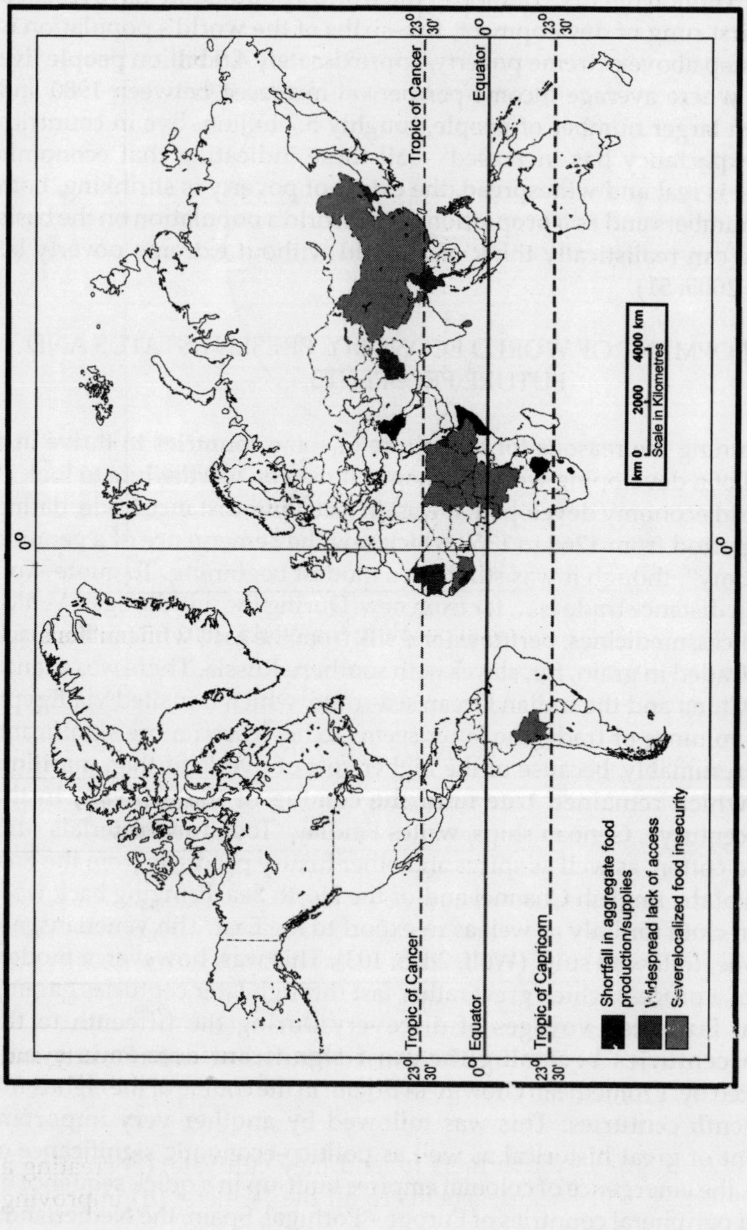

Fig 2.3: Countries facing food crises, 2008

Source: FAO, 2008

access to technology for the small producers; and (e) developing coping strategies so that they can manage risks may have a significant effect on the life and activities of the small farmers cushioning them against challenges affecting food security (United Nations 2007: 98).

In explaining 'why some countries fail to thrive', Sachs (2005) observes that roughly 5 billion out of 6.3 billion of the world's population 'have reached at least the first rung of development, five-sixths of the world's population is at least one step above extreme poverty, approximately 4.9 billion people live in countries where average income per person increased between 1980 and 2000, an even larger number of people, roughly 5.7 billion, live in countries where life expectancy has increased' – all these indicating that economic development is real and widespread, the extent of poverty is shrinking, both in absolute numbers and as a proportion of the world's population on the basis of which we can realistically think of a world without extreme poverty by 2025 (Sachs, 2005: 51).

C. DEVELOPMENT OF WORLD ECONOMY: PRESENT STATUS AND FUTURE PROSPECTS

Before examining the reasons for the failure of some countries to thrive in a background of a developing world economy, it may be worthwhile to look at how the world economy developed. It was, in fact, long-distance trade, dating back to the period from 1260 to 1350 which saw the "emergence of a genuine world economy", though it was still then a modest beginning. To quote Wolf (2005),"Long distance trade is ... far from new. During the middle ages, Venice imported spices, medicines, perfumes and silk from the east, while in the Black Sea, Genoa traded in grain, fur, slaves with southern Russia. There was a land-route from China and the Indian Ocean sea-route, which transited via Egypt. In terms of volumes of trade, the latter seems to have been more significant. This was, presumably, because of the higher cost of overland than maritime transport, which remained true until the coming of the railways in the nineteenth century". Genoan ships, writes Findlay, 'Took raw materials such as alum, and cotton, as well as spices and other luxury products from the East to the ports of the English Channel and/or the North Sea, bringing back wool and woollen cloth for Italy as well as re-export to the East. The Venetians, not to be outdone, followed suit" (Wolf, 2005: 103). This was, however, a modest beginning of a process which grew rather fast through later centuries gaining vastly from European voyages of discovery during the fifteenth to the eighteenth centuries becoming the most significant economic events supplemented by 'Promethean Growth' in Britain in the course of the eighteenth and nineteenth centuries. This was followed by another very important development of great historical as well as politico-economic significance of that period, the emergence of colonial empires built-up in a quick sequence by a number of peripheral countries of Europe – Portugal, Spain, the Netherlands, Britain and France – achieving control over much of the world seeking wealth through "plunder and trade" (often armed), though in the long run commerce proved more enduring and fruitful. The empires ultimately dissolved with the costs of control rising and rewards dipping but the result was the incorporation of much of the world into an economic system with its centre in Europe until the twentieth century after which that shifted to Europe and North America, and dominated by the West even today, but in which the advanced east Asian countries, specifically Japan, have since taken a role. Meanwhile, the European conquest of the Americas was a great event giving the Europeans the "ghost acreage", as it was called by Eric Jones, of 20 million sq. miles. This vast possession did not, however, benefit Spain or Portugal much besides giving them an opportunity for plunder, especially concentrated on silver and gold from Latin America in contrast to colonization and trade in North America under the aegis of the British, ultimately culminating in turning the economy of this British colony of North America consisting of the present United States and Canada three times larger than that of Latin America and the Caribbean at purchasing power parity, and more than five times in current US dollars.

The most spectacular success in getting overall control of the two Americas was maritime power achieved in developing a trade in sugar, tobacco, indigo (natural dye) and later cotton produced with the help of African slaves out of which grew the 'triangular trade' of the seventeenth and eighteenth centuries linking Europe with Africa and the 'New World', exchanging manufactures from the first (Europe) to slaves from the second (Africa) and commodities from the third (New World) (Wolf 2005: 104).

Angus Maddison, writing in his, '*The World Economy: A Millennial Perspective*' in 2001, mentions three periods of rapid growth in income per head in the world as a whole since 1820: 1870-1913; 1950-1973, the so-called 'golden age', and 1973-1998. Interestingly, in each of these three periods rapid growth of world exports in volume terms also happened in the following order: 3.4 per cent a year between 1870 and 1913; 7.9 per cent between 1950 and 1973, and 5.1 per cent between 1973 and 1998. Particular dynamic areas lie behind these periods of rapid world economic growth. The most 'dynamic parts' of the world between 1870 and 1913 were the former British colonies: the United States, Canada, Australia and New Zealand. Incomes per head grew at 1.8 per cent a year when exports grew at 4.7 per cent a year. Western Europe came to be the most dynamic region of the world economy – though, in fact, Japan was the fastest-growing individual economy – between 1950 and 1973, enjoying a growth rate of real incomes per head at 4.1 per cent a year with exports having grown at extraordinarily high rates of 8.4 per cent a year. But the picture changed after this between 1973 and 1998 when the fastest growing part of the world economy was non-Japan Asia where real incomes per head increased at 3.5 per cent a year while total exports from Asia, including Japan, grew at 6 per cent a year during this period (Wolf, 2005: 108). In this connection Table 2.2 gives a comparative picture of the growth of merchandise in exports for various regions of the world.

Table 2.2: Growth in Volumes of Merchandise Export: World and Major Regions, 1870-1998
(Annual Average Compound Growth Rates in percentage)

Region	*1870-1913*	*1913-1950*	*1950-1973*	*1973-1998*
Western Europe	3.24	-0.14	8.38	4.79
Western Offshoots*	4.71	2.27	6.26	5.92
Eastern Europe and former USSR	3.37	1.43	9.81	2.52
Latin America	3.29	2.29	4.28	6.03
Asia	2.79	1.64	9.97	5.95
Africa	4.37	1.90	5.34	1.87
World	3.40	0.90	7.88	5.07

Source: Wolf, 2005 (Ref. Maddison, *The World Economy: A Millennial Perspective*)
* United States, Canada, Australia and New Zealand

Thus, over a period of 250 years, 1700-1950, 'a global division of labour developed and intensified with industrialization' in which the newly industrialized countries of the West, led by Britain, some West European countries and later the United States took a dominant role involving dramatic changes with some of these economies experiencing a progressive decline, particularly in the late nineteenth and early twentieth centuries. As Fig. 2.4 shows, while there has been deep decline of the GDP in Asia during the period 1700-1950, sliding down from nearly 60 per cent of the world GDP in 1700 to less than 20 per cent in 1950, the United States maintains a record of a steady rise of the GDP through each of the periods between 1700 and 1950. Both India and China exhibit features similar to that of a constant decline. However, world economy has performed better in the last half century as compared with past, increasing six times from 1950 to 1998 with an average growth of 3.9 per cent a year compared with 1.6 per cent from 1820 to 1950, and 0.3 per cent from 1500 to 1820 (Dicken, 2007: 35). Some twenty years ago, the Japanese Management writer Konichi Ohmae coined the word 'global triad', to mean that the 'world economy is now essentially organized around a tripolar, macro-regional structure where three pillars are North America, Europe and East Asia, and on the basis of data on production, trade and Foreign Direct Investment (FDI), the three macro-regions contain 86 per cent of world GDP and 82 per cent of the total world merchandise exports and 98 per cent of outward stock of foreign direct investment (Dicken, 2007: 38).

A Global Triad: Concentration of World GDP and Exports

North America: GDP: $12,431.15 billion, or, 34.1 per cent of World GDP
Total Exports: $ 1,162 billion, 15.9 per cent of the world's total
FDI: Inward Stocks: 22.0 per cent; Outward: 24.7 per cent
FDI: Inward Flows 35.5 per cent: Outward: 46.4 per cent

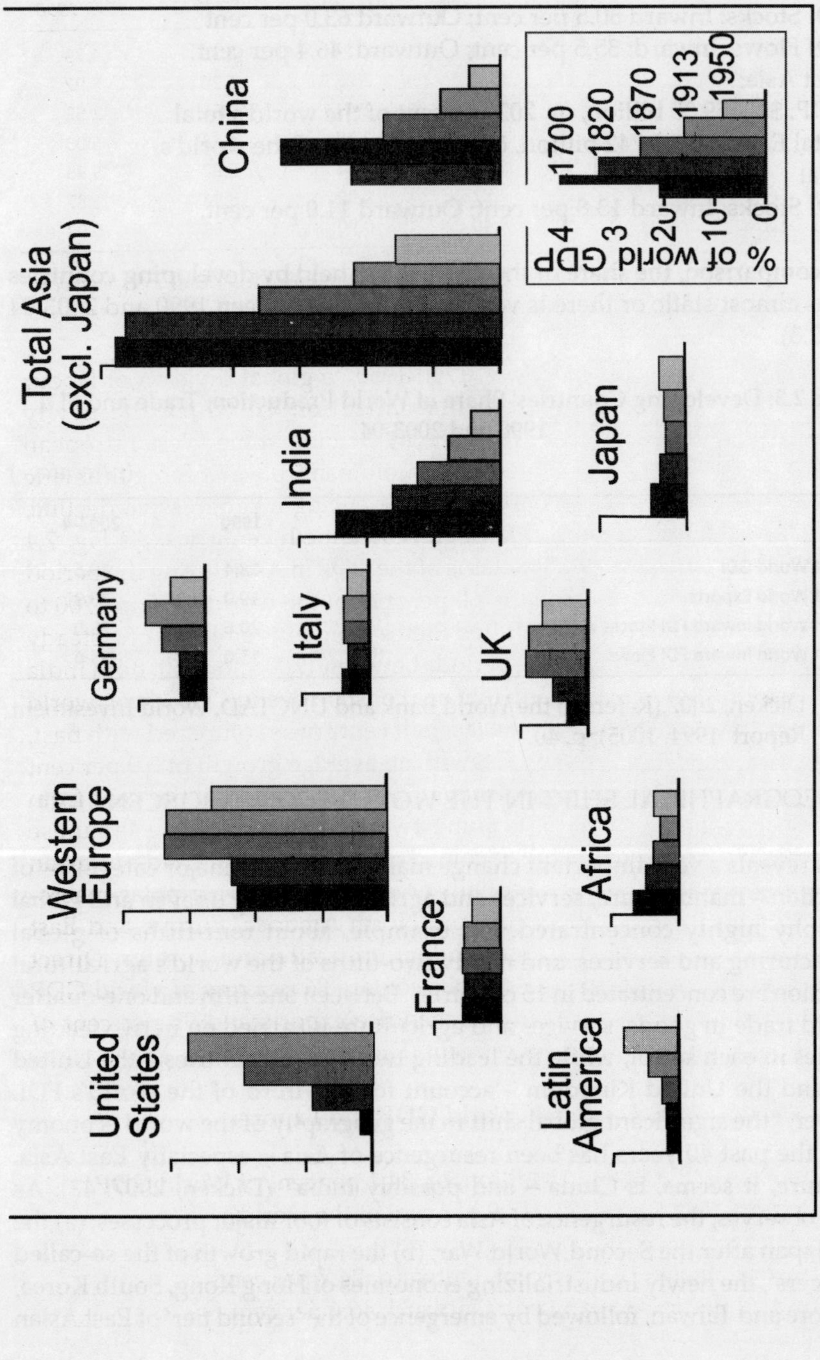

Fig. 2.4: Global shifts GDP, 1700-1950

Source: Peter Dicken, 2003

Europe: GDP $ 11,526.45 billion, or, 31.6 per cent of world GDP
Total Exports: $ 3,337 billion, or, 45.7 per cent of the world's total
FDI Stocks: Inward 50.5 per cent; Outward 63.0 per cent
FDI Flows: Inward: 35.5 per cent; Outward: 46.4 per cent
East Asia:
GDP: $ 7,369.35 billion, or, 202 per cent of the world's total
Total Exports: $ 1,742 billion, or, 20.2 per cent of the world's total
FDI Stocks: Inward 13.8 per cent; Outward 11.0 per cent.

In comparison, the share of the global GDP held by developing countries remains almost static or there is very little change between 1990 and 2003-04 (Table 2.3).

Table 2.3: Developing Countries' Share of World Production, Trade and FDI, 1990 and 2003-04

(percentage)

	1990	2003-4
Share of World GDI	18.4	19.5
Share of World Exports	19.0	26.3
Share of World Inward FDI Stocks	20.6	25.0
Share of World Inward FDI Flows	15.0	35.6

Source: Dicken, 2007 (Refers to the World Bank and UNCTAD, World Investment Report 1994. 1005): p. 40.

D. GEOGRAPHICAL SHIFT IN THE WORLD ECONOMIC SCENARIO

All this reveals a very important change making the three major categories of production – manufacture, services and agriculture – very uneven and global geography highly concentrated. For example, about four-fifths of global manufacturing and services, and nearly two-fifths of the world's agricultural production are concentrated in 15 countries. Between one-fifth and one-quarter of world trade in goods, services and agriculture is carried on in two leading countries in each sector, while the leading two source countries – the United States and the United Kingdom – account for one-third of the world's FDI. However, "the significant global shift in the geography of the world economy during the past 40 years has been resurgence of Asia – especially East Asia. The future, it seems, is China – and possibly India" (Dicken, 2007: 43). As Dicken observes, the resurgence of Asia consists of four major processes: (a) the rise of Japan after the Second World War; (b) the rapid growth of the so-called 'four tigers', the newly industrializing economies of Hong Kong, South Korea, Singapore and Taiwan, followed by emergence of the 'second tier' of East Asian developing economies, mainly Indonesia, Malaysia and Thailand; (c) emergence of China – the 'dragon' – a 'powerhouse whose size and efficiency few competitors can match', and (d) the 'potential economic dynamism of India".

Among these new powerhouses of world economy, the postwar growth of Japan's economy is spectacular with its manufacturing growth averaging 13.4 per cent per year in the 1960s, making it the second largest economy by 1980 and in spite of a few slumps, it later recovered enough to maintain its position of second largest economy preceded only by the United States. The countries joining this race of manufacturing growth, such as South Korea and Taiwan registered high rates, the annual rates in South Korea averaging 18 per cent in the 1960s, 16 per cent in the 1970s, 13 per cent in the 1980s and 7 per cent in the 1990s (to 2003); Taiwan's manufacturing industry grew at the rates of 16, 14, 8 and 6 per cent respectively during the same periods. Later, extremely high rates of manufacturing growth were also attained by Malaysia, Thailand and Indonesia. Similarly, the rate of East Asian countries in exports of manufacturing is equally unique. The seven East Asian Countries – South Korea, Hong Kong, Singapore, Taiwan, Indonesia, Malaysia and Thailand – jointly increased their share of world manufacture exports from a mere 1.5 per cent in 1960s to almost 20 per cent in 1990s. But compared to all this China's achievement is unmatchable. Between 1980 and 2003, China's growth rate (of GDP as a whole and manufacturing) was 'the highest in the world', with an annual average of about 10 per cent and an annual growth rate of merchandise export increasing from 13 per cent in the 1980s to 14 per cent between 1990 and 2003. All this turned China into the world's fourth largest manufacturing producer, the second largest agricultural producer, the fourth largest exporter of merchandise (to displace Japan from the third place in a short while), and the third largest importer. China is very likely to emerge as the second largest economy and the second largest exporter in the world in 2010 (*See* Note D). India, on the other hand, has exemplified spectacular growth in one specific type of economic activity: the outsourcing of IT services, such as software, data processing, cell centres, etc. Meanwhile, India's GDP growth rate between 1980 and 2003 though was higher than the world average of 5 to 6 per cent, it remained at half that of China during the same period, but increased to 9.4 and 9.6 per cent respectively in the years 2005-06 and 2006-07, and is likely to be maintained at 9 per cent during the period 2007-08 to 2011-12 (Government of India, 2008: 3).

India is the world's 13th largest manufacturing economy while China is the fourth largest. India is not among the 15 top merchandise exporters; China is the fourth largest. India does not have a strong export base in manufactures. China's merchandise exports are eight times larger than India's, the latter taking ninth place among the nine leading East Asian countries in merchandise export. However, the situation seems to have largely improved for India in merchandise exports, having grown continuously at more than 20 per cent since 2002-03 and was expected to reach 22.6 per cent in 2006-07. Further, the value of merchandise exports increased by 21.6 per cent in April-December 2007, valued at US $ 111 billion which was likely to increase further to US $ 160 billion in 2007-08. Manufacturing industries in India have grown by 9.2 per cent in 2004-05, 9.1 per cent in 2005-06 and 12.5 per cent in 2006-07, and was likely to maintain growth at more than 9 per cent in 2007-08 (Government of India, 2008: 182). At the same time, India's performance in the production of food crops increased at an average rate of 2.5 per cent between 1950-51 and 2006-07 compared to the growth of population averaging 2.1 per cent during this period though it witnessed a steep slump between 1990 and 2007 when the rate of growth declined to 1.2 per cent, lower than the growth of population, averaging 1.9 per cent.

Further change in the geography of world economy took place from 1989 onwards in a part of the world till then dominated by the erstwhile USSR, the collapse of which gave rise to 'transitional economies', affecting an important change in the situation. The USSR accounted for almost 10 per cent of world manufacturing output which came down to about 1 per cent for the Russian Federation. Since then it along with Poland, the Czech Republic and Hungary representing the most significant transitional economies, accounted for most of the manufacturing production and exports for this group. Poland, the Czech Republic and Hungary increased their exports by over 10 per cent a year while growth was 7 to 8 per cent for the Russian Federation.

Latin American countries, on the other hand, are among the most resource-rich in the world but show very modest economic performance with average export-growth significantly lower than East Asian countries during the past two decades though in economic growth there had been wide differences between countries, with Chile and Mexico maintaining relatively faster growth than the comparatively slower rate of economies in Argentina and Brazil.

Most of the countries of Africa, parts of Asia and part of Latin America in contrast constitute the 'troughs' of the global economic map with Sub-Saharan Africa forming the largest single area of 'economic peripherality', entangled in 'deepest poverty and deprivation', posing great social challenges.

However, it is further relevant to note the slower global economic growth in 2007 following a 'solid broad-based growth in terms of the growth of world gross product (World Gross Product, or, WGP is "calculated using country weights for gross domestic product, or, GDP in dollars at market prices") considered for three consecutive years' going down from 3.8 per cent in 2006 to 2 per cent in 2007. As noted in the report of the United Nations, the main reason for this global slowdown is the 'weakening housing market to a rate of 2.2 per cent in 2007' in the United States when "no other developed economy is expected to emerge as an alternative engine for the world economy, as growth in Europe is forecast to slow to around 2 per cent and in Japan to below 2 per cent in 2007" (United Nations, 2007: iii). As a matter of fact, Japan fared worst among the leading developed countries recording negative growth in 1998 and 1999, and its growth was one of the lowest in 2007 as in Germany. In contrast to this situation, strong economic performances have continued unhindered in the developing countries and in the Economies in Transition during 2006, reaching average growth rates of 6.5 and 7.2 per cent respectively (Table 2.4). The performance has been much better in this regard in countries with 'Economies in Transition' among which Azerbaijan provides an instance of exceptionally high rate of growth, remaining above 20 per cent in each of the year between 2005 and 2007 as an exporter of oil. Growth is expected to remain strong in these regions with a slight fall to 5.9 per cent in the developing economies and 6.5 per cent in Economies in Transition as indicated by 'strong demand for and higher prices of energy and primary commodities'. At the same time the performance of the 'Least Developed' countries remains strong, averaging 7 per cent in 2006. In the poorest countries as well growth continues to remain strong. Interestingly, growth in the developing countries is largely due to rapid industrialization besides growth in services in countries like China, India (with growth remaining consistently above 10 per cent for each year between 2003 and 2006 in China, and above 7 per cent between 2003 and 2007 in India, and exceeding 8 per cent in 2004 and 2005) and a few other emerging economies, such as those in Latin America, become highly dependent on the production and exports of primary commodities. Nevertheless, the situation for developing countries may not remain the same given the economic slow down in the developed countries, such as the United States and those in Europe as well as Japan, since the developed countries take the lion's share of the global demand for primary commodities exported by the developing countries, and any prolonged recession in the developed countries curtailing their demand for commodities they import will have a consequent effect on the economies of the developing countries in the long run.

E. REASONS FOR FAILING ECONOMIC GROWTH

As poverty and hunger are the results of economic failure of a country, one needs to look for the actual reasons leading to such a situation. Sachs (2005) refers to eight major types of problems that can cause an economy to stagnate or decline adversely affecting the life of the people.

1. Poverty itself is considered to be a trap in the poorest countries where in the absence of necessary capital and infrastructural growth in the form of paved roads, transport, power and irrigation facilities no developmental work can take place. The people are too poor to save to generate capital for investment in income and employment generating activities. On the contrary, the poor cause depletion of natural resources by cutting down trees, exhausting soils of their nutrition doing what little farming they do in the absence of fertilizers, mining their minerals to exhaustion and overfishing, ultimately leaving no prospect for a developed

Table 2.4: Rates of Growth of Real GDP, 1997-2007

(Annual percentage change)

Selected Economies	1997-2005	1997	1998	1999	2000	2001	2002	2003	2004	2005	2006	2007
Developed Economies	2.5	3.2	2.5	3.0	3.6	1.2	1.2	1.9	3.0	2.5	2.9	2.2
United States	3.2	4.5	4.2	4.5	3.7	0.8	1.6	2.5	3.9	3.2	3.2	2.2
Canada	3.5	4.2	4.1	5.5	5.2	1.8	2.9	1.8	3.3	2.9	2.8	2.6
Japan	1.1	1.4	-1.8	-0.2	2.9	0.4	0.1	1.8	2.3	2.6	2.5	1.7
Australia	3.6	3.8	5.2	4.4	3.4	2.4	3.9	3.3	3.3	2.9	2.5	3.6
European Union	2.3	2.7	2.9	3.0	3.9	2.0	1.2	1.3	2.4	1.7	2.7	2.4
France	2.2	2.1	3.2	3.1	3.9	1.9	1.0	1.1	2.3	1.2	2.3	2.3
Germany	1.4	1.8	2.0	2.0	3.2	1.2	0.0	-0.2	1.3	0.9	2.2	1.7
Ireland	7.3	11.7	8.5	10.7	9.2	5.8	6.0	4.3	4.3	5.5	5.8	5.3
Luxembourg	4.9	5.9	6.5	8.4	8.4	2.5	3.8	1.3	3.6	4.0	5.5	4.5
U.K.	2.8	3.0	3.3	3.0	3.8	2.4	2.1	2.7	3.3	1.9	2.6	2.7
Economies in Transition	4.6	0.9	-2.7	3.9	7.9	5.7	5.1	7.0	7.7	6.4	7.2	6.5
South-eastern Europe	3.0	-0.7	0.3	-1.5	3.5	4.8	4.9	4.7	6.7	4.7	5.9	5.5
Commonwealth of Independent States[a]	5.0	1.4	-3.6	5.4	9.1	6.0	5.2	7.6	7.9	6.8	7.5	6.7
Russian Fed.	4.7	1.4	-5.3	6.4	10.0	5.1	4.7	7.3	7.2	6.4	6.7	5.8
Azerbaijan	11.2	5.8	10.0	7.4	11.1	9.9	10.6	11.2	10.2	26.4	32.5	28.0
Developing Countries[b]	4.6	5.4	2.1	3.5	5.7	2.7	3.9	5.2	6.9	6.4	6.5	5.9
Africa	3.8	3.2	3.4	2.8	3.2	3.7	3.4	4.7	4.8	5.4	5.6	5.6
North Africa	4.0	2.7	4.5	3.5	2.8	3.5	3.2	5.4	50	5.0	6.1	5.5
Sub-Saharan Africa (excluding Nigeria and South Africa)	4.2	4.5	4.3	2.6	2.4	4.9	4.0	3.3	5.7	5.9	5.9	6.6
East Asia and South Asia	5.9	5.9	1.2	6.3	7.1	4.5	6.5	6.9	7.8	7.5	7.4	6.9
Latin America & Caribbean	2.7	5.5	2.5	0.5	4.0	0.2	-0.8	2.0	5.9	4.5	5.0	4.2
Least Developed Countries	6.4	5.4	4.9	5.6	5.8	6.6	6.4	6.5	7.9	8.3	6.9	7.1
Major Developing Countries												
Brazil	2.2	3.3	0.1	0.8	4.4	1.3	1.9	0.6	4.9	2.3	3.3	3.5
China	9.0	9.3	7.8	7.6	8.4	8.3	9.1	10.0	10.1	10.2	10.2	8.9
Hong Kong SAR[c]	3.8	5.1	-5.5	4.0	10.0	0.6	1.8	3.1	8.6	7.3	6.2	6.2

(Contd.)

Table 2.4 (*Contd.*)

Selected Economies	1997-2005	1997	1998	1999	2000	2001	2002	2003	2004	2005	2006	2007
India	6.2	5.2	5.6	6.9	4.7	4.9	4.7	7.4	8.5	8.0	7.7	7.9
Indonesia	2.2	4.7	-13.1	0.8	4.9	3.8	4.4	4.7	5.1	5.6	5.3	6.0
Korea, Rep. of	4.2	4.7	-6.9	9.5	8.5	3.8	7.0	3.1	4.7	4.0	5.0	4.3
Malaysia	4.0	7.3	-7.4	6.1	8.9	0.3	4.4	5.4	7.1	5.0	5.5	6.0
Pakistan	4.3	1.8	3.1	4.0	3.5	2.5	4.1	5.7	7.1	7.2	6.5	6.2
Taiwan	4.2	6.4	4.5	5.7	5.8	-2-2	4.2	3.4	6.1	4.0	4.1	4.2
Thailand	2.4	-1.4	-10.5	4.4	4.8	2.1	5.4	6.9	6.4	4.5	4.2	4.8

Source: Compiled from United Nations: World Economic Situation and Prospects 2007: Tables A.1-A.3.

Notes

a. CIS consisting of 12 independent states;
b. Developing countries covering countries that account for 98 per cent population of all developing countries;
c. Special Administrative Region (of China)

economy. Any ameliorating process could be initiated by the government doing away with corruptive practices inviting people's participation.

2. Physical geography plays a major role in determining economic development on the basis of available natural resources, good soils, sufficient rainfall, navigable rivers, and being endowed with a coastline encouraging trading activities. A look at many of the countries steeped in extreme poverty can be explained largely by the lack of these resources or their presence being not enough to ensure necessary development. Any success in this regard nevertheless depends very much on good governance free of corruption and however large may be the population of a country or countries, absence of education and health services keeping people illiterate and unhealthy make human resource rather a burden than an asset. Physical setting like landlocked situation giving rise to high transport costs make poverty a persistent feature in them, as it happens in the case of Bolivia, Ethiopia, Kyrgyzstan, Nepal or Tibet. A comparative assessment of certain facilities available in Africa and Asia will reveal the difference and cause of greater hardship. For example, only 6 per cent of the arable land in Africa is irrigated compared to 40 per cent in Asia; only 2.2 kgs. of fertilizer per hectare of arable land in Africa (0.8 kg. in Sub-Saharan Africa) is utilized which is less than 15 per cent (and 7 percent in Sub-Saharan Africa) of the level in Asia and Latin America; the number of tractors per 1000 hectares of arable land is 3 times greater in Asia and 8 times greater in Latin America than in Africa; road density is more than 2.5 times higher in Latin America and 6 times higher in Asia than Africa (United Nations, 2007: 123). Further, arid conditions with low agricultural productivity and vulnerability to prolonged droughts crippled the economy of many African and Asian countries. Ecological conditions favouring killer diseases like malaria, schistosomiasis, dengue fever, and dozens of others have retarded economic progress in tropical countries for a long time and among them malaria is held primarily responsible in slowing down Africa's economic development throughout history.

All the same, tackling geographical disadvantages is not beyond human skill. There are many ways of meeting the challenges, doing a great deal in solving a difficult situation. Roads can be built from a landlocked country to connect the ports in another country. Tropical diseases can be controlled, as is being done for HYV/AIDS and malaria, the latter though a very much treatable disease, surprisingly claims up to 3 million lives per year, overwhelmingly in Africa (90%), the rest of the deaths occurring in tropical regions of Asia and the Americas. Arid climates can be overcome with irrigation as is being done in Israel and many Arab countries. Steps can be taken in advancing educational and similar other services though the overhead costs become sometimes quite prohibitive in certain cases which might lead the country through debts into a poverty-trap.

3. The third major problem is the fiscal trap implying that the government is not resourceful enough to invest in the development of public goods and services, such as roads, power grids, ports, health and education when the population is impoverished to pay enough taxes for creating capital to be utilized by the government, the latter may already be carrying a huge load of debt for which the limited tax revenue may have to be used to pay off the debt rather than finance new ventures to the benefit of the people, and lastly, the government may be corrupt and inefficient, incapable of deriving a larger revenue through taxes imposed on the people. Debt cancellation on the part of the donor country/countries in such cases is supposed to be the only way to give the country a fresh start.
4. According to Sachs (2005), the primary requirement for economic development is a government 'oriented toward development' for which it has to identify and finance the areas of higher priority infrastructure projects making it along with social services available to all the people without any selective approach. At the same time greater emphasis should be laid on creating an environment conducive to private investment with openness in safeguarding the interests of private investors.
5. Cultural barriers in the form of cultural or religious practices blocking the role of women in the society, is actually depriving the society of the services of half the population when women have no political or economic rights. Denying education makes a great effect on the fertility rates changing the structure of population particularly in the poorer countries where the poor households have a larger number of children. Cultural barriers may similarly apply to religious or ethnic minorities preventing them from gaining access to public services like education, health facilities or job-training. In extreme cases it may turn into ethnic cleansing as happened for Indians in East Africa or what is happening now in some other countries of Africa – all of them having a very dampening effect on the local economy.

6. Sachs (2005) calls it 'geopolitics' to explain trade barriers erected by developed countries to impede a poor country's economic development. Trade sanctions imposed by a powerful country on a 'regime' it does not like apart from weakening or toppling a regime, will impoverish the population of that country even though the regime does not get toppled. Such a situation is far from acceptable as the increasing animosity between opposing governments creates a greater damage to the people suffering from impoverishment.
7. In the absence of a proper market in the poor countries bringing sufficient profits for an innovation, investment in research and development is very low. In contrast, the rich countries having a large market increasing the incentives for innovation bring new technologies to the market raising productivity and expand the size of the market for fresh innovations (what the economists call endogenous growth), which has a very significant role to play as innovation increases the size of the market, while in turn the larger market raises the incentives for innovation.

   Advanced countries of North America, Europe and East Asia invest 2 to 3 per cent or over of their GDP, amounting to hundreds of billions of dollars each year in research and development activities. In contrast, the innovation process does not even get started in poor countries which makes production often of low standard and unresponsive in creating jobs. Thus, the existing inequality of innovative activity magnifying inequality of global incomes makes the situation worse. As a result, the low-income countries as home to 37 per cent of the world's population and 11 per cent of the world's GDP (adjusted for difference in purchasing power), accounted for less than 1 per cent of all the US-registered patents taken out by investors in the year 2000.

   The innovation gap, therefore, explains why the poorest of the poor do not have a foothold on growth. But they are not entirely barred from access to new technology which is easily accessible through computers and the like, imported through consumer goods, text books or in many other ways. There is, however, some limitations as many high-level technologies developed in rich countries relevant for particular ecological conditions may not successfully operate in the tropical, arid or mountainous countries where so many of the extreme poor live today (Sachs, 2005: 56-64). In seeking newer locations in the developing countries for investments, the advanced countries have so far kept close to sites lying on the Asia-Europe trade-route, such as Penang Island

(Malaysia), Singapore, Taiwan, Hong Kong and Mauritius – each of them being an island enjoying a coastal location; even Shanghai, China's leading economic city, is on the coast at the mouth of Yangtze River. In contrast, hinterlands have failed to attract the development of new industries with large investments.

8. An equally valid reason for poverty is the demographic trap when poor families choose a large family though they cannot afford to invest in nutrition, health and education of each child. And as long as women are not empowered through education to make a choice of their own in this regard, the rate of fertility will remain high.

The answer to the question – why some countries grew while others declined may be found in the fact that out of 58 poor countries not producing oil (with a per capita income below $3000), 22 experienced decline. Even some poor countries having high cereal yields per hectare using high levels of fertilizers have experienced economic growth. As revealed by data available from the World Bank (2004), there is a big difference between Africa and East Asia, among the developing or poor countries, in per capita food production, which has been high and rising in East Asia in recent decades but slow and declining in Africa (Table 2.5).

Table 2.5: Some Socio-Economic Facts: East Asia and Sub-Saharan Africa in 1980

	East Asia	*Sub-Saharan Africa*
Crop Yield (kg/hectare)	2,016	927
Irrigated Land (% of cropland)	37	4
Share of Crop area Planted to modern varieties (%)	43	4
Adult Literacy Rate (%)	70	38
Infant Mortality Rate (per 1000 live births)	56	116
Total Fertility Rate (births per woman)	3.1	6.6

Source: Sachs, 2005 (Ref. *The World Bank*, 2004)

The presence of an extensive network of roads in East Asia with a dense population is helpful for carrying fertilizers to rural farms and bringing farm output to the market. Use of fertilizer and irrigation in Asia make food yields very high and donor agencies have given ample support to the introduction of high-yielding variety (HYVs) seeds in Asian countries resulting in the famous Green Revolution benefiting farmers with higher returns. In contrast, rural Africa is much less densely populated, with an absence of roads to carry fertilizers and crops. The farmers do not use fertilizers on crops, but depend on rainfall making the situation uncertain for donors to take much interest in changing agricultural performance through the introduction of new varieties of seeds for increasing production. Hence, farmers with low productivity cannot escape poverty, a condition aggravated further by low rates of adult literacy and high rates of infant mortality and fertility. Investors from outside do not feel attracted to a situation where returns are not expected to be high, while larger investments in infrastructure, such as roads and power supplies, seem more likely to derive greater benefits with larger and more densely populated economies.

F. GROWTH, INEQUALITY AND POVERTY

In measuring success achieved so far in eradicating poverty, hunger and malnutrition, it becomes more relevant to make a comparative assessment of the impact of economic growth on equality and poverty.

'Inequality is a measure of relative income'. Looking at what happened in this context it is interesting to note that according to the World Bank, between 1980 and 2000, the Chinese average real incomes rose by about 440 per cent when it rose per head by only 60 per cent in the United States. In Purchasing Power Parity (PPP), the ratio of Chinese real incomes per head to those of the United States rose, accordingly, from just over 3 per cent in 1980 to just over 12 per cent in 2000 which is a 'big reduction in relative inequality'. "On all measures, global inequality rose until about the early 1980s. Since then, it appears, inequality among individuals has declined as a result of the rapid growth of much of Asia and, above all, China. But it is also important to understand what drove the long-term trend towards global inequality over almost two centuries. It is the consequence of dynamic growth that spread, unevenly, from the UK in the course of the nineteenth and twentieth centuries. In the process a growing number of people became vastly better off than any one had ever been before, but can have become worse off. Such dynamic growth is bound to be uneven" (Wolf, 2005; 157). Because of better climates, resources and policies some countries, quite naturally, could take better advantage of the new opportunities for growth, such as China and India. In this perspective, it is better to focus on poverty than on inequality.

An income of one US dollar a day at 1985 international prices (PPP) has been defined by the World Bank as extreme poverty, Bourgingnon and Morrison use the same figure to analyse extreme poverty since 1820 (Wolf, 2005: 157). On this basis of classification, the number of people in extreme poverty increased from 900 million in 1820 to a peak of 1.3 to 1.4 billion between 1960 and 1980, before falling slightly to 1.3 billion in 1992. As mentioned earlier in Chapter I, the proportion of the world's population living on less than a dollar a day fell from 80 per cent in 1820, a time when living on the margin of subsistence was the norm, to about two-thirds at the beginning of the twentieth century, to about 50 per cent by 1950, and 32 per cent in 1980 and finally, to 24 per cent by 1992, during which period the world's population increased from a little over a billion in 1820 to 1.7 billion by 1910, and by 1992 it had risen to 5.5 billion (Wolf, 2005:157-158).

Undoubtedly, the proportion of people in extreme poverty, however defined, has fallen enormously over the last two centuries, and the decline has certainly continued since 1980. The number in extreme poverty has fallen in East Asia particularly during the past two decades, both China and India showing very large declines as a result of accelerated economic growth: about 100 million in India between 1980 and 2000, and 220 million in China between 1978 and 1999 despite large increases in population in these two countries.

Reasons behind the changing scenario with declining numbers of people living in extreme poverty may be found in growth in food production, having substantially outpaced that of population, increasing the 'average daily supply of food per person by 24 per cent between 1961 and 1999 globally'. In developing countries it increased by 39 per cent, touching 2,684 calories per head against the average need of an adult being between 2,000 and 2,310 calories as per the FAO's estimate. In China, the average food supply had gone up 82 per cent, or 3,044 to 2,417 calories, from 1,635 calories in 1950-51. As a result, the food supply in the developing countries has on average changed from inadequate to adequate, though hunger still persists. According to the estimates of the FAO, the number of people suffering from chronic malnutrition declined from 920 million in 1969-71 to 790 million in 1997-99, from 35 to 17 per cent. Elsewhere, the figure of undernourished people is given as a little over 820 million for the developing countries in 2001-03, comprising 17 per cent of their total population (FAO, 2006:118/Table A2). However, the situation in sub-Saharan Africa was far worse where the population of undernourished people fell from 38 to 34 per cent between 1979-81 and 1997-99, but absolute numbers increased in a rapidly growing population, from 168 million to 194 million (In another estimate FAO 2006 gives it at 206.2 million in 2001-03, constituting 35 per cent of the region's total population).

A clearer view may, perhaps, be obtained about the recent trends of regional economic development as reported in "World Economic Situation and Prospects 2007" by the United Nations. According to this report, economic growth in development in recent years has remained strong, increasing slightly from 6.4 per cent in 2005 to 6.5 per cent in 2006, sustained through 'rising commodity prices and strong external demand' particularly for manufactures and especially for great benefits for East Asia from exports of the latter while Western Asia gained primarily from rising oil prices. In spite of that, however, barring Latin America, the unemployment situation has not improved much in some parts of East Asia and scarcity of sufficient number of jobs and increasing unemployed labour force make poverty a great challenge, particularly in Africa and Western Asia.

It may be observed that Africa's GDP has increased by 5 per cent in 2006 with its momentum continuous for the previous three years and was likely to be maintained in 2007, but growth 'highly concentrated in a narrow range of activities' makes Africa's economy highly vulnerable to such factors as 'weather conditions, terms of trade development and aid flows' and growth has not been accompanied by job creation, affecting hard the continent's ability to reduce poverty. The fact is, in many countries "growth rates have not been strong or sustainable enough to generate efficient labour demand' for the increasing labour. Besides,"the high volatility of GDP growth" leads to uncertainty in future profitability, hampering job creation. Further, the shift of economic activity from agriculture into more capital-intensive sectors, such as mining and oil production with commensurate very low absorption of labour has not been accompanied by any increase in agricultural productivity which ultimately has led to persistent high underemployment in the rural sector. As a result, the economy has not become robust enough to compete in the global market and even its exports with concentration in a few commodities (without diversification) have slowed down growth and the deterioration of the terms of trade offers it a small share of world exports, the latter showing a declining trend over the past few decades. In many African countries the amount spent on debt-repayment exceeds expenditure on social services like education and health undermining the continent's growth prospects. Finally, the current economic global recession (2008-2009) may "significantly hamper the region's prospect, as exports are a key engine of growth in the continent" (United Nations, 2007: 108).

In East Asia, on the other hand, economic growth peaking at 7.6 per cent in 2006, surpassed that of 2005, with China setting the pace for this soaring growth. The pace was maintained during the first half of the year (2006), with robust export and investment boosting the Chinese economy when other countries as well enjoyed strong external demands for their exports. But eventually the momentum slowed down during the second half of the year resulting from the tightening measures started taking effect in China in the form of controlling investment growth, raising interest rates in banks, change in land use policies and restrictions for foreign investments in the housing market which was to continue in 2007 as well as 'the government... putting increased emphasis on a more equitable and sustainable growth path'. Similarly, weakening of the global electronic cycle and deceleration of export growth, the latter having been at the root of economic growth, slowed down the economies of newly industrialized countries of East Asia, causing decline in the domestic investments in countries like Thailand and the Philippines of which in the former 'investment growth fell from a three-year average of 15 per cent per annum to 3.6 per cent in 2006. Interestingly, several countries, such as Hong Kong, Taiwan, Singapore including China are experiencing "jobless" economic growth showing 'no reduction in unemployment rates since 2003' which may have been the "result of the continued restructuring, strong productivity increases and internal labour migration" (United Nations, 2007: iii), and this happens to be the reason for the same situation in four of the ASEAN countries (Indonesia, Malaysia, the Philippines and Thailand).

In 2006, the aggregate economic growth in South Asia, led by India, has been strong at 6.7 per cent, India maintaining it around 8 per cent since 2004 (Table 2.4) Despite robust economic growth in other economies of this region, such as Bangladesh, Sri Lanka and Nepal, unemployment rates in South Asia remained due to the inability of the formal sector to absorb the rapidly growing labour force.

However, if the current economic growth is sustained, the MDG goal of reducing extreme poverty by half by 2015 is likely to be achieved. On the whole, growth in the region of South Asia has been broad-based, provided by a good agricultural production, except in Pakistan where crop output fell, while the manufacturing sector, particularly the textile and ready-made garments industry, continuing to enjoy a strong external demand, the service sector deriving benefit from growth in the other sectors and turning especially in India into 'an important driver of growth'. However, a trade deficit occurred in most countries in 2005 and 2006, imports exceeding exports, owing to the increase in oil prices and simultaneous growth in domestic consumption and investment demand.

In contrast, the surge in oil revenues as a consequence of rising oil prices has led to an estimated growth of 5.7 per cent in the regional economy in Western Asia but political instability in some countries, such as Iraq, Palestine, Lebanon, etc, came to be a drag on economic growth. On the whole, the strong growth in the region could not reduce high rates of unemployment. The rapid growth of population and labour force along with increasing female participation and the steady stream of migrants seeking hard jobs have caused excess labour supplies where the public sector as the major employer of the domestic labour force "remains oversaturated". At the same time the lack of economic diversification apart from some diversification into other energy production along with limited scope of industrial diversification is likely to hinder more sustainable growth in future with retarded possibility of employment generation.

The economy in Latin America and the Caribbean countries grew by about 5 per cent in 2006, but is likely to slow down to 4.2 per cent in 2007, both rates happening to be the lowest in the developing world, though making the region less vulnerable to external shocks as domestic demand has become the main driving force for economic growth supported by a strong growth in export volumes due to consistent external demand and with the resulting economic activity spurred further by continued increase in import demand. A broad-based growth in the region as a whole is its hallmark and this ultimately has contributed to the increase in employment opportunities by 0.5 percentage points, decreasing unemployment to around 8.7 per cent. Formal employment increased in almost all the countries throughout the region in 2006, making for an increase in real wage for the first time in several years. The strong growth in export volumes and their increasing prices — especially of metals, minerals and hydrocarbon - accelerating capital flows have strengthened local currencies as a result of which external debts for the region as a whole have started decreasing since 2006 in terms of GDP. Nevertheless, the situation remains far from encouraging as is revealed hereafter.

It may hence be observed that the economic performance of the developing countries has been remarkable in sustaining the growth of GDP in almost all the countries though with a visible lacuna in the area of employment generation which has so far been much less impressive everywhere, including China, with the sole exception of Latin America and the Caribbean region where job creation as well as rising rates of real wage for labour have become an important achievement

G. SCOPE FOR EMPLOYMENT AND DECENT WORK FOR ALL

The recent trends of favourable global economic growth are not fully reflected in commensurate growth in employment opportunities. Global unemployment reached a record of 195.2 million in 2006, the rate of unemployed remaining at 6.3 per cent unchanged from the previous year, and for the first time the global unemployment in the service sector – 40.0 per cent – rose above that in agriculture – 38.7 per cent. Looking ahead, there is going to be a great demand for employment with the population aged 15 to 64 years increasing by 175 million in South Asia, by 100 million in South-East Asia, Eastern Asia and Sub-Saharan Africa, and by 50 million in Latin America.

Despite remarkable changes in employment structure in the developing countries, most of them are still predominantly agricultural economies. Nearly 70 per cent of the labour force in the lowest-income countries is employed in agriculture compared with 5 per cent in high-income developed countries. Even in developing countries belonging to the upper-income-group where most industrial development has taken place, agriculture employs almost 20 per cent of the labour force. In spite of the fact that the relative importance of agriculture as employer has declined in all categories of countries, the balance shifting towards other sectors – industry and services (Fig. 2.5), in absolute terms the number employed in agriculture continues to grow.

As a matter of fact, although the employment structure in developing countries has undergone remarkable changes, the sectoral changes in employment needs to be seen in the 'broader context of growth in the size of the labour force'. As clearly evidenced (Table 2.6), agriculture employed 69 per cent of the labour force in 1990 (and which was above 90% in some countries) as compared with 15 per cent in industry and 16 per cent in services for the low-income countries. But the proportion of labour employed in agriculture falls quite dramatically and increases simultaneously in the other two sectors of economy – industry and service – with rising incomes so that the picture almost reverses in high-income countries which keep 64 per cent of the total labour force employed in services and 44 per cent in industry (rising to 69% in some countries) in contrast with 5 per cent (falling to 1%) in agriculture (Dicken, 2003: 556).

The labour force is growing at more than 2 per cent each year in the developing countries which means it has doubled roughly every 30 years

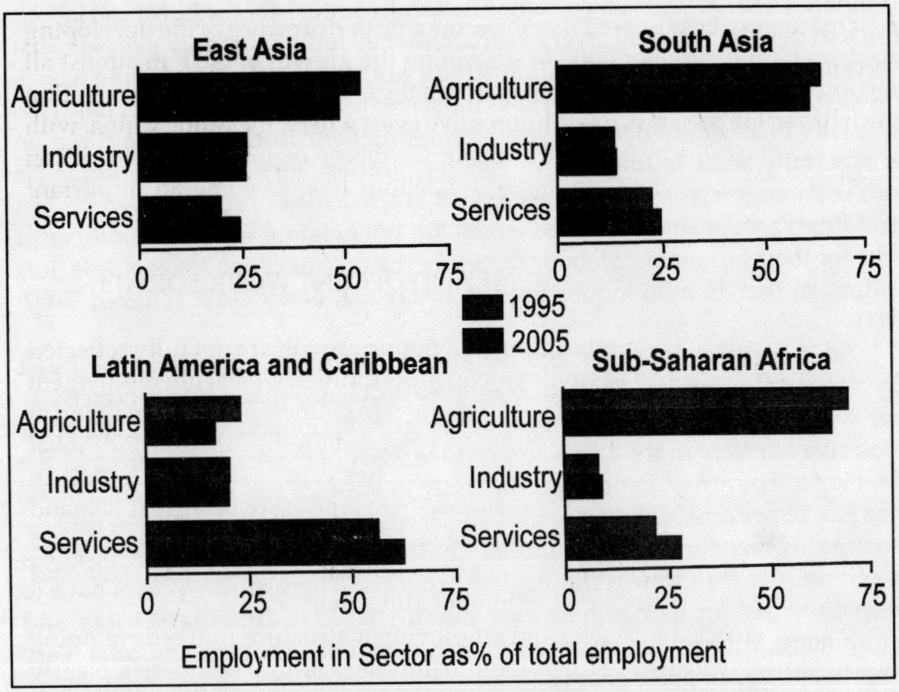

Fig. 2.5: Geographical Variation in Employment Structures in the Developing World

Source: Peter Dicken, 2003

Table 2.6: The changing Structure of the Labour Force in Developing Countries 1960-1990

Country Group	*Percentage of the Labour Force (mean and range) in*					
	Agriculture		*Industry*		*Services*	
	1960	*1990*	*1960*	*1990*	*1960*	*1990*
Low-income	77	69	9	15	14	16
	54-95	18-94	1-18	2-31	3-37	N/A
Lower Middle-income	71	36	11	27	18	37
	39-93	13-79	2-26	7-48	5-39	N/A
Upper Middle-income	49	21	20	47	31	52
	8-67	6-51	9-52	16-46	16-69	N/A
High-income	18	5	38	31	44	64
	4-42	1-18	25-50	25-38	27-57	N/A

Source: Peter Dicken, 2003 (4th Ed): Global Shift: Reshaping the Global Economic Map in the 21st Century (based on The World Bank 1995: Tables 1 and 3)

compared with 90 years taken in the nineteenth century for the European labour force to double. "Overall, every year another 80 million or more people reach working age and join the 2.1 billion-strong workforce in the developing world. Mexico, Turkey, and the Philippines must create 500,000 to 1 million new jobs annually to accommodate the youths entering their workforce. The situation is especially acute in low-income Asian countries like Bangladesh and India and some countries in South-East Asia. It is also a major problem even for fast growing East Asian economies, which are not creating sufficient number of jobs for their burgeoning labour forces. In the case of China for example, it is estimated that 15 million jobs need to be created every year" (Dicken, 2007: 505).

As it appears, however, the developing countries, especially the Least Developed Countries (LDCs), face another problem – that is low productivity per worker, resulting in slow productive growth. On average, in 2002-03, it took five workers in the LDCs to produce what one worker produces in the developed countries. Labour productivity in the LDCs as a group in 2000-03 was just 12 per cent higher than in 1980-83, while it increased by 55 per cent on average in other developing countries. The low labour productivity in the LDCs is primarily due to slow technological progress, where most workers have to earn their living using their raw labour, with rudimentary tools and equipments, little education and training, poor infrastructure and often very poor health causing less energy. Besides, there is widespread unemployment. In 2006, 86 per cent of the working men and women lived in families with a per capita income of less than US $2 a day – a rate virtually unchanged from a decade earlier. The number of people seeking employment outside agriculture is increasing, but labour productivity outside agriculture is declining. In the most LDCs, the most likely explanation is that employment opportunities in formal sector are not expanding fast enough to absorb the growing non-agricultural labour force and due to this employment in the informal sector as a share of non-agricultural employment is increasing while labour productivity within the informal sector is on average lower than labour productivity within the formal sector.

It requires major efforts to generate productive employment and decent work leading to a faster pace of reduction of poverty and hunger. "Pursuing such a strategy will require improved collaboration between international development actors in support of national development strategies together with a major improvement in the quantity and quality of international development assistance, continued reduction in debt burdens and a favourable outcome to international trade negotiations" (United Nations, 2007: 203-204). With almost 40 per cent of the 3 billion labour force in agriculture, technical change and rising productivity in agriculture created a large surplus of labour force which tended to migrate to urban areas but urban jobs in manufacturing not being able to absorb this surplus, diverted them to fall into the sector of last resort – services – most in self-employment.

Hence, what is needed as a solution to slack labour market in both rural and urban areas of the developing countries is suggested as follows:

(a) The rural non-urban sector needs to be developed so that all the rural surplus labour does not migrate to urban areas straining urban labour markets, urban infrastructure and urban services.
(b) Public investment in infrastructure is necessary to increase investment, productivity and value-added goods in rural and urban sectors
(c) Basic social security is an important safety net for the vulnerable in the labour market (United Nations, 2007:206).

As observed in case-studies conducted in 23 countries, Viet nam 'stands out as a very positive exception where economic opportunities have increased, and poverty has declined substantially in the last ten years due to changes in economic and social policies of the government. An emphasis on building assets and development of secondary sources of income such as livestock, gardening, tree cultivation and trading, as well as an extensive network of credit provision along with development of markets, land allocation to households etc. has helped people generate incomes. In several communities in Bangladesh, India and Sri Lanka, among the 23 countries studied, groups speak of greater opportunities, but they consider access to such openings are sometimes limited to those who are better off. In almost every community visited for the India study, the poor perceive that "interventions by NGOs and the advent of self-help groups and village development committees have improved the social status, livelihood security and availability of livelihood alternatives, but they do not think these gains are shared among some of the poorer groups ... In Thailand people report that recent investment in 'water, electricity and transportation has vastly improved, making work easier. Poor people make similar observations on better living and work conditions in some of the favelas (slums or squatter land) of Brazil" (Narayan, *et al.*, 2000: 63-64).

In recent years, employment opportunities have tended to favour skilled workers over the unskilled so that the poor, with their limited training, have benefited least. Nevertheless, some progress has been made as, among the employed, the number of those earning less than US $ 1 a day fell by over 10 per cent between 2000 and 2006 (Table 2.7). Most of these workers earn less than US $ 2 a day, whose number rose from 768 million in 2000 to 846 million in 2006. Young people are estimated to represent 44 per cent of the world's unemployed, many of them, especially those living in poverty in developing countries, find it difficult to get work because of inadequate education and skills or the absence of information about job opportunities. Governments should provide both formal and non-formal education for young people so that they can acquire skills that enhance their job opportunities and strengthen their entrepreneurial and innovative skills, for which a Youth Employment network needs to be established as has been done in many developing countries to help create new opportunities for young men and women. One possible alternative to land-based security is to enable the land-poor to engage in those self-employment opportunities not critically dependent on land. As has already happened and may be encouraged further, the non-farm activities occupy a significant position in the occupational structure of South Asia, accounting for 34 per cent of the rural labour force in Bangladesh, 46 per cent in Sri Lanka and 19 per cent in India. In Bangladesh, of all the gainful activities – both primary and secondary – pursued by workers owning no more than 0.5 acres of land, 50 per cent belong to the non-farm sector while agricultural labour accounts for only a third of the total (Osmani, 1991: 319).

As the Table 2.7 shows, in most developing countries, economic growth, one of the conditions for reducing poverty, has strengthened since 2001, averaging over 6 per cent per year between 2004 and 2006. The proportion of people living on less than US $ 1 a day has appreciably declined from 31.4 per cent in 1990 to 22.3 per cent in 1999 to 18.4 per cent in 2004 as a result of economic growth converting into poverty reduction, affected particularly by such development in China and other parts of East Asia. Despite the fact that elsewhere the number of extremely poor has mostly increased or remained roughly constant, the Millennium Development Goal of halving the proportion of people living in extreme poverty by 2015 seems to be achievable except in Sub-Saharan Africa.

Table 2.7: Selected indicators of progress in Developing Countries, 2000-2006

	2000	*2001*	*2002*	*2003*	*2004*	*2005*	*2006*
Economic Growth							
Growth of output (Percentage per year)	5.6	2.7	3.9	5.2	6.9	6.5	6.9
Poverty (millions)							
Population living on Less than US $ 1 a Day (millions)	1120[a]	1067	—	—	986	—	—
Proportion of Population (percentage)	223[a]	—	20.4	—	18.4	—	—
Working Poor (US $ I per day)	570	568	560	545	528	515	503
Employment (millions)							
Employment (Female)	787	800	816	834	847	862	878
Employment (Male)	1255	1276	1299	1325	1347	1371	1394
Education (Percentage)							
Primary Gross Enrolment Rate Female	78.9	79.5	79.2	81.3	83.7	—	—
Primary Gross Enrolment Rate Male	85.5	85.6	85.4	85.6	87.9	—	—

Source : United Nations, 2007 (Ref. World Economic Situation and Prospects 2007, UN Publication; World Bank, Global Monitoring Report 2007; Millennium Development Goals Data Base).

Note : a = 1999

H. FACTORS CONTRIBUTING TO POVERTY REDUCTION

As an indicator for assessing the decline in poverty and hunger, life expectancy takes an important place indicating the measure of improvement. There has been remarkable progress in life expectancy in the developing world as a whole, rising by 'four months each year after 1970, from 55 years in 1970 to 64 years in 2000'. In South Asia it increased from 49 years in 1970 to 62 years in 2000 and in East Asia the increase was from 59 years to 69 years although two countries, North Korea and Zimbabwe, have a very poor record in this respect largely due to AIDS epidemic or the gross incompetence of governments. Improvement in life expectancy meant a decline in global inequality.

Infant mortality is another indicator for ascertaining improvement in life in general as a direct consequence of improvement in income and basic services as happened in many developing countries with growing economies. Infant mortality rates have declined in the developing world as a whole from 107 per thousand in 1970 to 87 in 1980 and 58 in 2000. In East Asia, the region with the fastest growing economy, the rates of infant mortality have fallen from 56 in 1980 to 35 in 2000. Similar declines have also been experienced by South Asia, the rate falling from 119 in 1980 to 73 in 2000. Even Sub-Saharan Africa has registered a quite impressive fall in this regard, from 116 in 1980 to 91 in 2000.

Better education appears to be at the root of declining infant mortality rates making people more conscious about the benefit of small families offering children better scope for education and health. Greater awareness in this regard is visible in putting emphasis on adult-literacy in developing countries, the rate rising from 53 per cent in 1970 to 74 per cent in 1998. By 2000, adult illiteracy came down to 8 per cent in East Asia, though it was still higher in Sub-Saharan Africa with 30 per cent and in South Asia with 34 per cent. In fact, adult female illiteracy was more widespread than that of men, but it was improving, falling from 29 to 21 per cent between 1990 and 2000 in East Asia, while in South Asia it fell from 66 to 57 per cent, and in Sub-Saharan Africa from 60 to 47 per cent during the same period. The rate is likely to continue its fall in future as illiteracy is much lower among the young population (Wolf, 2005: 145).

Similarly, reduction in fertility rates has been quite remarkable over the recent years, the births per woman having fallen in the developing world, in general, from 4.1 per cent in 1980 to 2.8 per cent in 2000. In East Asia, the fertility rate having fallen from 3.0 per cent to 2.1 per cent is already at close to the replacement rate (when half the world, including all the rich countries, is at or near the so-called replacement rate of fertility which means that each mother is raising one daughter on average to "replace" her in the next generation). The replacement rate is two children, one of whom, on average is a girl, though, in fact, the replacement rate is a little more than two, taking into account the possibility that the daughter will not survive to reproductive age. In Latin America, the fertility rate has gone down from 4.1 to 2.6 per cent and in South Asia it has registered the slowest progress, with a fall from 6.6 per cent in 1980 to 5.2 per cent in 2000. This is quite high by all standards affecting improvement in lifestyle since a smaller family is always welcome, indicating women's empowerment in deciding about the size of the family which comes from education on the one hand, while on the other it assures larger parental investment for each child toward his/her education, health and survival, contributing to greater welfare for the society.

One of the most important indicators of persistent poverty in the developing countries is child labour, elimination of which is essential to step up the welfare-status of the people. Approximately, 200 million children between the ages of five and fourteen work in developing countries, which undoubtedly is a far lower proportion of the world's children than ever before in history. Almost 70 per cent of them are engaged in agriculture and about 10-15 million in export-industry, predominantly in South Asia. These children work under the compulsion of poverty to supplement the family income. The decline in birth rates making for smaller families in the developing world has a direct impact in increasing the scope of education for children in these families. The population growth-rate in middle-income developing countries is likely to come down to 0.9 per cent a year between 2000 and 2015. According to the World Bank, the proportion of children aged 10 to 14 in the labour force has fallen from 23 per cent in all developing countries in 1980 to 12 per cent in 2000. In East Asia, the rate of fall is highly impressive, from 26 to 8 per cent while in South Asia it has declined from 23 to 15 per cent during the same period. In Sub-Saharan Africa, the decline has been less impressive, from 35 to 29 per cent. China's record in this respect seems to be spectacular, bringing it down from 30 per cent in 1980 to 8 per cent in 2000, while India's performance appears less noteworthy with a fall from 21 to 12 per cent (Wolf, 2005: 164-166). According to a report of the Government of India, the total number of child workers declined from 11 million in 1991 to 3.6 million in 2001. Similar decline is also evident from reports of the NSSO (National Sample Survey Organisation) which estimated the total number at 6.2 per cent of children in the age of group of 5-14 in the years 1993-94 and 3.4 per cent in 2004-05. They, together with out-of-school children, represented about 18 per cent of the children, 15 per cent boys and 21 per cent girls, signifying a high measure of deprivation (Government of India, 2008: 100-101).

Another factor to be taken into consideration for its close relationship with relieving people from poverty is protection of the environment so that promoting growth does not become inimical to environmental sustainability. It is a highly complex issue, no doubt, and as Brown (2006) observed, any strategy for eradicating poverty will not succeed if an economy's environmental support system is collapsing. The health of an economy cannot be separated from that of its support system. In his words, "If croplands are eroding and harvests are shrinking, if water tables are falling and wells are going dry, if rangelands are turning to desert and livestock are dying, if fisheries are collapsing, if forests are shrinking, and if rising temperatures are scorching crops, a poverty-eradication programme– no matter how successfully drafted and well implemented – will not succeed" (Brown, 2006: 142). Talking of environmental sustainability, the two major elements of sustainable development cannot be ignored, that is, first, meeting the needs, in particular, of those who have been left far behind in a century of extraordinary growth, and second, the limits which the world society must now impose to protect the resource base of the world both locally and globally, the latter imposing standards of consumption to be met within the bounds of ecological possibility. The Burtland Commission in its report, 'Our Common Future', 1987, defined sustainable development as "development that meets the needs of the present without compromising the ability of future generations to meet their own needs" (World Commission on Environment and Development 1987). The World Commission equated sustainable development with "more rapid economic growth in both industrial and developing countries" because "economic growth and diversification ... will help developing countries mitigate the strains on the rural environment" (WECD, 1987:87), the latter being more impoverished than the urban areas. Achievement of sustainable development depends on broader participation in decision-making where people need to be directly involved in new forms of multilateral cooperation; the extension and sharing of new technologies; increased international investment; an expanded rate of transnational corporations; the removal of artificial barriers to commerce, and an expanded global trade (Hamm and Muttagi, 2001:21).

The sense and responsibility for general welfare of all the people living in a country is a product of the industrial revolution making people more aware of their rights to social benefits. Hence, the concept of welfare state and its development remained confined for a long time to the industrially developed countries. Ironically, many of them building their colonial empires did not pay much heed at any time of their rule of colonies to the welfare of the ruled. The mindless policy of continuous exploitation of the natural resources went on in an unrelenting manner harming the environment and as a result the present state of environmental degradation in many of the developing countries, once colonized, is a legacy of colonial rule, aggravated further during the post-colonial period largely in the absence of any damage-control efforts. One needs to realize the stark reality of the fact that the emergence of these countries – comprising the developing world today – from colonial rule did not offer immediately a promising situation. Rather, the inheritance of a shattered economy posed a great challenge on every front with a growing population exacerbating environmental degradation further in the form of soil erosion, droughts and floods, lack of irrigation and fertilization, poor transportation, lack of marketing and communication facilities, lack of education and technological improvement along with health and sanitation problems. The resulting ecological and social collapse have reinforced each other in a downward spiral into poverty, environmental degradation, social injustice, disease and violence.

As answer to meeting such environmental challenges and restoring ecological sustainability, conserving and rebuilding soils together form a major step to ensure greater sustained production. Preserving biological productivity of highly erodible cropland can be accomplished by planting such land in grass or trees before it becomes wasteland. Terracing, a time-old method for dealing with water erosion, has proved to be very effective throughout the mountainous regions of Asia. On less steep sloping land contour strip farming is very useful. Another method found very effective for conservation of soil is conservation tillage consisting of both no-till and minimum tillage. "In addition to reducing both wind and water-erosion this practice helps retain water, raises soil carbon content and reduces energy needed for crop cultivation (Brown, 2006: 150).

Instead of preparing the seedbed by ploughing, harrowing etc., farmers simply 'drill seeds directly through crop residues (of the previous crop) into undisturbed soil, controlling weeds with herbicides'. In this way 'the only soil disturbance is confined to the narrow slit in the soil surface where the seeds are inserted, leaving the remainder of the soil undisturbed, covered by crop residues and thus resistant to both water and wind erosion'. In the United States the no-till area has increased from 7 million hectares in 1990 to 25 million hectares in 2004, used for the production of corn and soybeans, and this practice is rapidly spreading in the western hemisphere, already covering 24 million hectares in Brazil, 18 million hectares in Argentina, and 3 million hectares in Canada. In Australia the no-till cultivation covers 9 million hectares.

Slow desertification is another type of loss of cropland faced by many countries in the world. Algeria is one example. In 2000, she adopted a policy of concentrating her orchard vineyards in the southern part of country in a bid for halting the desertification of the cropland. Similar concerns about the expansion of the Sahara desert on its southern boundary has prompted Nigeria to propose planting of a 'Great Green Wall' of trees, five kilometres wide and extending 7,000 kilometres across the continent in an effort to arrest further advance of the desert. Similar efforts are being made in China using a Great Green Wall to check the progress of desert besides paying farmers in threatened provinces to plant trees on their croplands, thus covering 10 million hectares of grain land, amounting to almost one-tenth of the country's grain land area. Further, for releasing pressure on grazing lands, the government has asked farmers to reduce their stock of sheep and goats by 40 per cent – though practicing it may not be easy where people live in perpetual poverty. But overgrazing of the two-fifths of the grazing lands of the world needs to be controlled, particularly for checking soil erosion induced by the pulverization of soil by the hooves of sheep and goats. The only viable action, as shown by India, is to keep the animals in enclosures and feed them there (Brown, 2006: 149-152). In Mongolia, planting of desert shrubs on abandoned cropland has succeeded in stopping soil erosion on 7,000 hectares of land — a first attempt at reclamation which is being expanded.

Meeting water needs makes another crucial factor for environmental stability and sustainable development as measures for challenging human poverty. The growing population and increasing demands for water are intimately connected with improvement in living standards, but demands for water promise a gloomy future. 'Scarcity at the heart of the global water crisis is rooted in power, poverty and inequality, not in physical availability'. Today, about 1.1 billion people in the developing countries have inadequate access to water, and 2.6 billion lack basic sanitation. The 1.8 million child deaths each year are related to unclean water. However, the fact is that the apparent deficit of water in developing countries is "rooted in institutions and political choices. Not in water's availability" (UNDP, 2006: 2). As a matter of fact, household needs make a very small fraction of total water needs for the society, usually less than 5 per cent. But variation in its availability as commanded by circumstantial factors is immense. For example, in the high-income areas of cities in Asia, Latin America and Sub-Saharan Africa, people enjoy access to several hundred litres of water a day delivered into their homes at low prices by public utilities. In contrast, the slum dwellers in the same countries get less than 20 litres of water a day per person, which is the minimum social threshold requirement. Besides, women and young girls in developing countries carry a double burden of disadvantage, as they have to sacrifice their time and education for collecting water, quite often walking long distances.

While the scarcity of water is a widespread phenomenon, sometimes it is the result of political clout and institutional support and not a physical one, as in India, wherein water-stressed parts irrigation pumps extract water from aquifers '24 hours a day for wealthy farmers', while neighbouring small holders depend on uncertain rain. As a matter of fact, there is more than enough water in the world for domestic uses, for agriculture and industry and the existing scarcity is created by political processes and institutions disadvantaging the poor.

Not having access to water and sanitation is one of the most critical forms of deprivation threatening life, livelihood and undermining human dignity. The Millennium Development Goal of halving the proportion deprived of access to clean water happens to be a key target as clean water and sanitation would save the lives of millions of children, support progress in education and liberate people from illness keeping them in poverty. But we have to keep in mind that even if the targets are achieved there will still be over 800 million people left without water and 1.8 billion people without sanitation by 2015. We need to remember that sweeping reforms in water and sanitation made possible with rising income through industrialization brought a drastic change to the life of people living in present day rich countries where not even a hundred years ago London, New York and Paris were the centres of infectious diseases, with diarrhoea, dysentery and typhoid affecting public health with child death rates as high as in much of Sub-Saharan Africa today. England (with 80 years of life-span in London today) was far from a picture of health which, "in 1564 ... had a population of between three and five million – much less than three hundred years earlier – when plague used to take a continuous, heavy toll. Then the number of Britons was actually in retreat. The previous decade had seen a fall in population nationally about 6 per cent. In London as many as a quarter of the citizenry had perished. But plague was only the beginning of England's deathly woes. The embattled populace also faced constant danger from tuberculosis, measles, rickets, scurvy, two types of small pox... dysentery, and a vast amorphous array of fluxes and fevers... These were, of course, no respectors of rank. Queen Elizabeth was nearly carried off by small pox in 1562...Leprosy, one of the great dreads of the Middle Ages, had likewise mercifully abated in recent years (perhaps around 1550s) never to return with vigour. But no sooner had these perils vanished than another violent fever called 'the new sickness' swept through the country, killing tens of thousands in a series of outbreaks between 1556 and 1559" (Bryson, 2008: 22-23).

According to an estimate, water purification alone explains almost half the reduction in mortality in the United States in the first third of the twentieth century, and the expansion of sanitation increased life expectancy by 15 years in the four decades after 1880 in Great Britain (Bryson 2008: 48). The basic needs for water vary but the minimum threshold is about 20 litres per head a day while 1.1 billion people lacking access to clean water use about 5 litres a day in contrast to people in Europe using more than 200 litres and in the United States more than 400 litres. In this context it is highly dismaying to note that a person in Europe uses more water in flushing a toilet or an American taking a shower than is available to hundreds of millions of individuals living in urban slums or arid areas of developing countries. Further, "dripping taps in rich countries lose more water than is available each day to more than 1 billion people" (UNDP, 2006:6). It is, however, encouraging to note that progress in access to water and sanitation is improving with rising incomes, though showing large variations around the world. As observed in the UNDP 2006 Report, one may find different results: "some countries – such as Bangladesh and Thailand in sanitation and Sri Lanka and Vietnam in water – do far better than would be expected solely on the basis of income. Others, such as India and Mexico for sanitation do far worse" (UNDP, 2006:6). Factual evidences show that global aggregates for water and sanitation coverage conceal large differences between regions. For example, Sub-Saharan Africa has by far the lowest coverage rates (55%) for water, while most people without access to clean water live in South Asia. But deprivation in this context appears to be more evenly spread for sanitation, its coverage being almost as low in South Asia as in Sub-Saharan Africa, with two of every three people in both regions lacking access. Half of the people in East Asia and a quarter in Latin America do not have access to even the most basic sanitation. About 40 developing countries provide clean water for less than 70 per cent of their population and 54 countries provide safe sanitation for less than half of the population. According to the World Bank, 2002, India becomes an underwater stress zone by 2025 and an underwater scarce zone by 2050 which sounds quite chilling.

The human development costs resulting from deprivation in water and sanitation are immense: (i) Almost 1.8 million children die each year of preventable diseases such as diarrhoea, making unclean water and lack of sanitation the second biggest killer of children in the world, besides the loss of 443 million school days each year from water related diseases. Research for this report (UNDP, 2007)) shows that access to safe water reduced child death rates by more than 20 per cent in Cameroon and Uganda. In Egypt and Peru, the presence of a flush toilet in the houses reduces the risk of infant death by over 30 per cent, (ii) The suffering of nearly half of the people in developing countries at any time from a health problem is related to water and sanitation deficits, (iii) Millions of women and grown up girls spend several hours a day to collect water for domestic use and (iv) Illness and lost education opportunities in childhood for millions of people lead them to poverty in adulthood.

As pointed out in the 'United Nations World Water Development Report 2003', "our primary source of water is the runoff diverted for use in irrigation, in industry and in homes, both rural and urban and for waste disposal. It is the water of evapotranspiration that mainly supports forests, rainfed cultivated and grazing land, and a variety of ecosystems. Despite a withdrawal of only 8 per cent of total renewable freshwater resources, there is 26 per cent of annual evapotranspiration and 54 per cent of accessible runoff appropriated by humans. With increase in per capita consumption due to changes in lifestyle and population increases, the proportion of appropriate water is also increasing. This coupled with variations in water availability is putting pressure on water to produce food for human consumption, industrial processes and all other uses, making it scarce".

"It has been estimated that today more than 2 billion people are affected by water shortages in over 40 countries, 1.1 billion do not have sufficient drinking water and 2.4 billion have no provision for sanitation (WHO/UNICEF, 2000). The outcome can mean increases in disease, poorer food security, conflict between different users and limitations on many livelihood and productive activities. A current prediction is that by 2050 at least one in four people is likely to live in countries affected by chronic or recurring shortages of fresh water. At present many developing countries have difficulty in supplying the minimum annual per capita water requirement of 1,700 cubic metres (m^3) of drinking water necessary for active and healthy life for their people. The situation is particularly grave in many of the cities of the developing world... given predictions of a 60 per cent world urban population increase by 2020. At present, half the population of developing countries live in water poverty" (United Nations, 2003: 10).

According to another estimate, water stress is increasing, making today about 700 million people in 43 countries living below the water-stress threshold of 1,700 cubic metres per person and by 2025, is likely to reach 3 billion as water-stress intensifies in China, India and Sub-Saharan Africa. Already 538 million people in northern China alone live in an intensely water-stressed region. In the world today, about 1.4 billion people live in river basin areas where water use exceeds sustainable limits. The ecological stress is very pronounced in many parts of the world, further aggravated by the decline of river systems – starting from the Colorado River System in the United States to the Yellow River in China. No less detrimental to human development is the rapid depletion of groundwater in South Asia and in some parts of India. The groundwater table falling by more than 1 metre a year is a sure indication of its jeopardizing effect on agriculture in the near future. Global warming is understood to transform hydrological regimes determining the availability of water and there is overwhelming evidence to justify the apprehension that 'many of the world's most water-stressed areas will get less water, and water flows will become less predictable and more subject to extreme events'. Some of the outcomes of this, as predicted, will be of disastrous consequences. Such as – (i) There will be marked reductions in the availability of water in East Africa, the Sahel and Southern Africa resulting from decline in rainfall and rise in temperature with large productivity losses in basic food staples. Projections for rainfed areas in East Africa indicate losses of potential productivity of up to 33 per cent in maize and more than 20 per cent for sorghum and 18 per cent for millet, (ii) The disruption of food production systems will subject an additional 75-125 million people to the threat of hunger, (iii) Accelerated melting of glaciers will lead to medium-term reductions in water availability in a large group of countries in East Asia, Latin America and South Asia, (iv) Disruption of monsoon patterns in South Asia a potential for more rain and fewer rainy days and also for more people being affected by drought and (v) Rising sea levels resulting from glacial melt is likely to cause fresh water losses in river delta systems in Bangladesh, Egypt and Thailand (UNDP, 2006: 14-15) (See Note E).

In order to meet the challenge, among the most important strategies recommended are: (a) An integrated water management system keeping national water use levels within the limits of ecological sustainability with a coherent planning programme for all water resources and (b) Keeping equity and interests of the poor at the centre of integrated water resource management programme and making the latter an integral part of poverty reduction strategies. (c) There need to be appropriate pricing policies for establishing real value of water and withdrawal of harmful subsidies encouraging water overuse. (d) Increasing pro-poor supply of water, making provision of safe waste water available for productive use by separating industrial and domestic waste and working with farmers to reduce health risks, and (e) Increasing national and international aid for investment in water infrastructure including storage and flood control (UNDP, 2006: 16) (*See* Note F).

I. BUDGET FOR EARTH RESTORATION

Lester Brown (2006) has made an attempt at estimating the approximate costs to reforest the earth, protect the earth's top soil, restore rangelands and fisheries, stabilize water tables and protect biodiversity. South Korea provides an example of a big success in the reforestation of its denuded mountains and hills carried out for the last four decades using local labour. China has tried extensive reforestation though mostly confined to arid areas with much less success. Turkey has an ambitious NGO-led grassroots reforestation programme, relying largely on volunteer labour. In Kenya the women's group led by Nobel Peace Prize winner Wangari Mathai have planted 30 million trees. The major focus of reforestation will be on developing countries where the growing demand for fuel wood will require roughly about 55 million additional hectares of forest land. For stabilizing soil and restoring hydrological stability would require another 100 million hectares located in hundreds of watersheds in developing countries, besides an additional 30 million hectares to produce lumber, paper and other forest products.

In conserving topsoil affected by erosion, the highly erodible land amounting to about one-tenth of the world's cropland, representing about half of all erosion that cannot sustain cultivation needs to be retired. The cost of keeping this land out of production will come to about $2 billion as total payments to farmers to plant this land in grass or trees under 10-year contracts. The second initiative taken on adopting conservative practices on the remaining land subject to excessive erosion where it exceeds the rate of new soil formation, include steps, such as giving incentives to farmers 'to adopt conservation practices like contour farming, strip cropping, minimum or no-till farming'. For this the cost is estimated to be about $16 billion. On the other hand, the annual rangeland protection and restoration including such measures as improved rangeland management, financial incentives to eliminate overstocking, and revegetation with approximate rest periods when grazing is banned, will amount to $ 9 billion. The restoration of oceanic fisheries involves establishment of a worldwide network of marine reserves, covering roughly 30 per cent of the ocean's surface on an annual expenditure amounting to about $ 13 billion (Brown Lester, 2006 : 158-162).

J. INCREASING FOOD PRODUCTION AS A DEVELOPMENT PRIORITY

China's landmark achievement in almost eradicating hunger as a consequence of drastic reduction in poverty is associated with its eight-fold expansion in economic reforms of 1978 and the 50 per cent increase in its grain harvest between 1977 and 1986. Grain output increased from 305 to 407 million tons between 1978 and 1984. The per capita GDP more than quadrupled from 1978 to 1995, rising from 657 to 2,970 yuan, or by 9 per cent per annum. Over the period 1978 to 1996, the value of exports and imports increased by almost 16 per cent per year, about 6 percentage points higher than in the pre-reform period, 1952-1978. Economic growth followed by the opening up of Chinese economy brought about a much higher living standard increasing per capita disposable incomes nearly four times in the rural areas and more than three times in the cities, a period during which urban-rural inequality got substantially reduced (Yao, 1999: 107 and 109).

While hunger has been disappearing in China, it has been spreading in parts of the Indian subcontinent and in Sub-Saharan Africa, increasing the number of people in hunger from 820 million in 2000 to 852 million in 2002. One of the major ways of combating hunger is naturally an increase in grain production. The three-fold increase in grain production since 1950 in the developing countries was through the adoption of high-yielding wheats and rices developed in Japan and hybrid corn from the United States combined with a 'tripling of irrigation area and nine-fold increase in fertilizer use in the World. But the efforts to raise cropland production are slowing as evident from the fall in the rate of grain yield production per hectare which fell from 2.1 per cent a year between 1950 and 1990 to 1.2 per cent from 1990 to 2000. Multiple cropping where soil moisture permits seems to be an answer to raise grain production. As a matter of fact, tripling world grain production since 1950 was due to multiple cropping in Asia resulting from crop combination, such as wheat and corn in northern China, wheat and rice in northern India, double or triple cropping of rice in southern China, southern India and nearly all the rice-growing countries in southeast Asia.

The double cropping of winter wheat and corn in the north China plain raised China's grain production to US levels beginning two decades ago. Winter wheat grown there gives a yield of nearly 4 tons per hectare; corn averages 5 tonnes, and grown in rotation they give 9 tonnes of grain per hectare per year while double-cropped rice in China yields 8 tonnes per hectare. (*See* Note G). Brown (2006) points out that, forty years ago wheat used to be the only crop in north India, but with the advent of the earlier maturing varieties of wheats and rices, wheat could be harvested in time to plant rice. The wheat/rice combination is now widely used throughout the states of Punjab, Haryana and parts of Uttar Pradesh. However, India lags behind China in yield per hectare; it being 3 tonnes per hectare for wheat and 2 tonnes per hectare for rice. And dismayingly enough, the production of food declined from about 212 million tonnes in 2003-04 to about 204 million tonnes in 2004-05 in India which is not adequate for feeding its 1.1 billion population. Further, it is predicted that by 2028, the country will face a deficit of 28 million tonnes per year with its growing population. Two-thirds of the net-cropped area is under dry-land farming, accounting for 42 per cent of total production. India needs to increase her annual growth rate to 4.7 per cent from the present growth rate of 1.5 per cent per year to meet the challenge. Progress in industrialization may see declines in multiple cropping in both China and India, pulling away labour from agriculture as happened in Japan where grain harvest peaked in 1960 at nearly 5 million hectares giving two harvests, and cropland dropped to 2 million hectares because of the conversion of cropland to non-farm uses. "Even a rice support price four times the world market level could not keep enough workers in agriculture to support extensive multicropping" (Brown, 2006: 165). Similarly, the harvested grain area peaking in 1965, has shrunk by half because of the decline in multicropping. In Taiwan, it has declined almost two-thirds since 1975. In the United States, on the contrary, double cropping has increased since 1996 following a combination of winter wheat with soybeans grown as a summer crop, giving six per cent of the soybeans production. Western Europe, with its mild winters and high-yielding wheat, may be able to double crop yield in combination with summer grain, such as corn or oilseed. Elsewhere, Brazil and Argentina having a longer frost-free growing season supports extensive multicropping, usually wheat or corn grown with soybeans (Brown, 2006: 16) (*See* Note H).

In comparison, due to lack of consistency in the degree and direction of priorities, African agriculture is today one of the most under capitalized in the world with minimum facilities for irrigation, use of fertilizers and tractors for tilling the soil, access to transport, energy and communications, or extension services and help from any research institutions. In all these aspects, as mentioned earlier in this chapter, Asia fares far better than Africa on all counts. "The combined effect of all these are reflected in stagnating or declining agricultural productivity, weak backward and forward linkages between agriculture and other sectors, loss of competitiveness in world markets, increased food insecurity and natural resource and environmental degradation" (United Nations, 2007: 124).

The importance of land in Africa, central to the achievement of the Millennium Development Goals, is very much relevant to the development of human rights and security. Land is the main source of livelihood for the majority of rural Africans. Almost 60 per cent of all Africans work in agriculture and three-fifths of their farms function at subsistence level. The state of agriculture is poor, facing declining productivity due to poor soil fertility, poor access to improved seeds and planting materials, and dependence on rain-fed agriculture and poor management. In spite of continuous food imports and recourse to annual food aid ($ 2 billion) about one-third of Sub-Saharan people suffer from hunger, and the continent's share of global agricultural exports has fallen sharply from 15 per cent in the mid-1960s to 5 per cent in 2000. Experts in the field do believe that responding effectively to the cycle of food crisis in Africa requires a major transformation of the agricultural sector in the form of the 'Green Revolution' that took place in Asia in the 1950s. The African Union Heads of State and Government in 2003 called for an increase in budgetary allocations to agriculture and rural development to at least 10 per cent of total national public expenditure by 2008. A successful agricultural transformation process depends on farmers getting access to the markets through agro-industrial and agro-business sectors. This demands that both agro-industrial and agro-business development should closely collaborate with the transformation process. It implies the crucial importance of agriculture-industry-trade nexus in addressing the agricultural transformation challenge in Africa.

To meet the requirement of a larger food production, Africa needs a Green Revolution as happened in India in the mid-sixties of the twentieth century and to make it a success Africa requires strong, consistent, and holistic government support as India got. 'The farmer must face a realistic set of conditions – in input supplies and prices, available technologies, supportive infrastructure, and output prices – to facilitate the adoption and diffusion of

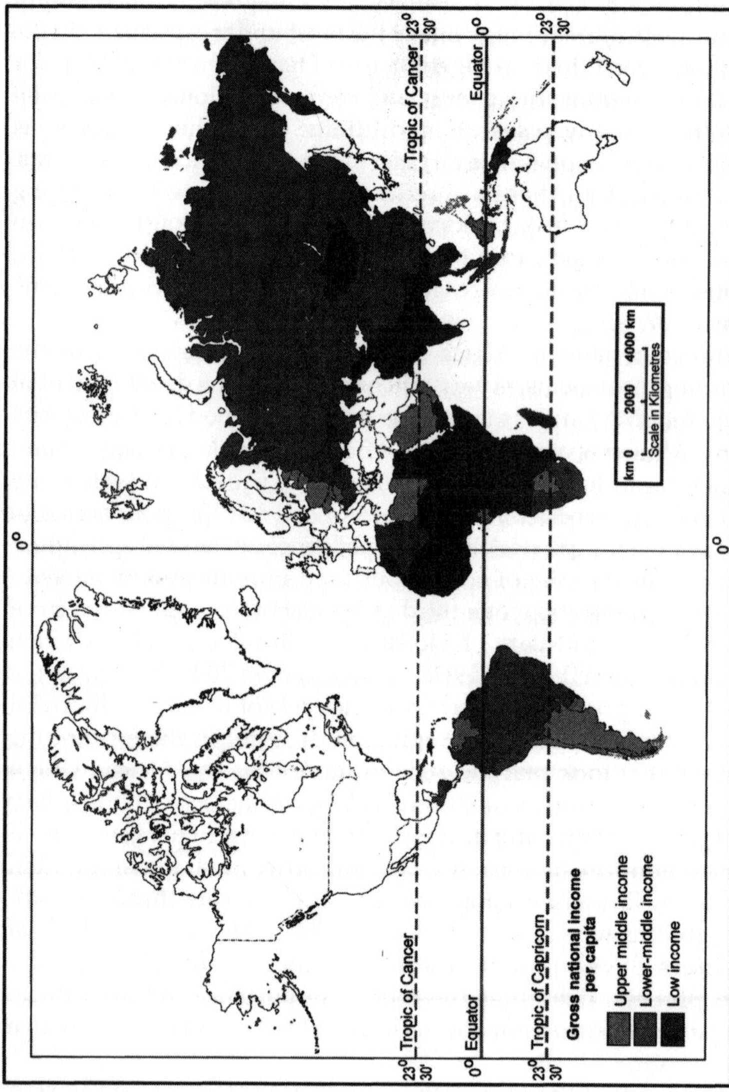

Fig. 2.6: Distribution of Upper-middle, Lower-middle and Low-income Countries in the Developing World

Source: Peter Dicken, 2003

the new technology. Solutions for water management, seeds and fertilizers should be viewed in an integrated manner'. Markets play an important role, no doubt, but in the case of the poorest of the poor, the government should ensure a strongly supportive environment for the success of Green Revolution. At the same time foreign assistance is crucial. "The new alliance for an African Green Revolution (AGRA), spearheaded by the Gates and Rockfeller Foundations, is likely to mark an holistic sea-change in the global approach" (United Nations, 2007: 131), as Sachs, Special Advisor to the United Nations Secretary-General, puts it. The said 'alliance', in fact, indicates 'a new readiness of the world to engage in the challenge of raising African food production' in which donor governments – the UK, Ireland, Brazil, Luxembourg, Norway, and others – are taking an active role. Yet what might impede the progress as well as success of Green Revolution in Africa are the land policies eroding the rights of the rural farmers to land, forcing poor farmers to become tenants to land merchants and speculators. Among the worst affected are pastoralists, hunter-gatherers, fishermen and small-scale rural and peri-urban food gatherers. Most African countries are endowed with a variety of natural and human resources, and poverty and hunger can be overcome through proper utilization of them. What is actually lacking is a strong political will of leaders and peoples in full deployment of public resources supporting productive activities. Land tenure, as claimed by experts, is a critical factor for achieving success in Green Revolution in Africa where the poor are often displaced by the elite, are forcibly evicted and cannot manage long-term sustainable practices. Implementation of appropriate land policies seems to be the most effective way of meeting this challenge (*See* Note I).

3
Issues Relevant to Human Deprivation

A. LIFE EXPECTANCY AND OLD AGE

Till the beginning of the twentieth century, the death of infants and children was a common feature all over the world, primarily caused by diarrhoea, malaria, measles, pneumonia, small pox, tuberculosis and various forms of undernutrition making life, in general, short, the estimated global average life expectancy remaining at only 30 years while the world population numbered around 1.65 billion. Concerted efforts to control infectious diseases, such as malaria, yellow fever and hookworm started early in the twentieth century, and carried on with greater emphasis in many of the developing countries becoming independent of colonial rule during the 1950s and 1960s. The progress achieved in health services was reflected in the lengthening of life expectancy which had risen to 47 years by 1950. Apart from malaria control happening during this period, eradication of small-pox by 1977 made a very strong impact on life expectancy for the world as a whole, which increased to around 65 years by 2000.

In terms of life expectancy at birth, the females seem to play a much stronger role than the males, surviving longer, in general, than the males according to data provided by the United Nations for 2004. Nevertheless, it does not convey the real facts since the females living longer than the males, in general, are those surviving death at infant stage, and death through maternal mortality mostly happening at the time of child birth and domestic or other types of violence and abuse in most of the societies irrespective of developed or developing countries though with large variations.

As observed in a report of the United Nations, "the primary demographic consequences of fertility decline, especially if combined with increases in life expectancy, is population aging, a process whereby the proportion of older persons in the population increases and that of younger persons declines" (United Nations, 2006:7).

According to the United Nations report in 1950, just 8 per cent of the world population was aged 60 years or over and by 2005, the proportion had risen to

10 per cent, but is expected to reach 22 per cent in 2050 (Table 3.1). As a result, the number of older persons – aged 60 years and above – will nearly triple in the world, increasing from 673 million in 2005 to 2 billion in 2050. In sharp contrast to this, the number of children (below age 15) is projected to decline from 1.84 billion in 2005 to 1.82 billion in 2050 with their share of the total population dropping from 28 per cent in 2005 to 20 per cent in 2050, making a global population less young. In the geographical distribution of the global population in both 2005 and 2050, the most striking feature in the concentration as well as in the proportion of older population is the sharp division between the developed and developing regions. Developed regions have much less young people (0-14 age group) as percentage share of the total (17%) against a much higher percentage share of them in the developing regions (the highest share being 41.5% in the Least Developed Regions in 2005 with Africa in the lead with 41.4% which drops to 28% in Asia). Europe has the lowest share of young people among the developed regions, 14.9 per cent, followed closely by Northern America (20.5%). Almost a similar trend is predicted to be maintained in 2050, forty-five years later, in regard to the geographical concentration of the younger population though on a moderate scale in direct response to a lower fertility rate, the percentage share of younger population dropping to 28.2 per cent in the Least Developed Regions, Africa still leading with 28 per cent while it is expected to go down to 18 per cent in Asia in 2050 (Fig.3.1). However, as a result of the high rate of fertility still prevalent in the Least Developed countries, 'eight least developed countries will have the youngest population on earth, with median age at or below 24 years. They are – Afghanistan, Angola, Burundi, the Democratic Republic of Congo, Guinea-Bissau, Liberia, Niger and Uganda – all in Africa except Afghanistan (United Nations, 2006:7). On the other hand, in geographical distribution the picture is just the reverse in the concentration of ageing population (60 and above), the developed regions having the highest percentage of them, as mentioned earlier (20.1%), Europe leading among them with 20.6 per cent. This is going to increase further in the following decades with greater healthcare, better food and living conditions, with the ageing population constituting over 32 per cent of the total population (or more than 42% including people in the age-group of 80 and over). Interestingly, by 2050 nearly a tenth of the total population in Europe will be more than 80 years old. Comparatively, significant changes in the concentration of ageing population is also predicted to happen in the developing world, where 'population ageing had been considerably slower till 2005 due to higher fertility rates. But the slow decline in fertility rates resulted in a visible decline of younger population even earlier, as demonstrated by a fall in their percentage share from 37 per cent in 1950 to 31 per cent in 2005, the proportion of older people having increased from 6 to 8 per cent during this period, and is likely to rise to 20 per cent by 2050, equaling the proportion of population of the younger category in LDCs (Less Developed Countries) though those belonging to the latter category are likely to be a much higher proportion in the LDCs as well as in Africa – both sharing 28 per cent or

a little more than that of the total population. Interestingly, the proportion of older population (60 and above including 80 and above) is going to increase two times or more than that in the developing world by 2050 (Table 3.1). This is clearly evidenced in the proportion held by Africa, Asia and Latin America individually where the rise is exactly the double in Africa and over two-and-a-half times in both Asia and Latin America. At the same time, in comparative assessment the fall in the proportion of younger population in the developing world is much steeper than in the developed world in general – being from 17.0 per cent in 2005 to 15.2 per cent in 2050 in contrast to 30.9 to 20.6 per cent in the former, the proportion of children falling from 41.1 in 2050 to 28.0 per cent in 2050 in Africa.

Table 3.1: Percentage Distribution of the Population by broad Age-group for the World The Major Development Groups and the Major Areas, Medium Variant

Major Area	*Percentage Distribution in 2005*				*Percentage Distribution in 2050*			
	0-14	*15-59*	*60+*	*80+*	*0-14*	*15-59*	*60+*	*80+*
World	28.3	61.4	10.3	1.3	19.8	58.3	21.8	4.4
More Developed Region	17.0	62.9	20.1	3.7	15.2	52.2	32.6	9.4
Less Developed Region	30.9	61.0	8.1	0.8	20.6	59.3	20.1	3.6
Least Developed Region	41.5	53.4	5.1	0.4	28.2	61.5	10.3	1.1
Other Less Developed Countries	29.1	62.3	8.6	0.9	18.4	58.7	22.9	4.3
Africa	41.4	53.4	5.2	0.4	28.0	61.7	10.4	1.1
Asia	28.0	62.7	9.2	1.0	18.0	58.3	23.7	4.5
Europe	15.9	63.5	20.6	3.5	14.6	50.9	34.5	9.6
Latin American and the Caribbean	29.8	61.2	9.0	1.2	18.0	57.8	24.3	5.2
Northern America	20.5	62.7	16.7	3.5	17.1	55.6	27.3	7.8
Oceana	24.9	61.0	14.1	2.6	18.4	56.9	24.8	6.8

Source: UN: World Population Prospects, The 2006 Revision : Table 3

What finally emerges for the demographic structure of the two disparate worlds is, first, a 'definite trend of declining population of children in the total volume of population through the decades, making it more aged with persistent increase in the proportion of older population; second, the rate of change in population structure is quite at variance between the two worlds', but in the long run the ratio between the old and the young is going to be much higher for the developed world where life expectancy is going to exceed 60 years for a quite a large section of the total population with 42 per cent in that age-group by 2050 when in the developing world it will remain at 20.1 per cent.

Significantly, while the total number of children is likely to decrease from 1,840 million in 2005 to 1,824 million in 2050, the total number of old people is going to make a sharp increase, from about 673 million to about 2,006 million between 2005 and 2050. This gives an entirely different perspective since the

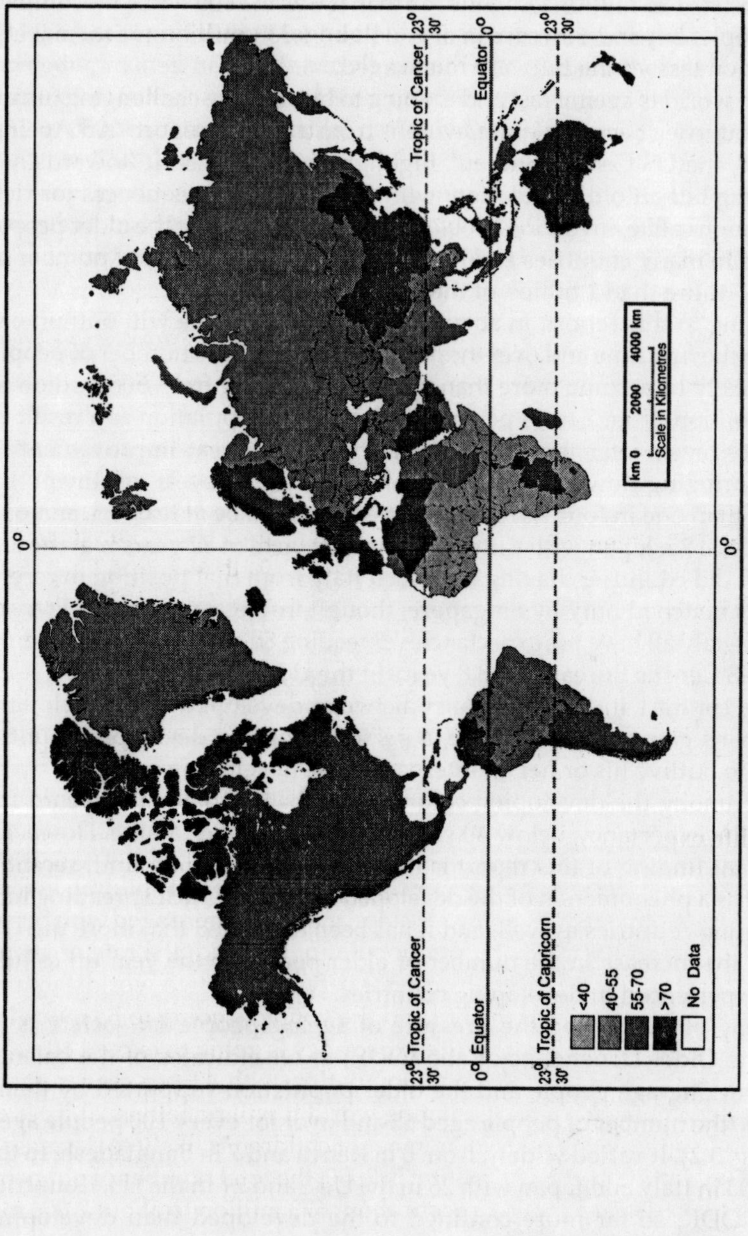

Fig. 3.1: Variation of Life Expectancy in the World

Source: **Jeffrey Sachs**

fall in the proportion of children from 28.3 per cent of the total population in 2005 to 19.8 per cent in 2050 masks the fact that the actual reduction in their number is far less – about 21 million, whereas for the old population the increase in actual number will be as much as1,333 million extra heads (comprising 14.6% of the total population). In other words, the total world population in 2050 is likely to consist of 21 million less children than it was in 2005 while the number of older people, beyond 60 years in age, will be 1,333 million more, making the population less young but very much aged.

That the world is ageing fast and is going to face a more challenging future for many countries becomes quite evident from a recent report, 'An Ageing World, 2008, the US Census Bureau', highlighting a huge shift 'towards not just an ageing but an old population, with formidable consequences for rich and poor nations alike, involving problems like how to care for the older people living alone in many countries and how to pay for unprecedented number of pensioners – more than 1 billion of them by 2040'.

According to this report, in about 10 years older people will outnumber children for the first time and over the next three decades the number of people over 65 is likely to become more than double, increasing from 506 million in 2008 to 1.3 billion, from 7 to 14 per cent of the world population as a result of high fertility levels after the Second World War and great improvement in healthcare bringing down death rates. Europe is the 'greyest' continent. By 2040, more than one in four Europeans are expected to be at least 65, and one in seven at least 75. Japan with a life expectancy at birth of 82 years is at the top of the list of old countries, having displaced Italy from that position in recent years and is matched only by Singapore, though in Western Europe, France, Sweden and Italy all have life expectancies exceeding 80 years which, according to USCB (US Census Bureau) is 78.8 years in the U.K. (*See* Note A).

But the contrast in life expectancy between developed and developing world remains glaring. For example, a person born in a developed country can expect to outlive his or her counterpart in the developing countries by 14 years. And among the developing countries Zimbabwe holds the record for the lowest life expectancy, below 40 years due to AIDS and famine. However, the important finding of this report is that the wave of ageing until recently considered as a phenomenon of the developed countries is fast spreading into the developing countries as well, and it has been estimated that more than 80 per cent of the increase in the number of older people in the year up to July 2008 was experienced in developing countries.

One way of measuring the pressure of ageing people on society is to consider the Older Dependency Ratio (ODR) as an indicator of the balance between working-age people and the older population supported by them. The ODR is 'the number of people aged 65 and over for every 100 people aged 20 to 64 (Fig. 3.2). It varied widely, from 6 in Kenya and 7 in Bangladesh, to the highest of 33 in Italy and Japan with 26 in the U.K. and 21 in the US. Countries with high ODR, so far more confined to the developed than developing countries, are already creaking under the intense pressure of funding prolonged retirement for their older population. Besides, women living on average seven years longer than men, means more of them living alone which creates another problem for the society to tackle. For example, by 2040, there will be 96 female pensioners in Italy for every 100 women aged between 20 and 64. China is another of the fast-ageing countries in the world, as its fertility rate has remained below replacement level since 1991. Within a decade its labour force will begin to shrink reducing China's basic advantage in the global economy – cheap labour. The grave concern about a lack of young workers to support its ageing population has prompted the Shanghai authorities to urge eligible couples to have a second baby, after years of following a one-child policy, which already included a series of exemptions including for ethnic minorities and couples who are both only children. Earlier, in 2009, the US-based 'Centre for Strategic and International Studies' warned that China would have more than 438 millions over 60s by 2050, each having just 1.6 working-age adults to support them compared with 7.7 in 1975 (Fig. 3.3).

Age and life expectancy have changed further during the last thirty years or so during which period the world has become more urbanized – from about 38 per cent of global population being urban to over 50 per cent or 3.3 billion people in 2008 becoming urban residents, likely to increase to almost 5 billion people by 2030. Most of the growth will be in the smaller cities of the developing world and in the metropolises of unprecedented size in southern and eastern Asia. Healthcare services are obviously better in urban than rural areas but 'the enormous social and economic stratification within urban areas results in significant health inequities... within the developing countries, the best local governance can help produce 75 years or more of life expectancy, with poor urban governance, life expectancy can be as low as 35 years" (WHO, 2008: 7). This is primarily because one-third of the urban population today – over one billion people – lives in slums which are prone to many hazards including 'pollution, accidents, workplace hazards and urban violence'. "Loss of social cohesion and globalization of unhealthy lifestyles contribute to an environment that is decidedly unfavourable for health" (WHO, 2008: 7-8), particularly in situations where a large section of the world's 200 million international migrants live, constituting at least 20 per cent of the population in 41 countries. Quite naturally, migrants having minimum access to local healthcare benefits run greater health risks. "This mobile and urbanized world is ageing fast and will continue to do so. By 2050, the world will count 2 billion people over the age of 60, around 85 per cent of whom will be living in today's developing countries, mostly in urban areas. Contrary to today's rich countries, low- and middle-income countries are ageing fast before having become rich, adding to the challenge" (WHO, 2008: 8).

Naturally, such a situation is likely to affect the society hard because of the effect on the working population, as the number of persons in the working age determines the potential for economic growth. The proportion of population of working age – 15 to 59 – appears to have decreased from 61per cent in 1950

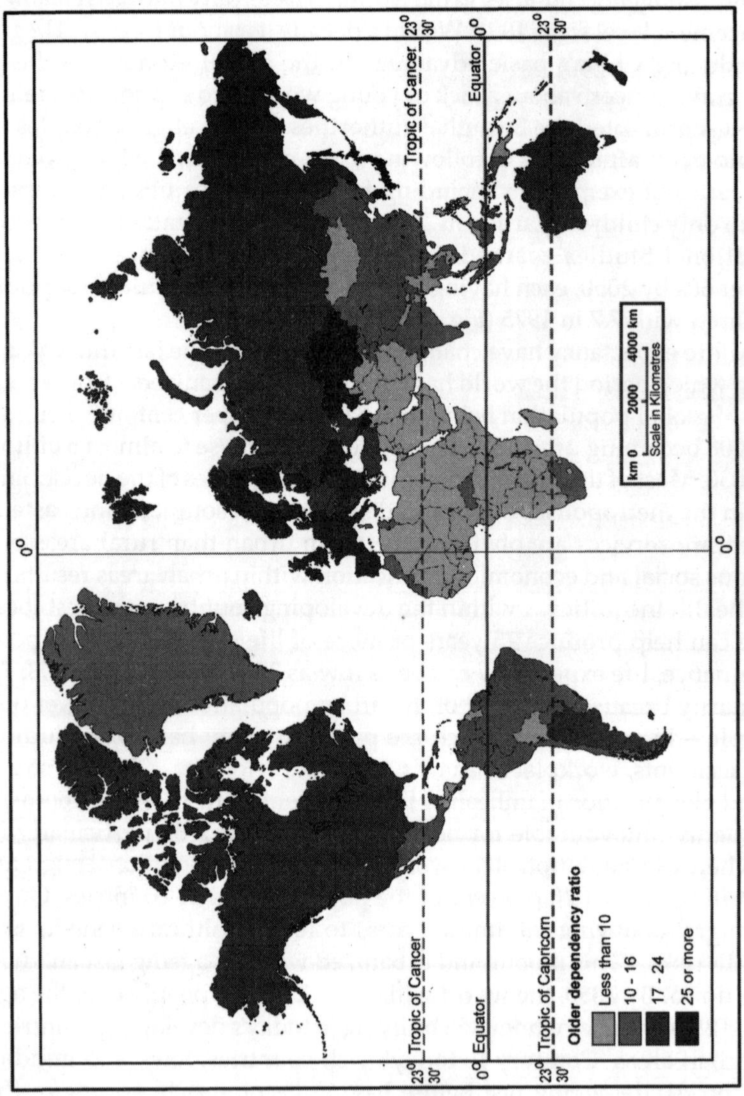

Fig. 3.2: Old Dependency Ratio in the World, 2010 and 2040

Source: The Guardian, The UK

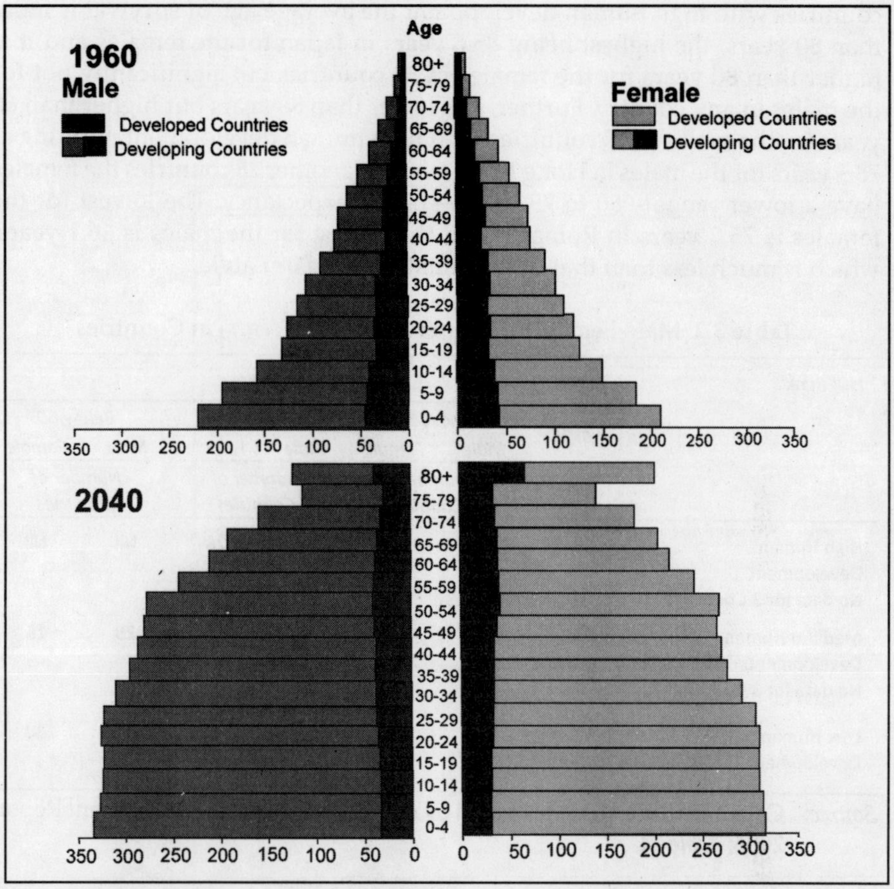

Fig. 3.3: The Number of People by Age and Sex, 1960 and 2040

Source: *The Guardian*, The UK, 2009

to 59 per cent in 1970, but increased steadily to 63 per cent in 2005 though the trend is not going to be maintained afterwards, and is likely to decline to 52 per cent by 2050. In the developed regions, on the other hand, the proportion of population in the working age is 'expected to decline slightly' from 61 per cent in 2005 to a little over 59 per cent in 2050. In contrast, in the Least Developed countries it will rise from 53 per cent in 2005 to 61 per cent in 2050 (Table 3.1).

As Table 3.2 shows, the data available for a total of 177 countries, divided according to HDI (Human Development Index) values into three groups, High Human Development (HHD), Middle Human Development (MHD) and Low Human Development (LHD) countries, there are wide differences in life expectancy with each group of countries, For example, countries with HHD, consisting of 68 countries with wide geographical locations, consistently have the highest level of life expectancy compared with the other two groups. In countries with high human development the average age of survival is more than 80 years, the highest being 85.6 years in Japan for the females and it is higher than 80 years for the females in 32 countries but significantly not for the males in any country. Further, it is lower than 80 years but higher than 60 years for the males in 60 countries with high human development, peaking at 78.9 years for the males in Hong Kong while in another 28 countries the females have a lower range – 60 to 79 years – of life expectancy. The lowest for the females is 75.2 years in Romania and the lowest for the males is 66.1 years, which is much less than that of the females found in Latvia.

Table 3.2: Male-Female Life Expectancy Rate (Years) in Countries

HDI Rank	*Total Number of Countries*	*Years*					
		Above 80		*60-79*		*Below 60*	
		Male	*Female*	*Male*	*Female*	*Male*	*Female*
		Number of Countries		*Number of Countries*		*Number of Countries*	
High Human Development No data for 3 Countries	63	Nil	32	60	28	Nil	Nil
Medium Human Development No data for 3 Countries	83	Nil	Nil	61	66	20	15
Low Human Development	31	Nil	Nil	Nil	1	31	30

Source: Calculated from data provided by the UNDP, Human Development Report 2006: *Table* 24

Almost a similar trend in this regard prevails in the countries with medium human development. Consisting altogether of 83 countries situated across the world, none has a life expectancy exceeding 80 years, but averages between 60 and 79 years for 61 countries for the males and 66 countries for the females. There are 47 countries within this group where life expectancy enjoyed by people exceeds 70 years on average and Albania among them has the highest average reaching 76.99 years for the females but this goes down to 71.8 years for the males in Syrian Arab Republic. In fact, the length of life falls below 60 years for the males in 20 countries and in 15 countries for the females, falling to the lowest of 31.3 years for both the females and the males in Swaziland.

The low human development countries, 31 in total, have a more modest record of life expectancy. People living in them have a shorter life, the males surviving less than 60 years on average as a common rule while the females in only one country – Yemen – cross the threshold of 60 years (62.4 years). On the whole, in all the 31 countries the males survive for less than 60 years and the females in 30 countries enjoy life not exceeding that limit. The highest average longevity (59.7 years) for the males is found in Yemen while both the lowest of 34.0 years for the males and 36.2 years for the females occur in Lesotho.

Significantly, the HDI has a very close relationship with the span of life and geographically the shortest span occurs, in general, in the African countries. Besides the prevailing trend, the females have longer life expectancy than males in 172 countries. Just five countries have longer life expectancy among males - Maldives (Males: 67.4 years and females: 66.6 years); Botswana (Males: 34.9 years and females: 34.8 years); Zimbabwe (Males: 37.2 years and females: 36.0 years); Zambia (Males: 38.2 years and Females: 37.1 years); and Malawi (Males: 40.0 years and Females: 39.6 years). The scientific explanation for this disparity favouring females given by Murray and Lopez (1997) is as follows: "The life expectancy at birth is substantially longer for women than for men, as disability-adjusted life-expectancy at birth, and the percentage of life-span lived with disability is higher for men than women. Rates of years lost due to mortality and years lived with disability are higher in men than in women. These differences seem to be much greater than those expected purely on the basis of biological differences in maximum life-span and health potential. Much of the difference can, however, be explained by the combined effects of tobacco, alcohol, and other addictions or exposures that lead to life injuries. Narrowing this gap will be a major challenge for health policy" (Murray and Lopez, 1997: 1441).

Table 3.3 provides an insight into the changes in life expectancy over the period 1960 to 2002, in different regions divided into two income-levels. The

Table 3.3: Levels and Changes in Life Expectancy, 1960-2002, by World Region

Region	*Life Expectancy (Years)*			*Rate of Change (Years per decade)*	
	1960	*1990*	*2002*	*1960-90*	*1990-2002*
Low and Middle-Income Countries	44	63	65	6.3	1.7
East Asia and the Pacific	39	67	70	9.3	2.5
(China)	(36)	(69)	(71)	(11)	(1.7)
Europe and Central Asia	—	69	69	—	0.0
Latin America and the Caribbean	56	68	71	4.0	2.5
Middle East and North Africa	47	64	69	5.7	4.2
South Asia	44	58	63	4.7	4.2
(India)	(44)	(59)	(64)	(5)	(4.6)
Sub-Saharan Africa	40	60	46	3.3	-3.3
High-Income Countries	69	76	78	2.3	1.7
World	50	65	67	5.0	1.7

Note: Entries are the averages of male and female life expectancies (— not available)

Source: Dean T Jamison, 2007: Table 17.1 (Following World Bank 2004: CD-ROM Version)

progress in life expectancy between 1960 and 2002, for a period of a little over four decades, shown separately for the first three decades followed by a little over one decade clearly shows that the length of life increased faster during the period 1960-1990 than later during 1990-2002. Apart from Sub-Saharan Africa, there has been gain in all the regions though at a slower pace between 1990 and 2002 but in Sub-Saharan Africa there has been a significant loss largely due to an increase in mortality rates caused by HIV/AIDS. However, very interesting cases are provided by China and India, in the former the length of life almost doubled between 1960 and 1990 while in India the rate of increase was equally high during the entire period, rising at nearly five years per decade. In contrast, Eastern Europe and Central Asia recorded no gains (Jamison, 2007: 296-297).

B. GENDER EQUALITY

Before making any conclusive observation it is worth ascertaining how the premise of gender equality — inequality being a clear manifestation of deprivation - is supported by the ratio in the status of men and women in the society, its trend and the determining factors setting the trend. According to the statistics provided by the United Nations, 1991, 50.3 per cent of the world's population in 1985 was male; meaning there was about 101 males for every hundred females according to *Demographic Year Book* 2007, males constituted 50.4 per cent against 49.6 per cent females among world population, making a ratio of about 98 females for every 100 males in 2007, indicating a downward trend for the female segment of the world population (*Demographic Year Book*, 2007). The ratio in 1985 however, differed between countries, such as at that time there were 95 to 97 males for every 100 males in Europe, the United States and Japan; 99 males per 100 females in Africa; and 105 males per 100 females in Asia where the situation remained almost unchanged in 2007 when there were about 95.3 females for every 100 males while it seems to have improved in Africa with about 100.6 females for every 100 males; 102.4 females per 100 males in North America and nearly 108 females per 100 males in Europe as shown in *Demographic Year Book*, 2007). Difference in masculinity ratio in a population depend on three factors: difference in the ratio at birth, difference in gains and losses through migration, and difference in mortality rates. In 24 countries in Europe, "the differences in the ratio of males to females at birth being small are unlikely to be the source of difference of more than 1 or 2 per cent in the masculine ratio of the populations, and in any case cannot account for interregional differences of 10 per cent or more ... between Europe and Asia" (Coale, 1991: 517). The major male-female ratio is also influenced by in-, and out-migration. In general, the masculinity of the urban population is increased by larger male than female rural to urban migration while in some countries female rural to urban migration is heavier. As Coale points out, immigration into the United States was predominately by males in the nineteenth and early twentieth centuries, so that in 1910, about 14.7 per cent of the population in the United States was foreign-born, and the ratio of males to females among foreign-born was 1.31, compared to 1.017 among the native-born population. On the whole, the net effect of immigration prior to 1970 was to lower the proportion of males in the population of the United States which was not very significant. The effect of sex differences in mortality is another factor affecting male-female ratio of population. Coale maintains that, "in societies that provide unbiased health conditions and more or less equal nutrition to males and females, male mortality rates are higher at every age from zero to the highest age attained. Such has been the case in mortality schedules for European populations at least since the middle of the nineteenth century, over a range of expectations of life at birth from about 35 to nearly 80 years" (Coale, 1991: 518).

It is generally believed that gender inequality results from the way the productive and reproductive roles of men and women develop in the society. The existing arrangements 'tie production to reproduction in certain ways'. Hence, gender equality involves a social change largely independent of the process of economic growth. "A clear distinction needs to be made between women whose work participation goes up in response to new opportunities (usually those who are educated urban, and very often young enough not to find family responsibilities a constraint) and those whose work intensity goes up as a result of excessive demands on time. The latter is especially likely to happen to poor and older women, who are not mobile, usually illiterate and unskilled, who are unable to find alternative employment in a situation of restructuring. Such women may find themselves moving to less and less productive work with longer hours and probably increasing the quantum of unpaid work in an effort to save household resources. Greater mobility will benefit the former group but not the latter" (Krishnaraj, *et al*. 1998:7).

Gender discrimination is among the greatest and most pervasive of inequalities. South Asia is the region with the most visible female discrimination in the world as manifest very strongly in educational differences between the two sexes, in employment and earning. They are revealed very clearly in parts of South Asia where India especially takes centre-stage giving highly shocking examples of discrimination between male and female children. For example, the rate of infant mortality among the children aged 1 to 5 is 50 per cent higher for girls than boys. "These girls, deprived of life because they were born with two chromosomes, are among the circa100 million missing women in South Asia. The higher mortality rates among girls and women from birth to about age 30 inverts the normal demographic gender balance, pointing to structural inequalities, nutrition, healthcare and status" (UNDP, 2005: 15). Amartya Sen (1989 and 1990) estimates a much higher number of missing women, a great many more than 100 million, by assuming that the low "masculinity ratios in Europe, North America, and Japan would prevail in Asia if there were no excess female mortality" (Coale, 1991: 521-522). Sen and Coale define "missing women" as "the number of expected women in a country minus the number of women actually alive, holding the number of men in the country constant"

(Klassen, 1061). Coale (1991) provides some data on "missing women" and names the countries primarily involved in this great loss (Table 3.4), resulting largely from higher rates of fertility and female mortality. However, estimates of missing women have differed widely between authors depending on different 'assumptions used', the number varying between 65 and 105 million (Klassen and Wink, 2009), and following significant improvements in the situation lying at the root of this problem India seems to have now the largest share of missing women in the world, "while improved female education and employment opportunities have generally reduced female mortality" (as the major reason for missing women), "the increased use of sex-selective abortions: (preference for male child) "has had the opposite effect of raising the number of missing women, particularly in China, but increasingly also in India" (Klassen and Wink, 2009: 297). Klassen and Wink have further included countries like Afghanistan, South Korea, Sri Lanka, Iran, Algeria, Tunisia and Sub-Saharan Africa (SSA) in the list of countries affected by missing women, and according to them, the global figure of missing women 'shows an increase ... in the number from 87 million to 95 million' though the share of them 'as a share of women alive has fallen from 7.7 per cent to 7 per cent with a considerable drop in Bangladesh, Pakistan and, West Asia, accounting for '2-3 percentage points', while 'in India and Egypt it dropped by some 0-2 percentage points' in contrast to a rise of '0-4 percentage points in China, the latter adding another 6.3 million missing women" (Klasen and Wink, 2009: 305) (*See* Note *B*).

Table 3.4: Estimated Deficits in Female Populations caused by Excess Female Mortality: Selected Areas

Country or Region	*Ratio of Males to Females*		*Number of Females (millions)*	*Percent of Females Missing ***	*Number of Missing Females (millions)*
	Actual Ratio	*Expected Ratio**			
China, 1990	1.066	1.010	548.7	5.3	29.1
India, 1991	1.077	1.020	406.3	5.6	22.8
Pakistan, 1981	1.105	1.025	40.0	7.8	3.1
Bangladesh, 1981	1.064	1.025	42.2	3.8	1.6
Nepal, 1981	1.050	1.025	7.3	2.4	0.2
West Asia, 1985	1.060	1.030	55.0	3.0	1.7
Egypt, 1986	1.047	1.020	23.5	2.6	0.6

Source: Coale, 1991

As stated by Coale –

* Based on a model stable population incorporating levels of fertility and mortality prevailing some ten years earlier and assuming 1.059 as the sex ratio at birth.

** Females 'missing' as a per cent of the reported female population shown in column 3.

Son preference becomes almost a pathetic issue existing widely in India for a long time among people belonging to lower-middle to low-income classes though not entirely absent among people of the higher echelons of the society. Female foeticide as a practice still now is not unheard of. What, however, really matters as appalling and crooked is the deprivation of the girl child from birth, missing on care, food and other normal facilities including education and healthcare in such families where she is largely discriminated as an intruder into family budget. A girl in such a family is expected, in reality is subjected to even before reaching puberty to give a hand in helping her mother in her domestic chores, such as cleaning, washing, cooking, looking after younger siblings, collecting water quite often from long distance, foraging for fuel and making it (cowdung cakes in India) sometimes sacrificing school and any sort of recreation. A girl child gets less than two-thirds of the recommended calories intake. A son in such families enjoys priority in everything, in being served with a glass of water when he needs it by sister or sisters, younger or older does not matter, as it is by custom a long practice yielding the place of hierarchical superiority to the boy-child in preference to the girl-child. This pervasive son-preference, however, does not exist among the tribals of India though they are considered to be the most culturally backward communities in the country.

An interesting insight into the society living in Edinburgh and Glasgow (Scotland) in the early twentieth century gives almost a similar picture of deprivation for the females in their experiences as girls where within the family girls especially were required to help in the home and care for younger siblings. Flora MacDonald, like many women, has no memory of her brothers helping in the home. She had three brothers and four sisters. As the girls got older they did more work. Brothers were treated like gentlemen – that was general rule… 'Mother used to carry the red carpet for brothers'. Alex Kellock, growing up as one of eleven children in the 1920s, does not remember even having to do 'as mush as boil an egg' as he had older sisters who did everything for him and his seven brothers, even making their beds…. The harsh circumstances of daily life (particularly in tenement houses consisting of one or two rooms in which majority of the urban population lived) meant that many women were prematurely aged and relied on their children's help; girls might be kept off schools to help at home, particularly on wash day… Girls had often to juggle their school studies with home commitment and parents could rarely afford to let them stay on at school past their minimum leaving age … Educating girls beyond their need was anyway considered by many parents a 'waste of time and unnecessary burden', as Phemie Anderson remembers… leaving school at fourteen without a leaving certificate (Clark and Carnegie, 2006: 27-28 and 35, 42).

Even though there is no conclusive evidence to confirm discrimination takes place on the basis of modern diagnostic tools to determine the sex of the child during pregnancies, the Census data revealing an unusually high proportion of male births and male children under five in Asia, particularly in China and India suggest sex-selective foeticide and infanticide in practice in spite of serious attempts at its eradication.

Further, all-India data clearly reflects the gender bias, having an increasing proportion of males in the juvenile population with the exception of Kerala. As Krishnaraj *et al.* observe, it is very difficult to conclude on the attitude of the parents on having girl child, obviously in the economically constrained families which is not very encouraging as in most of the cases marrying off a girl, considered to be very much tied up with the prestige of a family in the society, becomes a major issue involving financial position on the one hand, and the prevailing system of dowry on the other, as a "status marker". In many ways the dowry determines the fate of a girl since a higher demand from the bridegroom's family makes the girl a liability to parents and much of the violence against married girls happening quite often, culminating at many instances to bride killing is related to non-fulfilment of dowry demands, sometimes even falsely claimed (*See* Note *C*).

In India, the sharp decline in sex-ratio in 0-6 age-group results from the practice of female foeticide which has been very adversely criticized by the Supreme Court of India, urging the Union and State Governments to enforce laws banning sex discrimination and female foeticide. Despite legal ban, over 6 million girls have been aborted in India during the past 14 years and among the states the major part of this barbaric action takes place in Punjab, Maharashtra and Rajasthan (*See* Note *D*). Among them in Maharshtra alone more than 100,000 such fetuses are aborted every year. Incidentally, a study of a clinic in Mumbai found that 7,999 to 8,000 aborted fetuses were females. The sex-ratio among aged 0-6 has declined from 945 females for 1,000 males in 1991 to 927 females in 2001, and according to the UNICEF, 7,000 fewer girls are born in India everyday than the global average. Punjab has the lowest sex-ratio of 798 females per thousand males, in Maharashtra it is 805, in Haryana it is 819 and in Delhi it is 868. The decline is particularly acute in Punjab, Haryana, Himachal Pradesh, Madhya Pradesh, Mahasrashtra, Chandigarh and Delhi. According to a report (Gupta, 2009), 'the general preference for a son is stronger in Punjab and Haryana, rather shockingly among the literate than unlettered, rejecting the general notion that girl child as an economic liability is a major concern with poor families'. The situation is aggravated further by early marriage of the girls, and 'as reported by The International Planned Parenthood Federation, 3 million girls in India are married off before 18. Almost 50 per cent of the girls in Bihar, Rajasthan, Madhya Pradesh and Uttar Pradesh are married off even earlier, at or before the age of 16 (*See* Note *E*). This is partly corroborated by a report of the Asian Development Bank, submitting that "about 23 per cent of females between the age of 15 and 19 are married, compared with 5 per cent of the male population in the same age group" in Pakistan and "30,000 women die during delivery each year" (Sayeed, 2009: 144) (Fig. 3.4).

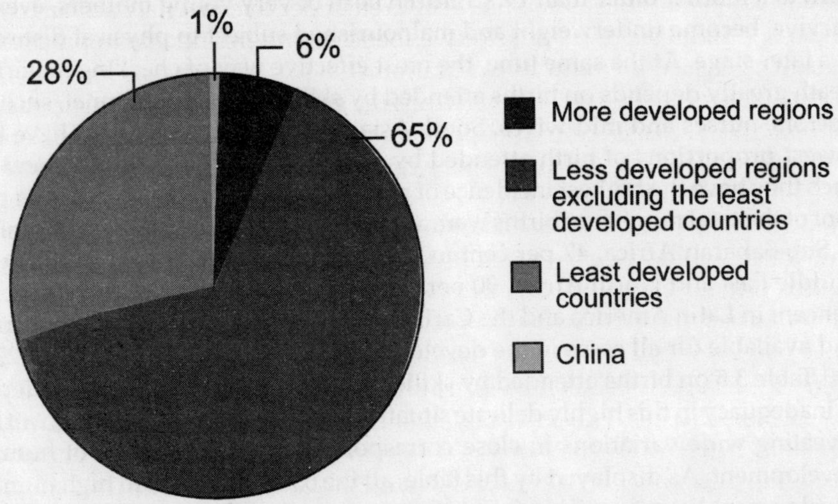

Fig. 3.4: Global Births by mothers aged 15–19, 2000-2005

Source: UNICEF, 2007

Referring to Sen (1992 and 2001) and Miller (2001), Smith and Byron (2009) point out "Perhaps the most disturbing manifestations of son preference are female infanticide (relatively rare) and an increasingly common 'high tech' form of discrimination- sex-selective abortion – leading to 'natality inequality' between the genders… The end result of gender discrimination among both adults and children in South Asia is a high rate of excess female mortality, unnaturally low life expectancies for females relative to males, and population sex ratios skewed disproportionately in favour of males, the latter which documents at an aggregate level evidence of millions of 'missing women' in the region" (Smith and Byron, 2009: 116-117). Early marriage of girls very closely linked with serious consequences still now remains as a persistent problem. In Asia, 18 per cent, in Africa 16 per cent and in Latin America 8 per cent of the girls are married off at the age of 15 which results in early motherhood bringing greater stress on teenage mothers, giving rise to the birth of underweight babies.

Interestingly, in global perspective, child marriage turns out to be an important phenomenon when 36 per cent of women aged 20-24 are found to be married before reaching 18 years of age and most of them happened in South Asia and Sub-Saharan Africa leading to premature pregnancy and motherhood at a very early age. "As estimated 14 million adolescents between 14 and 19 give birth each year' and in consequence girls under 15 are five times more likely to die during pregnancy and childbirth than women in their twenties.It has further been estimated that 'if a mother is below 18, her baby's chance of survival in the first year of life is '60 per cent' less than that of a baby born to a mother older than 19. Children born of very young mothers, even if survive, become underweight and malnourished subject to physical disorder at a later stage. At the same time, the most effective way of checking maternal death greatly depends on births attended by skilled health personnel, such as doctors, nurses and mid-wives. South Asia and Sub-Saharan Africa have the lowest proportions of birth attended by skilled healthcare personnel and as such they have the highest incidence of maternal death. In 2000-06, only 44 per cent of the total number of births were attended by skilled healthcare personnel in Sub-Saharan Africa, 47 per cent in South Asia, but this was 84 per cent in Middle East and North Africa, 90 per cent in East Asia and Pacific region, 91 per cent in Latin America and the Caribbean, 97 per cent in CEE/CIS countries and available for all cases in the developed world (UNIFEM 2008/2009: 127).

Table 3.5 on births attended by skilled health personnel reveals the depth of inadequacy in this highly delicate situation across the countries of the world, revealing wide variations in close correspondence with the level of human development. As displayed by this table, all the 63 countries with high human development have provision for birth attendance by skilled health personnel covering 81 to 100 per cent of delivery cases, making it available for 100 per cent women in 32 countries while in another 15 countries it is available for 99 per cent of the pregnant women and in some countries, such as Kuwait (98%), Bahrain (98%), Malaysia (97%), Oman (95%), Libya (95%) and Saudi Arabia (91%), quite a very high percentage of women have access to this service. In comparison, however, the picture is much less assuring for the middle human development countries where only 43 out of a total of 85 countries provide similar service which varies from 81 to 100 per cent of the pregnant women out of which only in 6 countries 100 per cent women have access to this, another 5 countries provide this service to 99 per cent of the women, latter including Kazakhstan, Thailand, Fiji etc. but while China makes it available for 97 per cent of the pregnancies, in India only 43 per cent of pregnant women gets this benefit. The situation is far more distressing for the low human development countries, 22 in all, and access to this service in 9 of them extends to less than 40 per cent of the cases, the lowest going down to 6 per cent in Chad in Africa

Table 3.5: Births Attended by a Skilled Health Personnel in % for 1997-2005

HDI	*Total number of Countries*	*0-10*	*11-20*	*21-40*	*41-60*	*61-80*	*81-100*
High Human Development Without data = 7 Countries	70	0	0	0	0	0	63
Medium Human Development	85	0	4	6	12	20	43
Low Human Development	22	1	2	6	10	3	0

Source: Calculated from data provided by UNDP Human Development Report 2007-2008, Table 6

and the highest is 68 per cent in Cote d'Ivoire, which including the Democratic Republic of Congo and Benin makes only three countries where access to skilled healthcare services are available for just over 60 per cent of the pregnant women.

Undoubtedly, the deplorable state of the availability of services from skilled personnel in its wide variation and increasing inadequacy with level of human development, is a reason for higher incidence of maternal mortality which is available with greater attention to redeem their paucity. As reported in UNDP 2002 report, every year more than 500,000 women die in pregnancy and child birth, the situation being worst in Sub-Saharan Africa where a woman has a 1 in 13 chance of dying in pregnancy or childbirth. "Increasing the number of births attended by skilled health personnel is key to reducing maternal mortality ratios and again there is wide variation – with as few as 29 per cent of births attended by skilled personnel in South Asia and 37 per cent in Sub-Saharan Africa" (UNDP, 2002: 26).

I. Maternal Mortality

Maternal mortality presents another very discouraging feature of female life, which gets permanently terminated because of various reasons, and prominent among them are malnutrition of the mother during the period of pregnancy, larger number of births than mother's health-condition permits, lack of skilled delivery system and lack of proper care during post-natal period. As mentioned earlier, every year more than 500,000 women or roughly one woman every minute dies as a result of pregnancy and childbirth, such deaths being 100 times more likely in Sub-Saharan Africa than in high-income OECD countries (Table 3.6)."Some 99 per cent of all maternal deaths occur in developing countries, with over 90 per cent of those in Africa and Asia. Two-thirds of maternal deaths in 2000 occurred in 13 of the world's poorest countries. The same year, India alone accounted for one quarter of all maternal deaths. One out of every 16 Sub-Saharan African women will die as a result of pregnancy or childbirth, compared to just 1 out of every 4000 in industrialized countries. Moreover, motherless newborns are between 3 and 10 times more likely to die than newborns whose mothers survive" (UNICEF, 2007: 5).Great disparities in the risks of maternal mortality across and within regions are quite remarkable. Such as, the lifetime risks of maternal mortality is higher than 1 in 100 women in 59 countries, where more than one third of the world's population lives, making it 13 times higher than in developed countries. It is 1 in 76 women in developing countries, more than 1 in 50 in 44 countries containing about 12 per cent of the world's population, less than 1 in 1000 in 61 countries with 41 per cent of the global population but only 1 in 8000 in developed countries. The situation, as shown in UNDP, 2002 Report, seems to be worst in Sub-Saharan Africa, where a woman has an 1 in 13 chances of dying in pregnancy or child birth, indicating virtual absence of medical care at the time of delivery or during the period of pregnancy. The lifetime chances for a woman to survive the ordeal of procreation, 1 in 13 to 1 in 283 – makes a wide variation in the scale of such death occurring for a pregnant mother in the developing world which is highly regrettable compared with pregnant mothers in the developed countries where 1 out of 797 to 1 in 4,085 women face this grim future (Table 3.6). Needless to say, increasing the number of births attended by skilled health personnel is key to reducing these untimely maternal deaths. However, in recent years there has been decrease in maternal mortality ratio, falling by 7 per cent between 1990 and 2005, the global maternal mortality ratio decreasing from 430 deaths per 100,000 live births in 1990 to 400 per 100,000 in 2005. But this rate (roughly less than 0.4 per cent per year at the global level) is much behind the required level of 5.5 per cent annual reduction to achieve the global target. It is proportionately high in Sub-Saharan Africa where maternal mortality ratio is 920 deaths per 100,000 live births in 2005 which is 500 in South Asia, decreasing from 650 in 1990, while, by contrast, in the developed world only one woman out of 800,000 dies from pregnancy related causes (UNIFEM, 2008-2009: 126-127). Further, it is of gravest concern for the developing countries that 'some 99 per cent of all maternal deaths occur here, with over 90 per cent of them in Africa and Asia. In 2000, two-thirds of the maternal deaths occurred in 13 of the world's poorest countries (Fig. 3.5).

Table 3.6: Maternal Mortality in Different Regions

Region	*Lifetime Chance of dying in pregnancy of childbirth*
Sub-Saharan Africa	1 in 13
South Asia	1 in 54
Middle East and North Africa	1 in 55
Latin America and the Caribbean	1 in 157
East Asia and the Pacific	1 in 283
Central and Eastern Europe and CIS	1 in 797
OECD	1 in 4,085

Source: UNDP, 2002 (Refers to UNICEF, 2002)

Maternal mortality rate and fertility rate are the two most important components defining the status of women in the society, of which the first is defined "as the death of a woman while pregnant or within 42 days after the termination of that pregnancy regardless of the length and the site of pregnancy, due to any cause related to or aggravated by pregnancy or its care but not due to accidental or incidental causes" (UNDP, 2007: 369). As such, maternal death is bound to affect demographic structure of the society with untimely death of mother mostly within her reproductive period. On the other hand, fertility rate defined as the number of children born to a woman also have a positive impact on the demographic structure, Table 3.7 on maternal mortality rate shows two figures – one, reported by the national authorities, and the other

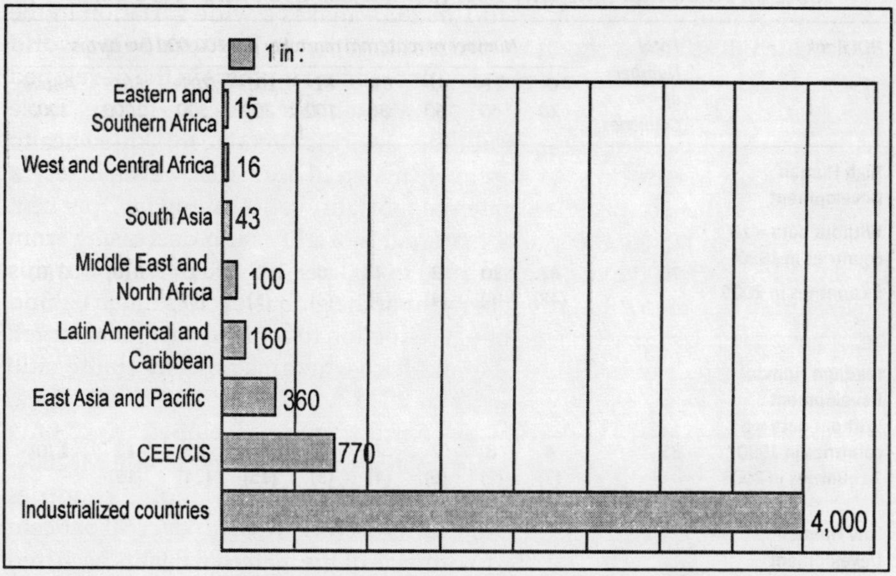

Fig. 3.5: Lifetime risk of maternal death, 2000

Source: UNICEF, 2007

(shown in parenthesis) is the adjusted one based on reviews made by different international sources. Figures for countries belonging to three separate groups of ranking in terms of the level of human development, such as high, middle and low, show that the rate of maternal mortality was significantly low in majority of countries with high human development, staying below 20 per cent per 100,000 live births in 42 out of 60 countries with data. This corroborates further with the 'revised' figures for 2005, when in 48 out of a total of 65 countries (with data) had it below 20 per cent per 100,000 live births, going down to one digit figure in 30 countries according to both 'reported' and 'adjusted' figures. Importantly, there were at least 4 countries between 1990 and 2005 - Luxembourg, Cyprus, Brunei and Barbados – from where the national authorities reported having no maternal deaths though according to the 'adjusted' data rate of maternal death was 1 per 100,000 recorded by Ireland alone in 2005. However, the rate ranges between 21 and 40 per 100,000 live births in 10 countries with high human development as reported, but it is only in 6 countries, according to adjusted data. While the highest of such deaths for the High human development countries reached 250 for 100,000 live births in Saint Kitts and Nevis in 1990-2005 among the 'reported' countries, the highest 'adjusted' figure in 2005 was 110 per 100,000 in Brazil.

In comparison, the rate of maternal deaths appears to be higher for a large number of countries with middle human development, remaining above 100 per 100,000 live births, as exhibited by both 'reported' and 'adjusted' figures in

Table 3.7: Maternal Mortality Rate (per 100,000 live births) 1990-2005

HDI Rank	*Total Number of Countries*	*Number of maternal mortality per 100,000 live births*								
		0-20	*21-40*	*41-60*	*61-80*	*81-100*	*101-200*	*201-500*	*501-1000*	*Above 1000*
High Human Development Without data = 10 countries in 1990 5 countries in 2005	70	42 (48)	10 (6)	3 (4)	4 (3)	0 (3)	0 (1)	1 (0)	0 (0)	0 (0)
Medium Human Development Without data = 5 countries in 1990 7 countries in 2005	85	4 (3)	8 (2)	7 (6)	7 (4)	8 (5)	16 (15)	17 (24)	12 (19)	1 (0)
Low Human Development Without data = 3 countries in 1990 None in 2005	22	0 (0)	0 (0)	0 (0)	0 (0)	0 (0)	0 (0)	4 (1)	10 (11)	5 (10)

Source: Calculated from data provided by UNDP Human Development Report 2007-2008 Table 10

Note: Two sets of data:
(i) Reported by the National Authorities shown outside brackets.
(ii) Adjusted data for 2005 shown within brackets (data based on reviews made by UNICEF, WHO and UNFPA to account for well-documented problems of underreporting and miscalculation.)

46 out of 80 countries (with data) in 1990-2005, and in 58 out of 78 countries (with data) in 2005 it was above 100, reaching the highest of 1,100 in Zimbabwe according to the national authorities, and the highest of 1,000 in Cameroon as 'revised' data. The lowest of the 'reported, figure on maternal death was 1 for 100,000 in Granada but was 10 as 'revised' figures occurring in Ukraine.

Similarly, the rate is mostly on the higher side as reported for the countries with low human development, being higher than 200 per 100,000 live births in all the 19 out of 22 countries (with data), becoming still higher when 'adjusted' for all the 22 countries in 2005, exceeding 200 in 4 countries and going above 500 in 21 countries. According to the 'reported' data, Sierra Leone had the highest number of maternal deaths, both the reported and adjusted data for this country becoming as high as 2,100 per 100,000 live births in 2005.

The indication is clear enough that the level of human development has a determining effect on the rate of maternal mortality, making it obvious that this human tragedy is largely the result of factors not very difficult to control.

II. Fertility Rates

Since maternal mortality is directly linked with pregnancy, the rate of fertility may be considered as a mirror reflecting how far women are considered safe in bearing the burden of giving birth to children. Table 3.8, giving fertility rates for two periods, 1970-75 and 2000-05, shows that of the total 67 countries with high human development having data, in 1970-75, 51 had a fertility rate below 4 number of births per woman, and in 16 countries the rate was over 4, reaching the highest of 7.6 children per woman in Libya Arab Jamahiriya, with two other countries – Saudi Arabia and Oman – having 7 children per woman. However, in 2000-05, all 67 countries (with data) have fertility below 4 and in 48 countries it is below 2 per woman, the highest being 3.8 children per woman in Saudi Arabia and the lowest of 0.9 in Hong Kong China (SAR) (in comparison to the lowest of 1.6 children per woman in Finland and Germany).

Compared with this, the picture reverses for the middle human development countries, where the rate of fertility was above 2 per woman in all the 84 countries (with data) in 1970-75, exceeded 4 per woman in 79 countries, ranging as high as between 6.1 and 8.0 children per woman in a total of 47 countries, and going above 8 children per woman in one country – Yemen – which had an actual score of 8.7. In 2000-05, 52 countries had fertility rates between 2.1 and 4.0 children per woman but this exceeded in two countries with 6.1 to 8.0 children, with the highest of 7.0 in Timor-Leste. Interestingly, the lowest for both the periods – 1970-75 and 2000-05 – is recorded by the same country, Ukraine, with 2.2 per woman in 1970-75 and 1.2 in 2000-05, though both are higher than the lowest record during these two periods in High human development countries.

The same trend of the larger number of countries holding higher rates of fertility prevails in the Low human development countries, exceeding 6 children per woman in 21 out of a total of 22 countries, the maximum rising to 8.3 children per woman in Rwanda in 1970-75, whereas the fertility rate exceeds 6 children per woman in 9 of the countries in 2000-05, the highest being 7.4 in Niger. The rate did not go down below 5 children per woman in either of the two periods, being lowest at 5.7 in Central African Republic in 1970-75 and 5.1 in Cote d'Ivoire in 2000-05.

It emerges from a comparative study of the data available for the three groups of countries – HHD, MHD, and LHD – showing the fertility rates across the world in 1970-75 and 2000-05 that there is an inverse relation – the lower is development the higher is the rate of fertility. Further, the Muslim countries tend, in general, to have a higher level of fertility rate irrespective of their rank in human development. Lastly, the higher rates of fertility seem to be closely linked to the rate of maternal mortality, the larger number of children born leading to a larger number of deaths among women during or following

Table 3.8: Fertility Rate (births per woman) 1970-75 and 2000-05

HDI	*Total Number of Countries*	*Number of Children born per woman*									
		0-2.0		*2.1-4.0*		*4.1-6.0*		*6.1-8.0*		*Above 8.0*	
		1970-75	*2000-05*	*1970-75*	*2000-05*	*1970-75*	*2000-05*	*1970-75*	*2000-05*	*1970-75*	*2000-05*
High Human Development Without data = 1970-75 : nil 2000-05 : 3	70	13	48	38	19	9	0	7	0	0	0
Medium Human Development Without data = 1970-75 : 1 2000-05 : 1	85	0	0	5	52	33	0	45	22	1	0
Low Human Development	22	0	0	0	0	1	13	19	9	2	0

Source: Calculated from data provided by UNDP Human Development Report-2007/2008, Table 5

pregnancy indicating less resistance to the strain of childbirth undergone by mothers having many children, besides inadequate medical care primarily at the time of delivering children.

One of the major factors influencing fertility rates is the prevalence of measures for family planning, or prevention of unplanned pregnancies adoption and practice of which could reduce maternal deaths by around "one quarter, including those result from unsafe terminations. "According to the United Nations Statistics Division, 137 million women in the world still have an unmet need for family planning, and contraceptive prevalence has increased from 55 per cent in 1990 to 64 per cent in 2005. Another 64 million are using traditional methods of contraception, which can have high failure rates" (UNIFEM, 2008/2009: 127).

III. Underweight Infants and Children under Age 5

Maternal mortality rates have a direct relationship not only with fertility rates but also their impact is felt on two other areas of demographic structure, infant mortality and mortality of children under the age of five. Undoubtedly, the major reasons for high rates of maternal mortality are repeated child birth at short intervals making women physically weak, lack of necessary medical care, and high malnutrition before and during the period of pregnancy, all of which render women less resistant and impact heavily on the health of the child, born quite often underweight and not recuperating later as the newly-born are mostly deprived of maternal nutrition as well as immunity provided by breast milk, flow of which usually dries up prematurely due to poor health of the mother. The shortening of the lactating period proves highly risky with the infants, making them dependent upon unbalanced market-food, often with quick exposure to fatal ailments or reduction in their weight compared with any normally nourished child. A look at Table 3.9 reveals that, in 65 out of a total of 66 high human development countries with data, less than 20 per cent of the infants are short of normal weight, with just one country Trinidad and Tobago more than 20 per cent (23%). Among the 83 medium human development countries with data, in 74 of them the infants are short of normal weight by up to 20 per cent and in 9 other countries short by 20 per cent or more, including Bangladesh, Yemen, Sudan and India, reaching the highest deficit of 36 per cent in Bangladesh. In contrast in China only 4 per cent of children have less than normal weight of an infant. Among the 22 low human development countries in 4 of them deficit in weight is over 20 per cent below normal. Sierra Leone has the highest record of 23 per cent less than normal weight and the lowest is 9 per cent in Rwanda (*See* Note F).

The most striking fact revealed in this table (3.9) is that between 1998 and 2005, low birth weight for infants is, rather, a universal phenomenon than to be identified with any particular group of countries or regions and the rate of birth weight appears to be low even in some of the leading industrial countries,

Table 3.9: Infants with Low Birthweight, 1998-2005

HDI Rank	*Total Number of Countries*	*Low Birthweight (in %)*				
		0-20	*21-40*	*41-60*	*61-80*	*81-100*
High Human Development	70	65	1	–	–	–
Without data = 4 countries						
Medium Human Development	85	74	9	–	–	–
Without data = 2 countries						
Low Human Development	22	18	4	–	–	–

Source: Calculated from data provided by UNDP Human Development Report 2007-2008, Table 7

such as it is 5 per cent low in Norway, 7 per cent in Germany, France and Australia, 8 per cent in the United States, the United Kingdom and Japan while situation in this regard is better with Tunisia with 5 per cent and China with 4 per cent. Nevertheless, deficit in birth weight for infants does not exceed 9 per cent in the industrial countries whereas it is, in general, comparatively much higher in most of the developing countries, obviously for diverse reasons. However, the low birth weight of infants in the advanced countries may, perhaps, be attributed to the growing incidence of early marriage especially among girls even before they are financially established in life, and may also be due to the intolerance of figure-conscious young mothers for breast-feeding, affecting proper development of the infant. On the other hand, prevalence of higher rate of underweight infants in the developing countries is primarily due to economic reasons though social reasons (such as large number of births among both the Hindus and Muslims as dictated by religious taboos/beliefs) as well play a deciding role in expanding the family-size, ruining the mother's health.

There is very likely to be a link between underweight infants and underweight children under age five but what is discouraging is the absence of any data for this for all the advanced and some of the developing countries, making it difficult to arrive at any firm conclusion. But as underweight itself is another form of deprivation, the prevalence gives an idea about the future demographic prospects of those countries where children lacking in normal growth become prone to disease and other connected deficiencies (even hampering brain development). Figures presented in Table 3.10 for the high human development countries, show that 52 of them provide no data for underweight children under five while in 15 out of 18 countries with data, less than 20 per cent of their children under age five were underweight, in 3 countries there were more than 20 per cent, reaching the highest of 39 per cent in Albania. The lowest per cent of underweight children (under 1%) are found in Macedonia. Interestingly, while in the United States 2 per cent of the children under age five were underweight, all countries in Europe, barring Czech Republic with 3 per cent, Albania (39%), and Macedonia (1%), and such industrial countries as Canada, Australia and Japan provide no information in this regard which does not necessarily mean that all children in those countries are having normal weight. On the other hand, the percentage of underweight children is much higher in many of the developing countries. For example, in 54 out of 74 medium human development countries providing data, the range of underweight children under age five exceeds 20 per cent, reaching the highest of 6o per cent in Yemen, and is over 50 per cent in 7 countries. However, China has a better record in this with only 9 per cent of the children being underweight. The record for the 22 low human development countries is comparatively far worse, with 15 out of 22 countries having more than 41 per cent children under age five underweight and in 6 it is 50 per cent or more, of which Burundi has the highest record of 60 per cent while the lowest is 20 per cent underweight children found in Senegal.

Table 3.10: Underweight Children under age five, 1996-2005

HDI Rank	*Total Number of Countries*	*Underweight Children (%)*				
		0-20	*21-40*	*41-60*	*61-80*	*81-100*
High Human Development	70	15	3	—	—	—
Without data = 52 countries						
Medium Human Development	85	20	34	20	—	—
Without data = 11 countries						
Low Human Development	22	1	5	15	1	—

Source: Calculated from data provided by UNDP Human Development Report 2007-2008, Table 7.

Though, logically underweight at infant stage is likely to lead to underweight among children up to age five, there does not seem to exist any direct relationship – one leading to the other – between the two stages as displayed by data in Tables 3.9 and 3.10 apart from the fact that the occurrence of similar conditions are more prevalent in developing than in developed countries, showing the level of economic (and social as well) development acts as a determining factor for making infants and children under five in age underweight. Further, "according to study conducted by the International Food Research Institute, if men and women have equal influence on decision-making, the incidence of underweight children under three years old in South Asia would fall by up to 13 percentage points, resulting in 134 million fewer undernourished children in the region; in Sub-Saharan Africa an additional 1.7 million children would be adequately nourished" (UNICEF, 2007: !6).

Admittedly, infant mortality (Table 3.13) and mortality of children under the age of five (Table 3.14) are two strong indicators of deprivation, depriving life to a high percentage of infants and children under five, indirectly influenced by both fertility rates and the rates of maternal mortality. For example, high fertility rates leading to early death of mothers (many of them dying quite at an early age of reproduction/motherhood) raises the incidence of maternal mortality while having a serious impact on the health of the child born, as physically weak mothers not having enough nutrition themselves during the pre-, or post-natal period usually give birth to children in many instances far ahead of the normal delivery date and that too happening most often without proper medical help, in unhygienic conditions. Children born in such situations are very likely to be physically weak to resist any infection or disease, causing premature death.

In general, undernutrition is implicated in up to 50 per cent of all deaths of children under five, reduced resistance making them easy prey to various ailments. As estimated, "almost 40 per cent of all deaths under five occur during the neo-natal period", or first five months of life, caused by a variety of complications, and around 26 per cent, representing 10 per cent of all under five deaths, occur due to severe infections, such as pneumonia and sepsis (a serious blood-borne bacterial infection). The latest evidence is that 4 million babies die each year in their first month of life and half of these die in their first 24 hours – a child is about 500 times more likely to die in their first day of life than at one month. According to available statistics, around 2 million children under five die from pneumonia each year – "around 1 in 5 deaths globally". In addition, up to 1 million more infants die from severe infections including pneumonia, during the neo-natal period... Diarrhoeal diseases account for 17 per cent of under-five deaths. Malaria, measles and AIDS together, are

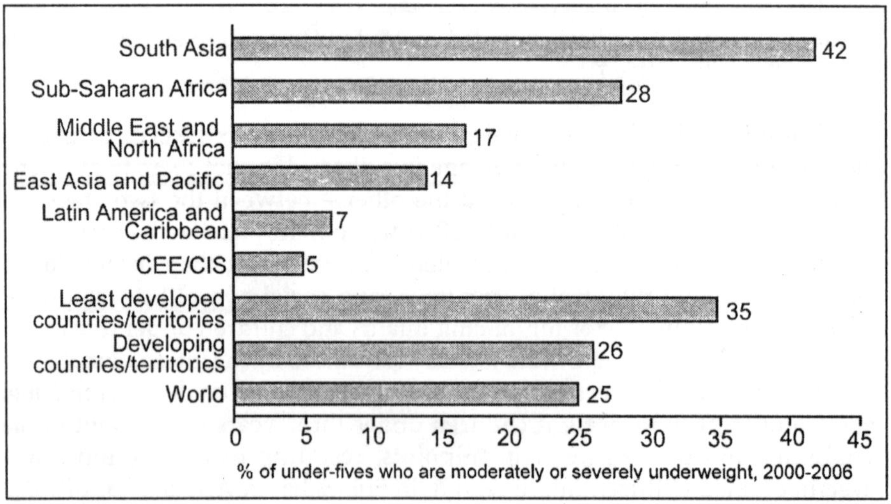

Fig. 3.6: Level of Undernutrition among children under five around the World
Source: UNICEF, 2008

responsible for 15 per cent child deaths (UNICEF 2008: 8). Among them measles alone happens to be a great killer, with 90 per cent of people (345,000 in 2005) dying of measles being children. Neo-natal mortality accounts for almost 40 per cent of all deaths under-five and for nearly 60 per cent of infant (under one) deaths. The largest absolute number of new-born deaths occurs in South Asia – India contributes a quarter of the global total – but the highest national rates of neo-natal mortality occur in Sub-Saharan Africa. A common factor in these deaths is the health of the mother – each year, as mentioned earlier, 500,000 women die in childbirth or from complications during pregnancy. In this connection it needs to be emphasized that, "according to the latest estimates for 2000-06, at present in the developing world, one quarter of pregnant women do not receive even a single visit from skilled health personnel (doctor, nurse, midwife); only 59 per cent of births take place with assistance of skilled attendant; and just over half take place in a healthy facility" (UNICEF 2008: 4). As mentioned in the *Lancet*, the world's leading medical journal, in 2005, 3 million of the 4 million neo-natal deaths could be prevented each year with proper intervention for reducing child mortality. South Asia has the highest level of undernutrition, affecting 42 per cent of the under-five children, followed by Sub-Saharan Africa with 28 per cent of them, decreasing to 17 per cent in Middle East and North Africa, 14 per cent in East Asia and Pacific, 7 per cent in Latin America and the Caribbean and 5 per cent in the CEE/CIS countries (UNICEF 2008: 11) (Fig. 3.6).

Undeniably, the data given here is far from giving the actual picture as many such deaths go unrecorded or uninvestigated in many developing countries where child birth is not properly and systematically recorded. For example, among the developing countries Sub-Saharan Africa holds the highest record of birth registration (66% of births are duly registered, as claimed), the actual number of registered births is much higher in South Asia (because of much larger population) with 64 per cent registered, but the rates are very low in Afghanistan, Bangladesh, the United Republic of Tanzania, and Zambia due to the absence of effective as well as properly functioning systems of birth registration. "In Bangladesh and Zambia, UNICEF estimates that only 10 per cent of births are registered, while in the United Republic of Tanzania the registration rate is just 8 per cent" (UNICEF, 2008: 23).

In this connection, the data provided by Jamison (2007) on the causes of death of children under five reveals the nature and severity of diseases. Table 3.11 gives the causes of death of children under age five and underlines the fact that apart from children dying due to reasons not clearly known, comprising about 39 per cent of the total of about 13.9 million cases investigated, respiratory infection happens to be the major killer, taking away 14 per cent of children below five years of age. This is followed by those dying of low birth-weight or malnutrition, making over 8 per cent of victims while malaria alone takes a total toll of 8 per cent. Interestingly, the picture changes when deaths of children at neo-natal stage are considered, showing almost 85 per cent of the deaths due to low birth-weight occurs during the earliest period, at an age

Table 3.11: Causes of Under Five Children's Mortality, Worldwide in 2001 (in 000)

Cause	*Total*	*Age 0-4*	*As % of the Total**	*Neonatal (age 0-27 days) (% of the total shown within brackets)*	*Stillbirths*
HIV/AIDS	340	340	2.4		
Diarrhoeal Diseases	1,600	1,600	11.5	116 (1.3%)	
Measles	557	557	4.0		
Tetanus	187	187	1.3	187 (100%)	
Malaria	1,087	1,087	7.8		
Respiratory Infection (and sepsis)	1,945	1,945	14.0	1,013 (52.1%)	
Birth asphyxia and birth trauma	739	739	5.3	739 (100%)	
Low birth weight	1,301	1,301	8.4	1,098 (84.4%)	
Congenital Anomalies	439	439	3.2	321 (73.1%)	
Injuries	310	310	2.2		
Other	5,375	2,101	38.74	446 (8.3%)	3,274
Total	13,874	10,600	76.40	3,900	3,274 (23.60%)

Source: Dean T Jamison, 2007: Following Mathers, Murray and Lopez (2006), Jamison (2006), Jamison *et al.*, (2006)

Note: (a) Of the estimated 13.9 million under five deaths only 0.9 per cent occurred in high income countries. Thus, deaths mentioned in this table refer to low and medium income countries.

(b) "Stillbirths" refer to fetal loss in the "third trimester of pregnancy. About 33 per cent of stillbirths occur after labour has begun".

* computed.

between 0 and 27 days and hundred per cent of the deaths from tetanus and birth asphyxia and trauma before the child crosses its 27th day of birth. Congenital anomalies similarly take a very high toll, accounting for over 73 per cent of such deaths at the neonatal stage. Respiratory infections as well appear to be a great killer at this earliest stage, terminating the lives of 1,013,000 children or over 52 per cent of them dying from this. Taking all deaths of children under five, 390,000, or more than 28 per cent, die at the neonatal stage while another 23.6 per cent lives are lost as stillbirths which seems to be exceptionally high but preventable with adequate care during pregnancy and at the time of delivery. Undoubtedly, the situation prevailing in the low, and medium-income countries is far from good for dealing with pregnant women in the absence of adequate medical facilities supplemented with nutritional intake.

As pointed out by Jamison (2007), it becomes more worrying to observe that the role of many of these curable diseases does not diminish as killers of children even at the age of five or above. For example, estimates given for

Table 3.12: Causes of death among Children aged 5 and above in low and middle income countries, 2001

	Deaths (in millions)	*% of the Total*
1. Communicable, maternal, perinatal and nutritional conditions		
TB	1.5	4.0
AIDS	2.2	5.8
Respiratory infections	1.5	4.0
Maternal conditions	0.5	1.3
Other	2.5	6.6
Sub-total	8.2	21.7
2. Non-communicable diseases		
Cancers	4.9	13.0
Diabetes	0.7	1.9
Ischaemic and hypertensive heart disease	6.5	17.2
Stroke	4.6	12.2
Chronic Obstructive Pulmonary disease	2.4	6.3
Other	6.1	16.1
Sub-total	25.2	66.7
3. Injuries		
Road Traffic Accidents	1.0	2.6
Suicides	0.7	1.9
Other	2.7	7.1
Sub-total	4.4	11.6
Total	37.8	100.0

Source: Dean T Jamison, 2007 (aggregated from Mathers, Murray and Lopez, 2006; (126-31)

nearly 38 million of them dying at this age in low- and medium-income countries in 2000 (Table 3.12), the largest proportion of such deaths, nearly 64 per cent of the total, resulting from various diseases are caused by non-communicable diseases, including cancers, diabetes, heart problems, etc., in comparison to about 32 per dent of the children dying from communicable diseases or maternal, parental and nutritional conditions. Intriguingly, 0.7 million children aged five and above committed suicide while road accidents took a toll of another 1 million children. It may be noted further that heart disease is more deadly for older children compared with children at a lower age level, causing as much as 17 per cent deaths among the older group, involving 6.5 million children in 2001, the largest cause, and another 4.6 million dying of stroke, together making over 11 million deaths among children aged

Table 3.13: Infant Mortality Rate (per 1000 live births), 1970 and 2005

HDI Index	*Total Number of Countries*	*0 - 20*		*21 - 40*		*41 - 60*		*61 - 80*		*81 - 100*		*Above 100*	
		1970	*2005*	*1970*	*2005*	*1970*	*2005*	*1970*	*2005*	*1970*	*2005*	*1970*	*2005*
High Human Development Without Data: 3 countries	70	16	66	26	3	14	—	6	—	2	—	3	—
Middle Human Development Without Data: 22 countries in 1970	85	0	17	1	26	8		15	6	15	9	39	3
Low Human Development Without Data: 1 country in 1970	22	—	—	—	—	—	1	0	3	0	5	21	13

Source: Calculated from data provided by UNDP Human Development Report 2007/2008: Table 10

Table 3.14: Under-Five Mortality Rate (per 1000 live births), 1970 and 2005

HDI Index	*Total Number of Countries*	*0 – 20*		*21 – 40*		*41 – 60*		*61 – 80*		*81 – 100*		*Above 100*	
		1970	*2005*	*1970*	*2005*	*1970*	*2005*	*1970*	*2005*	*1970*	*2005*	*1970*	*2005*
High Human Development Without Data: 3 countries	70	9	63	27	6	12	—	6	—	6	—	7	—
Middle Human Development Without Data: 22 countries in 1970	85	—	11	1	29	1	4	5	16	4	4	52	21
Low Human Development Without Data: 1 countries in 1970	22	—	—	—	—	—	—	—	1	—	—	21	21

Source: Calculated from data provided by UNDP Human Development Report, 2007/2008: Table 10

five and above, maintaining almost a parity with children under five, 14 per cent of them dying of heart disease. On the other hand, cancers come second as a killer with death-toll of almost 5 million older children, representing 13 per cent of the total. AIDS and TB together take away another 3.7 million older children more severe than for children under five, who become greater victims of malaria, a killer of nearly 8 per cent of the total number of children falling prey to various diseases.

A comparative assessment of the prevalence of infant mortality and mortality of children under five (Tables 3.13 and Table 3.14) show that the rates for both are much higher in countries with low, and medium human development level, the highest infant mortality, amounting to 225 per every 1000 live births dying at infancy in Mali (LHD), followed by 202 in Yemen in 1970 (UNDP 2006; Table 10) with 123 in Equatorial Guinea and 165 in Sierra Leone in 2005, all of them in Africa, while the highest elsewhere was 126 in 1970 in Oman, in the Middle East. In contrast, the lowest infant mortality occurred in Sweden (11 per 1000 live births) in 1970, falling to 3 in 2004 and 2 per 1000 in Iceland in 2004 in comparison to the lowest of 9 in Belarus in 2004 for an MHD country and the lowest of 61 per 1000 live births in 2005 for a LHD country and the highest of 165 in Sierra Leone in 2004, another LHD country. Significantly, the highest 1970 figures for mortality for under fives was 400 per 1000 live births again in Mali with its highest record of infant mortality for that year, and it was 331 per 1000 in Gambia, a LHD, country, with the highest infant mortality for an HHD country of 200 in Oman. In 2004, Equatorial Guinea as an MHD country, on the other hand, accounted for the highest record of under-five mortality of 205 per 1000 and for an LHD 283 in Sierra Leone. In comparison, the rate of under-five mortality came down from 14 per 1000 in 1970 to 3 in 2004 in Iceland. Interestingly, the majority of the African countries maintained high rates of both infant and under-five mortality making that continent quite distinct in comparing data between 1970 and 2004.

The close connection between infant mortality and mortality under the age of five on the one hand, and fertility rate and maternal mortality rate on the other, becomes further clear from a comparison between the two sets of data, which exhibit a close resemblance. For example, it is in the African countries that the highest rates of fertility as well as maternal mortality occur, the fertility rate remaining above 5 births per woman in all the LHD countries (all of which, incidentally, happen to be located in Africa). Even births exceeding 7 per woman are encountered in 10 of them, the highest rising to, as mentioned earlier, 8.7 in Yemen in 1970-75, 8.3 per woman in Rwanda, both LHD countries, while it was 7.6 per woman in Libya, an MHD country, rising to 7.9 in Niger in 2000-05 as the highest fertility rate for any country, remaining at 7.8 in Timor-Leste as, the highest for an MHD country during that period. Interestingly, the lowest rate of fertility was 1.6 in Finland and Germany in 1970-75, going down further to 0.9 in Hong Kong SAR, and it was 1.1 for a woman in Ukraine, the lowest for an MHD country in 2000-05.

The rates of maternal mortality may further strengthen the idea of a direct relationship not only with fertility rate but also about its impact on two other aspects of demographic structure, infant and under-five mortality rates. For example, the highest rate of maternal mortality amounting to 1,800 per 100.000 women as reported by the national government and the 'adjusted' figure of 2100 for 2000 in Sierra Leone (UNDP 2007-2008), one of the LHD countries of Sub-Saharan Africa, along with the highest mortality of 1100 per 100,000 women as 'reported' data against 1000 as the 'adjusted' data for 2005 occurring in two other Sub-Saharan countries – Zimbabwe and Cameroon – are MHD countries. Important to note that there were 22 countries reported to have maternal deaths exceeding 500 per 100,000 women which, after adjustment in 2005 was found to be the record for 30 countries, and it exceeded 1000 maternal deaths for

Table 3.15: Comparative Data on the rates of fertility/maternal death/infant mortality (per 1000 live births) and under-five mortality (per 1000 live births) in 2005 in countries mentioned under A, B and C groups.

Countries with Higher Rates	*Fertility Rate (Births per woman)*	*Maternal Mortality Rate (per 100,000 live births)*	*Infant Mortality Rate (per 1000 live births)*	*Under-five Mortality Rate (per 1000 live births)*
1. Nigeria	5.8	1,100	100	194
2. Rwanda	6.0	1,300	118	203
3. Angola	6.8	1,400	154	260
4. Malawi	6.0	1,100	79	125
5. Burundi	6.8	1,100	144	190
6. Congo	6.7	1,100	129	205
7. Chad	6.5	1,500	124	208
8. Niger	7.4	1,800	150	256
9. Guinea-Bissau	7.1	1,100	124	200
10. Sierra-Leone	6.5	2,100	165	283
11. Cameroon	4.9	1,000	87	149
Average	6.4	1,327	122	206.6
Countries with Medium Rates				
1. Ghana	4.4	560	68	112
2. Mauritania	4.8	820	78	125
3. Congo	4.8	740	81	108
4. Madagascar	5.3	510	74	119
5. Haiti	4.0	670	84	120
6. Kenya	5.0	560	79	120
7. Djibouti	4.5	650	88	133
8. Togo	5.4	510	78	139
9. Yemen	6.0	430	76	102
10. Uganda	6.7	550	79	136
11. Gambia	5.2	690	97	137
12. Equatorial Guinea	5.6	680	123	205
13. Timor-L'este	7.0	380	52	61
Average	5.2	567	81.3	124

Table 3.15 (*Contd.*): Comparable data on the rates of Fertility (births per woman), maternal death (per 100,000 live births), infant mortality (per 1,000 live births) and under-five mortality (per 1,000 live births) in 2005

Countries with Low Rates	*Fertility Rate (Births per woman)*	*Maternal Mortality Rate (per 100,000 live births)*	*Infant Mortality Rate (per 1000 live births)*	*Under-five Mortality Rate (per 1000 live births)*
1. Hong Kong SAR	0.9			
2. Korea (Rep. of)	1.2	14	5	5
3. Slovakia	1.2	6	7	8
4. Slovenia	1.2	6	3	4
5. Czech Rep.	1.2	4	3	4
6. Latvia	1.2	10	9	11
7. Belarus	1.2	18	10	12
8. Ukraine	1.2	18	13	17
9. Japan	1.3	6	3	4
10. Spain	1.3	4	4	5
11. Italy	1.3	3	4	4
12. Germany	1.3	4	4	5
13. Greece	1.3	3	4	5
14. Hungary	1.3	6	7	8
15. Poland	1.3	8	6	7
16. Bahrain	1.3	32	9	11
17. Lithuania	1.3	11	7	9
18. Croatia	1.3	7	6	7
19. Bulgaria	1.3	11	12	15
20. Romania	1.3	24	16	19
21. Bosnia-Herzegovina	1.3	3	13	15
22. Russian Federation	1.3	28	14	18
23. Armenia	1.3	39	26	29
Average	**1.2**	**265>12.0**	**185>8.4**	**222>10.0**

Source: Compiled and calculated from UNDP 2007/2008: Tables 5 and 10.

every 100,000 women in 6 countries as reported but following adjustment altogether 11 countries in 2005 had maternal deaths exceeding 1000 per 100,000 women. All these countries with low human development are geographically located in Africa. They are – Nigeria (1100), Rwanda (1300), Angola (1400), Malawi (1100), Burundi (1100), Congo, the Democratic Republic (1100), Chad (1500), Niger (1800), Guinea-Bissau (1100) and Sierra Leone (2100). Cameroon with 1000 represents an MHD country. This is corroborated by high rates of fertility, ranging between 6.1 and 8.0 births per woman in 45 out of a total of 84 MHD countries in 1970-75, the rate remaining unchanged in 22 of them in 2000-05.

The linking between the four major aspects of demographic character of population of countries gets fully reflected in Table 3.15 focusing on the rates of maternal deaths, fertility rates, rates of infant mortality and under-five mortality bringing out how far they are correlated, one affecting the other. Of the 11 countries with high rates included in the table, the first ten fall under the category of low human development while the last one comes under medium human development. The highest rate for each of these demographic characteristics is recorded in one or the other of the first ten countries, and of them Sierra Leone holds the record of the highest rates in the world in three areas – maternal mortality, infant mortality and under-five mortality. Another country, Niger, holding the second highest rates of maternal mortality, comes first among all the countries in the world in fertility rate and takes third position in under-five mortality rate. Significantly, the rate of maternal mortality for all the first ten countries and also that of Cameroon, the eleventh country, exceeds that of any other country elsewhere in the world which as well applies partly to the rate of fertility, being almost uniformly high, remaining above 6 children per woman in all of the ten out of 11 countries along with the rate of infant mortality remaining above 100 per 1000 live births in all except two countries (Malawi and Cameroon) followed by higher rates of under-five mortality exceeding 200 per 1000 live births in 7 out of 11 countries maintaining higher rates on all the four counts. It is further interesting to note that all these countries are geographically located in Africa.

That fertility takes a pivotal role in determining the rates of maternal mortality, infant mortality and under-five mortality, is substantiated further by data available for countries with medium and low scores in the rates, as exhibited by the second and third groups of countries in Table 3.15, the former including 13 and the latter 23 countries. Among the 13 countries, the rate of fertility varies between the highest of 6.7 per woman to the lowest of 4.0 which is considerably less than 6.4 per woman on average in countries with higher rates while the rate of maternal deaths has decreased appreciably from an average of 1327 to 567. Decrease in the rate of infant mortality as well as under-five mortality between the two groups of countries is remarkable, falling from an average of 122 deaths per 1000 live births of infants and about 207 deaths of under-five among the first group of countries (with higher rates) to respectively 83 and 124 among the second group of countries (with medium rates). The reduction is more than two times in maternal mortality, infant mortality and a little less in under-five mortality when the average rate of fertility or the number of babies born to a woman has dropped from 6.4 to 5.2. Markedly, a still lower rate of fertility (average 1.2 children per woman) as recorded in 23 countries (with lower rates) is accompanied by further reduction in the average rate of maternal mortality (12 per 100,000 live births), infant mortality (8.4 per 1000 live births) and under-five mortality (10 per 1000 live births) which seems to be quite fast, pointing out very clearly that this has been possible due to a drastic fall in fertility rate, from the highest average of 6.4 to 1.2 children per woman between countries with higher and lower rates. Fertility playing a determining role in saving mothers, infants and under-five children from premature death strengthens the view that repetitive births ruining the health of the mothers not only hastens their death but at the same time jeopardizes the health of the infants they give birth to and those who survive death as infants fail to recoup later at tender age, resulting in the termination of life of both the infants and the under-five children. Another interesting fact may be observed in a higher rate of mortality for the under-fives than infants in all these countries which is not easy to explain apart from the fact that surviving mothers cannot provide them with proper nourishment for increasing their sustainability. We cannot ignore the fact that all the 11 countries with higher rates, 11 out of 13 countries with medium rates are located in Africa while 18 out of a total of 23 countries with lower rates are in Europe where the difference between the average rates of infant and under-five mortality is much less pronounced (8.4 against 10 respectively) in comparison to countries with higher rates (122 against 267) and medium rates (81 and 124).

IV. Gender Discrimination and Domestic Violence

(i) Gender Abuse

Every year, an average of 22,489 women and children are reported missing in India. Of these an average of 5,452 women and 11,008 children are not traced. The 'Action Research on Trafficking in Women and Children in India' reports (2002-2003) that many of the missing persons are not really missing but have instead been trafficked. Female infanticide is important in this connection, being prevalent among some of the higher castes in India as records show 'large groups of villages in Rajasthan and Gujarat, comprising several hundred upper-caste households, where no female child had been allowed to survive for many generations', being a strategy to acquire further landholding for consolidating their socio-economic status (*See* Note G). Female foeticide has been reported in modern time in Tamil Nadu, an estimated 3,174 girl children killed in 1975, the causes for which are shown as 'increasing landlessness and poverty, an escalating dowry system, higher gender differentials in wages, low education among women and few economic opportunities'. Female infanticide also happens to be widely practiced in rural north India and an author (Adithi, 1995) estimated that the total number could be as many as 1,632,000 each year in seven contiguous and culturally similar districts of Bihar. What is highly disturbing is that "whereas previously only upper castes such as Rajputs and Brahmins practised female infanticide, the custom has now spread to all other groups in the rural spectrum, including Scheduled Tribes, Christians and Muslims...emulating upper-class customs" (Sinha and Rajan *et al.* 2001:181).

Inequality in marriage affecting women has also other sides limiting women's freedom as "marriage confers unequal rights and responsibilities between men and women. Women's moral responsibility for homework and maintenance of chastity (not usually applicable to men/husband) "is part of the conjugal contract. Non-fulfilment of this contract according to expected standard entitles a man to punish his wife. In this context family planning programmes have been insensitive to sexual relationships in Indian marriages.

Reduction in unwanted pregnancies or unwanted births certainly lessens the reproductive burden for women but this very freedom threatens male authority, where wives have undergone sterilization. Rao cites cases of wife beating due to suspicions about their fidelity" (Krishnaraj *et al.* 1998: 12).

"Every year thousands of women are maimed or killed by rejected suitors in many countries. A landmark World Health Organization (WHO) multi-country study on women's health and domestic violence against women reveals that of those interviewed, 37 per cent of women in a Brazilian province, 56 per cent of women in a province in the United Republic of Tanzania, and 62 per cent women in a province in Bangladesh reported having experienced physical or sexual violence by an intimate partner. The pattern is broadly similar for industrialized countries. According to another key report from the same organization, the 'World Report on Violence and Death', studies show that 40 per cent to 70 per cent of female murder victims in Australia, Canada, Israel, South Africa and the United States were killed by their husbands or boy-friends—often within the context of an ongoing abusive relationship. In the United Kingdom, 40 per cent female homicide victims are killed by their intimate partners" (UNICEF, 2007: 23).

Gender abuse takes various forms and crosses the boundaries of race, culture, wealth and religion. Forced sex, rape on a massive scale particularly during wartime, domestic violence besides beating (by husband or in-laws) are some of the most common forms of discrimination or outrageous treatment met by women in different countries. More than men, women are the victims of war, as witnessed even in the recent clashes between different ethnic groups in some of the African countries. As reported by Diane Taylor about the Rwandan genocide, quoting Mirindi Euprazi, she states "almost every woman here" (Democratic Republic of Congo) "has been raped, some countless times, some have them so badly injured by repeated rapes by different militias that they are barely able to stand. Many of their children conceived through rape" (*The Guardian*, 2008). Leah Chishugi, a nurse living in London and a survivor of genocide in Rwanda, travels into the heart of Congo to record the testimony of about 500 rape victims – women and girls – abused by militias. She found victims a year old and as old as over 90 years, and hears of unspeakable atrocities perpetrated by the militia after rape of the victims. For example, 'nearly a quarter of 14-year olds have been forced to have sex or do something sexual they did not want to do, and one in four 16-year olds have been hit or hurt in some way by someone they were dating' in Britain and according to Nicola Harwin, the Chief Executive of Women's Aid, "it is shocking that 25 per cent of 14-year olds have been forced or coerced into having sex – and that's them admitting it – and even more who know someone who's has been hurt or humiliated" (*The Guardian*, 2008). (See *H*) According to WHO study, 150 million girls and 73 million boys under the age of 18 experienced forced intercourse or other forms of physical or sexual violence in 2002. An estimated, 1.8 million children are involved in commercial sex work (UNICEF, 2007: 5).

Prostitution or sex-work represents one of the vilest forms of violence degrading human dignity and freedom of women deeply rooted in the society for a very long time and called by various names in different situations. However named, prostitution happens to be the most debasing power exercised by man's world as a way of fulfilling its crude demand of physical (sexual) pleasure for which women are compelled directly or subjected to unavoidable circumstances while financially insecure. As stated by Sinha and Dasgupta (2009), the chilling facts are that "the tremendous trauma women of South Asian descent endure and the severe disintegration of self they experience after sexual violence cannot be dismissed simply as the result of middle class sensibility or rare instances of coerced sex. Sexuality seems to be at the core of South Asian women's identity (and perhaps all women), the disruption of which wreaks havoc with their sense of self" (Sinha and Dasgupta, 2009:6). Prostitution, called sex-trade, as it is a commercial activity allowing women involved to earn their living out of it, seems to be quite a large involvement, engaging as many as 23,000 workers in Kolkata (India) alone not including "college girls" and "flying prostitutes", i.e. part-timers etc., often working in secret for some income. Taken together, the total number of performing prostitutes or sex-workers may amount to 'at least 60,000'. Trafficking of persons, especially women and children (young girls and boys as well) is a global phenomenon, operating through countries to satisfy man's sexual urge and no nation is immune to human trafficking and sexual slavery. In fact, there has been an increase in sex commerce in parts of Australia, Netherlands and Germany where this has already been legalized. "For example, in the fundamentalist Islamic nation of Iran, trafficking is a highly lucrative business and buying and selling of women and girls is thriving. Traffickers in Iran seem to target girls between 13 and 17, who are sold off to Arab countries, Pakistan and European nations. To supply a new global trade, every ten minutes a woman or a girl somewhere in the developing world enters into a web of indented labour... According to the United Nations, trafficking is a 30 to 40 billion dollar per year business worldwide and is closely tied to organized crime. As estimated, 2 million people are trafficked every year to which number South Asia contributes nearly 150,000. Although the figures include labour trafficking, the number of women and girls transported across national boundaries makes up an astounding 80 per cent of the trade" (Sinha and Dasgupta, 2009: 32-33) (*See* Note I). An estimated 500,000 women are trafficked each year from Eastern Europe and the CIS to Western Europe. An estimated 15,000 Russian and East Europeans work in Germany's red-light districts. In the Netherlands, 57 per cent of the trafficked women are under 21 (UNDP, 1999: 89).

In many countries, strangely enough, steps towards legalizing sex trade are demanded by sex-workers – for women already in trade it can be a sure means of livelihood in the absence of any other work for them. This is happening in Bangladesh and also in Thailand, and in the latter it is already an annually 4.3 billion dollar industry. Undeniably, the sex industry has reached the stage of a flourishing international business of extraordinary proportion "instigated by ever-increasing demands of women and children to populate the sex trade, a trade that is conducted by organized criminal elements in society" (Sinha and Dasgupta 2009: 181).

It is true that the sex trade is not always done under force. In many cases women do it voluntarily on false premises of employment and thus end up in the flesh trade. For example, many of the Sri Lankan women transported by labour agencies on agreements for domestic duties, had to provide sexual services demanded by the employers. The US State Department estimates that around 12,000 Nepali girls and women are trafficked into Kolkata to work in the sex market while trafficked women from West Bengal villages are transferred as prospective brides to the northern states of India with dwindling female population. Available facts strongly indicate violence as a part of sex trade where the customers are the main source of that with endless instances of prostitutes 'raped, beaten, robbed, threatened and injured with weapons used by the customers'. "Furthermore, customers do not necessarily stop at brutalizing the sex- workers, but at times physically and sexually abuse their children" (Sinha and Dasgupta 2009: 185).

Shoma A. Chatterjee, writing for the *Statesman, Sunday Magazine*, Kolkata Issue, (2009) mentions about trafficking of women and children as prostitutes and slaves under the socially approved 'devadasi system', prevalent in some parts of India. The so-called '*devadasi*' (meaning slave or servant of God) system extant in different names in some of the Indian states, such as Tamil Nadu, Kerala, Maharashtra, Karnataka, Andhra Pradesh and Assam, is nothing but a system of prostitution in disguise, originally meant as entertainers of gods in temples, the practice traced back to the Third century A.D. in some places. Thousands of young girls work in this capacity and according to a source, the '*devadasi*' accounts for 25 per cent of the total number of prostitutes in India (Chatterjee, 2009). However, according to Pande (2008), "The '*devadasi* or temple girls played a centralized role in the religious and cultural life of India from 10th century onwards, as conveyed through numerous texts and inscriptions. Emerging from the concept of divine fecundity as an integral part of agrarian societies, this institution continued to flourish and is in existence even in the 21st century ... Historically, the temple dancers of India were influential women who were not only acclaimed for their artistry, but also for their untold wealth and influence in politics. In some instances they enjoyed a very high status and often shared betel leaf with the wives of kings and conversation with royalty as equals (Paes, 1985). They were honoured by the public in the past, and even offered seats alongside the figures of royalty. Modern classical dance forms, such as *Bharatnatyam*, originally *Sadir*, *Mohini Attam* and *Odissi* owe their origin to these temple dancers. The *devadasis* never married a mortal man and they were regarded as ever auspicious ones...for they were married to gods, immortals and hence could never become widows. There is no conclusive proof that they practiced prostitution though it cannot be denied that, depending on their status, they cohabited with men of their class" (Pande, 2008: 102-103). The practice, as it seems, continues and according to an estimate, 6,000 women were dedicated to gods as '*devadasi*' just in one state, Karnataka. Quoting an M.P. (Member of Parliament in India) Vinita Deshmukh reported that in 2004-05, at least 200 devadasis' practiced prostitution in two districts of Maharashtra. Though the system has been declared illegal in many states of India, such as Andhra Pradesh, Karnataka and Maharashtra, it continues stealthily in many areas (Sinha and Dasgupta, 2009: 21-22). Pande refers to the fact that "many of these women earn their livelihood" (today) "through street performance, begging or prostitution". On being requested by The National Commission of Women made to various state governments in India for providing it with the number of devadasis in their states, Maharashtra informed about the presence of 2,479 devadasis enjoying maintenance allowance; in Tamil Nadu the government identified 14,339 devadasis and in Karnataka 16,560 were identified as devadasis (Pande, 2008: 106-107).

Chatterjee (2009) further refers to the selling of little girls in upper Sind, in Pakistan, purchased by animal herders coming from farther away lands."Local Sindhis exchange women and girls across and through generations, even pledging girls yet to be born... The girl-child sees her life pledged, negotiated, bartered, sold long before she knows who she is, and sometimes years before she is born" (Chatterjee, 2009). Vast trading networks have developed and trafficking takes place across the length and breadth of Pakistan, girls taken to as far places as Quetta, Lorlai and Kandahar. As per an estimate of the Human Rights Commission in Pakistan, about 200,000 women and children are involved in this slave trade. Another 1,500 Bengali women are in Jail. Incidentally, Bangladesh announced a penalty in 1983 for the abduction of women, and similar national laws exist in other member countries of the South Asian Association of Regional Cooperation but they have not proved very effective.

The Fourth World Conference of Women in Beijing working on the UN platform for action against domestic violence against women made some headway in helping to enact laws in different countries including India to ensure justice to women and in this perspective spelled out that the onus of proof should be placed on the perpetrators of violence.

Domestic violence has attained serious proportions in many societies, in developing as well as developed world, indicating the sheer hollowness of the claim of these societies about gender equality and human dignity. For example, the Council of Europe has reported about domestic violence as the major cause of death and disability of women aged 16 to 44, accounting for more deaths and ill health than cancer or traffic accidents. In the United States, according to an estimate of the United Nations, women accounted for 85 per cent (671,110) of the victims of domestic violence in 1999. "Family violence have established clearly the link between women's inferior status in the family and society and the violence perpetrated on them...Family violence also prevents women from playing a role in decision-making within the household, at work and in the political, economic and social spheres and therefore directly influences their participation or the lack of it in public activities and hence, the exercise of their citizenship." (Elizabeth, 2009: 146-147). According to WHO, statistics available from around the world show that one out of every three women has experienced violence in an intimate relationship at some point in life. U.K. Statistics on Domestic Violence (1993) reveals that 'severe, repeated and systematic violence occurs in at least five of every hundred marriages in Britain; between 40 and 45 per cent of murdered women are killed by their male partners' (*See* Note *J*). In India, on the other hand, as reported by the Government of India, 'there has been a 71.5 per cent increase in cases of torture and dowry-related deaths between 1991-95 (Elizabeth, 2009: 153-154). In spite of the fact that the Dowry Prohibition Act originally passed in 1961 was amended three times in the 1980s, it failed to provide a clear definition of 'dowry', giving an opportunity to people to take advantage of the situation (*See* Note *K*).

As reported by Ranjana Kumari, Director, Centre for Social Research, New Delhi (India), 'one of the definite fall out of globalization and marketization has been growing incidence of domestic violence on women'. "In Asia, while national economies grew, the human costs of the gain included continued social stress and fragmentation. Increasing domestic violence, street crimes and suicides have been reported in Malaysia, Thailand and Korea. In Korea, a hot line for women received an escalating number of calls from women suffering from domestic violence – seven times as many as during the year 1998" (UNDP, 1999: 40). According to a WHO study, violence against women exists in 'epidemic proportions' in many countries across the world, the rates varying between 15 per cent in Japan and 71 per cent in Ethiopia, with about 20 per cent in the United States and Sweden and 23 per cent in Canada and Britain. The incidence of domestic violence in France appears to have increased in recent years, with 3 to 4.5 per cent of women reported to have been the victims of domestic violence once in their lifetime. Surveys conducted in various countries show that between 10 and 69 per cent women reported having experienced domestic violence, such as 10 to 35 per cent in Latin America, and 18 to 45 per cent of women in Sub-Saharan Africa having been victims of physical violence by intimate partners at sometime or other in their life (Kumari 2009:211). To cite cases from India, crimes of violence against women touched

Table 3.16: Incidence of Crime Against Women in India

Nature of Crime	*Incidence*					
	1998	*1999*	*2000*	*2001*	*2002*	*2003*
Dowry Death	6,975	6,699	6,995	6,851	6,822	6,208
Torture (cruelty by husband and other relatives)	41,376	43,823	45,778	49,170	49,237	50,703
Dowry Prohibition Act	3,578	3,064	2,876	3,222	2,816	2,684

Source: Kumari, Ranjana, 2009

a new high in 2003 with 132,720 cases registered of which dowry alone totalled 6,285 with 2,532 other cases registered under the Dowry Prohibition Act. There were 40,285 cases of cruelty by husband or relatives against the married women. Table 3.16 shows 'incidence of crime against women in India' for the period, giving an indication that the number of victims linked to dowry has remained above 6,000 a year during this period while the cases of cruelty has increased from about 41 thousand to over 50 thousand between 1998 and 2003. There has been no improvement in the situation over this period of about five years. In all probability the situation seems to be worse than in the figures as a large number of cases are never reported or even hushed up exercising control over the sources of giving out the truth. A study conducted by the Centre for Social Research, found that nearly 50 million married women in India suffer from domestic violence but only 1 out of 1,000 cases are reported, and in 2 out of 100 cases the accused get conviction under the Indian Penal Code. Records from the Special Cell for Women and Children in Mumbai, established by the Police Commission, revealed that 43 per cent of women endured domestic violence for 3 to 17 years before complaining to Police (not always sympathetic either). According to 2001 Census data, a total of 236,272, 617 women (including under-age girls) were married and if one out of five women suffers from domestic violence, one can infer that at least 47,354,523 must be suffering. (Kumari, 2009: 212-213).

Looking at the highly industrialized countries of the developed world with their high standard of living the situation in this regard does not appear as a consolation since nearly one-third of American women (31%) report being physically or sexually abused by husband or 'a boy friend' at some point in life. In Sweden, a country with the highest representation of women in its Parliament, there were 22,500 cases of domestic violence reported in 2007-08, an increase of 10 per cent since 2000, while as many as 80 per cent of the cases are not reported at all which is also the case in Japan where till 1998, the issue of domestic violence was treated as a family affair and 'no woman were allowed to speak about it in public'. Recent statistics available for Japan indicate that one in every twenty women surveyed felt her life was in danger of domestic violence from her husband or boy friend. The cost of domestic violence as estimated for some countries shows that in Canada it amounts to over $ 1 billion per year spent as costs for criminal justice, police, counselling and training (Kumari, 2009: 218-219). In Finland, the costs of domestic violence as estimated in 2002, amounted to 91 million Euro per year while in the U.K., a study undertaken by Walby (2004) for the UK Women's Equality Unit, put it at around 23 billion pounds in 2001 (about 34 billion Euro at that time). (*See* Note *L*).

Violence primarily 'rooted in unequal relations between men and women' affects 'at least one in three women and girls in the world'. Importantly, there seems to be a large gap between violence charged and reported as available data for some countries, indicated in Fig 3.7, showing that 'no more than one-third of the cases of violence are reported, as women are more likely to report crimes if they were not committed by a partner, and more likely to report non-sexual than sexual violence', resulting in wide gap between actual happenings and reporting. Nevertheless, as it appears, not even half of the cases face conviction by criminal justice system of the relevant countries, exposing the overall weakness of the entire system not safeguarding the rights of women.

Domestic violence may, perhaps, be eliminated to a great extent through (a) democratization of the family making women as equal partners in decision-making; (b) eliminating any discrimination between man and woman in the society; and, (c) making equal representation of women in state legislatures, parliament, etc, giving them equal power as men. In this respect, encouragingly many countries have come forward making legislation or amending the existing ones to combat violence against women and children in the footsteps of the United Nations which, in 1993, included domestic violence against women in the declaration on the 'Elimination of violence against women'.

Nevertheless, perhaps, nothing compares in severity and brutality of violence to women as the barbarity associated with the practice of widow burning in India, an age-old ritualistic torture though not frequent now (banned by law in early nineteenth century through the initiative of one of the foremost social reformers at that time in India – Raja Rammohan Roy: *See* Note *M*) except for very rare occurrences in remote rural areas. This happens to be the most offensive type of violence perpetrated on women among the highly orthodox Hindus on religious pretext where the young wives (age in not a consideration) obviously married to men much older in age are expected to join the burning pyre choosing self-immolation – 'shahamaran' – or dying together (See Amitabh Ghosh's Novel '*Sea of Poppies*', 2008) with the husband on his death, an act virtually forced on them. What happened on these occasions – as described in many reports/writings – is when the dead husband's body is burnt, the wife is expected to climb on to the flaming pyre and burn herself to death. She, declared a 'sati' (the word originally derived as the feminine version of 'honest', here means a wife intensely devoted to her husband), according to prevailing practice, is taken in a procession to the place of cremation, often heavily (forcibly) drugged to numb her consciousness, accompanied by very loud music intended to drown her voice if she cried out in protest, and is forcibly pushed on to the pyre. The practice strictly banned by law sometimes makes headlines taking place even now at places in remote rural areas ignoring legal bar.

Another form of humiliation and crude violence occurring against women across the world is the labelling and ostracizing of women as 'witches' - another gross violation of human rights, a right to live with dignity, happens to be a dreaded practice, representing one of the most cruel offences. The practice goes back to the thirteenth century or thereabout and the United Nations has named Africa, Asia and South America as high-incidence zones for witchcraft-related killings. Recently, there has been a sudden spurt in such cases in India, the highest number of incidents being reported from Andhra Pradesh, Bihar, Jharkhand, Chhattisgarh, Maharashtra, Orissa, Gujarat, Assam, West Bengal and Rajasthan where women branded as witches or '*dayan*'/*dynis*/*dakins* etc. as

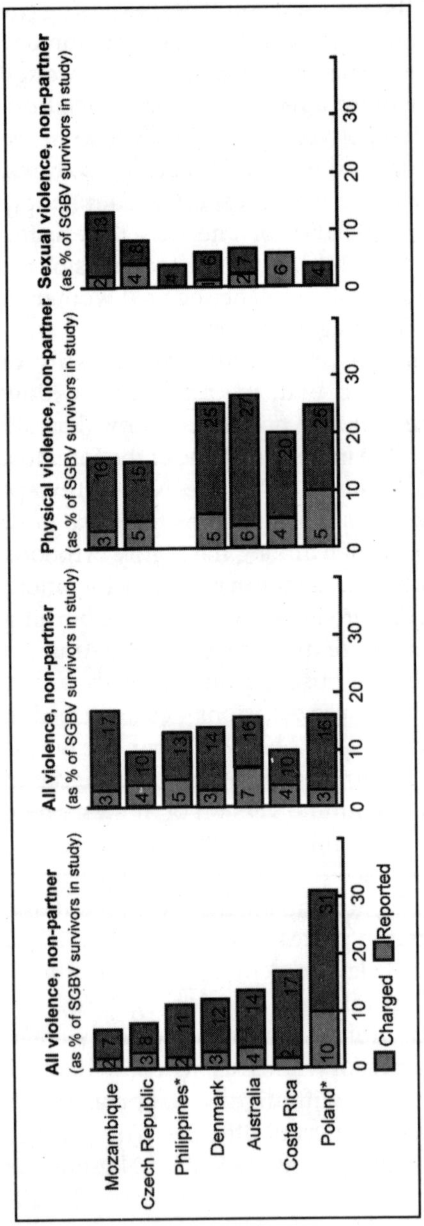

Fig. 3.7: Violence Against Women: Reporting and Charging Rates

Source: UNICEF, 2008-09.

they are locally called, are tortured to death on false or fabricated charges. "The form of torture in most parts of the world are somewhat similar in its vulgarity and cruelty. They include stripping and parading the victim, tonsuring, blackening the face and slashing her with knives or any other sharp instrument. Women are often made to eat human and animal excrement. They are also stoned, drowned, hanged or burnt to death, their teeth smashed, heads shaved and breasts chopped off" (Mathur, 2009: 286). According to an official estimate, about a thousand women were killed in the early nineteenth century in the plains of Central India alone while another source (Sarkaria, 1997) reports that the number of women killed as witches far exceeded those who died as 'satis'. According to official data, 2,556 women were branded and killed as witches in India between 1987 and 2003. In tribal-dominated Sundergarh district in Orissa, at least 50 women have been killed branded as witches since 2001. In Rajasthan, labelling women as witches is common, and in the state of Jharkhand too, 'the problem is widespread in tribal areas', witchcraft taking place on a large scale among the Santhal and Munda tribes. In the south of Baster, Chhattisgarh, 50 per cent of men in prison on murder charges are for witch killing and since 1996, 73 women have been murdered on charges as witches. According to a report of the "Free Legal Aid Committee', 536 women had been killed on this allegation in the past ten years in just two districts of Bihar. The practice of branding women as witches is very much prevalent in the southern regions of India, including Gujarat where accusations for 'witchcraft' are not uncommon and there are reports of witch-hunts from Maharashtra and West Bengal (Mathur, 2009: 287-88).

Interestingly, according to a field investigation in Rajasthan, the husband's family was found to be responsible for initiating the process of labelling or branding the woman as witch in 23.8 per cent of the cases based primarily on superstitions, property disputes, infertility and childlessness (the blame always put on wife but not on the husband in absence of medical advice), physical deformity etc., and once the allegation is made, the woman victim going through social exclusion were subjected to torture and violence ending in brutal murder (Mathur, 2009:293-299).

Female genital mutilation/cutting (FGM/C), forming another form of gender abuse, "involves partial or total removal of, or other injuries to female genitalia for cultural and medical reasons." The practice of FGM/C mainly occurs in countries in Sub-Saharan Africa, the Middle East, and North Africa and some parts of South-East Asia. It is estimated that "more than 130 million women and girls alive today have been subjected to FGM/C (which) can have serious health consequences, including the failure to heal, increased susceptibility to HIV infection, child birth complications, inflammatory diseases and urinary incontinence. Severe bleeding and infection can lead to death" (UNICEF, 2007 4). Mostly confined to the countries of Sub-Saharan Africa and the Middle East, where respectively 36 and 28 per cent of their women aged 15 to 49 and their daughters have the experience of undergoing genital mutilation/ cutting for social reasons. The data available for 27 countries (Table 3.17) show

Table 3.17: Female Genital Mutilation/Cutting among women aged 15 to 49 years and daughters, 2000-2006

(In percentage)

Country	*Women (15-49 years)*			*Daughters Total*
	Total	*Urban*	*Rural*	
1. Benin	17	13	20	6
2. Burkino Faso	73	76	71	25
3. Cameroon	1	1	2	1
4.Central African Republic	26	21	29	7
5. Chad	47	47	44	21
6. Cote d'Ivoire	36	34	39	9
7. Djibouti	93	93	96	49
8. Egypt	96	92	98	29Y
9. Eritrea	89	86	91	63
10. Ethiopia	74	69	76	38
11. Gambia	78	72	83	64
12. Ghana	4	2	6	1
13. Guinea	96	94	96	57
14. Guinea-Bissau	45	39	48	35
15. Kenya	32	21	36	21
16. Mali	92	90	93	73
17. Mauritania	71	65	77	66
18. Nigeria	2	2	2	1
19. Occupied Palestinian Territory	19	28	14	10
20. Senegal	28	22	34	20
21. Sierra Leone	94	86	97	35
22. Somalia	98	97	98	46
23. Sudan	90	92	88	58
24. Tanzania	15	7	18	4
25. Togo	6	4	7	1
26. Yemen	23	26	22	20
27. Zambia	1	1	1	—

Source: UNICEF, 2008, Table 9

Note: Y: data refer to years or periods other than those specified in the column etc.
— data not available

wide variations in the percentage of women/daughters having undergone genital mutilation or cutting, ranging up to 98 per cent of the total number of women and daughters. In seven of these countries, more than 90 per cent of the total number of women within the age-group of 15 to 49 have experienced this, with the highest rate of 98 per cent occurring in Somalia, 96 per cent in Egypt and Guinea, going down to below 5 per cent in four of the Sub-Saharan African countries, in two of them it is falling to 1 per cent of the total number of women/daughters involved. The four countries are Zambia (1%), Cameroon (1%), Nigeria (2%) and Ghana (4%). There are five countries where the percentage of women affected is higher in urban than in rural areas, but the most redeeming feature that can be noticed is that, in general, the percentage of daughters subjected to this practice seems to be distinctly lower than women and, though varying widely between 1 and 64 per cent, it has not exceeded 10 per cent in 9 countries, involving only 1 per cent of the daughters in 4 countries (Cameroon, Ghana, Nigeria and Tanzania) along with Zambia maintaining the lowest record of female genital mutilation for women as well. The parity between women and their daughters in the low incidence of this social practice confined to particular communities is indicative of a slowing down of this process, which is surely connected with the education and empowerment of women.

(ii) Gendered Educational Gap

It is widely estimated that women make up the majority of the world's poor, comprise nearly two-thirds of the people who are illiterate, and along with children represent 80 per cent of civilian casualties during armed conflicts. Gender bias is very clearly reflected in inequality in educational facilities available for girls. Gender disparities in education, especially in South Asia and West, Central and North Africa is very striking. Of the world's estimated 854 million illiterate adults, 544 million are women, and of the 113 million children not in primary school, 60 per cent are girls. In India, the enrolment ratio of boys aged 5 to 14 is 17 percentage points higher than that of girls at the same age. Yet in many developing countries, mostly in Latin America, girls have no disadvantage or even a small advantage. In other words, discrimination in educational facilities is very striking comparing figures, such as, for every 100 boys out of school, there are 115 girls in the same situation. Sadly, in spite of the gender gap closing steadily over the past few decades, 'nearly 1 of every 5 girls enrolling in primary school in developing countries does not complete primary education', getting deprived of the opportunity to develop her full potential. "The education gap is widest in South Asia where men, on average, spend 2.5 years more in school than women, declining to 1.3 years in Sub-Saharan Africa, 1 year in Latin America and the Caribbean" (UNICEF, 2007: 23). According to a recent UNICEF estimate, an average of only '43 per cent of girls of the appropriate age in the developing world attend secondary school, though as evident from facts girls derive multiple benefits from secondary education, such as it delays the age of marriage and also the age at which a young woman first gives birth, and it can enhance freedom of movement, maternal health, strengthens women's bargaining power within household and plays as a crucial factor in providing women's economic and social participation'. A UNICEF survey of selected countries across Latin America and the Caribbean, South Asia and Sub-Saharan Africa including India, found that "on average, children with uneducated mothers are at least twice as likely to be out of primary school than children whose mothers attended primary school" (UNICEF, 2007: 27). Another study found '73 per cent of children aged 7 and 14 years in Sub-Saharan African countries with educated mothers were in school, compared with only 51 per cent of children whose mothers lacked schooling. Evidence from Nepal and India shows that women's participation in household decisions decreases backwardness among children and reduces child mortality. Women with education prioritize healthcare and education in the family. Empowered women particularly prioritize girls' education which has a positive consequence spanning generations.

Table 3.18 on adult literacy among the males and the females in 1995-2005 brings home the glaring truth that there is a fundamental bias, showing much higher rates for the males than that of the females across the world. The countries divided into three different groups on the basis of human development, present wide differences in the scale of achievement in this respect, closely following their ranking in order of human development. Such countries with high indices of human development (including countries like Hong Kong SAR, the Republic of Korea, United Arab Emirates, Saudi Arabia, Brazil, etc.), 70 in all, 27 of which provide no data, have adult literacy rate exceeding 81 per cent for the males in all the 43 countries of which in 35 countries it is over 90 per cent, as against adult literacy going above 81 per cent for the females in 40 countries, and above 90 per cent in 35 of them while falling below 81 per cent in 3 countries. It falls to the minimum of 80.4 per cent for the males in Malta but goes further down to 73.5 per cent for the females in Oman. The situation, however, is more varied in the group of countries with medium human development, 85 in all, 9 of which provide no data. Countries, such as China, India, Venezuela, Peru, the Islamic Republic of Iran, South Africa, Pakistan, Bangladesh, Nepal, etc. fall within this group where the predominating position of the males in adult literacy remains unchanged, being higher than 81 per cent in 50 countries for the males in comparison to in 36 countries for the females. The rate, however, goes down for both the males

Table 3.18: Adult Literacy Rate, Males and Females, 1995-2005

HDI Rank	*Total Number of Countries*	*Number of Countries with % in adult literacy*									
		0–20		*21–40*		*41–60*		*61–80*		*81–100*	
		M	*F*	*M*	*F*	*M*	*F*	*M*	*F*	*M*	*F*
High Human Development Without data = 27 countries	70	–	–	–	–	–	–	–	3	43	40
Medium Human Development Without data = 9 countries	85	–	–	–	5	1	16	25	19	50	36
Low Human Development	22	_	5	2	7	8	7	10	3	2	–

Source: Calculated from data provided by UNDP, Human Development Report 2007-2008, Table 28.

and the females, ranging between 61 and 80 per cent for the males in 25 countries and in 19 countries for the females, dropping further to 41 to 60 per cent for the males only in 1 country (41.91% in Gambia), while in 16 countries it was between 41 and 60 per cent for the females, and goes down to less than 41 per cent in another 5 countries, with the lowest rate being 34.7 per cent in Yemen. The rate slides down further in the countries with low human development, 22 of them, all in Africa, including countries like Nigeria, Niger, Tanzania, Ethiopia, Zambia, etc., where, with the exception of 2 countries having an adult literacy rate higher than 80 per cent only for the males, the rate falls to 61 to 80 per cent for the females in 3 countries against in 10 countries for the males but goes down to 41 to 60 per cent in 8 countries for the males and in 7 countries for the females, falls further to 21 to 40 per cent for the males in 2 countries and in 7 countries for the females. It goes down to below 21 per cent for the females in 5 countries, reaching the lowest of 12.8 per cent in Chad, though the lowest for the males in low human development countries is comparatively much higher, 31.4 per cent in Burkino Faso.

Evidently, the disparity is large enough to deprive women of a comfortable place in the society since education empowers them to make them heard not only in the household but also outside. The existing literature confirms the positive relationship between female schooling and labour participation. Education is one of the determinants of female employment and it has been found that labour force participation of illiterate women is around 37 per cent (confined to agriculture and some informal sector activities where education is not required) while that of literate women is around 43 per cent (Psacharopoulos and Tzannatos, 1991: 275).

At the same time, the importance of maternal education as a determinant of child mortality, as mentioned before, cannot be ignored. Maternal education has proved to be one of the most important determinants of mortality in childhood since educated mothers breaking tradition may adopt many of the alternatives in child care and treatment available in the rapidly changing society. Further, an educated mother is more capable than an illiterate mother of demanding the attention of doctors and nurses. "The final and most important factor is that the education of women greatly changes the traditional balance of familial relationship with profound effect on childcare. In particular, an educated woman is more likely to assume personal responsibility for the care of the child" (Flegg, 1982: 443) and thus effectively slow down the rate of infant mortality. Hence, underdeveloped countries placing a low priority on improving women's education and in achieving a more equal distribution of incomes are unlikely to accomplish a rapid fall in their infant mortality rates, or, in other words, education among women has a greater positive role in reducing infant mortality than income (Flegg, 1982: 454). That education for women takes a vital role in reducing child mortality is supported by Caldwell (1986) who believes that changes in mortality rates over time and between regions could largely be explained by differences in maternal education and family income. According to him, it is education of the parents (including mothers) that convinces them to get modern health treatment facilities, even taking a long journey in rural areas, unlike uneducated parents who do not trust in modern treatment. Educated mothers understand better than illiterate ones the significance of hygiene, the implications of infection to keep children separated from people with infectious diseases as a defense against mortality risks. Since children when very young are the greatest concern for mothers, top priority should be given to education – particularly female education as a sure way to bringing down infant mortality (Caldwell, 1986:171-202). To be more specific, the status of women is very crucial for accurately assessing the state of the world's children, as pointed out by UNICEF (2007), a world fit for children is also a world fit fort women and it can emphatically be claimed that 'women's rights are good for children and ultimately good for the world'.

(iii) Women's Empowerment and Participation

Among different forms of deprivation, discrimination of women is highly crucial to the development of any society where women representing nearly or somewhat more than half of the total population have no role to play in building up the society. It is not only the loss of human power — considered as the most valuable and important of all resources on earth as human brain power is at the root of conversion of all other natural resources into human-use – itself as a resource makes a formidable loss for the society if this human potential is unutilized. Despite some progress, gender equality and full empowerment of women remain indispensable though both are yet unrealized goals. Gender equality is a vital part of global and national efforts to reduce poverty, not only of women themselves, but also of their families, particularly their children. All national development strategies should include programmes on gender equality and empowerment of women, ensuring equal access to education and training at all levels for women and girls as an important part of empowerment of women and reduction of poverty and hunger.

(a) Working Hours and Wage Differentials for Women

Discrimination against women is very strongly reflected in women working longer hours than men whether they live in developed or developing countries in rural or urban settings. Oxfam estimates that women work around 60 to 90 hours per week and, as revealed through time-use surveys conducted in several selected countries across Asia, Latin America and Sub-Saharan Africa, women's working hours exceed those of men often by a wide margin. For many women, again, the unpaid work in the household takes up the majority of their working hours leaving less time for remunerative employment. As has been found from data from urban areas in 15 Latin American countries, unpaid household work becomes the main activity for 1 in every 4 women, whereas it is so for 1 in every 200 men. Household work exceeds the total number of hours spent by women where they are in remunerative employment, such as in Mexico 33 hours each week is spent by a woman in household duties on top of her remunerative employment against 6 hours of domestic work done by men in a week. In India, it is 35 hours a week for women in caring for the family against 4 hours for men (UNICEF, 2007:37-38). Over 60 per cent of all family workers are women. A particular example further reveals how women's service to household can be a great burden on her daily time budget. For example, absence of water in or near the premises makes it a responsibility of the women or young girls to do the time-consuming job to bring and carry water often from a considerable distance and according to a research report, in Sub-Saharan Africa women spend some "40 billion hours a year collecting water" which "is equivalent of a year's worth of labour by the entire work force in France" (UNIFEM, 2008/2009:37-38). This huge wastage of human labour could be saved by making provision for accessible water (*See* Note *N*) (Fig. 3.8).

That women work longer hours than men across the developing countries is given in data provided by one of the UNICEF Reports (2007). *See* Table 3.19 which shows the estimated average daily difference in working hours between men and women in 9 selected countries.

By 2005, women accounted for roughly 45 per cent of the world's economically active population though this varied widely across regions, East Asia and the Pacific represented by the highest with 68.9 per cent of the economically active population, dropping to 62.3 per cent in Sub-Saharan Africa and 57.5 per cent in CEE/CIS countries (UNICEF, 2007: 39).

Discrimination against women is also very obvious in the earnings gap, women being paid, in general, less than men in similar positions, being roughly 20 per cent less than men. According to UNIFEM Report, 2008-2009, women earn 17 per cent less than men. "Estimates based on wage differentials and participation in the larger labour force suggest that women's estimated earned income is around 30 per cent of men's in the countries surveyed in the Middle

Table 3.19: Working hours of Women and Men in Selected Countries

Country	*How much longer do women work than men each day?*
Benin (1998)	2 hours 25 minutes more
Mexico (2002)	1 hour 45 minutes more
India (2000)	1 hour 6 minutes more
South Africa (2000)	59 minutes more
Republic of Korea (1999)	58 minutes more
Madagascar (2001)	51 minutes more
Mongolia (2000)	44 minutes more
Mauritius (2003)	24 minutes more
Average for 8 developing countries	1 hour 9 minutes more

Source: UNICEF, 2007.

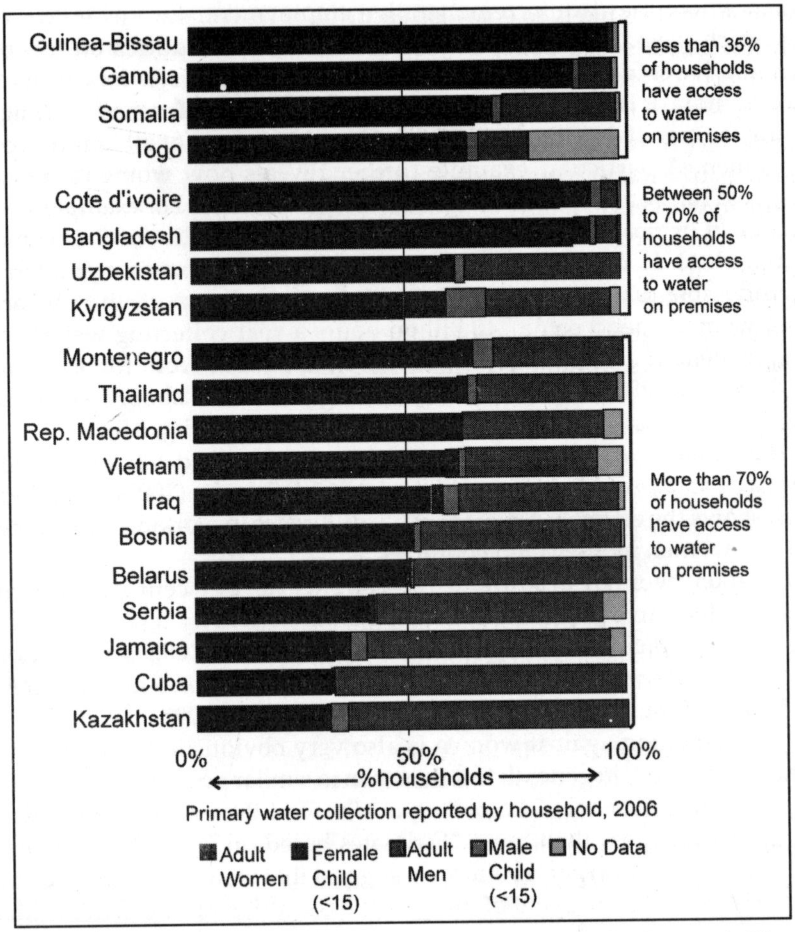

Fig. 3.8: Women Carrying Water from a Distant Source for Domestic Use

Source: UNIFEM, 2008-2009.

East and North Africa, around 40 per cent in Latin America and South Asia, 50 per cent in Sub-Saharan Africa and around 60 per cent in CEE/CIS, East Asia and industrialized countries" (UNICEF, 2007: 40). In other words, estimated earnings for women are substantially lower than for men and measured at thousands of US dollars at 2003 prices adjusted to PPP (purchasing power parity) women's earnings varied from 59 per cent of men's earnings in CEE/CIS to 57 per cent in industrialized countries, 40 per cent in Latin America and the Caribbean, 62 per cent in East Asia and the Pacific, 51 per cent in Sub-Saharan Africa, 39 per cent in South Asia and 28 per cent in Middle East and North Africa (UNICEF 2007, 41) (*See* Note O) (Fig. 3.9).

Another very important aspect affecting women joining the labour force is that it does not generally match that of men in improving their quality of life as women are less likely to enjoy job security, or financial security, earning less with few or no social benefits. Jobs are in many cases far from stable and working conditions in majority of the cases are deplorable – far from women-friendly – since in developing countries the majority of the women workers outside agriculture are concentrated in the informal sector and 'are more likely than men to be own-account workers, domestic workers, industrial out-workers or unpaid workers in family enterprises'. Nearly 60 per cent or more of women workers engaged in non-agricultural work in developing countries are employed in the informal sector, representing 86 per cent in India, Shoma A. Chatterjee, 2009 refers to a report which claims that the informal sector in India employs about 90 per cent of the country's workforce and 97 per cent of its women workers (Shoma A. Chatterjee, (2009)), 83 per cent in Kenya, 77 per cent in Indonesia, 69 per cent in El Salvador, 67 per cent in Brazil, 58 per cent in South Asia, 55 per cent in Mexico and 39 per cent in Tunisia. While in North Africa this is 43 per cent, Sub-Saharan Africa has one of the highest rate of informally employed women, rising to 84 per cent (UNICEF, 2007: 43-44).

(b) Women's Empowerment

The empowerment of women and their full participation in all social activities—economic as well as non-economic – need to be perceived as an essential partnership for the overall welfare of the society which requires, as Amartya Sen (2000) points out, that there should not be "a: restricted view" of the 'personhood of women'. In his words "The relative deprivations in the well-being of women were – and are – certainly present in the world in which we live, and are clearly important for social justice, including for women...That problem ("missing women") is unquestionably important for the well-being of women and in understanding the treatment of women as "less than equal". There are also pervasive indications of culturally neglected needs of women across the world. There are excellent reasons for bringing these deprivations to light and keeping the removal of these inequities very firmly on the agenda" (Sen, 2000: 190-91). Sen's observation in recent years "has brought out very clearly how the relative respect and regard for women's well-being is strongly influenced by such variables as women's ability to earn an independent income to find employment outside the home, to have ownership rights and to have literacy and be educated participants in decisions within and outside the family" (Sen, 2000: 191). In his opinion, working outside the home and earning an independent income have a deep impact on enhancing the social standing of a woman in the household and the society. Freedom to hold a job reduces women's "relative and absolute privation". Making monetary contribution to the prosperity of the family, she has a stronger voice in family decisions while education giving her more exposure acts as a way of emancipation. A study of nearly three hundred districts in India conclusively proves that women's education and employment are the two most important influences in reducing fertility rates (3.0 per woman in India as against 1.9 in China) obviously releasing her from 'constant bearing and rearing of children that plagues the lives of young women in many societies in developing countries'. Further, women's education and literacy have a great impact on reducing the mortality rates of children, as observed earlier, while at the same time women's empowerment has a strong influence in reducing gender bias, doing away with son preference.

Besides, gender inequality corresponding with higher female mortality of infants and children in early years exists in countries, such as India, Pakistan, Bangladesh, China, Iran and those in West Asia and North Africa and others, in contrast with countries of Europe or North America or Sub-Saharan Africa where female children have better survival rates. Among different variables female literacy has proved to have greater impact on the reduction of child mortality, and it has been observed that increase in the crude female literacy rate, from, say 22 per cent (the actual 1981 figure for India) to 75 per cent reduces the predicted value of under-five mortality of males and females combined from 156 per thousand (again, the actual 1981 figure) to 110 per thousand (Sen, 2000: 197-198). Similarly, female literacy tends to reduce fertility rates.

The result of women's economic participation, apart from providing such social benefits as reduction of child mortality and fertility rates, also reduces gender bias in societies, yielding women an important place in making family-decisions having a great impact on social changes in general. Incidentally, Bangladesh presents the most classic example of women's empowerment and participation with access to credit provided by the 'Grameen Bank' through its "visionary micro-credit movement' initiated and perpetuated by Mohammad Yunus (awarded Nobel Prize for the effort and exemplary success,) making a special effort to provide small credit to poor women borrowers, who proved more reliable than their male counterparts, where 98 per cent of the borrowers repaid the loans within stipulated period. "Behind the belief that 'loans for women' will lead to their economic and social advancement are a number of assumptions; that women will use loans for their own enterprise; that these enterprises will be successful; that women will control the profits derived from such enterprises and that greater involvement in economic activities will strengthen the social and political position of women in society" (Hulme and Mosley, 1996: 125). Similar emphasis has been placed on women's participation by BRAC, led by another visionary leader, Fazle Hasan Abed. As Sen remarks, social and economic movements like these have brought in great changes in the society with a sharp decline in fertility and with the increasingly higher involvement of women in social and economic affairs, two of the most important developments achieved, and this has made family planning programmes more accessible and popular even in the rural areas of Bangladesh.

On the other hand, social and political empowerment of women in every society of both developed and developing countries across the globe is not often as clearly reflected in this occupational involvement in the formal or informal sector as it is in their participation as members of the Parliament, as legislators, as senior officials and managers and professional as well as technical workers.

These are considered as the major determinants of female empowerment in any society, and indicates women's status in this regard undoubtedly varies in global perspective.

As all said and done, perhaps, the question of the identity of women is a more fundamental issue in expressing their independent, born-free status in the society than any other form of recognition earned from participation in various fields of activities. However, innocuous it may appear on the surface, there is a deep-rooted implication in changing the surname/family-name (family of origin) after marriage existing as a social custom (compulsion?) for women in many societies by which pre-marital status of a woman changes with a new identification when the woman on marriage is customarily required to adopt the family-name of her husband, but not the other way round. Because, this does not happen to men who retain the family-name untarnished (with, however, very rare exceptions where it is volunteered by man but is not imposed by the society), obviously indicating superiority/domination of masculine self rather than both 'sharing' equal status as a means of identification, and children born to them out of wedlock also carry the family-name of the father (father becoming the only person for identification of children at all stages of future activities, mothers having no role in them as individually mothers are never accepted for such purpose though father alone has that right), further establishing patriarchal domination. Notably, encouraging signs of deviation from this practice are visible, at some instances women retaining their original family-names even after marriage, quite often tagged with husband's family-name. There are also other areas where masculine superiority is very much dominant. For example, women following marriage are required, in particular communities, to wear red vermillion powder at the parting of hair, conch-shell bangles as signs of a married woman while in some communities they are required to cover their head and body by a loose garment in order to keep away from the eyes of other males. Also serving male members of the family, women eating (in many cases consisting of left-overs) after the male members of the family have consumed the better part of food. Women (as wife/daughters/mother/sisters, etc.) bear most of the load of domestic work (sometimes all of it), such as cooking, which includes quite often foraging for fuel, making cow-dung cakes to be used as fuel, washing clothes, along with at many instances carrying water for domestic use where it is needed, sewing/mending, shopping where it is allowed, working in the field as free labour, sweeping and feeding animals where they exist, and making beds for all in the household, etc. The females are often called the 'unpaid servants' of the family though they are generally the major carer of their children as well as other members of the family in illness. Ironically, in many societies a clear distinction is made in teaching curriculum in educational institutes for girls, making cooking and sewing as compulsory part of learning for girls though in the same societies when it comes to remunerative jobs, the overwhelming majority of chefs, cooks and tailors etc. are males.

Furthermore, in many countries, mostly in the developing world, women are generally excluded from taking any decision on household affairs involving large expenditure, such as purchase of "land, cars, livestock... crucial for families. Money spent on large purchases may be regarded as a wise long-term investment. However, the short-term cost of acquiring these can consume a large share of household income that might otherwise be used for more immediate household needs, such as medicine, school supplies and food ... "In Nigeria, for example, 78 per cent of women indicated their husbands have exclusive control over large purchases. Approximately 60 per cent of women in Egypt and over a third of women in Bangladesh and Nepal felt excluded from such decisions" (UNICEF, 2007: 21) (Fig. 3.10). There is, however, wide variations between countries in the share of women having no right in deciding on daily household expenditure, ranging between 5.8 per cent in Madagascar to the highest of 65.7 per cent in Malawi in Sub-Saharan Africa, and with only 2.4 per cent in Indonesia and 9 per cent in the Philippines. In a situation like this the mobility of women is likely to be highly restricted."Survey data suggest a high degree of male control over women's mobility in each of the regions surveyed. In Burkino Faso and Mali, for example, approximately 60 per cent of women reported that husbands alone decide when wives can visit with family or relatives. One-third of Bangladeshi husbands control their wives' mobility outside of the home. In Latin America and the Caribbean, data from Nicaragua show that 18 per cent of women require a man's permission before leaving home to visit friends and family" (UNICEF, 2007:21). This is 61.6 per cent in Mali; 59.7 per cent in Nigeria; 56.1 per cent in Zambia; 47.3 per cent in Tanzania; diminishing to 9.2 per cent in Madagascar while in the Philippines it is 10 per cent; in Indonesia 10.8 per cent; in Peru 12.5 per cent; in Bolivia 12.4 per cent; in Nepal 33.7 per cent; in Morocco 28.8 per cent and in Egypt 25.6 per cent of women depend on their husband's approval for visiting anybody outside home (UNICEF, 2007: 20). Similarly, male domination prevails in taking decisions concerning healthcare needs of women in the family which is closely related to the health and well-being of children as well. In South Asia and Sub-Saharan Africa, as surveys show, it is the husband who decides the issue. For example, in Burkino Faso, Mali and Nigeria almost 75 per cent women reported that husbands alone take the decision. In Bangladesh and Nepal, two countries surveyed for this purpose, around 50 per cent of the cases are decided by the husbands.

One of the reasons behind the prevalence of these practices, substantially dwarfing the role of the wife in the family, is the age gap between husband and wife which may vary in majority of the cases enormously. While the average age at marriage in Western Europe is estimated to be '27 for women and 30 for men', in developing countries the difference is much wider, such as "in South Asia ... husbands are approximately five years older than their wives; the gap rises to six years in Sub-Saharan Africa (excluding southern Africa)" (UNICEF, 2007: 22), and could be very much wider with girls married off at ages below 18 years, as is practised in many countries including India which 'affects the power that women have over household decisions'.

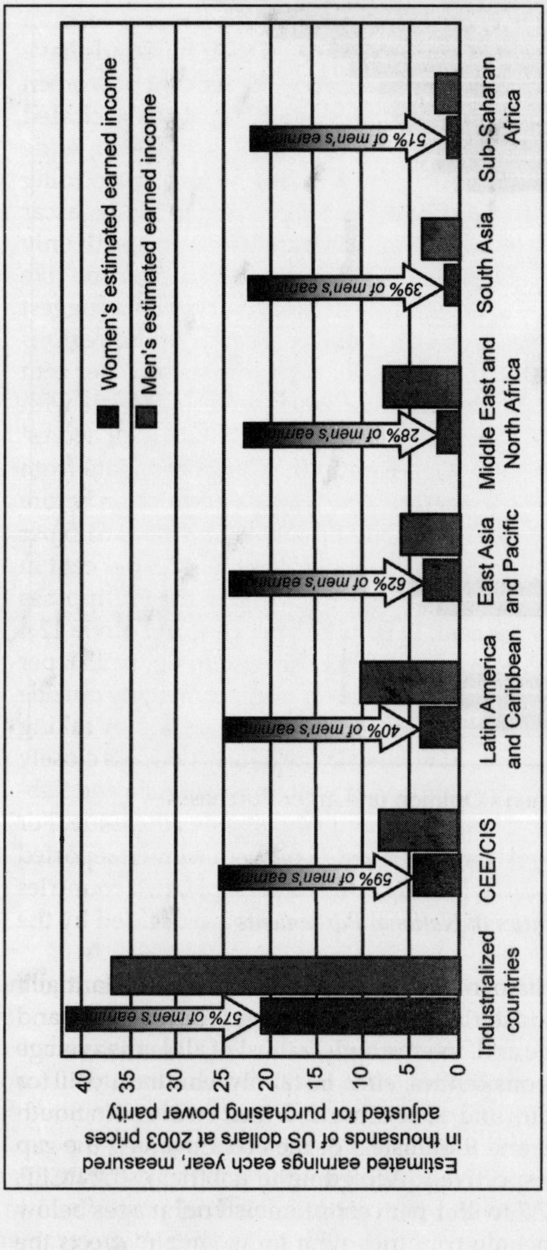

Fig. 3.9: Women earning less than Man

Source: UNIFEM, 2007

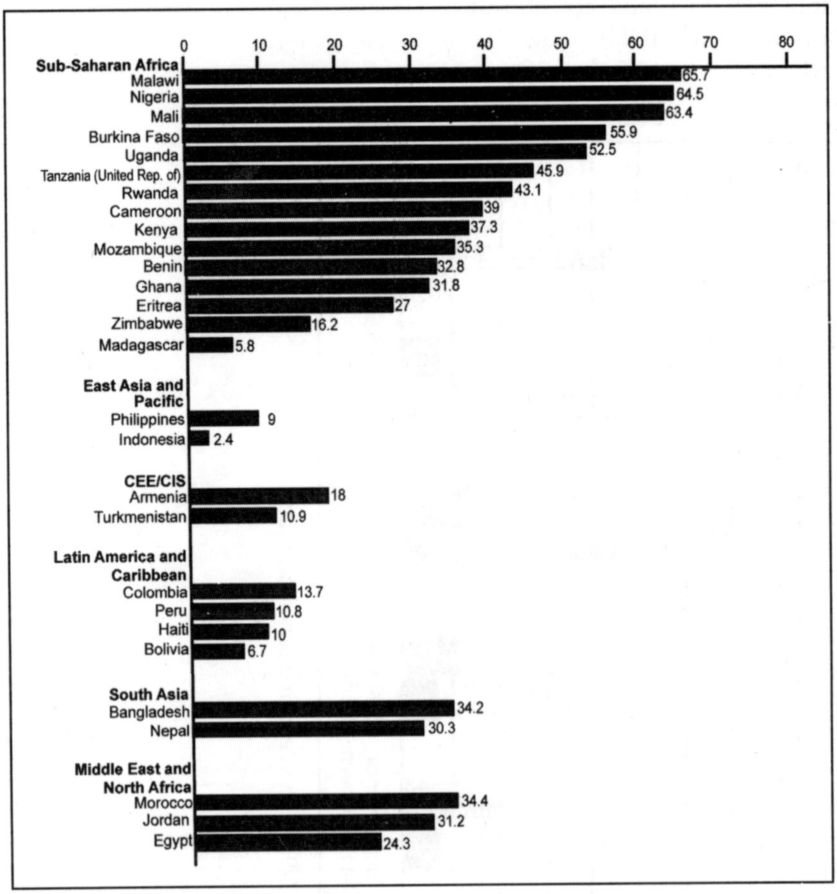

Fig. 3.10: Women's Opinion on Larger Purchases

Source: UNIFEM, 2007

(c) Women Sharing Responsibilities of National Parliaments

According to IPU (Inter-Parliamentary Union, 2005) women taken together for 183 countries held 858 ministerial posts, accounting for only 14.3 per cent of ministerial posts worldwide and, interestingly, a third of all ministerial posts fell in the area of minor responsibilities, such as family, children, youth and social affairs or women's affairs and education, and worldwide accounted for only 13 ministers of defence, and 9 ministers of the economy, or 1.5 per cent and 1 per cent of the total respectively. According to another estimate, as of 2008, women held between 7.7 to 28.1 per cent of ministerial posts, i.e., 1 in 3 at best, while five regions include countries with no women in government (UNIFEM, 2008-2009: 26). The role of women as members of Parliament is considered as the most important one among all responsibilities. The data in

Table 3.20, available for 68 out of 70 countries (2007) with High Human Development (HHD), show that in 37 countries women's representation in Parliament was less than 20 per cent of the total number of seats, including under 10 per cent in 8 countries among which Kuwait had the lowest representation, amounting to 3.1 per cent of the total number of seats in the Parliament. The highest share of women, accounting for 47.3 per cent, occurred in Sweden and, interestingly, all the three of a total of four successive highest shares (first, third and fourth) were in the Scandinavian countries, including Sweden (47.3%), Finland (42%) and Denmark (38.8%), followed by Iceland (31.7%) while the highly industrialized countries, like the United Kingdom with women sharing only 19.3 per cent, the United States with 16.3 per cent, France with 13.9 per cent, and Japan with only 11.1 per cent fell far behind with the exception of Germany where women's representation in the Parliament was comparatively better, having 30.6 per cent of the total number of seats. Female representation in the Russian Federation also gives a very poor impression, with 8.0 per cent share of the total and this slides down to an entirely negative position for the two Islamic states of Saudi Arabia and Qatar where women are unrepresented in the National Parliament. Three other Islamic states, Kuwait (3.1%), Libya (7.7%) and Oman (7.8%) stood almost at the bottom of the ladder in spite of scoring high in human development.

Among the 85 countries with Medium Human Development (MHD), 3 of them were without any representation of women in the National parliament, Fiji where the parliament remained dissolved at the time (2007) when data were collected, in Myanmar (Burma) where the Parliament was not convened in spite of a democratic election, and in occupied Palestinian Territory going without a parliament. Of the remaining 82 countries (3 without data), 39 had female representation of less than 10 per cent of the total number of seats in the parliament, with no representation in two of them – Kyrgyzstan and Solomon Islands. Further, it is not difficult to find out that women's

Table 3.20: Women Having Seats in Parliament, 2007
(in % of total number of seats)

HDI Rank	*Total Number of Countries*	*Number of Countries with seats in the Parliament (in % of the total)*				
		0-20	*21-40*	*41-60*	*61-80*	*81-100*
High Human Development Without data = 2 countries	70	37	29	2	—	—
Medium Human Development Without data = 3 countries	85	54	18	—	—	—
Low Human Development Without data = 1 country	22	15	5	1	—	—

Source: Calculated from data provided by UNDP Human Development Report 2007/2008, Table 29

representation is low in Islamic countries in general, such as in Yemen (4.7%), Iran (4.1%), Jordon (7.9%) etc. However, the highest representation of women in the National Parliament among the MHD countries remains much below 40 per cent, South Africa having 32.8 per cent, which happens to be the only country where it exceeds 30 per cent.

In comparison, in all 22 countries with Low Human Development (LHD), women do not seem to be doing badly. In half of the total number of LHD countries the membership exceeds 10 per cent of the total number of parliamentary seats and it is above 20 per cent in 5 countries, reaching the highest of 45.3 per cent in Rwanda which is, interestingly, much higher than not only in any of the 85 MDH countries but also exceeds that of two of the Scandinavian countries, Finland with 42.0 per cent and Denmark with 36.9 per cent and as such Rwanda acquires the second position in the world for parliamentary representation of women.

Comparing this picture with that of 2006, some important changes in the overall scenario may be detected. For example, as reported in 2006, ten countries had no women parliamentarians, and in more than 40 countries women represented less than 10 per cent of the legislators. The Scandinavian countries had the highest rate of representation, around 40 per cent, while the Arab States ranked the lowest with a regional average of less than 8 per cent. However, it is encouraging to find a rising trend in women's participation in parliament – the number of national parliaments where women accounted for 30 per cent or more of the legislators, taken as the critical yardstick of women's participation as recognized by 1995 Beijing Platform of Action (and why not 50 per cent ?), having already registered four-fold increase in the past 10 years. The most dramatic changes in this regard, however, had taken place in conflict-ravaged countries, like Afghanistan where women, excluded from politics, now (2007) hold 27.3 per cent representation. Similar instances are provided by war-torn countries like Burundi and Timor-Leste where women account for 30.5 per cent and 25.3 per cent respectively.

At the same time, women as heads of the state is another very strong indication of their increasing empowerment. For example, Sri Lanka had the first female Prime Minister in the world with Mrs. Bandaranaike taking that position as early as 1959 followed by other countries, in India Mrs. Indira Gandhi became the Prime Minister for the first time in 1966 (to be in the same position again later), in the UK Mrs. Margaret Thatcher became the Prime Minister in 1979 and in Pakistan Mrs. Benajir Bhutto took that position in 1988. Further, the election of women as the President of a state happened in Liberia in 2005, then in Chile in early 2006, thus marking a new opening in the 'history of women's political leadership'. In Europe, Latvia was the first country to elect a female President, followed by Finland and Ireland. Other countries with women leaders include the Philippines, Bangladesh (as Prime Minister), Jamaica, New Zealand, Mozambique, Netherlands Antilles and the Republic of Korea (UNICEF, 2007: 51-57), and later are Argentina, Panama, Nicaragua and Germany. The latest to be elected in 2010 is Laura Chinchilla as President of Costa Rica and Dilma Rousseff as Brazil's President who became the eighth elected female President in Latin America and the Caribbean.

Nevertheless, the rate of improvement in women's share of seats in national Parliaments has been very slow, so far, increasing "from 11.6 per cent in 1995 to 18.6 per cent as of May 2008" which had been still lower during the previous two decades, 1975 and 1995, "rising less than one per cent". At the current rate of increase, few countries will reach a critical mass of 30 per cent by 2015, and the parity zone – meaning neither sex holds more than 60 per cent of seats – will not be reached by the developing countries until 2047 (UNIFEM, 2008/2009: 21).

According to the same source (UNIFEM, 2008/2009), (a) Women remained unrepresented till 2008 in 9 out of a total of 194 countries, Belize, Micronesia, Naura, Oman, Palau, Qatar, Saudi Arabia, Solomon Islands and Tuvalu. Interestingly, no women candidate was elected in the 2008 election in Kuwait, but two women were appointed to the 16-member cabinet sworn in June 2008, (b) Women occupied less than 10 per cent Parliamentary seats in 41 countries, including Egypt (1.6%), India (9.1%), Iran (2.8%), Japan (9.4%), Sri Lanka (5.8%), Ukraine (8.2%), and Yemen with as low as 0.3 per cent. (c) Out of 187 countries with data nearly 38 per cent of the total number had women's representation accounting for 10 to 20 per cent of the total number of seats in the parliament and it was 18.2 per cent in France, 11.1 per cent in Hungary, 13.3 per cent in Ireland, 14 per cent in Russian Federation, 16.8 per cent in the United States and 19.5 per cent in the United Kingdom, (d) Comparatively higher in the ladder were another 67 countries in which the share of seats for women varied between 20.1 per cent and above. There were at least 12 countries – such as Bahrain (2.5%), Egypt (1.6%), Haiti (4.1%), Kiribati (4.3%), Kuwait (3.1%), Lebanon (4.7%), Papua and New Guinea (o.9%), Sao Tome and Principe (1.8%), Vanuatu (3.8%), Yemen (0.3%), Chad (5.2%) and Sri Lanka (5.8%) – in which the share of women's seats remained below or close to 5 per cent in contrast to only 4 countries (Argentina: 40.0%; Cuba: 43.2%; Finland: 41.5%; and Rwanda: 48.8%) where the share exceeded 40 per cent, Rwanda among them heading the list. Strikingly, three Scandinavian countries — Finland, Denmark and Norway are far ahead of all other industrial countries, including the United States, the United Kingdom, France and Japan in this respect, undoubtedly indicating that these countries (including China with 21.3 per cent women's representation) are lagging far behind some socially and economically backward countries in progressive attitude towards women's empowerment.

(d) Role of Women as Legislators, Senior Officials and Managers

The next important portfolio exercising power and preference in the society constitute the role of women as legislators, Senior Officials and Managers at public and private levels, and in this 8 out of a total of 70 HHD countries with data (10 of the total number of High Human Development (HHD) countries do not provide any data in this regard) have less than 20 per cent women's representation, in 6 of them it is falling below 10 per cent of the total number of such positions available in the country. However, a large majority of the countries with HHD – altogether 44 maintain a slightly higher representation, exceeding 20 per cent (Table 3.21), and in 8 countries the representation is above 40 per cent, reaching the highest of 47 per cent in Bahamas. Interestingly, female representation in such countries as legislators, senior officials or as bureaucrats, and managers, exceeds 40 per cent only in one large industrial country, the United States, while the remaining seven countries with more than 40 per cent representation are small and far from industrial in comparison to industrial countries like the United Kingdom (34%), with much lower female representation.

On the other hand, among the 85 countries with Middle Human Development (MHD), there is no data for 49 of them, and out of remaining 36 countries providing data in this context, female representation is below 20 per cent in 11 of them, in 5 of which it falls below 10 per cent, 4 of them are Islamic countries including Pakistan, the latter having the lowest representation of just 2 per cent. However, altogether in 19 countries women's share of these positions exceeds 20 per cent of the total and in 6 of them the share exceeds 40 per cent, rising to above 50 per cent in 3 countries, the Philippines having the highest at 58 per cent which is considerably higher than the highest score of 47 per cent of these positions made by women in a HHD country (Bahamas). The overall situation in regard to female representation in the higher ranks of occupations appears to be better in the MHD countries than that of the HHD countries, implying the fact that women enjoy better opportunities in many of the former than that of the latter.

Table 3.21: Female Legislators, Senior Officials and Managers, 2007 (as % of the total)

HDI Rank	*Total Number of Countries*	*Number of Countries with Female Legislators, Senior Officials and Managers (as % of the total)*				
		0–20	*21–40*	*41–60*	*61–80*	*81–100*
High Human Development Without data = 10 countries	70	8	44	8	–	–
Medium Human Development Without data = 49 countries	85	11	19	6	–	–
Low Human Development Without data = 20 country	22	1	–	1	–	–

Source: Calculated from data provided by UNDP Human Development Report 2007/2008, Table 29.

In contrast, the 22 countries with Low Human Development (LHD) have data for only 2 countries – Ethiopia and Tanzania – of which the former, amounting to 20 per cent, but it reaches 49 per cent in Tanzania, surprisingly making it the second highest among all the countries with different human development levels, and is higher than that of Bahamas having the highest for MHD countries.

(e) Women as Professionals and Technical Persons

The third very important as well as quite explicit indicator of women's empowerment is constituted by their role in the society as professional and technical personnel, the data for which compiled by the United Nations for 2007 are presented for countries with different levels of human development (Table 3.22). Among the 70 countries with ratings of high human development, 10 do not provide any data, and among the remaining 60 countries in only one does the representation of women in such activities fall below 20 per cent - South Africa with 6 per cent, while it rises to between 20 per cent and 40 per cent in 8 countries, and exceeds 40 per cent as the share of women in 44 countries, becoming higher than 60 per cent in 7 states. In other words, altogether in 51 countries the participation of women in such responsible positions is higher than 40 per cent, and becomes still higher, going above 50 per cent in 34 countries, exceeding 60 per cent and above in altogether 9 countries. Incidentally, the countries having more than 60 per cent and above representation of women in available jobs in these sectors are Hungary (62%), Poland (60%), Lithuania (67%), Estonia (70%: the highest), Latvia (65%), Costa Rica (60%), Seychelles (62%), Bulgaria (60%) and Russian Federation (65%). Interestingly, 7 out of 9 of them are East European countries and, barring Russian Federation, all of them (8 in all) are rather small to very small countries with not a very robust economy as compared with that of the highly industrialized countries like the United States (56% shared by women), Germany (50%), the United Kingdom (47%), France (47%) and Japan (46%) where women seem to play a more subdued role in spite of their strong economies.

Among the countries with medium human development only 38 out of a total of 85 countries provide data, and women's involvement in positions mentioned above falls below 20 per cent in 3 of them, going down to the lowest of 12 per cent as women's share in Bangladesh. However, it is between 20 per cent and 40 per cent in 11 MHD countries, between 40 and 60 per cent in 18 of them, and becomes more than 60 per cent in 6 countries. In fact, 24 countries have female representation of more than 40 per cent while it exceeds 50 per cent in 18 countries, going higher up in Venezuela (61%), the Philippines (61%), Georgia (62%), Ukraine (64%), Moldavia (66%) and Kazakhstan (67%), the last one having the highest percentage of female participation, naturally far exceeding that of men in each of the 24 countries with more than 50 per cent share of female participation in professional and technical positions available

Table 3.22: Professional and Technical Positions held by Women, 2007 (as % of the total)

HDI Rank	*Total Number of Countries*	*Women holding professional and technical positions (as % of the total)*				
		0–20	*21–40*	*41–60*	*61–80*	*81–100*
High Human Development	70	1	8	44	7	–
Without data = 10 countries						
Medium Human Development	85	3	11	18	6	–
Without data = 47 countries						
Low Human Development	22	–	2	–	–	–
Without data = 20 countries						

Source: Calculated from data provided by UNDP Human Development Report 2007/2008, Table 29 (Calculated on the basis of parliamentary seats and occupational data from IPO 2007 and ILO 2007 respectively)

in respective countries, which also applies to all the Scandinavian countries. In contrast, China with the largest population of the world involves only 17 per cent of such positions for women while no data is provided by India on this count.

Similarly, there is no data for 20 out of 22 LHD countries and here again only Ethiopia (30%) and Tanzania (32%) - both in Africa – show a moderate level of female participation.

To summarise, some problems emerge from the study of female representation as an expression of women's empowerment and participation in particular socio-economic activities and among them their representation in Parliament may be taken as the most important as well as symbolic recognition of women's status in societal perspective. Arguably, the situation does not present a very encouraging picture taking the ground realities into consideration since in majority, in more than half of the total number of countries, to be specific, with data for the HHD countries, women elected to the Parliament comprise less than 20 per cent of the total number of seats which very clearly demonstrating men's dominance in the most powerful ruling body of the country. This is highly objectionable, and it appears men in highly industrial countries, advanced in every consideration, are pathetically lagging behind in accommodating more capable women for equity's sake, at least, sharing power and responsibility equally between the two indivisible components of the society. Lack of this is clear in the United Kingdom with women's participation amounting hardly to 19.3 per cent, 13.9 per cent in France and 11.1 per cent in Japan though it rises to 30.6 per cent in Germany. In contrast, the situation seems to be comparatively better in Scandinavian countries with 47.3 per cent female representation in Parliament in Sweden and 42.0 per cent in Finland. Significantly, the Scandinavian countries took the lead in opening the doors of Parliament to women and the first-ever country in the world had been Finland to elect women in 1907 followed by Norway in 1911 (in the latter country though she was appointed but not elected), the United States in 1917, Denmark, the United Kingdom and Ireland in 1918; Germany, Austria, Luxembourg, Poland and Estonia in 1919; Lithuania and Belgium in 1921 to be followed by other countries doing it much later. In some of them women had to wait till 2002 to acquire this right, such as happened in Bahrain, and Kuwait only in 2005, and that too not as elected but as appointed members of the house of Parliament in both countries while, at worse, women never earned this right of access to Parliament in Saudi Arabia and Brunei Darussalam till this date. Interestingly, the United States had allowed women the right to contest election to the Parliament as early as 1788, preceding any country in the world doing it, but very strangely the first time women were elected to the Parliament was in 1917 (UNDP, 2007/2008: Table 33) – a long lapse of almost 129 years women had to wait to make it a reality, making the United States the third country in the world admitting women to the National Parliament, the gap between the two dates, perhaps, reveals the extremity of male aversion and resistance to yield this fundamental human right to the female sex. Similarly, the right to contest election was allowed in 1902 in Australia but the first woman to be elected happened as late as 1943.

On the whole, countries in Europe have been more prompt in recognizing female participation in Parliament with the lone exception of Switzerland where it happened as late as in 1971, in sharp contrast to many other countries where the recognition of women's right of access to National Parliament came after 1950 in 132 out of a total of 194 countries (17 countries providing no information) listed by the United Nations for this purpose, and in 44 countries it did not happen till 1980 or thereafter, the 1990s becoming remarkable in this context as in all the central Asian countries, namely Kazakhstan, Turkmenistan, Uzbekistan, Kyrgyzstan and Tajikistan women earned access to National Parliament though this right was acceded to them as early as in 1918 in Kyrgyzstan under the Soviet rule, but they were elected for the first time in 1993 after the dissolution of the Soviet Union. However, in 24 (out of 132) of these countries, allowing females' access to Parliament, they were appointed rather than elected and in another 9 countries, they were both elected and appointed. Some of the countries in which the female members were appointed are – Norway, Belgium, Lithuania, Mexico, Thailand, Colombia, Nepal, Kuwait, Bahrain, Turkey, Jordon, Lebanon, Algeria, Indonesia, Ghana, Lesotho, Uganda, etc. while, in countries like Iran, Swaziland, Kenya, Zimbabwe, Zambia, Guinea-Bissau, etc., both the processes of election and appointment were involved in giving women access as members to the Parliament.

Unlike women's entry into Parliament, no definite pattern emerges from their role as legislators, senior officials and managers in different countries apart from the fact that while their overall participation is far from high (though slightly higher than their participation in Parliament), the highest being 47 per cent, and this exceeds 40 per cent in only 8 countries out of a total of 116 countries with data available, the average for small countries is found to be higher than that of the industrially advanced countries, such as the United States, the United Kingdom, France and Japan.

The only area where women fare comparatively better is in professional and technical positions since it is only here that women take a better or even leading role in at least 43 countries, taking up over 50 per cent of the total opportunities available. The majority of these countries where women take the lead are, however, small states. Seven of the countries having 60 per cent of the positions held by the females are located in Europe (including the Russian Federation spreading beyond European border) and also in 5 highly industrial countries – the United States (56%), Germany (50%), the United Kingdom (47%), France (47%) and Japan (46%) –female participation varies between 46 and 56 per cent.

On the whole, however, the picture does not give a very comfortable feeling about women's empowerment and achievement through participation in the most socially sensitive areas of activities. As a matter of fact, all these areas are by and large the domains of men controlling the society in every possible manner, often sidetracking women's interests and treating them not as 'equals'. A few instances cited here may open a window on the persistent discrimination (belying all assurances for its removal) in different countries, becoming quite evident from an observation made recently by the UNESCO that while women make up 58 per cent of Saudi Arabia's student population, they represent only 16 per cent of the work force. Segregation for women practiced in that country is very palpable in the establishment of the world's largest 'women-only' university to start functioning in 2010 on the outskirts of Riyadh, the Saudi capital, taking up over 8 million square metres of area to accommodate students complete with facilities, such as library, conference centres, 15 academic faculties, laboratories and a 700-bed hospital, a school, a kindergarten and theme parks offering facilities for research into nanotechnology, bioscience and information technology. Segregation in Saudi Arabia also takes many different forms: "Women cannot open bank accounts for children, enroll them in schools... travel with them (children) without written permission from the child's father. They cannot drive cars or ride bicycles, but are allowed to fly planes – only if they are driven to the airport by a male guardian or chauffeur... To be eligible for government scholarships to study abroad, the ministry of higher education requires female students to be married and accompanied by their husband or a male guardian" (Butt, 2008).

Similarly, looking at the participation of women in Senior Management positions worldwide, the difference between men and women is highly significant, indicating almost overwhelming masculine control. As shown in Table 3.23, the share of women enjoying senior positions in public or private organizations worldwide ranges between a little over 3 to below 13 per cent across the world, remaining below 8 per cent in the developing world except in Latin America and the Caribbean where it reaches the highest, and in the CEE/CIS countries. Compared to this, women's share as full-time workers seems to be slightly better, rising to nearly 40 per cent in East Asia and Pacific countries but falls to the lowest of about 18 per cent in Middle East and North Africa – mainly Islamic countries – with the second lowest of about19 per cent in South Asia and Sub-Saharan Africa. In comparison, women's share in ownership of firms, many of them run on family basis depending largely on traditional skill, happens to be the lowest in South Asia (12.6%), followed by that in Middle East and North Africa (17.4%) while Sub-Saharan Africa with nearly 29 per cent of the total share enjoyed by women is ahead of women in CEE/CIS countries, but superseded by women living in Latin America and the Caribbean in this respect. That the position of women is highly discriminated against in all of these regions is further exposed in their share of senior managerial positions which happen to be very much lower compared with that of men, ranging from a maximum of 1 for every 26 women full-time workers getting the position of a senior manager in Latin America and the Caribbean countries as well as in Sub-Saharan Africa against 1 male senior manager for each 8 or 9 male full-time workers in respective regions. The discrimination deepens further, increasing in scale to the highest of 1 senior manager for each 62 full-time women workers in East Asia and Pacific whereas the males have the prerogative of 1 senior manager for each 6 full-time male workers. In other words, while the number of full-time workers required for attaining a position as senior manager varies between 6 and 9 for the males it varies between 26 and 62 for the females in similar situations, showing high gender discrimination in promotion operating in these regions.

Further, very poor representation of women may be found in the proportion of women judges on the Supreme Courts of various countries and as judges on International courts. According to UNIFEM, 2008/2009 Report, less than 25 per cent judges on Supreme Courts are women, with wide variation, from

Table 3.23: Share of Women in Senior Positions (in %), Full-time Female Workers, Firms with Female Participation in Ownership (in %) and Senior Management positions (numbers), 2002-2007

Region	*Women in Senior Positions (%)*	*Full-time Female Workers (%)*	*Female Ownership in Firms (%)*	*Number of Full-Time Male-Female Workers in Senior Management positions (1 in every -)*	
				Males	*Females*
South Asia	3.4	18.8	12.6	8	55
Middle East and North Africa	4.7	17.8	17.4	9	38
East Asia and Pacific	6.4	39.7	23.3	6	62
Sub-Saharan Africa	7.3	18.8	28.6	9	26
CEE-CIS	11.8	49.3	26.5	6	42
Latin America and the Caribbean	12.6	32.2	35.8	8	26

Source: Compiled from UNIFEM, 2008/2009, Fig. 4.10

being 0 per cent in India to 53 per cent, the highest, in Honduras. In industrial countries, such as Japan (7%), Germany and Luxembourg (19%), and Norway (38%), there is far from equality. Comparing with this, women's representation in the International Courts is equally discouraging, only 'one third' of the existing 'international and regional courts' having participation of women as judges higher than 30 per cent of the total number of positions available, and some of them having none. For example, the 'International Tribunal for the Law of the Sea' with 21 judges has no representation of women (Table 3.24). Interestingly, in five instances, women's participation varies between a ratio of 1 to 14 males and 1 to 3 females while the maximum of slightly more than half (7 against 11) happens only in one out of all the 13 International Courts and regional Forums of Justice.

Table 3.24 shows how poorly women are represented as judges on either international or regional courts. Besides, seeking justice becomes a very complicated issue particularly for women since in many cases "her evidence does not have the same weight as the men's; she may face time limits for bringing a case forward; she may fear retribution for standing up against perceived male prerogatives; or she may not be able to reach the courts because of the distances involved or because the costs may be prohibitive ... In terms of rules of procedure and evidence it is still the case that, in some countries, courts weigh a woman's evidence as equivalent to one half of a man's" (UNIFEM, 2008, 2009: 78). All these make justice very limited in practice and in many cases barriers like time limits for bringing a case to court for sexual

Table 3.24: Number of Female and Male Judges on International Courts and Regional Courts, 2006

Name of the Courts	*Status*	*Number of Female Judges*	*Number of Male Judges*
International Tribunal for the Law of the Sea	IC	0	21
International Court of Justice	IC	1	14
International Criminal Tribunal for the former Yugoslavia	IC	1	8
Inter-American Court of Human Rights	RC	1	6
Caribbean Court of Justice	RC	1	6
European Court of Justice	RC	6	29
Andean Court of Justice	RC	1	3
European Court of Human Rights	RC	12	33
Appeals Chamber	IC	2	5
Special Court of Sierra Leone	IC	3	7
International Criminal Tribunal for Rwanda	IC	3	6
Court of First Instance	RC	9	16
International Criminal Court	IC	7	11

Source: Compiled from UNIFEM, 2008/2009, Fig. 5.5

Note: IC — International Courts
RC — Regional Courts

abuse like rape is highly prejudicial as strongly objected to by Women's Rights activists who argue that "justice systems should recognize rape as a crime for which there is no time limit, similar to murder and kidnapping cases" (UNIFEM, 2008, 2009:79). Despite a United Nations' resolution made in 2006, at the meeting of the 'United Nations High Level Panel on Wide Conference', arriving at the conclusion that "the UN needs ... a dynamic unity focused on gender equality and women's empowerment" (UNIFEM, 2008, 2009:105), and the same reiterated in 2007 and again in 2008, but gender equality remains a far cry as evidenced by the figure of employment of UN Professional Staff and the appointment of female professional staff in the UN. For example, considering the distribution of male and female professional staff of the United Nations, women seem to be at the 'bottom of power hierarchy' at every stage comprising maximum representation of about 43 per cent for three out of eight levels of employment, going down below 30 per cent in four others (Table 3.25). The degree of greater inequality, however, persists in the representation of women constituting the professional staff of various UN agencies, remaining below 25 per cent in agencies like the UNU, UPU, IAEA and UNDO. It increases to 29 per cent in FAO and WHO, rising to 41 per cent in ILO, to 46 per cent in UNESCO, UNAIDS, reaching 47 per cent in UNICEF, and crosses 50 per cent in UNFPA and UNITAR (UNIFEM, 2008, 2009: 105). Gender bias in utilizing the productive potential of female labour force is reflected further in the fact that in 2007, there were 1.2 billion women in paid work as against 1.8 billion men in spite of more than 200 million women labour joining the work force in the last decade. Besides, women having so far been concentrated more in the informal sector, including subsistence and vulnerable areas make a highly deplorable underuse of the real female potential.

The changing pattern of women's employment in the recent years show

Table 3.25: Distribution of UN Professional Staff by Sex and by Grades, 2006

Grade	*Total Number of Staff*	*Females*	*Males*	*Females as % of Males*
Ungraded*	217	45	172	20.7
D2	509	119	390	23.4
D1	1679	430	1249	25.6
P5	5225	1521	3704	29.1
P4	7337	2554	4783	34.8
P3	6474	2775	3699	42.9
P2	3112	1342	1770	43.1
P1	155	66	89	42.6

Source: Compiled and calculated from data provided by UNIFEM, 2008/2009, Fig. 6.8

Note: Ungraded* — Ungraded staff include positions of the Under Secretary-General, Deputy Secretary-General,etc.

that women's share of non-agricultural paid employment was 50 per cent or more than that in only 17 out of a total of 110 countries with data for 2002, coming down to 25-49 per cent in another 76 countries and in 11 out of 13 countries in the Middle East and North Africa women's share was below 25 per cent of the labour force employed in paid non-agricultural jobs. Further, of the 111 countries with data for both 1990 and 2002, women's share of non-agricultural employment rose in 81 countries but declined in 29 countries, most of the declines occurring in Europe and Central Asia, Middle East and North Africa (Table 3.26). The informal jobs constitute one-half to three-quarters of non-agricultural employment, and the share is increasing fast. "In Sub-Saharan Africa, Latin America, and India the share of informal employment in total non-agricultural employment is higher for women than men. In several countries – Benin, Chad, Guinea and Kenya – most of the female non-agricultural labour force is in informal employment" (Table 3.27 from UN Millennium Project, 2005: 90). On the other hand, though still important, agriculture is a declining source of employment for poor women in developing countries, employment for them falling from 34.6 per cent of the total agricultural employment in 1990 to 29.7 per cent in 2000, recording the largest drop in Sub-Saharan Africa followed by East Asia and Pacific. In comparison, it rose only in Latin America and the Caribbean during this period.

Ironically, however, according to the estimate given in UNIFEM 2008, 2009 Report, women are half of the world's total migrant population, comprising nearly 100 million, and they dominate the category of migrants with tertiary education as demonstrated by Africa, Oceania, Latin America and Caribbean countries and Asia, besides Europe, in each of which women represent the larger section of the migrants (Table 3.28) so that with the exception of North America, in all other regions women take a leading position, representing a 'female brain drain'. "For some women, whether migrating for domestic or

Table 3.26: Share of Women in Non-Agricultural Paid Employment, by Region, 1990 and 2002

Region	*High (more than 50%)*		*Medium (25%-49.9%)*		*low (less than 24.9%)*	
	1990	2002	1990	2002	1990	2002
Developed Countries	3	5	20	18	0	0
East Asia and Pacific	1	1	17	10	1	0
Europe and Central Asia	8	9	18	15	1	1
Latin American and the Caribbean	1	1	26	23	0	0
Middle East and North Africa	0	0	4	3	15	11
South Asia	0	0	1	3	8	2
Sub-Saharan Africa	3	1	23	4	14	4
Total	16	17	109	76	39	18

Source: UN Millennium Project, 2005, Table 7.1

Table 3.27: Informal Employment in Non-Agricultural Jobs by Sex and Region 1994 and 2000 (%)

Region/Country	Women	Men
North Africa	43	49
Algeria	41	43
Morocco	47	44
Tunisia	39	53
Egypt	46	57
Sub-Saharan Africa	84	63
Benin	97	87
Chad	95	60
Gúinea	87	66
Kenya	83	59
South-Africa	58	44
Latin America	58	48
Bolivia	74	55
Brazil	67	55
Chile	44	31
Colombia	44	34
Costa Rica	48	42
Dominican Republic	84	63
El Salvador	69	46
Guatemala	69	47
Honduras	65	74
Mexico	55	54
Venezuela	47	47
Asia	65	65
India	86	83
Indonesia	77	78
Philippines	73	71
Thailand	54	49
Middle East and North Africa		
Syria	35	43

Source: Millenium Project 2005, Table 7.2

professional work, migration offers the chance of economic independence and empowerment... World Bank estimates for 2004 indicate that remittances received by developing countries amounted to US $ 125 billion, almost twice the amount of Official Development Assistance (ODA) and approximately 75 per cent of the total foreign direct investment ... Women's contribution to these remittances ... in some countries ... is significant. In the case of Dominicans working in Spain, for example, as much as 78 per cent of all remittances were sent by women, even though they accounted for 61.4 per cent of migrants,. In the Philippines, 97 per cent of migrants send at least some money home, with women sending about 45 per cent of their income on average" (UNIFEM 2008/2009:58).

Table 3.28: Brain Drain as led by Women (Adults with tertiary education in around 2000)

Region	*Emigration Rate (%)*	
	Female	*Male*
Africa	27.7	17.1
Oceania	23.8	16.5
Latin America and the Caribbean	21.1	17.9
Europe	10.4	9.9
Asia	7.3	6.1
North America	3.5	4.1

Source: Compiled from UNIFEM 2008/2009, Fig. B

C. CHILD LABOUR AND CHILD ABUSE

The existence of about 250 million child labourers in the world, as reported by the United Nations, consisting of 140 million boys and 110 million girls, is a convincing proof of deprivation taking another form. Asia accounts for 153 million of them and 80 million of them are found in Africa. Millions of these children act as domestic workers, subjected to physical and psychological abuse (UNDP, 2000: 41) (Table 3.29). Encouragingly, the Convention on the Rights of the Child (CRC), convened by the United Nations, adopted in 1989, and entered into force in 1990, ratified by 191 countries recognizes 'the need for specific attention to protecting and promoting the rights of children, to support their growth, development and becoming worthy citizens of the world'.

Sen considers child labour as a terrible issue related to the question of freedom of choice. In his words, "The worst violations of the norm against child labour come typically from the virtual slavery of children in disadvantaged families and from that their being forced into exploitative employment (as opposed to being free and possibly going to school)" (Sen 2008:30). In his opinion, child labour, as prevalent in Pakistan, India and Bangladesh involves "slavery and bondage, since many of the children working exacting tasks are forced to perform them." The situation arises "from economic deprivation of the families from which they come" as "sometimes the parents are themselves under some kind of bondage" with the employers, one or generations back remaining unpaid have multiplied fabulously with added interests for which papers are often fabricated by the employers taking advantage of the illiterate loanee. Apart from offering labour "there is also the barbarity of children being forced to do things" leaving no choice for the children to go to school or "deciding what they want to do" (Sen, 2008: 114-115).

Besides the countries mentioned in Table 3.29, most countries in Latin America and Africa have similar problems regarding child labour which, though much smaller in magnitude, is also present in North America and

Table 3.29: Child Domestic Workers in Selected Countries, 1990

Country or City	*Thousands*
Philippines	766
Jakarta, Indonesia	700
Dhaka, Bangladesh	300
Haiti	250
Lima, Peru	150
Sri Lanka	100

Source: UNDP, 2000, Table 2.5

Western Europe where, however, working mostly part-time they are in a position to attend school (Bajpai, 2009).

Despite acceptance of international standards and commitments on restricting the use of child labour, the national child policy and various regulations at the national and state levels in India, millions of children are engaged in work, often under hardship and dangerous conditions. As reported in Censuses, there were 10.75 million child workers in the age-group 5 to 14 years in 1971, 13.64 million in 1981 and 11.28 million in 1991. The decline in the absolute number of child labourers in 1991 is attributed partly to 'a significant decline in population growth along with improvement in overall literacy rates, in general, and school enrolment/attendance in particular' (Government of India, 2002: 94-95). The ILO, on the other hand, on the basis of data collected in 1996 put the figure at '23.17 million in India, out of which 12.67 million' were full-time workers. According to the NSSO (National Sample Survey Organization, India) child workers constitute 3.4 per cent of the total labour force in India and of this agriculture accounts for more than 75 per cent. Interestingly, about 78 per cent of the male child workers are concentrated in the agricultural sector and over 20 per cent in the non-agricultural sector while of the female child workers about 52 per cent is engaged by the agricultural sector with more and more of them getting involved in low-paid jobs as compared to male child labour (Bajpai, 2003: 151-153). Helen Sekar (2001) points out a host of reasons responsible for pushing children to work at an early age. They are: poverty; parental illiteracy and ignorance; tradition of making children learn the family skills; absence of universal compulsory primary education; non-availability and non-accessibility to schools; irrelevant and non-attractive school curriculum; social and cultural environment; informalization of production; employer's preference to children for their cheap labour and inability to organize against exploitation; family work; level of technology; and apathy of trade unions' (Sekar, 2001: 107).

In urban areas of India, around 35 per cent of the boys and girls were employed in household and non-household industries of which the boys were likely to work more in non-household industries while the girls worked mainly in household/domestic work. Besides, a much larger proportion of the boys worked in trade and commerce than girls when nearly one-fourth of the working girls and one-sixth of the working boys in urban areas were engaged in the service sector, including domestic work. Poverty is stated to be the most important factor for children to enter into labour market often prompted by their parents to augment family income. "Children belonging to bottom 30 per cent of the households, account for 36 per cent of the working children (37% for boys and 35% for girls) in rural areas. In urban areas this ratio is 49 per cent (48% for boys and 53% for girls)" indicating greater participation of children in labour force (Government of India 2002: 97) (*See* Note P). As it is pointed out in the government report, child labour is closely related to the rates of school drop-outs so that wherever drop-out rates are high at the primary and middle levels, the volume of child labour is also commensurately high. Another factor contributing to the creation of child labour, especially in big cities in many countries is the presence of street children joining the labour force. This happens in countries like Brazil, Kenya and India. In Mumbai, for example, 7 per cent of the city's population, or almost 980,000 people live on pavements of whom about 40 per cent are children (Bajpai, 2009: 428). According to Burra (2006), "street children constitute the third category of children who work for survival, and are largely found in semi-urban and urban metropolitan centres. These children find themselves on the street because they have either run away from their families, or have been abandoned or kidnapped. They usually live in public places, such as railway stations, bus stops and footpaths and are without the protection of their families. They may be self-employed, such as shoe-shine boys, newspaper vendors, rag-pickers, hawkers and vendors, or they may work in establishments like '*dhabas*' (small way-side eating places), as domestic servants and as '*coolies*' (porters), or as casual labourers on construction sites, as helpers in shops, and so on... It is estimated that in Delhi, 400,000 children are child workers and 100,000 are street children (as reported by UNICEF). In the city of Hyderabad ... there are 30,000 street children (as reported by Rao and Malik 1992 in an article, 'Street Children of Hyderabad: A Situational Analysis), between 75,000 and 100,000 on the streets of Calcutta (Kolkata) (as estimated by A. Ghosh in 1992: Street Children of Calcutta: A Situational Analysis), ... an estimated 25,000 street children in the city of Madras (Chennai), (as estimated by J. Arimpoor in 'Street Children of Madras: A Statistical Study in 1992) ... about 45,000 street children in Bangalore city (as estimated by Nandana Reddy in 'Street Children of Bangalore: A Statistical Analysis in 1992) (Burra, 2006: 27-28).

Working children are often associated with migrant families providing labour at construction sites, brick-kilns, mines and plantations. Significantly, involvement of child labour in non-agricultural occupations covers a wide canvas giving an insight into the real nature of the problem. For example, the carpet industry in India happens to be one of the largest employers of child labour. This industry located in Mirzapur-Bhadodi-Varanasi region of Uttar Pradesh, producing almost 80 per cent of carpets in India, employs 200,000 people, 150,000 of whom are children, excluding those working as part of family labour while the carpet indiustry in Kashmir employs another 100, 000 child labour (Burra, 2006: 191-192). In fact, the number of child labourers engaged in carpet looms has made a quantum jump, from 10,000 in 1975 to 420,000 in 1989 (Bajpai, 2009: 201). Interestingly 600 carpet-weaving training camps have been set up by the Government of India, training annually about 30,000 children as they are always preferred by employers for their 'nimble fingers'. About 50 per cent of the child labour force has been sold to the loom-holders, and as observed by Neera Burra (2006), about 100,000 child workers employed in the Carpet industry work "in conditions of slavery". According to an estimate, nearly 400,000 female children aged 7 to 14 years are employed in the cottonseeds fields, of which 250,000 are employed in the fields of Andhra Pradesh alone. "The figure is greater than the total number of children working in the so-called 'hazardous' industries sector, such as carpet weaving, glass bangles, diamond polishing, gem polishing, and limestone kilns and quarries put together (Weiner, 2006: xxxiii). Among them the glass bangles industries in Firozabad (U.P.), making 99 per cent of India's glass bangles using furnaces in which temperature ranges between 700 and 1800 degrees centigrade (in tank furnaces) employ 40,000 to 50,000 children (Weiner, 2006: 27-29). Similarly, children preferred for their 'nimble fingers 'are also employed by another hazardous industry, the match factory in Sivakasi in Tamil Nadu, alone employing 45,000 children of whom approximately 90 per cent are girls (Burra, 2006: 193 and 205). The gem industry in Jaipur (Rajasthan) employs another 13,000 children below the age of 14 years, working long hours from 8 am to 6 pm while the brassware industry of Moradabad (U.P.) engages about 40,000-45,000 children. Pottery is another industry making use of about 5,000 children as a source of cheap labour in Khurja (U.P.). Further, at least 50 per cent of the total labour force employed in spray painting in the lock industry, mainly concentrated in Aligarh (U.P.), are exposed to inhaling 'paint and paint thinners' (Burra, 2006: 56-59, 79, and 81; 110-111, and 163).

One needs to investigate the situation in regard to child labour prevailing across the world and its concentration in specific regions for a comparative assessment. As indicated by Table 3.30, the largest percentage of children engaged in offering labour are in Sub-Saharan Africa, involving 35 per cent of the total number of children, increasing further to 36 per cent in the eastern and southern countries of this region. The percentage, however, drops to 9 per cent in Middle East and North Africa, and it is 13 per cent in South Asia. Data for the world as whole for children involved in work is not available.

Among the 119 countries listed by UNICEF, 2008, 21 provided no data, including countries like China, Pakistan, Congo, South Africa etc., and there are wide disparities in the geographical distribution of child labour. For example, in 12 countries 36 per cent or more of children are labouring, the figure rising to the highest rate, 53 per cent in two of them – Chad and Ethiopia, followed by Somalia with 49 per cent and Sierra Leone with 48 per cent. Some of the countries with high percentage of child labour are Afghanistan (30%), Belize (40%), Cambodia (45%), Democratic Republic of Congo (32%), Cote

Table 3.30: Child Labour (5-14 years) (1996-2006) and Child Marriage (1987-2006) in different regions of the World

(in percentage)

Regions	*Child Labour*			*Child Marriage*		
	Total	*Male*	*Female*	*Total*	*Male*	*Female*
Sub-Saharan Africa	35	36	34	40	24	47
Eastern and Southern Africa	36	35	33	36	20	44
Western and Central Africa	34	33	34	44	27	53
Middle East and North Africa	9	10	8	17	11	19
South Asia	13	—	—	45	30	53
East Asia and Pacific	10	11*	10*	19*	12	25
Latin America and the Caribbean	11	12	10	26	24	31
CEE/CIS	5	5	5	11	7	9
Developing Countries	16	19	17	34	23	45
Least Developed Countries	29	31	28	49	37	57
World	—	—	—	—	—	—

Source: UNICEF, 2008, Table 9

Note: (a) A child is considered to be involved in child labour (1) Children 5 to 11 years old who, during the week preceding the survey, did at least 1 hour of economic activity or at least 28 hours of domestic work, or (2) children 12 to 14 years old who, during the week preceding the survey, did at least 14 hours of economic activity or at least 28 hours of domestic work.

(b) Child Marriage means percentage of women 20 to 24 years old who were married or in union before they were 18 years old.

* excludes China.

d'Ivoire (35%), Ghana (34%), Guinea-Bissau (39%), Mali (34%), Republic of Moldavia (32%), Nepal (31%), Nigeria (38%), Rwanda (35%), United Republic of Tanzania (36%), and Uganda (36%). Interestingly, of the 14 countries having more than 30 per cent child labour, 10 are in Africa, and there are at least two countries – Romania and Trinidad and Tobago –where the number is 1 per cent. Industrialized countries, such as the United States, The United Kingdom, Germany, France, Japan Canada and Australia have not been included in this list presuming, perhaps, that they have no child labour or provide no data on child labour though there are instances of reference to their existence in some of them.

Child labour is a brutal fact which reminds us about the criminal injustice involved in keeping children deprived of their childhood associated otherwise under normal circumstances with playing, learning and exploring the wonders of the slowly unfolding world around them, all of this denied to them when the children at tender age are compelled to work, engaged as labour as a means of livelihood supplementing family income. The joyless life in a gruelling work environment devoid of any freedom for the child to feel the boundless enjoyment derived in better (normal) circumstances is nothing but a curse for being born of poor parents. Another equally important aspect making childhood extremely unhappy and highly vulnerable to external pressure is child marriage. As mentioned briefly earlier in this chapter, child marriage happens to be practiced in many communities as a form of social injustice done to female children having no opinion in the matter. In this case, Saudi Arabia takes the lead, marrying off 45 per cent of women within the age range of 20-24 before they were 18 years old. Sub-Saharan Africa follows with 40 per cent, which rises to 44 per cent in its western and southern states while in the Least Developed Countries, taken together, almost half (49%) of the married women (20-24 years old) were married before reaching the age of 18 years.

Naturally, the overall rate of child marriage is higher in rural than the urban areas – the highest being 57 per cent in the rural areas of the Least Developed nations. Interestingly, among the 119 countries listed in UNICEF 2008, from which the industrialized countries are excluded, 25 have no data on child marriage and out of the remaining 94 countries, in 23 the rate of child marriage is 40 per cent or more, and exceeds 50 per cent of the married women in 11 of them, such as in Bangladesh (64%), Central African Republic (61%), Chad (72%), Guinea (63%), Malawi (51%), Mali (51%), Mozambique (56%), Nepal (51%), Niger (75%), Sierra Leone (56%), and Uganda (54%). Among those 11 countries, besides Bangladesh and Nepal, all other countries are in Africa. Niger having three-quarters of her women's marriages executed in childhood heads the list closely followed by Chad with nearly three-quarters. Contrastingly, the lowest record of 2 per cent of the girls married off in their childhood happens in an African country – Algeria, followed by another African country, Djibouti with 5 per cent while the second lowest of such early age marriages occurs in the former Yugoslav Republic of Macedonia with 4 per cent of the young girls getting married in childhood There is no data for China on child marriage. In India, 45 per cent of the marriages are recorded as child marriage which is considerably higher than that of some of its neighbouring countries like Pakistan (32%) and Sri Lanka (12%).

It is highly objectionable as well as discouraging to note that, "in Rajasthan on '*Akshay Tritiya*' day, which is popularly known as '*Akha Teej*', hundreds of child marriages are openly performed. *Akha Teej* is regarded as the most auspicious day for celebrating marriages. On this day, even infants, who have just been born or are only a few years old and cannot even sit or walk, are married. The child brides or the bridegrooms do not understand the solemnity of these ceremonies, but for elders it is the safest and most tested way of keeping property and money within the family and of preserving the 'chastity' of their daughters... In rural India, the girls are married off at 12-14 years even before they are physically, psychologically, and emotionally mature' (Bajpai, 2009:212-214). Table 3.31 gives a very uncomfortable feeling in this regard, exposing further the wickedness of many people in Indian society in their false pretence of religious compulsion. The data (Table 3.31) as compiled by the National Family Health Survey (NFHS) show that the median age of first cohabitation with the husband is as low as 16 years, in some states going down to as little as 15.4 years as in Andhra Pradesh, while the median age at first marriage is 14.7 years in Madhya Pradesh and is around 15 years in Uttar Pradesh (15.0), Rajasthan (15.1), Bihar (14.9) and Andhra Pradesh (15.1). While the highest median age at first marriage is 23.2 years in Goa, and 20 years in five other states, namely Mizoram (22.0), Manipur (21.7), Kerala (20.2), Nagaland (20.1) and Punjab (20.0). Five of the states — Goa, Mizoram, Kerala, Nagaland and Punjab – have non-Hindus as the dominating population which partly explains (along with the higher rates of female education in Kerala, Goa and Mizoram) comparatively the higher age of marriage for girls. The minimum age of marriage in India according to existing law is 18 years. As a matter of fact, average practice exceeds 18 years in only 13 out of 25 States in India with the North-Eastern Region standing out with all of its 7 states (along with Sikkim) keeping the age of marriage of their girls above 18 years. Barring Assam, in all other states religion acts as a factor in explaining this. However, as the registration of marriage has not been made compulsory till recent times in many states, it is not possible to have a correct picture, and it is presumed that actual figures are much higher. "The percentage of Indian girls getting married by 13 years of age is 15.4 per cent, by the age of 15, 33.3 per cent" which, together make up nearly half (48.7%) of the girls getting married before reaching the stipulated age of 18, the legal age. "By 18 years this (the percentage of married girls) figure reaches 64.6 per cent, at 20 it is 79.4 per cent, at 22 it increases to 87.9 per cent, and at 25 it is 93.7 per cent. Early marriage (a more common phenomenon) is a form of sexual abuse of the girl child as they have to become mothers" (Bajpai, 2009: 214) when at an early age they are not at all capable of caring for newly borns.

Speaking of work conditions at most of the places they are far from safe for children. For example, surgical instruments bought by British hospitals from Pakistan are produced there in 'dangerous working conditions… using child workers as young as eight, each paid as little as 170 rupees a day (about 2 British pounds), the workshops producing scalpels, clamps and scissors to be used in British National Health Service (NHS) operations. According to an estimate of the ILO, as many as 5,800 children are working in surgical instrument manufacture in Sialkot (Pakistan) alone (Randerson, 2008).

The fact that more than 30, 000 children are dying everyday in the world from preventable diseases is quite alarming. Looking at the present trend, the Millennium Development Goal (MDG) of reducing child death by two-thirds by 2015 seems to be unattainable. The reasons for this may be found in the fact that 'the poor in most countries account for a far larger share of child death than commensurate with their share of population'. In other words, 'the children of the poor are overrepresented among the victims of child death – heavily so in many countries'. For example, "in China 36 per cent of child deaths occur among the poorest of the population, while 7 per cent occur among the richest". In Zambia, child mortality among the richest 20 per cent came down by 6 per cent in a year in the second half of the 1990s and "it was three

Table 3.31: Age at First Marriage by State in India 1998-99

State	*Median age at first marriage*	*Median age of first cohabitation with husband*
India	16.4	17.0
North		
Delhi	17.97	18.48
Haryana	19.0	19.3
Himachal Pradesh	16.9	17.9
Jammu and Kashmir	18.6	18.8
Punjab	20.0	20.1
Rajasthan	15.1	16.4
Central	14.85	16.15
Madhya Pradesh	14.7	16.0
Uttar Pradesh	15.0	16.3
East	16.4	17.03
Bihar	14.9	16.6
Orissa	17.5	17.6
West Bengal	16.8	16.9
North-East	19.4	20.01
Arunachal Pradesh	18.7	18.8
Assam	18.1	18.2
Manipur	21.7	21.8
Meghalaya	19.1	19.3
Mizoram	22.0	22.0
Nagaland	20.1	20.2
Sikkim	19.8	19.8
West	19.07	19.37
Goa	23.2	23.2
Gujarat	17.6	18.2
Maharashtra	16.4	16.7
South	17.7	17.88
Andhra Pradesh	15.1	15.4
Karnataka	16.8	17.0
Kerala	20.2	20.3
Tamil Nadu	18.7	18.8

Source: Bajpai, Asha, 2009 (Taken from National Family Health Survey, 1998-99)

times as fast for the poorest 20 per cent" (UNDP, 2005: 61). Further, the rate of child mortality is declining much more slowly among the poor than the average rate of decline for most of the countries, the rate of decline among the poorest 20 per cent being half the average rate of decline, so that the mortality gap between rich and poor children is widening with time. According to an estimate, 'closing the gap in child mortality rates between the poorest 20 per cent and the national average would reduce child deaths by 60 per cent, thus saving about 6.3 million lives a year, making it easier to achieve the Millennium Development Goal target. It is pointed out that the failure of the national and the international community in removing inequalities in income costs the lives of more than 6 million children a year (UNDP, 2005:62). Arguably, if the average child mortality rate could be brought down to the average rate of the richest 20 per cent, it would result in very large declines in child deaths, reducing the overall total by more than one-half in India alone, saving half-a-million lives of children in the first month of life in just three countries – Bangladesh, India and Nepal. It is further claimed that reduction in gender inequality 'would have a catalytic effect on cutting down child deaths and the effect would very much be pronounced particularly in South Asia, where gender inequality is very deeply entrenched. For example, if the gender gap in mortality between girls and boys aged 1-5 could be closed, it would save an estimated 130,000 lives, reducing the child mortality rate by 5 per cent. Besides, the indirect link between maternal nutritional deprivation and child mortality is a hard reality and as the percentage of underweight women, suffering from nutritional and vitamin deficiencies particularly during (and also before) the period of pregnancy is four times higher in South Asia than even in Sub-Saharan Africa, it is intimately connected with child deaths which are far greater. Although South Asia has lower levels of poverty and higher average incomes than Sub-Saharan Africa, the child malnutrition rate is 20 per cent higher in South Asia as compared with that of Sub-Saharan Africa, and as such half of the world's underweight children are in South Asia, as a strong reminder of the association with gender inequalities.

Naturally, a greater degree of gender equity would have a great impact on child mortality, and as estimated by the International Food Policy Research Institute, equity in access to education, nutrition, income and property rights could reduce the underweight rate among children less than 3 years old, vulnerable to early death, by 13 percentage points in South Asia, reducing their number by 13.4 million. In Sub-Saharan Africa, it will be 3 percentage points, or 1.7 million less undernourished children.

An International Food Policy Research Institute study in Latin America, South Asia and Sub-Saharan Africa on hunger and nutrition linking up children's nutritional status and women's decision-making power shows that children are likely to be undernourished where women have low status and are deprived of decision-making power in households. For example, in South Asia 'where between 40 and 60 per cent of women are underweight, almost 45 per cent of children were born with low birthweight making it the highest incidence of underweight births in the world'. On the other hand, as evidence from India and Nepal shows, 'women having greater influence in decision-making can promote better healthcare practices for the family'. This is clearly demonstrated by instances of the impact of women's decision-making on children's development in female-headed households. "It was estimated in 1998 that roughly 20 per cent of households worldwide are headed by women'. This "accounted for 24 per cent of all households in Latin America, 22 per cent in Sub-Saharan Africa, 16 per cent in Asia, and 13 per cent in the Middle East and North Africa". Contrary to general belief, female households do not always represent the poorest of the poor: Research on Latin America has shown that female-headed households may even generate higher incomes, or have more income earners than their male equivalents due to more effective use of household labour. (Another) study based on data from 17 countries "in which at least 15 per cent of children lived in female–headed households reveals that single mothers manage to raise their children with outcomes similar to those of two-parent families, despite the numerous challenges they face" (UNICEF, 2007: 28-29). A child benefits in a household with female-head of the family through the mother having full control over allocation of income. Evidence from rural Bangladesh indicates that children suffer less from malnutrition in a female-headed than a male-headed household, with more money being spent on food and healthcare services, with a more varied diet, and a female with one year of secondary education has been found to contribute to the better nutritional status of children, ultimately reducing child death at tender age. Furthermore, another very revealing fact is that over 50 per cent of the children in seven countries in Sub-Saharan Africa (Burkino Faso, Cameroon, Ghana, Kenya, Mozambique, Nigeria and the United Republic of Tanzania) having lost their fathers usually live with their mothers in contrast to fewer than half of the children who lost their mothers continued to live with their fathers, which makes it quite evident that women, in general, are more likely to care for orphaned children, irrespective of whether they have lost their mothers, fathers or both parents. In such cases often the grandmother, in particular, takes care of orphaned children, and by the end of 2005, there were 12 million such children orphaned by AIDS in Sub-Saharan Africa, (roughly 1 in every 5 children aged 12-17, and 1 in every 6 children aged 6-11), and the number of them was likely to rise to 15.7 million or 12 per cent of all children by 2010 in Sub-Saharan Africa due to various causes, one-quarter of them by AIDS (UNICEF, 2007: 30-31).

Flegg observes that "income per head as a determinant of infant mortality "is likely to influence it indirectly with its "effects on the rate of consumption of items affecting health, such as food, housing, sanitation, medical care and education. Moreover, certain diseases are thought to be dependent largely upon standards of living and not readily amenable to reduction through specific medical and public health measures (e.g. diarrhoeal diseases which are influenced strongly by nutrition and personal sanitation)" (Flegg, 1982: 442-443). According to him, there are many positive determinants of infant mortality besides income and inequality, such as maternal education, the availability of medical care and fertility.

Child Abuse

According to a report of the UNICEF (2007), every year a staggering number of as many as 275 million children are subjected to domestic violence, suffering the 'full consequences of a turbulent home life' across the world. Interestingly, according to this report, studies conducted in some of the largest countries in the developing world, such as China, India, Colombia, Egypt, Mexico, the Philippines and South Africa show that 'there is a strong correlation between violence against women and violence against children.

In a recent report published in Britain it is mentioned that 1 in 10 children suffer physical, sexual, emotional abuse or neglect, and as stated by some researchers from the University College of London's Institute of Child Health, between 4 and 17 per cent of children suffer physical abuse, such as hitting, punching, beating and burning. All the figures in this respect come from high-income countries, including the U.K. Besides, 'some 5-10 per cent of girls and 1-5 per cent boys have been subjected to penetrative sex, usually by a family member or relative'. 'Around 10 per cent of children suffer emotional abuse every year, including persistently being made to feel worthless, defined as the failure of their parents or carer to meet the child's basic emotional or physical needs or ensure safety' (Bosely, S., 2008) The same source reports about 15 per cent of girls and 5 per cent of boys by age 18 become exposed to sexual abuse and 13 per cent of mistreated girls and 8 per cent of mistreated boys fall into prostitution. In another report it is pointed out that children from poorer families are far more likely to die or be harmed than those from better-off families. The WHO estimates that 1.9 children in every 100,000 die yearly from injury or accidental poisoning among higher managerial and professional families, a ratio that rises to 25.4 in every 100,000 for children whose parents have never worked or are long-term unemployed. The overall rate of that for Kazakhstan is 26.98; in Moldavia it is 23.34; in Kyrgyzstan 22.19 and in Belarus the rate falls to 23.13 per 100,000 children. According to WHO, every year across the world "830,000 children die from unintentional injuries, sometimes called accidents. It is like wiping out the whole child population of a city like Chicago or Marseilles" (Bosley, S., 2008). Millions more children suffer permanent damage, and need long-term hospital care and rehabilitation. The leading cause of childhood injury and death is road crashes which kill 260,000 children a year globally, and injuring about 10 million. In fact, road traffic injuries in developing countries have reached epidemic proportions. This apart, of the 3.6 million people killed, over 90 per cent of which were civilians, and injured many more millions in violent armed conflicts – 57 major ones of them occurring during 1990-2001 in 45 locations, Sub-Saharan Africa having been hit the hardest, and at least half of the victims were children (UNDP, 2003: 46). At the same time, growing up in violent homes may become psychologically harmful for children even if they are not directly abused. Children with experiences of growing up in a violent atmosphere suffer from traumatic stress disorder, such as bed-wetting, or nightmares, allergies, asthma, gastrointestinal problems, depression and anxiety, leading many of them to drugs or alcohol at later stages.

Sexual abuse of children is a universal phenomenon occurring all over the world across all social, economic and cultural borders (including educational_ levels without being deterred by caste or religion or language, bearing a "sameness" in all situations). The incidence of sexual violence in domestic settings is a very common feature involving, according to a recent multi-country study by WHO, up to 21 per cent of children victims of sexual violence, affecting larger number of girls than boys. Sexual and gender-based violence often takes place beyond homes, in schools and colleges, especially in case of girls, or at work places from employees, with consequences spanning generations. For example, the effects of violent behaviour have a lasting influence on children, whether boys or girls, the former having grown up in a family watching father becoming violent to mother or to himself is twice as likely to become violent to his wife in future life. And girls experiencing their mothers physically violated by fathers, are likely to accept violence in marriage from their husbands.

A study conducted by the International Statistics on Child Abuse found that at least two out of four girls and one out of six boys are victims of child abuse. Another study made by Kennedy and Manwell (1992) on the pattern of child sexual abuse in Northern Ireland, reveals that in 85 per cent of cases, all the abuses were committed by relatives or persons known to the victim, done on faked pleas or threats which a child cannot understand or overcome at his/her age. Such abuses, as sometimes reported in the press (though very rarely) may continue for years, and increasing in severity as the child trusts the person. Hence, children, more girls than boys, are not safe even at home. According to an International Study, statistics on child sexual assault show that at least one out of every six boys is victim of sexual abuse. Another study in the U.K. (1983) discloses that 79 per cent of the victims already knew their abusers and 58 per cent were abused by family members. The picture in the U.S.is more or less the same. Finkelhor "on the topic of 'sexually victimized children (1979), reported that of a group of 530 college students, 17 per cent had been abused before puberty and out of those approximately 50 per cent had been abused by a relative" (Phansalkar-Joshi, 2009: 162). In its 1999 report, WHO mentions that one out of every ten children in India is sexually abused at some time of life. Figures available for Maharashtra in 2004 show a 10 per cent increase compared to data for 2003, of which more than 45 cases were reported of girls below the age of 18; and in 85 cases out of a total of 1,392, the girls were below 10 years. The girl child is never safe even from her infancy to puberty or thereafter. She is always in danger of being defiled or deflowered from which even handicapped and deaf and dumb child are not spared, making highly shocking as well as disturbing. In a study of about 100 reported cases made by P-Joshi it was confirmed that "the common categories of perpetrators are the father or uncle, cousin or family friend and in other cases, either he is a school teacher, principal or even the priest, the neighbour, or servant... The 99 per cent of such cases remain unreported to any one (on account of saving the prestige of the family) and hence unknown to the public, therefore, the myth prevails that the child is safe in the family with relatives, neighbour or friends" (Phansalkar-Joshi, 2009: 162-163)

The irony of social protection in the form of law and enactment hardly touches the extent of the vulgarity and abuse of human rights perpetrated by offenders mostly going scot-free in majority of the instances, obtaining easy acquittal (a sort of social protection) and the majority of the cases remaining hidden under the wrap of social ignorance multiplies the risks the children experience at tender age, leaving quite indelible mark on their immature mind carried all through the rest of life as a sign of loss of trust on elders. No protection, it seems, is adequate enough to face this heinous challenge of perversion, making children highly vulnerable to sexual and physical abuse unless society is sufficiently conscious to make children aware of their rights to dignity, violation of which is highly condemnable and punishable at the court of law. Children need this training to awaken their consciousness about the different forms of provocation coming from near and dear ones leading to abuse. Perhaps, schools could be roped in by the government making it a compulsory part of training from early age and parents concerned involved in this process. Ultimately, however, these highly base human instincts cannot be fought against as long as sanity does not prevail with the larger section of the society.

The problem itself has many dimensions: psychological, social (family background), economic and also political. Apart from individual harms inflicted by different forms of abuse, emotionally the most frightening and challenging issue for any society today is the immensity of the sex-trade involving children across the world. For example, "nearly 200 Nepali girls (as has been mentioned before) under 14 years of age are working in the sex industry in India; 15,000 children were sold to brothels in Cambodia between 1991 and 1994; possibly 300,000 Bangladeshi children are engaged in sex-trade in India; and approximately 10,000 children between 6 and 14 years of age work as prostitutes in Sri Lanka. The government in Thailand estimates that 60,000 children have been sold into prostitution but the unofficial number is close to 800,000...After the fall of Soviet Union (1989) and war in Yugoslavia, Europe has experienced a surge in child prostitution" (Sinha and Dasgupta, 2009: 27-28). Sinha and Dasgupta squarely blame the entire system saying that, "by encouraging tourist industries as a means of development in various poverty-stricken countries, World Bank and other financial institutions have also indirectly provided encouragement to local prostitution, particularly the child prostitution business. Sara Andrews writes pithily (*The Journal of Criminal Law and Criminology*, 2004) that "each year foreign travellers from predominantly western countries pump billions of dollars into the economies of developing nations when they purchase sexual services'. According to a 2002 estimate, as many as ten million children worldwide may be working as prostitutes. Philippines is arguably the worst offender with nearly 100,000 children involved in prostitution but South Asia is not far behind. India itself may have between 400,000 and 575,000 children exploited in the commercial sex-trade" (Sinha and Dasgupta, 2009: 27-28).

Children suffering from various types of physical disabilities make a serious problem for any country in dealing with proper rehabilitation of the such children. According to the report of the Rehabilitation Council of India on

Manpower Development 1996, estimated on the basis of the NSSO report, the following are the population of children with disabilities in the age group of 5-14:

Locomotor Handicap	: 8.94 million
Hearing Handicap	: 3.24 million
Speech Handicap	: 1.96 million
Visual Handicap	: 4.01 million
Mental Retardation	: 9.00 million
Cerebral Palsy.	: 3.00 million (Bajpai, 2009: 420).

Undoubtedly, however, it is highly shocking to note that deprivation of women in all societies (a more universal phenomenon than any other form of deprivation) still prevalent in various forms crossing all boundaries of economy or social classes, happens to be the most degrading type of human behaviour affecting the security and dignity of women. Attempts made, so far, in different countries (both developing and developed), however positive, have not yet been fruitful in eradicating the wide-ranging disparities between man and woman across the world. As a result, human rights for man and woman as equals in all spheres of life has remained largely elusive for women, making unaccountable loss of the huge human potential as a resource that could be provided by fully developed women comprising almost half of the global population.

At the same time, deprivation of children, almost 250 million of them still wasting life at tender age engaged in physical labour and subjected to concurrent abuse in different ways, is one of the greatest disgraces for any society, particularly affecting the developing countries, as a great blot on them, often making false, hollow promises for stopping the barbaric practices. The colossal wastage results in another great loss of human capital barring these children from growing up to their full potential with proper education and necessary healthcare.

4
Deprivation in Other Forms; Education, Health, Water, Sanitation

Education has a pivotal role in eliminating or reducing deprivation among people, particularly poverty, hunger, and all that without which life is not worth living in the basic sense which means comfort, security, dignity, inclusion in societal activities, and being treated as equals, with freedom, but not necessarily affluence. The expectations of people suffering from various forms of deprivation are for a good quality life, including "material well-being, which is often expressed as having enough" where "ill-being included material lack and want (of food, housing and shelter, livelihood, assets and money); hunger, pain and discomfort; exhaustion and poverty of time; exclusion, rejection, isolation and loneliness; bad relations with others including bad relations within the family; insecurity and vulnerability, worry, fear and low self-confidence; and powerlessness, helplessness, frustration and anger" (Narayan *et al.* 2000: 21). In other words, elimination of these deprivations is likely to make a full man/woman, satisfied with life taking a deep, positive interest in the development of the society. Sen considers that "The well-being aspect is essentially important ... as social security, poverty alleviation, removal of gross economic inequality, and in general in the pursuit of social justice... The well-being aspect of a person has great importance of its own for the analysis of social inequality and the assessment of public policy. Problems of social injustice and inequity between different classes and groups relate strongly to extensive disparities in well-being – including the freedom we respectively enjoy to achieve well-being" (Sen, 1995: 72) and 'inequality in well-being freedom' leads to 'disparities in well-being achievement'.

The potential value of education in contributing to the well-being of those deprived, giving them freedom of choice along with social security has no match in thwarting deprivation since as a single factor, education alone ultimately becomes the most decisive one for any person in exposing and utilizing his/her full potential with a bargaining capability. For example, a person cannot satisfy all his/her immediate needs to have a comfortable, secure life living with dignity and freedom of choice and expression as long as he/she is not capable of earning them with his/her own efforts (which is) only possible through being sufficiently qualified (equipped) through proper education. Putting it differently, every individual needs to be adequately equipped with education and training to acquire this purchasing (bargaining) power for obtaining a position – better to call it an employment opportunity – to stabilize his/her status in the society making it possible for him/her emerge from a status of 'ill-being' to that of well-being. This cannot be done overnight. It involves a long-time process entailing an enormous amount of effort from various fronts. But what is aimed at is finding a dependable, lasting means of ultimate eradication of all forms of deprivation people are steeped in various situations across the world. Apart from education, all other means will prove hollow, insubstantial in the long run since, to sustain their performance and progress they depend directly on how far people for whom it is done can keep it going in the absence of educational expertise and training for required knowledge. The financial or technical help coming from outside offered by foreign donors, cannot transform the society unless it is in a position to recreate a situation from inside for tackling the challenge of deprivation in its multiple facets and this needs most of all the capability earned with education. This is a great challenge given that there are 325 million children out of school at the primary and secondary levels, 183 million of them girls.

One may feel encouraged to know that worldwide, primary education enrolment rose from 80 per cent in 1990 to 84 per cent in 1998; that 51 countries with 41 per cent of world's population have achieved or are on the way to achieving universal primary enrolment; that 90 countries, with more than 60 per cent of the world population, have achieved or are on the way to achieving gender equality in primary education by 2015, and more than 80 countries in secondary education. However, the gloomier side in this respect is that 113 million school-age children are not in school, 97 per cent of them in developing countries. Besides, it is very difficult to get a correct picture in this regard since 93 countries, having 39 per cent of the global population do not have data showing trends in primary enrolment. Moreover, 60 per cent of children not in a primary school worldwide are girls, and of the world's estimated 854 million illiterate adults, 544 million are women (UNDP, 2002: 11). Nevertheless, many countries seem to have good prospects for achieving the Millennium Development Goal of making primary education universal by 2015, but many countries have not, so far, made satisfactory progress in this respect, given the number of countries (93) without data. Of the 21 Sub-Saharan countries with data, for instance, 14 are far behind the target or having deteriorating enrolment. But the crux of the whole issue is that "enrolling children in primary schools is only half of the battle, because it is meaningful only if they continue it – which requires that they and their families be able to resist the pressures of foregone income and work in the home" (UNDP, 2002: 22). The data on completion of primary education show that, the record is again worse for Sub-Saharan Africa where six countries are on way to achieve the target, 5 are behind or slipping back, and 33 countries are without data (and are presumed to be poor performers).

The data made available by UNDP, 2002, giving an insight into the present status of achievement in gender equity in primary education show it to be still far from satisfactory, though the goal is elimination of gender disparity in primary education by 2015 while in secondary education the goal is to reduce gender disparity to 80 per cent by 2015 (UNDP, 2002: 22-23).

Table 4.1 shows the achievement of MDG (Millennium Development Goal) in respect of gender equity in education which has not at all been impressive. Only 20 out of a total of 168 countries listed in the Table 4.1 have achieved the target so far while 70 more countries are on way to achieve it unless something goes wrong meanwhile with the process of progress, the 90 countries together representing 63 per cent of world population. At least this represents a significant increase for girls' education. Over the past 30 years, gross primary enrolment rates for girls in low-income countries have risen from 52 per cent to over 90 per cent. However, "gender disparities remain, not only at the primary and secondary levels, but also in tertiary education, where a mere 5 to 10 per cent of students in low-income countries are female" (UNICEF, 2007: 71). Further, the existing wide variation in the enrolment of girls at primary level between developing countries is quite striking. While, on the whole, 87 per cent of the girls in the developing countries are reported by the UNIFEM, 2008/2009 to have been enrolled for primary education, that figure comes down to 30 per cent in Djibouti, 33 per cent in Niger, 45 per cent in Mali, 60 per cent in Ethiopia, 70 per cent in Senegal, 75 per cent in the United Arab Emirates and 80 per cent in Kenya and Maldives (UNIFEM, 2008/2009: 37: Fig. 3.3). Significantly, besides Maldives, and Pakistan (with 59%), all other countries having a very low profile in this respect are in Africa. Though there has been an impressive gain in narrowing the "gaps between boys' and girls' primary school completion rates in low-income countries [the gap dropping] from 18 per cent in 1990 to 13 per cent in 2000. In Gambia, girls' enrolment rates more than doubled between 1980 and 2000, while in Guinea they jumped

Table 4.1: Achievement of Millennium Development Goal in Gender Equity in Primary Education (Number of Countries)

Regions	*Achieved*	*On Track*	*Lagging*	*Far Behind*	*Slipping Back*	*No Data*
Sub-Saharan Africa	5	15	0	8	1	15
Arab States	1	12	0	1	0	3
East Asia and Pacific	5	7	0	1	0	6
South Asia	0	5	0	0	0	3
Latin America and the Caribbean	3	13	0	1	0	16
Central and East Europe and CIS	4	16	0	0	0	5
TOTAL	20	70	0	13	1	64

Source: UNDP, 2002: Page 25.

from 19 per cent to 63 per cent between 1990 and 2001" (UNIFEM 2008/2009: 38-39).

In India, the enrolment ratio of boys aged 6-14 years is 17 percentage points higher than that of girls. Yet, especially in Latin America, girls have no disadvantage, or even a small advantage. Still, of the 113 million of the world's children not in primary school, 60 per cent are girls. Though the gaps are closing in primary, it is less so in secondary enrolments. Perhaps, the most surprising is the performance of the Arab States, generally associated with higher gender inequality. Except for one, all other countries with data, are on way to meet the MDG for primary enrolment but, again, Sub-Saharan Africa has made least progress in this regard.

An examination of educational status of countries across the world throws light on the existing condition assessed on the basis of 'net primary enrolment', 'net secondary enrolment', and extending it further into an assessment of the 'number of researchers engaged in research and development' in every country during 1991 to 2005.

Taking the countries according to the level of human development (HDI) for this purpose, it may be found that those with high human development, 70 in all, 57 of them providing data for 1991, and 65 for 2005, exhibit some form of similarity in enrolment at the primary level of education in 1991 and 2005, with most countries maintaining high rates (Table 4.2). The rate does not go below 49 per cent, recorded as the lowest by Kuwait in 1991 followed by Saudi Arabia with 59 per cent while it was 100 per cent in 8 countries in 1991—Iceland, Norway, Sweden, Italy, France, Spain, Japan and the Republic of Korea. In 5 countries (Australia, United Arab Emirates, Russian Federation, Estonia and Bahrain) it was 99 per cent, and altogether in 52 countries the enrolment at primary level was above 81 per cent, exceeding 90 per cent in 40 of them. It was the highest in 6 European countries (along with Japan and the Republic of Korea) but dipped slightly in the United Kingdom with 98 per cent and in the United States with 97 per cent. The major difference in the level of enrolment after 14 years, that is between 1991 and 2005, is that only Japan was holding on to its previous record of 100 per cent while in the other seven countries the rate of enrolment slides down from its earlier position, (if only slightly), though 13 countries had 99 per cent enrolment in 2005 in comparison to 5 countries in 1991, and in a total of 54 countries the rate of enrolment was 90 per cent or more in 2005 compared to 40 countries in 1991. But, though the European countries along with industrial nations elsewhere maintained, in general, a higher rate of enrolment than other countries, improving their positions further in this regard from 1991 to 2005 (such as, in Canada, Ireland, Switzerland, Netherlands, the United Kingdom, Belgium, New Zealand, Germany, Greece, Cyprus, Czech Republic etc.), the rate came down in some of the advanced countries, namely the United States, Hungary, Poland, etc. In sharp contrast the enrolment rate rose from 49 to 87 per cent in Kuwait and from 59 to 78 per cent in Saudi Arabia.

Table 4.2: Net Primary Enrolment Rate, 1991 and 2005

HDI Rank	*Total Number of Countries*	*0 – 20*		*21 – 40*		*41 - 60*		*61 – 80*		*81 – 100*	
		1991	*2005*	*1991*	*2005*	*1991*	*2005*	*1991*	*2005*	*1991*	*2005*
High Human Development: Without data: 13 countries in 1991 5 countries in 2005	70	0	0	0	0	2	0	3	3	52	65
Medium Human Development: Without data: 27 countries in 1991 9 countries in 2005	85	0	0	5	1	6	4	15	14	32	57
Low Human Development: Without data: 1 country in 1991 4 countries in 2005	22	1	0	7	1	12	6	1	8	0	3

Source: Calculated from data provided by UNDP Human Development Report 2007/2008: Table 12

A similar trend in general increase in the rate of primary enrolment is observed in countries with middle human development. In all, 27 out of a total of 85 countries of this group provided no data for 1991, but that number was down to 9 by 2005. As in the countries with high human development, number of countries with the higher rates of primary enrolment increases from 1991 to 2005, with the highest rates being 98 per cent in two of them – Ecuador and Myanmar. Further, in 20 of these countries the level of enrolment rises up to 90 per cent and above, but it falls below 50 per cent in 5 countries, including the lowest in Haiti with 22 per cent preceded by Pakistan (35%), Mauritania (35%), Sudan (40%) and Gambia (48%), making the difference rather sharp as compared with the level of enrolment in countries with High Human Development (HHD). Both China and India provide no data for 1991. Incidentally, Cambodia happens to be the only country to achieve the distinction of reaching 99 per cent enrolment, the highest for the MHD group of countries in 2005, while in 33 countries (out of a total of 76 countries with data in 2005) the rate of enrolment went above 90 per cent which, in all consideration, appears to be quite an impressive progress. In the MHD group the lowest rate of enrolment was 33 per cent in Djibouti. India presents an encouraging picture with 89 per cent enrolment in 2005 while China provides no data.

The situation, however, is different in countries with Low Human Development (LHD), with larger number of countries having low percentage of enrolment at primary level. In 1991 out of 22 countries, in 7 countries it ranges between 21 and 40 per cent, in 12 countries it is 41 to 60 per cent, and only in one country it rises to 66 per cent in Rwanda, while the lowest is 16 per cent in Eritrea. Altogether, in 6 countries enrolment level is above 50 per cent. By 2005, a larger number of LHD countries have higher rates of enrolment at primary level, ranging between 41 and 60 per cent in 6 countries, increasing to 61 to 80 per cent in 8 countries and exceeding that in 3 countries, with the highest rate being 95 per cent in Malawi, followed by Tanzania with 91 per cent. Despite the lowest rate being 40 per cent in Niger, the overall performance seems to be better in 2005 with larger number of children having the scope of primary education.

The emerging trend of enrolment at primary level indicates two prominent aspects, first, the occurrence of higher rates of enrolment decrease, in general, with decrease in the level of human development, between high and medium, and between medium and low, and, second, the number of countries with higher rate of enrolment is higher in 2005 compared with 1991 in each of the groups (high, medium and low) of countries which, undoubtedly, is a positive indication of progress.

The rate of enrolment at secondary level reflects achievement of a country in the continuation beyond primary level of education. However it is difficult to make a proper judgment of comparative secondary enrolment because of the lack of adequate data (see Table 4.3). For High Human Development countries, 38, or more than fifty per cent out a total of 70 countries do not provide data for 1991, and 15 countries are without data for 2005. However, despite this major handicap, some definite conclusions may be drawn from available data for 1991 and 2005, which shows that out of 32 countries with data for 1991, 26 had higher rates of enrolment at secondary level, in 12 of them it ranged between 61 and 80 per cent, in 14 countries between 81 and 100 per cent, while in the remaining 6 countries it came down below 60 per cent with the lowest of 17 per cent in Brazil preceded by Saudi Arabia with 31 per cent. The highest was in Japan with 97 per cent in 1991 followed by Finland with 93 per cent enrolment at secondary level and Canada (89%), Norway (88%), Denmark and Belgium (87%), the Republic of Korea (86%), Sweden, the United States and Bahrain (85%), and the United Kingdom (81%), firmly indicating that though, in general, the rate of enrolment at secondary level was invariably higher in highly industrialized countries, though some non-industrial countries, like Mexico (98%), Qatar (90%) and Argentina (99%) and Bahrain (85% in1991, 90 per cent in 2005) did not lag behind in this respect

On the other hand, although there was no country with 100 per cent enrolment at secondary level in 1991 (among those providing data), at least one country – Japan – had achieved this distinction in 2005 followed by Sweden and France with 99 per cent, and Spain with 98 per cent. In 21 HHD countries out of a total of 55 countries with data for 2005 –15 other countries, including Germany, could not provide data – the rate of enrolment at secondary level was 90 per cent and above, including both the United States and the United Kingdom having improved their score by this time (2005) to 95 per cent (81%

Table 4.3: Net Secondary Enrolment Rate, 1991 and 2005

HDI Rank	*Total Number of Countries*	*0 – 20*		*21 – 40*		*41 - 60*		*61 – 80*		*81 – 100*	
		1991	*2005*	*1991*	*2005*	*1991*	*2005*	*1991*	*2005*	*1991*	*2005*
High Human Development: Without data: 38 countries in 1991 15 countries in 2005	70	1	0	2	0	3	1	12	13	14	41
Medium Human Development: Without data: 68 countries in 1991 22 countries in 2005	85	4	2	7	17	4	11	2	26	0	7
Low Human Development: Without data: 19 countries in 1991 8 countries in 2005	22	3	8	—	6	—	0	—	0	—	0

Source: Calculated from data provided by UNDP Human Development Report 2007/2008: Table 12

Note: — no data available.

in 1991), and 89 per cent (85% in 1991) respectively. The lowest rate of enrolment among HHD countries at secondary level in 2005 was in the United Arab Emirates with 57 per cent preceded by Panama (64%), Mexico (65%) and Saudi Arabia (66%), compared with the lowest of 31 per cent in 1991. On the whole, improvement in the rate of enrolment at secondary level has been quite remarkable between 1991 and 2005, and though the major achievement in this regard was, again, confined to the industrialized countries, it was not much less impressive so far as the non-industrial countries are concerned.

Lack of data on a very large scale for the 85 countries with middle human development in 1991, with no data for 68 countries, makes assessment very difficult as well as incomplete. Nevertheless, with the majority of the countries providing data on enrolment at secondary level for 2005, 63 out of 85, we can get a clear insight into the existing situation, revealing wide differences in achievement between 1991 and 2005. For example, in 1991, in 11 out of 17 countries with data, the rate of enrolment did not exceed 40 per cent, and in only 2 countries it was above 60 per cent, Guyana having the highest of 67 per cent while the lowest went down to 15 per cent in Lesotho-Logo, and with another 3 countries having under 20 per cent enrolment. Overall, it was a very poor show which, however, seems to have improved by 2005, 26 out of 63 countries with data having a rate of enrolment ranging between 61 and 80 per cent, and in 7 countries exceeding 81 per cent. The Palestinian Territory held the highest rate of enrolment at 95 per cent, followed by Kazakhstan with 92 per cent. In the remaining 30 countries the rate fell below 60 per cent, going down to the lowest of 15 per cent in 2 countries, Uganda and Mauritania. China and India do not provide any data for both 1991 and 2005. The rate being above 61 per cent for majority of the countries - 33 in all – indicates a remarkable improvement of the situation between 1991 and 2005.

There is no data for 19 out of 22 low human development countries for 1991 and for 8 countries for 2005. The 3 countries with data on enrolment at secondary level in 1991 present a very poor record, the rate falling to the lowest of 5 per cent in two of them – Niger and Mali — the highest being 7 per cent in Rwanda. There has been no remarkable change from this position by 2005 when in 8 countries out of 22 countries providing data, the rate of enrolment did not exceed 20 per cent, the lowest being 7 per cent in Mozambique and 8 per cent in Niger. In 6 countries the rate was above 21 per cent, the highest being 28 per cent in Ethiopia, followed by Nigeria with 27 per cent. Thus secondary school enrolment in low human development countries does not present a very encouraging situation.

A comparative assessment of the status of secondary education across the countries of the world makes it clear that the rate of enrolment at secondary level generally increases with the level of human development of the country, and progress made in this respect between 1991 and 2005 appears to be confined more to raising the percentage further in countries of high and medium human development. This progress is especially in the high human development countries, where rates of enrolment ranging between 81 and 100 per cent occur in almost 75 per cent of them with data for 2005. This may be considered as a significant achievement for HHD countries, a great advantage for them in eliminating or rooting out inequality in many areas, unlike that in MHD or LHD countries where comparatively low achievement in rates of enrolment acts as a deterrent in coping with mounting problems.

Furthermore, the proportion of researchers engaged in research and development in a country is an unfailing indication of the prospects held in different areas of activities directly contributing to development in social and economic fields which enhances the welfare of the people. Sadly, lack of data for a large number of countries –17 out of 70 HHD countries lacking data – is a great hindrance to get a complete picture (Table 4.4). However, the data on the number of researchers per 1,000,000 people shows 21 countries including Tonga, Finland, Iceland, Sweden, Japan and Denmark, having more than 5,000 researchers for every 1,000,000 people of these countries, or in other words, 5 per thousand of the total population of each of the six countries are engaged in research work, and one of them, Tonga, keeping the highest of 45,454 persons directly engaged makes a rather doubtful exception in view of its small population (a Polynesian island in the midst of Pacific, the projected population of Tonga is going to be 107,000 in 2015). The proportion of researchers per 1,000,000 people of the others are Finland (7,832), Iceland (6,807), Sweden (5,416), Japan (5,287), and Denmark (5,016), indicating the dominance of Scandinavian countries. There are 11 countries in which the proportion ranges between 2001 to 5,000, and 4 of them — Norway, the United States, Luxembourg

Table 4.4: Researchers Engaged in Research and Development 1990-2005

(per million people)

HDI Rank	*Total Number of Countries*	*0-0*	*51-100*	*101-200*	*201-500*	*501-1000*	*1001-2000*	
High Human Development:	70	1	1	0	8	5	5	
Without data:						2001-5000	Above 5000	
17 countries						11	21	
Middle Human Development:	85	0-50	51-100	101-200	201-500	501-1000	1001-2000	2001-5000
		0	7	5	7	6	3	4
Without data:							Above 5000	
53 countries							0	
Low Human Development:	22	3	0	0	0	0	0	0
Without data: 19 countries							0	

Source: Calculated from data provided by UNDP Human Development Report 2007/2008: Table 13

and Singapore having more than 4,000 head for each of their million population engaged in research, but this goes down to below 2,000 though exceeds 1,001 in 5 countries, and Seychelles, another small island country in the Indian ocean, has the lowest of 15 heads per million engaged in research and development.

In comparison, countries with medium human development present a greater hurdle to make a correct assessment since 53 countries provide no data. Of the remaining 32 countries, the majority maintain a much lower profile, the number engaged in research and development ranging from 51 to 500 in 19 countries, falling below 101 per million population in 7 countries while in only 4 countries it exceeds 1,001, the highest of which,1,927 heads, occur in Jordon, followed by Uzbekistan with 1,754 heads and Iran with 1,279 heads. Both China (708) and India (119) lag far behind with Pakistan (75) trailing further behind, and Madagascar with 15 heads for every million people of its country rests at the bottom of the pile.

Lack of data for 19 out of 22 countries with low human development pose similar problem in assessing their position in this regard. However, the available data for three countries – Mozambique (0.6), Burkino Faso (0.2) and Zambia (0.0) – present the most pathetic picture of an almost non-existent participation in this regard.

In short, involvement in research and development remaining largely confined to the people in countries with high human development, with greater concentration in the advanced countries, those in Scandinavia taking the lead as in many other areas, is a clear reflection of apparent domination of these countries in enrolment in primary and secondary levels of education, indicating a healthy continuing education in majority of the cases, keeping a parity between research on one side and primary-secondary education on the other, both remaining inextricably interconnected. In contrast, a very small fraction of total population engaged in research and development elsewhere seems to be also in parity with similar small number of enrolments at primary-secondary levels in countries with medium and low human development holding back human welfare.

However, so far as gender parity is concerned at all the three levels of education – primary, secondary and tertiary (science and development) – it seems to be quite promising as shown in Table 4.5. Parity in primary education has already been achieved in Latin America and the Caribbean, East Asia and Pacific, CEE/CIS countries and the developed regions on the whole. In contrast, in Sub-Saharan Africa, South Asia and Middle East and North Africa, the females still lag behind the males in enrolment at all levels of education – primary to tertiary except for the developed region in tertiary education, but parity has been achieved between three other regions and the developed regions (barring the CEE/CIS) at the secondary level, while leadership of the females in 'politics, the economy and administration' has made it more advantageous for them to take the lead almost at all levels of education in Latin America and the Caribbean, East Asia and Pacific, CEE/CIS and the developed regions resulting in a large 'brain drain' of females as mentioned earlier.

Interestingly, "research on education and economic growth has shown that failing to invest in girls' education lowers Gross National Product (GNP)... Hill and King estimate that, everything else being equal, countries in which the ratio of girls' to boys' enrolment in primary and secondary education is less than 0.75 can expect levels of GNP that are roughly 25 per cent lower than countries in which there is less gender disparity in enrolments. More recently, Abu-Gharid and Klassen (2002) report that countries that fail to meet the goal of gender parity in education will face considerable costs, both in foregone economic growth and in reductions in fertility, child mortality and malnutrition. They estimate that countries off-track in female primary and secondary enrolment might lose 0.1 – 0.3 percentage points in annual income growth

Table 4.5: Gender Parity in Education at Primary, Secondary and Tertiary Levels: Female to Male Ratio in 2005

Region	*Primary Education*	*Secondary Education*	*Tertiary Education*
Sub-Saharan Africa	0.9	0.8	0.6
South Asia	0.9	0.8	0.7
Middle East and North Africa	0.9	0.9	1.0
Latin America & the Caribbean	1.0	1.1	1.1
East Asia and Pacific	1.0	1.0	0.9
CEE/CIS	1.0	0.9	1.2
Developed Regions	1.0	1.0	1.2

Source: Compiled from data provided by UNIFEM, 2008/2009: Fig. MDG 3.4

between 1995 and 2005 and an average of 0.4 percentage points between 2005 and 2015" (UN: Millennium Project, 2005: 47)

A. EDUCATION AND DEVELOPMENT

Human rights enshrined in international law, most importantly in the Universal Declaration of Human Rights admits that "everyone has the right to a standard of living adequate for the health and well-being for himself and of his family, including food, clothing, housing, medical care and necessary social services, and the right to security in the event of unemployment, sickness, disability, widowhood, old age or other lack of livelihood in circumstances beyond his control". Furthermore, it is also admitted that "everyone has the right to education. Education shall be free, at least in the elementary and fundamental stages. Elementary education should be compulsory. Technical and professional education shall be made generally available and higher education shall be equally accessible to all on the basis of merit" (Sachs, 2005: 253-254).

There may be some ambiguity in defining the 'standard of living'. As Amartya Sen *et al.*, observe,"there are many fundamentally different ways of seeing the quality of living, and quite a few of them have some immediate possibility" (Hawthorn, 1998:1). But the standard of living included as one of the major themes in the Declaration of Human Rights alludes to a level of living standard achieved by people which enables them to attend to their basic needs, as mentioned above, fulfilling requirements of a life giving comfort and security including family members, with facilities for education available on a wide front taking a very important place as a positive factor equipping family members for utilizing their potential. Sachs (2005) thinks that certain measures taken by the society are likely to strengthen efforts for advancing the cause of education. For example, 'meals for all children at the primary school level could improve the health of the school children, the quality of education, and the attendance in school'. As already has been experimented in India, a measure like this has proved to be highly successful, arresting the number of drop-outs in an effective way. Second, 'expanded vocational training for the students could teach them the skills of modern farming (for example, using improved fallows and fertilizer), computer literacy, basic infrastructure maintenance (electrical wiring, use and maintenance of diesel generator, water harvesting, bore-well construction and maintenance), carpentry, and the like'. Further, in the rural areas the adults could be trained in hygiene, malaria control, HIV/AIDS, mobile phone use along with a number of other technical enormously pressing issues, coming directly for the use of the rural population. The role of barefoot doctors in China could be emulated seriously in developing countries where the ratio between patients and physicians is abnormally low, falling to 2 per 100,000 people in Tanzania and Malawi (UNDP, 2007: 249-250), in the absence of proper educational facilities churning out more physicians.

Education can be considered as a "form of wealth" and educated people as 'human capital' because their 'skills are valuable' and therefore "a country is helped by having a critical mass of educated people. The countries of bottom billion are already desperately shorn of qualified people, and the situation is likely to get worse. The flight of the skilled is at its most rapid in precisely those bottom-billion environments where there is most scope for change" (Collier, 2008: 93-94). This refers to the exodus of educated people needed most in countries they are leaving for better opportunities elsewhere, affecting countries of their origin, such as, India, Sri Lanka, Pakistan, Thailand, and so on.

Education reinforces the socio-economic dynamics of a society towards equality in attainments and opportunities for its people, making it to be the best social investment and this is proved by instances available from various countries. According to a report of the United Nations, made in 1992, a study conducted in 31 countries established that a farmer completing four years of elementary education can increase his productivity 8.5 per cent higher on average than that of a farmer without education. Evidence of higher production in agriculture is available from India where the adoption and spread of the Green Revolution, in the early years, was faster among the educated farmers. In industry, most evidence suggests that at enterprise level educated workers are more productive. A study of 37 middle-income and 29 low-income countries showed that a 1 per cent difference in primary enrolment ratios was associated with 0.035 difference in per capita income growth rates. Another study has shown that an increase in literacy rates from 10 to 60 per cent made a 2.8 per cent increase in the income share of the poorest 40 per cent of the population. Interestingly, the poor countries get much greater rates of return than rich countries from investments in education, making primary education the source of highest rates of return. Evidence available from African countries indicates that the estimated rate of return on primary education was 26 per cent in comparison to 17 per cent for secondary education and 13 per cent for higher education (Government of India, 2002: 49).

Significantly, "tertiary education reflects the level of human capital formation, resulting in higher returns to education, thereby also to high rates of economic growth. The poorer regions of the world form a very small percentage of those enrolled for tertiary education and not surprisingly, university graduation ratios are the highest in North America and Western Europe (NAWE) at 33.1 per cent followed by Central and Eastern Europe (CEE) at 29.3 per cent, and some countries in East Asia and Pacific (EAP) region in 2004. However, women from NAWE and EAP account for 30 per cent each and in CEE about 18 per cent, indicating gender gaps in regions with generally higher educational attainment" (Viswanathan, 2007: 377). There are also significant gaps in the type of tertiary education attained indicating that "women are largely enrolled in non-science courses (as is evident from instances observed in admitting female students in Oxford University, mentioned earlier) such as education, social sciences, business and law and

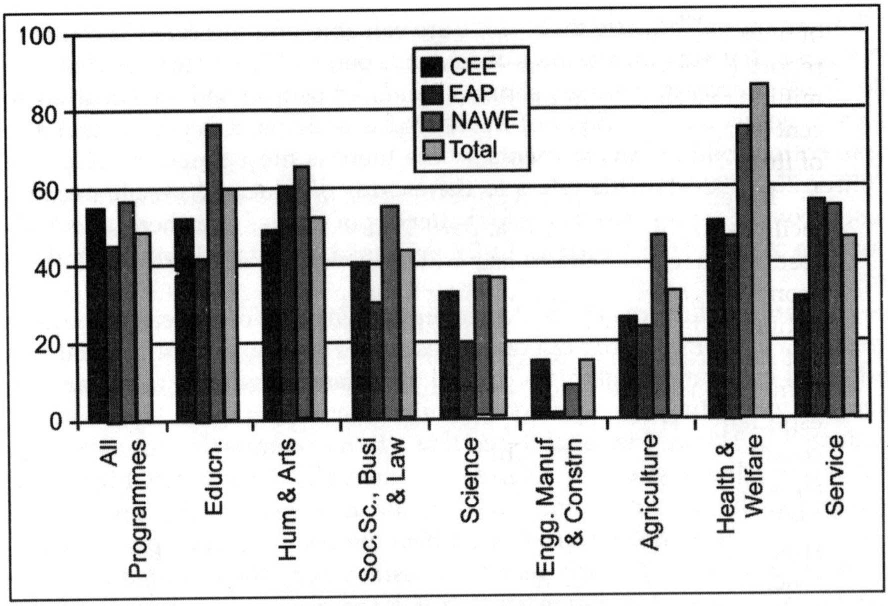

Fig. 4.1: Females as Per cent of Total Tertiary Enrolment, 2004

Notes: CEE: Central and Eastern Europe;
EAP: East Asia and Pacific
NAWE: North America and Western Europe

Source: Brinda Viswanathan, 2007

health services accounting for 90 per cent for the total of all countries (Viswanathan, 2007: 377) (Fig. 4.1)

Despite the claim made by the Planning Commission, Government of India 2002, that there has been a significant reduction in inequalities in educational attainments across gender, castes, income level and between rural and urban areas, it is, rather, very disappointing to note that even after more than sixty years' efforts, nearly one-third or close to 300 million children aged 7 years and above are illiterate (Government of India, 2002: 51). Free and compulsory education for all children until they complete the age of 14 years was enshrined as a right in the Indian Constitution some 60 years or so ago, but could not be achieved even at the end of the Tenth Five Year Plan (2002-2007) – making setting elementary education as a target for all by 2010 afresh. According to the estimate of the government, made in 2003, out of approximately 200 million children (which is rather conflicting with its earlier estimate) only 120 million were in school and the net attendance at the primary level was only 66 per cent of the enrolment (Government of India, 2003: 17). According to another report of the government "the total number of enrolment at elementary education level increased from 159 million in 2001-02 to 182 million in 2004-05, an increase of over 23 million". The number of out-of-school children, according to this report, (Eleventh Five Year Plan) has reduced from 32 million in 2001-02 to 7 million in 2006-07 and the Eleventh Plan (2007-2012) targets to achieve 80 per cent literacy rate, and reduce gender gap in literacy to 10 per cent (Government of India 2008:2, 13) (*See* Note *A*).

In Sen's opinion, social opportunities, in the form of education and health facilities 'facilitate economic participation'. Japan provides a very authentic case in this regard where basic education had taken a great role in 'enhancing economic growth', the same thing happening in other countries in East Asia. In his words, "while pre-reform China was deeply skeptical of markets, it was not skeptical of basic education and widely shared health care. When China turned to marketization in 1979, it already had a highly literate people, especially the young, with good schooling facilities across the bulk of the country. In this respect, China was not very far from the basic educational situation in South Korea or Taiwan, where too an educated population had played a major role in seizing the economic opportunities offered by a supportive market system. In contrast, India had a half-literate adult population when it turned to marketization in 1991, and the situation is not much improved today... The social backwardness of India, with its elitish concentration in higher education and massive negligence of school education... left that country poorly prepared for a widely shared economic expansion" (Sen, 2008: 41-42)

An adequately supportive social background, including 'high level of literacy, numeracy and basic education' have done remarkably well in spreading economic opportunities in many East and South East Asian countries, such as Japan, South Korea, Taiwan, Hong Kong, Singapore, post-reform China, Thailand and other countries among which India presents a case highly diverse in terms of human development, with some regions, especially Kerala, having much higher levels of education than Bihar, Uttar Pradesh, Rajasthan, and Madhya Pradesh (including present Chhattisgarh), having experienced a faster reduction in income poverty than other states.

Further, general expansion of education and literacy facilitates reduction in fertility and mortality. The impact of education, especially female education, is great on reducing fertility rates, as high fertility rates are detrimental to the quality of life, particularly of young women who, with their empowerment obtained through education and often with outside employment can influence family decisions on the size of the family, directly related to the number of children they are going to have. For instance, Iran has achieved one of the world's fastest transitions to low fertility, with its total fertility rate falling from 6.6 between 1980 and 1985 to 2.5 between 1995 and 2000. This has mostly been possible for a 'tremendous increase in girls' participation in school and literacy. Iran is not alone in education and demographic transition. Islamic countries, such as Egypt, Jordon, Morocco and Tunisia have all experienced tremendous increase in girls' schooling and significant decline in total fertility rates' (Sachs, 2005: 317), (*See* Note *B*).

The two neglected areas of social opportunity in India – gender equality and elementary education – are now receiving more attention, and as evidenced by a comparative study of nearly 300 districts in India, it emerges that women's education and women's employment are the two most important influences in reducing fertility rates and the mortality rates of children, with clear indications showing that 'the effect of female literacy on child mortality is extraordinarily large'. "For instance, keeping other variables constant, say, 22 per cent (the actual figure for India) to 75 per cent reduces the predicted value of under-five mortality for males and females combined from 156 per thousand (again, the actual 1981 figure) to 110 per thousand" (Sen, 2008:197-198). The great effectiveness of women's education and employment is conclusively proved in India, where the richest states of Punjab and Haryana have much higher fertility rates than southern states with much lower per capita income but with much higher female literacy and female employment. Further, women's improved educational status is thought to be associated with better bargaining power with effect on the fertility rate, with great significance in view of the facts revealed by the Family Health Survey conducted in the early 1960s that a large majority of pregnant women at that time did not want to be pregnant (Mukhopadhyay and Garimella, 2001: 99-101).

Referring to Sen (2008), we get an idea about a very interesting case of comparison between China and the State of Kerala in India in that Kerala's infant mortality rate is 'somewhat ahead of Chinese average', affected by high levels of basic education, healthcare and due to the fact that Kerala has some other favourable factors including women's empowerment and recognition by legal tradition of women's property rights. The infant mortality rate per thousand live births in Kerala (16 for girls 17 for boys) is much lower than China's (33 for girls and 28 for boys, as influenced by widespread abortion of female fetuses) even though both regions had similar infant mortality rates around the time, in 1979, when the one-child policy was introduced in China. In contrast, sex-selective abortion was absent in Kerala (Sen, 2008: 221)

Instances of education having positive effect on lowering fertility rates are provided by another state in India, Tamil Nadu, which has radically reduced fertility rates while many of the north Indian states – Uttar Pradesh, Bihar, Madhya Pradesh, Rajasthan – with much lower levels of education, especially female education, have high fertility rates, between 4.4 and 5.1 per woman in comparison to 1.8 in Kerala and 2.2 in Tamil Nadu.

There is substantial evidence to prove that schooling and healthcare are inseparable issues, and this is very closely found in cases of malnutrition in early life, affecting both 'cognitive and physical development' of a child which, though, can be reversed in later life through better nutrition. As has been found by some authors through investigation, children with malnutrition often start school late and complete fewer years of schooling. Nutritional supplements to pre-school children have shown permanent improvement in physical stature and cognitive development. A case study in Bolivia providing 'support for daycare, nutritional inputs, and pre-school activities for low-income children aged 6 months or for more than one year demonstrated "permanent gains in cognitive development"... which can raise lifetime earnings (Orazem, 2007: 249). Similar results were obtained in Jamaica on stunted infants and low-income rural households in the Philippines. It has also been found (Vermeersch and Kremer, 2005) that "producing free breakfast to pre-schoolers raised attendance by 30 per cent in Kenya... Deworming medication with an iron supplement for pre-schoolers in India also raised attendance and physical stature" (Orazem, 2007: 249) and in Kenya this increased school attendance by 0.15 years per student, implying a cost of $3.50 per child-year of schooling. National programmes are also beneficial to children at later stage, aged 13-15, and it is reported that giving iron supplements to them in a low-income country 'can raise cognitive abilities by 5 to 25 per cent, or the equivalent of 0.05 years of schooling'. Health reasons are not the primary one holding back children from attending school. Nevertheless, interventions in the form of nutritional supplements or deworming provision are likely to improve attendance to an appreciable measure and steps other than these needed to be taken towards making education universal are eloquently, though indirectly expressed in explaining the nature of the situation responsible for non-attendance (Table 4.6)

As is clearly evident from Table 4.6, 'lack of interest' seems to be the predominating reason, a dangerous signal, no doubt, irrespective of regions across the world for not attending school, which keeps back over three-fourths of the non-attendee children away from education in the urban areas of North Africa and Middle East, the same reason accounting for nearly two-thirds of non-attending school children in the urban areas of Central Asia and Europe. Poverty appears as the second most important reason keeping children away from school, the highest proportion being in South East Asia closely followed

Table 4.6: Reasons given for not attending school in urban and rural areas

Reasons	*All World Regions*		*Sub-Saharan Regions*		*North Africa & Middle East*		*Central Asia & Europe*		*South-East Asia*		*Latin America & the Caribbean*	
	Urban	*Rural*	*Urban*	*Rural*	*Urban*	*Rural*	*Urban*	*Rural*	*Urban*	*Rural*	*Urban*	*Rural*
Work outside home	7.4	4.2	3.3	1.8	0.7	0.7	9.3	7.8	8.7	4.4	18.3	10.0
Housework	7.3	11.5	5.3	7.9	5.6	9.9	6.3	9.3	10.7	19.7	11.7	17.9
Inadequate school supply	1.9	4.9	1.8	3.2	2.0	6.2	1.3	3.0	1.7	2.7	2.6	10.8
Poverty	18.2	18.1	24.1	23.9	4.6	3.4	1.3	0.8	24.2	26.3	11.9	11.3
Health reasons	6.3	5.0	7.9	76	1.2	0.5	0.7	0.4	1.5	0.9	9.4	4.2
Others	11.5	12.3	12.4	12.9	9.3	9.9	16.0	20.5	4.0	4.3	12.1	12.13
Total	100.0	100.0	100.0	100.0	100.0	100.0	100.0	100.0	100.0	100.0	100.0	100.0

Source: Peter F. Orazem, 2007. (Taken from compilations provided by Elizabeth King based on data from demographic and health surveys, various years)

by Sub-Saharan Africa while housework happens to be the third cause blocking the way to education for children, more in Latin America and the Caribbean and South Asia than in other regions. Apart from making education, at least, at the primary level free and compulsory through regulation, actual achievement of this goal depends largely on providing positive economic incentives to parents as a direct way of motivation in encouraging them to send their children to school which in a large measure depends on putting greater emphasis on awareness programmes for getting over the existing lack of interest involving a considerable section of the population of every region.

Needless to say, a child denied the opportunity of elementary schooling is 'not only deprived as a youngster, but also handicapped all through life'. Educated people, as observed earlier, are considered as 'human capital', and as observed by Sen "through education, learning and skill formation, people can become much more productive over time, and this contributes greatly to the process of economic expansion... If education makes a man more efficient in commodity production, then this is clearly an enhancement of human capital. This can add to the value of production in the economy and also to the income of the person who has been educated. But even with the same level of income, a person may benefit from education – in reading, communicating, arguing, in being able to choose in a more informed way, in being taken more seriously by others and so on. The benefits of education, thus, exceeds its role as human capital in commodity production" (Sen, 2008: 292-294).

Considering the overwhelming importance of education in uplifting human status with its ameliorating effects, greater attention needs to be placed on expanding the base of secondary and tertiary education and on improving the quality of education at all levels. While universal primary education have to be absolute, rather than relative, goals on raising the level of human capital, particularly of the poor, will increase the possibility of halting the intergenerational transmission of poverty and other connected inadequacies. At the same time, greater attention should be given, from the earliest level, to education in science and technology acting as primary tools in the proper utilization of the actual potential of human capability as a direct way to the removal of inequality.

B. HEALTH

Like education, lack of healthcare facilities is another form of deprivation for a very large section of people, mostly concentrated in the developing countries. It is very important to note that education, health, nutrition, safe water and improved sanitation complement each other, investments made in any one contributing to improve the situation in the others. This has a very interesting cumulative effect taking place in a manner of a chain reaction with positive impact on associated fields connecting one with the other. For example, while education promotes better hygiene and increases the use of health services, safe water and adequate sanitation improve health conditions with a reduction in infectious diseases which improve children's nutritional status, increasing their learning capabilities. "Together such interventions contribute to a health transition – from having communicable diseases account for most of country's disease burden to having chronic diseases as the main source. The health transition hastens the demographic transition from high to low birth and death rates. In addition, higher education levels are associated with better family planning. As more children survive, families reduce the number of children they have. Desired family sizes decline, a process helped by the ready availability of contraceptives. So, over time, lower infant and child mortality play a major role in falling fertility rates. This notion of synergies among social investments is central to reducing hunger, malnutrition, disease and illiteracy and to advancing human capabilities": (UNDP, 2003: 85).

Many countries have registered significant health progress during the recent decades though gains have been unevenly shared. For example, regarding under-five children's mortality, overall it has reduced by 6.7 million between 1978 and 2006, equivalent to saving 18,329 lives every day. The rate of decline has been much slower in low-income countries as a whole than in richer countries, and only Eritrea and Mongolia among all the low-income countries may be singled out for having reduced the rate of under-five mortality by as much as 70 per cent. This is still two-thirds or more of the 1975 level in 20 of the 25 countries of Sub-Saharan Africa. Fig 4.2 shows variation in the rate of reduction in under-five child mortality in selected countries between 1990 and

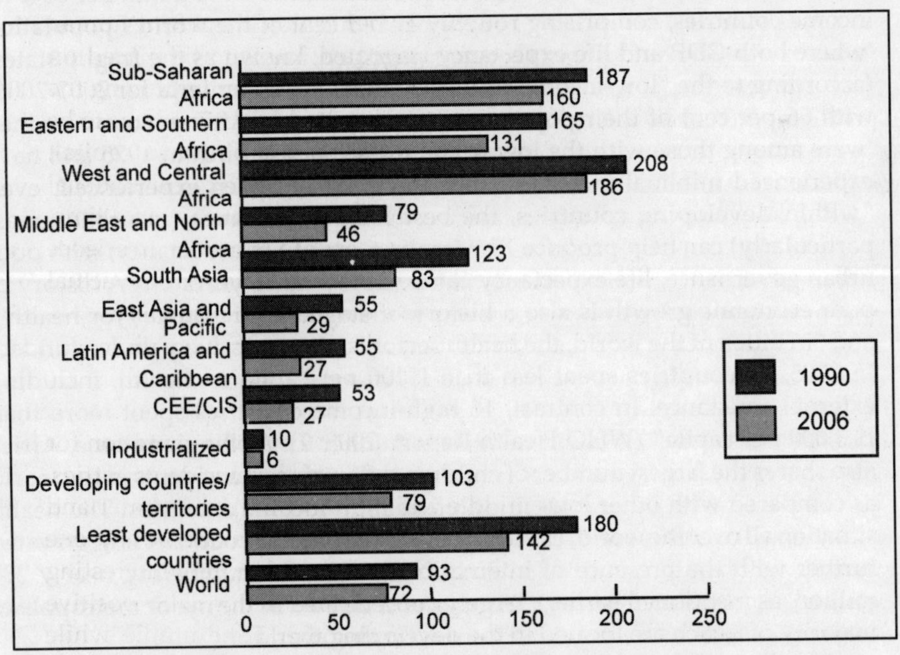

Fig. 4.2: Reduction in Under five Mortality between 1990 and 2006, by Regions

Source: UNICEF, 2008

2006 with corresponding per capita expenditure on health in 2006, ranging between the highest of $ 382 in Oman to the lowest of $ 35 in Madagascar. Disappointing progress in healthcare, particularly in Sub-Saharan Africa is evidenced in the vaccination coverage which, despite recent changes for better, happens to be significantly lower than that in the rest of the world. Further, while the use of contraceptives have increased to about 61 per cent during the past three decades in other developing regions leading to reduction in abortion rates, it 'remains as low as 21 per cent in Sub-Saharan Africa increasing the absolute numbers of abortions', performed mostly in unsafe conditions. Besides, Sub-Saharan Africa is also the only region in the world where childbirth care for mothers and newborns have not made any progress and less than half of all births are attended by skilled health personnel in 33 developing countries, going down to as low as 6 per cent in one country (WHO Health Report, 2008: 2-4).

The relation between life expectancy and economic growth during the last three decades exhibit distinct patterns, showing the most 'impressive relative gain in life expectancy at birth and considerable economic growth happening in a number of low-income countries in Asia (including India) Latin America and northern Africa 'with an increase of life expectancy at birth by 12 years, while GDP per capita was multiplied by a factor of 2.6'. Similarly, the high-income countries as well experienced a 'substantial economic growth and increased life expectancy' during this period in contrast to a number of low-income countries, comprising roughly 10 per cent of the world's population 'where both GDP and life expectancy stagnated, known as the fragile states' (according to the "low-income countries under stress" criteria for 2003-2006) with 66 per cent of their population concentrated in African countries they 'were among those with the lowest life expectancy at birth in 1975 and have experienced minimal increases since then'. As has been experienced, even "within developing countries, the best local governance (in urban areas, particularly) can help produce 75 years or more of life expectancy; with poor urban governance, life expectancy can be as low as 35 years"... "A history of poor economic growth is also a history of stagnating resources for health", and "in much of the world, the health sector is often massively under-funded. In 2005, 45 countries spent less than I$100 per capita on health, including external assistance. In contrast, 16 high-income countries spent more than I$ 3,000 per capita" (WHO Health Report, 2008: 7). At the same time, Africa also shares the largest number of children dying of road accidents in the world as compared with other low- middle and high-income countries. The health situation all over the world, especially in the developing countries is aggravated further with the presence of international migrants (numbering nearly 200 million, as mentioned earlier), largely concentrated in the major metropolises majority of which are located in the developing world.

The fast urbanizing world, which is already more urban than rural since 2008 with 3.3 billion residing in urban areas and by 2030, almost 5 billion of the global population are going to be concentrated in the urban areas when the major portion of the entire load is likely to be borne by the small cities of the developing countries and the metropolises of unprecedented size in southern and eastern Asia, the world itself is aging fast and will continue to do so. Around 85 per cent of this old population aged over 60 amounting to a total of about 2.6 billion by 2050, will be living in the developing countries, mainly concentrated in the urban areas. "Contrary to today's rich countries, low-, and medium-income countries are ageing fast... adding to the challenge" (WHO Health Report, 2008 :8).

Factors like urbanization, ageing and attendant globalized life-style have combined to 'make chronic and non-communicable diseases – including depression, diabetes, cardiovascular disease and cancer – and injuries' increasingly raising the level of morbidity and mortality, so much so that even in 'the industrialized countries, as many as 25 per cent of 65-69 year olds and 50 per cent of 80-84 years olds are affected by two or more chronic health conditions simultaneously'. The combined effects of globalization, urbanization and ageing will be compounded further by the effect of climate change and more adversely impact on the most vulnerable communities of the poorest countries such as Bangladesh and the like.

Inequity in health services is very much pronounced in comparing the situation across the world. While in Europe almost 93 per cent of the population derive benefit from national health service systems, it still has remained elusive in a country like the United States even though 80 per cent population strongly are in favour of its inclusion. However, similar opportunities are now available in some of the low-, middle-income countries, moving faster than many industrialized countries in providing universal coverage of health services. Countries having achieved this goal in recent years are Costa Rica, the Republic of Korea, Thailand and Turkey.

Estimates of the burden and prevalence of diseases, as mentioned in a recent study, are attributed to malnutrition, poor water supply, sanitation and personal and domestic hygiene, unsafe sex, tobacco use, alcohol, occupation, physical inactivity, use of illegal drugs and air pollution. "Developed regions account for 11.6 per cent of worldwide burden from all causes and disability and account for 90.2 per cent of health expenditure worldwide. Communicable, maternal, perinatal, and nutritional disorders explain 43.9 per cent; non-communicable causes 40.9 per cent; injuries 15.1 per cent; malignant neoplasms 5.1 per cent; neuropsychiatric conditions 10.5 per cent; and cardiovascular conditions 9.7 per cent of DALYs (Disability Adjusted Life Year) worldwide... 15.9 per cent of DALYs worldwide are attributable to childhood malnutrition and 6.8 per cent to poor water and sanitation and personal and domestic hygiene" (Murray and Lopez, 1997:1436).

According to the study mentioned above, in 1990, nearly 90 per cent of the worldwide burden of disease occurred in developing countries, where only 10 per cent of healthcare funds were spent. In terms of overall worldwide burden of diseases, Sub-Saharan Africa and India had the largest proportions

(21.4% and 20.9% respectively) but very small proportions (0.7% and 1.0% respectively) of health expenditure.

There were, however, wide regional variation in the contribution of the three major groups of causes leading to disease burden. Group I causes were 'communicable, maternal, prenatal, and nutritional disorder; Group II causes were non-communicable diseases; and Group III were injuries. In their regional distribution Group I disorders were dominant in Sub-Saharan Africa (65.9% of burden), but were much less important in market economies (7.1%). In China, Group I conditions accounted for 25 per cent of the disease burden, and the total regional burden of Group I for Latin America and the Caribbean, the Asian islands, and the middle-eastern crescent was 35 to 48 per cent. In contrast, Group II disorders were mostly concentrated in market economies, sharing more than 80 per cent of the total while it was less than 20 per cent in Sub-Saharan Africa. On the other hand, Group III diseases remained an important source of burden in all regions, accounting for 12-19 per cent of the total in each region.

As observed by Murray and Lopez (1997), on the basis of their study worldwide, '22.9 per cent of the DALYs were caused by infections and parasitic diseases, the proportion ranging from 2.7 per cent in formerly socialist economies of Europe' to 42.5 per cent in Sub-Saharan Africa. Respiratory infections contributed about 1.4 per cent of the DALYs in established market economies compared with 10.7 per cent in the middle-eastern crescent, and those from maternal disorders varied from 0.3 per cent in the former to 3.2 per cent in the latter. Among Group II disorders, cancers accounted for 5.1 per cent of worldwide burden with 13.7 per cent of the total occurring in developed regions. Worldwide, neuropsychiatric diseases accounted for 10.5 per cent of the DALYs, with 22.0 per cent of them happening in developing regions. Cardiovascular diseases comprising 10 per cent of the DALYs and respiratory and digestive diseases accounting for 3 per cent of the total DALYs worldwide, are two other types of important categories. The musculoskeletal diseases were also the major causes of burden, representing about 5 per cent of the DALYs in all regions except the middle-eastern crescent, established market economies and Sub-Saharan Africa. In comparison, intentional injuries had a wider range, from 1.5 per cent of DALYs in India to 6 per cent in Sub-Saharan Africa and the middle-eastern crescent.

Importantly, however, of the 30 leading causes of diseases, 3 of them held top positions, and they are lower respiratory infections, diarrhoeal diseases and perinatal or low birthweight and birth trauma. "The leading causes in developing regions were similar to leading causes worldwide, except that malaria was the seventh most important cause of burden" (Murray and Lopez, 1997: 1438)

According to their (Murray and Lopez, 1997) estimate in 1990, 819,000 deaths occurred worldwide from hepatitis B, and actually 2.8 million died from it. But malnutrition responsible for the greatest loss (15.9%) of DALYs, followed by the combination of poor water supply, sanitation, and personal hygiene causing 6.8 per cent while unsafe sex and consumption of alcohol, each caused 3.5 per cent of DALYs. Tobacco, on the other hand, was more responsible as a cause for death, killing about 1.8 million people in 1990, representing about 15 per cent of all deaths in the two regions of the established market economies and formerly Socialist economies of Europe alone, and about 12 per cent of DALYs. Tobacco-related mortality is predicted to increase from 3 million deaths in 1990 to more than 8 million in 2020, and worldwide the prediction for increase was just under 40 million in 1990 (2.6% of DALYs) to 120 million by 2020 (a little less than 9 per cent of all DALYs). The proportionate mortality from alcohol was higher in Sub-Saharan Africa than in established market economies. The impact was even greater in Latin America and the Caribbean countries where 5 per cent of the deaths and 8 per cent of the DALYs were from alcohol., causing about 750,000 more deaths worldwide than it averted, with more than 80 per cent of the deaths occurring in developing countries (Murray and Lopez, 1997: 1440).

I. The Killer Diseases and Global Situation

The study of diseases in populations, called epidemiology, focuses on mortality or death and its causes, or on morbidity, that is sickness which can include both disease and its illness, or a subjective experience of the patient to be diagnosed for a 'medically recognized disease' (Gatrell, 2002: 5). The term epidemiology, of Greek origin, "is the science and practice which describes and explains disease patterns in populations, and puts this knowledge to use to prevent and control disease and improve health… The central paradigm of epidemiology is that patterns of disease in populations may be analysed systematically to provide understanding of the causes and control of disease. The key strategy then, of epidemiology is to seek out the differences and similarities (compare and contrast) in the disease patterns of populations to gain new knowledge" (Bhopal, 2002: 3).

Whatever achievements have so far been made in ameliorating the living conditions of people, especially in the developing world, cannot mask the huge deprivations that remain in these areas. About a third of the children under five suffer malnutrition. Nearly 18 million people die every year from communicable diseases, about 30 million from non-communicable diseases. One of the most famous commitments made by the international community in 1978 was "Health for All by the Year 2000", but sadly enough, when 2000 arrived the world was severely gripped "with the AIDS pandemic, resurgent TB and malaria, and billions of the world's poor without reliable, or sometimes any, access to essential health services. At the World Summit for Children in 1990, the world pledged universal access to primary education by the Year 2000, yet 130 million or more primary-aged children were not in school by then" (Sachs, 2005: 213). The WHO Commission on Macroeconomies and Health (CMH) in its report (December, 2001), concluded that, 'poor health causes poverty and poverty contributes to poor health'. It identified "eight areas that accounted for the vast proportion of the disease burden in poor countries: AIDS, malaria, TB, diarrhoeal disease, acute respiratory infection, vaccine-preventable diseases, nutritional deficiencies, and unsafe child birth" (Sachs, 2005: 204).

One of the recent reports of WHO highlights the role of chronic diseases with greater impact in many low-, and medium-income countries which steadily growing, take a toll of 80 per cent of chronic disease deaths. Chronic diseases include heart disease, cancer, stroke, chronic respiratory diseases and diabetes. 'Visual impairment and blindness, hearing impairment and deafness, oral diseases and genetic disorders are other chronic conditions' and it was estimated that they were going to account for 35 million out of a total of 58 million global deaths or 60 per cent of all deaths occurring from all causes in 2005, such as cardiovascular deaths taking a toll of 17,528,000 people, cancer 7,586,000, chronic respiratory diseases 4,057,000 and diabetes 1,125,000 (WHO, 2005: 2-3) The scale of problem faced by India is strongly indicated by the fact that deaths from chronic diseases rising from 3.78 million or 40.4 per cent of all deaths, is likely to reach 7.63 million or 66.7 per cent in 2020 when the country will already be at the "top of the global league table" for the number of people suffering from diabetes which, from an estimated 19.3 million in 1995, is expected to rise to 57.2 million by 2025. China, similarly, is faced with the problem of chronic disease where "300 million" adult males smoke cigarettes and 160 million are hyper-sensitive, making chronic disease death rates among middle-aged population "higher than in some high-income countries". According to this report, globally 'each year at least 4.9 million people die as a result of tobacco use; 2.6 million die as a result of being overweight or obese; 4.4 million die due to raised total cholesterol levels and 7.1 million die of raised blood pressure. Significantly, there are 1.000 million people who are overweight and 388 million people will die in the next ten years of a chronic disease unless urgent action is taken (WHO, 2005: 6-7).

Fig. 4.3 shows the global distribution of the main causes of death in 2005, in which the share of death from diabetes is underestimated as usually living for years with diabetes often death is recorded to have been caused by heart disease or kidney failure both, as a matter of fact resulting from diabetes. On the other hand, Table 4.7 shows the rates of projected chronic diseases for four age groups of males and females, indicating the number of deaths as similar between males and females for all age groups except for those 70 years old or above among whom it becomes higher for females, reaching the highest for people of that age group. Further, people of that age-group are much more affected, increasing with age, such that, for age-group 0-29 years it is 0.7 per cent of the total (the total number of deaths is 135 times more than this), for the next age-group of 30-59 years its is 4.8 per cent (or the total is 21 times more of that), and it is 3.8 per cent for people 60-69 years (or the total is almost 30 times more of the oldest age-group - 70 and above years) which makes older people much more vulnerable to death from chronic diseases as a result of less physical resistance.

Table 4.7: Projected Chronic Disease Deaths Worldwide Numbers and Rates by Age and Sex, 2005

Age group	*Number of Deaths (Millions)*			*Death Rates per 100,000*		
	Males	*Females*	*Total*	*Males*	*Females*	*Total*
0 – 29	0.8	0.8	1.6	48	47	48
30 – 59	4	3	7	372	251	311
60 – 69	4	3	7	2328	1533	1911
70 and over	9	11	20	6981	6102	6467
TOTAL ALL AGES	18	17	35	556	543	549

Source: WHO, 2005. Preventing Chronic Diseases: 39

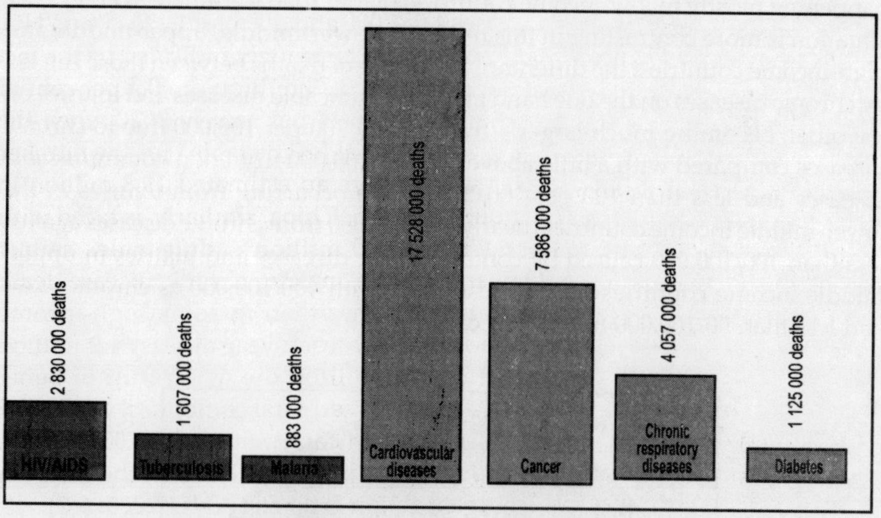

Fig. 4.3: Projected Global Deaths by cause for all ages, 2005

Source: WHO 2005

The geographical distribution of chronic disease deaths makes interesting revelations, such that 37 per cent of all these deaths will be concentrated in lower-middle income countries and 35 per cent in low-income countries, 20 per cent in high-income and 8 per cent in middle-income countries. In other words, "four out of five chronic disease deaths are in low- and middle-income countries… Many people believe that low- and middle-income countries should control infectious diseases before they tackle chronic diseases. In reality, low- and middle-income countries are at the centre of both old and new public health challenges. While they continue to deal with the problems of infectious diseases, they are experiencing a rapid upsurge in chronic disease risk factors and deaths, especially in urban settings. These risk levels foretell a devastating future burden of chronic diseases in these countries" (WHO, 2005: 8-9). As a matter of fact, poor people are much more likely than wealthy people to develop chronic diseases, poverty further deepening the crisis, and contrary to general belief, 'almost half of chronic disease deaths occur prematurely among people under 70 years of age and one-quarter of all chronic disease deaths occur in people under 60 years of age'.

Figures 4.4 and 4.5 indicate the concentration of the (projected) main causes of death in countries of different income groups, the rate per 100,000 progressively accelerating with fall in income, making it obvious that chronic diseases take a much larger role as compared with communicable diseases etc., and injuries as causes of death among all income groups as well as in global perspective. For example, while deaths taking place at the rate of 500 per 100,000 people from communicable diseases and such like, it is almost 750 per 100,000 from chronic diseases and a little more than 100 per 100,000 from injuries happening mostly in low-income countries, higher than the global average. The situation is more contrasting in this respect in lower-middle, upper-middle and high-income countries, the difference in the rate of deaths between those caused by chronic diseases on the one hand and communicable diseases and injuries on the other, becoming much larger – from nearly 700 per 100,000 due to chronic diseases compared with a little above 100 per 100,000 from non-communicable diseases and less than 100 per 100,000 people occurring from injuries in the lower-middle income countries (nearly 600/100,000 from chronic diseases against less than 100/100,000 caused by communicable diseases and injuries in upper-middle income countries contrasted further with 400/100,000 as chronic cases and less than 50/100,000 from other causes).

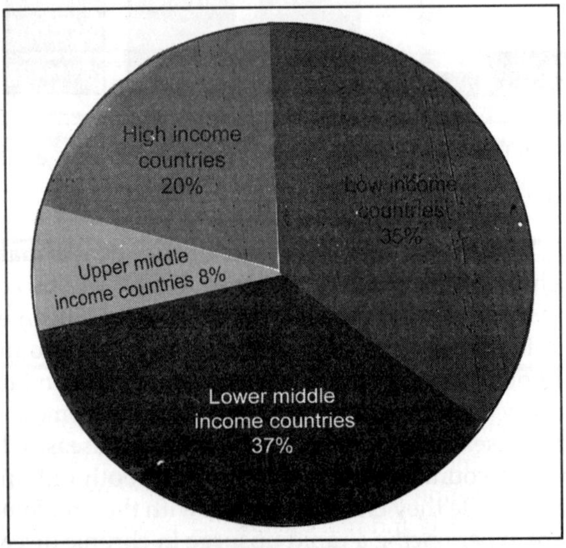

Fig. 4.4: Projected global distribution of Chronic disease deaths for Ages by Economic Group

Source: UNICEF, 2005

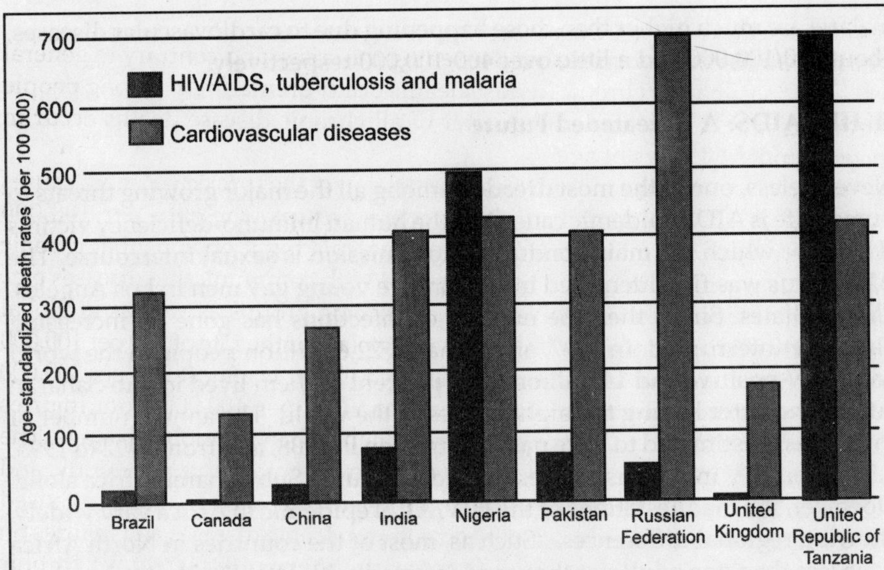

Fig. 4.5: Projected death rates by specific cause for selected countries for all ages, 2005

Source: UNICEF, 2005.

As pointed out in WHO, 2005 Report on health, the problem of chronic disease has a very serious impact on the society, affecting the quality of life of individuals, resulting in premature – often unexpected – deaths with severe economic consequences for the families as well as societies involved According to WHO, 2005 Report, China is likely to suffer a loss of national income "estimated at $558 billion over the next ten years as a result of premature deaths caused by heart disease, stroke and diabetes" (WHO, 2005: 5). As a matter of fact, the rate of death caused by chronic diseases such as cardiovascular diseases alone far outstrip the rate of deaths due to communicable diseases like HIV/AIDS, tuberculosis and malaria taken together as projected by WHO, 2005 though some selected countries are great victims of the latter. For example, apart from Nigeria and the United Republic of Tanzania, the rate of death from cardiovascular diseases is much higher in all countries, including China, India and Pakistan, remaining overwhelmingly ahead in each case, such as over 300 per 100,000 in Brazil against about 20 per 100,000 from communicable diseases; over 100/100,000 in Canada against hardly 5/100,000 from others; over 400/100,000 in India against 75/100,000; while the sharpest contrast occurs in the Russian Federation where cardiovascular deaths far outstrip deaths from HIV/AIDS, Tuberculosis and malaria taken together, amounting to about 700/100,000 against almost 50/100,000. On the other hand, Tanzania presents an entirely opposite picture where deaths from HIV/AIDS, Tuberculosis and malaria are much higher than those happening due to cardiovascular diseases, about 700/100,000 and a little over 400/100,000 respectively.

II. HIV/AIDS: A Threatened Future

Nevertheless, one of the most dreaded among all the major growing threats to human life is AIDS epidemic caused by the human immuno-deficiency victims (HIV), for which the main conduit of transmission is sexual intercourse. The AIDS virus was first identified in 1981 in five young gay men in Los Angeles, United States. Since, then the number of infections has gone on increasing almost uninterrupted. In 1997, an estimated 22.6 million people in the world were HIV-positive and 14 million or 63 per cent of them lived in Sub-Saharan Africa, the latter having the highest rates in the world. The annual number of infections is estimated to have passed 1 million in 1988, and from 1992 to 1995, 1.5 million HIV infections were estimated to occur in Sub-Saharan Africa alone. However, the manifestations of the HIV/AIDs epidemic in Africa vary widely, showing regional differences. "Such as, most of the countries in North Africa have less than one adult per thousand infected with HIV. Similarly, the island countries of Mauritius and Comoros have very low prevalence rates. In West Africa, most of the countries have intermediate levels of infection (less than 30 adults in a thousand). The countries with the highest prevalence rates (with some exceptions) are geographically clustered in what has been called the 'AIDS belt of Sub-Saharan Africa'. They are a chain of countries stretching from middle Africa through East Africa and Southern Africa and include the Central African Republic, Southern Sudan, Uganda, Kenya, Rwanda, Burundi Tanzania, Zambia, Malawi, Zimbabwe and Botswana" (United Nations, 1999: 447). Among them the five leading countries with higher rates of prevalence are Botswana (164.5 per 1,000); Zimbabwe (158.8 per 1,000); Zambia (156.6 per 1,000); Uganda (133.1 per 1,000); and Malawi (122.8 per 1,000) (United Nations, 1999: 449). The spatial clustering of countries with high levels of HIV infections is not fully understood though according to some authors (Caldwell and Caldwell, 1996), circumcision and prevalence of chancroid and STDs (sexually transmitted diseases) have some tenuous support from available data (United Nations, 1997: Fig. 133, p. 448).

According to the cumulated cases reported by the WHO, by region, in 1997, worldwide occurrence of AIDS gives a quite different impression in its geographical distribution, the Americas leading with 48.5 per cent of the cases in the world, followed by Africa with 35.1 per cent, Europe with 11.8 per cent, Asia with 4.3 per cent and Oceania 0.5 per cent of the total HIV infections (United Nations, 1997: 448). As the single largest infectious killer and the fourth leading cause of death in the world, HIV infection and AIDS were first detected in India in 1986, first in Maharashtra and then in TamilNadu. In the year 2000, the number of Indians infected with HIV was estimated at 3.86 million, exceeded only by South Africa, the prevalence rate of about 0.7 per cent of the general adult Indian population, quite low compared to the prevalence rates of 25 per cent and more in South Africa, Zimbabwe and Botswana. However, mortality due to AIDS is increasing in India, nearly 300,000 Indians are estimated to have died of AIDS in 1999 alone. In its geographical distribution, the data available for 2000 clearly demonstrate the great variation in the severity of the HIV epidemic among the states of India. Andhra Pradesh, Maharashtra, Karnataka, Tamil Nadu, Manipur and Nagaland are currently classified as high prevalence states where infection levels have crossed the one per cent-mark among antenatal women. Unprotected sex appears to be the predominant source of transmission, accounting for nearly 83 per cent of the total (Misra *et al.*, 2003: 87-88)

The results of various models suggest that HIV infection would increase sevenfold over a period of 25 years, and by 2015, there would be about 70 million cases in Sub-Saharan Africa. At the same time, the number of AIDS deaths as a proportion of deaths from all causes were to increase from 7.6 per cent in 1995 to 12.9 per cent in 2005. According to an estimate, between 1995 and 2000, AIDS would have caused about 620,000 deaths annually in Sub-Saharan Africa (United Nations, 1997: 462).

There is no denying the fact that the AIDS pandemic has ravaged Africa more than any other place in the world but yet the reasons for Africa's extremely high prevalence rates "are not well understood or agreed". The simplest explanation often given is that Africans have more sexual partners, and therefore, more risk of transmitting the disease. But this conclusion is refuted very strongly by 'The Lancet', a leading medical journal, on the basis of its recent surveys which found the number of casual sexual partners (five or more) in, say, Thailand and Rio de Janeiro, was more than that of African countries. Perhaps, the time of having multiple sexual partners or the number of African migrant male workers living away from their families for long periods or the extent of male circumcisions may explain such differences (Sachs, 2005: 322-323)

In terms of HIV and AIDS-related causes, WHO estimates that about 6.4 million people have died of AIDS since the epidemic began in 1981. The estimates given by the Harvard based HIV/AIDS experts at the 'Center of Health and Human Rights' are, however, different, the number of AIDS-related cumulated deaths put at 9.2 million worldwide as of January, 1996, and that 82 per cent of these, or 7.6 million, took place in Sub-Saharan Africa.

As reported in the UNDP Human Development Report, 2002, by the end of 2002 almost 22 million people had died from AIDS, 13 million children had lost their mothers or both parents to the disease and more than 40 million people are living with the dreaded HIV virus – 90 per cent of them in developing countries, 75 per cent in Sub-Saharan Africa. As mentioned earlier, in Botswana, the most affected country (with 164.5 per 1,000 persons) more than third of adults have HIV/AIDS and a child born today can expect to live only 36 years – about half as long as if the disease did not exist,. In Burkino Faso, the 20th most affected country, 330,000 adults are living with HIV/AIDS, and life expectancy has fallen by 8 years (*See* Note C).

The toll on life expectancy is only one way of sufferance. In Thailand, for instance, 'one-third of the AIDS affected rural families saw their 'incomes reduce by half because the time of farmers and those caring for them, was taken from the fields'. At the same time, medical expenses shoot up. In Cote d' Ivoire, caring for a male AIDS patient costs an average of $300 a year, a quarter to half of the annual net income of most small farmers affecting very hard on family consumption and expenses on schools for children. In the Caribbean as well HIV/AIDS is of great concern, Latin America having 1.3 million people with HIV/AIDS. Central and Eastern Europe and the CIS have fast-rising infection rates – 240,000 people are infected in Ukraine alone. And there are warnings that Asia is on the verge of an epidemic. In Ho Chi Minh city, Viet Nam, one sex-worker in five is HIV positive while there was none in the mid-1990s. It is pointed out that without preventive measures, as in Thailand, the epidemic could go out of control (UNDP, 2002: 27).

Beginning its ravages in early 1980s, the number of HIV infected people has reached 42 million by now, killing 22 million people and leaving 13 million orphans in its wake. As mentioned earlier, HIV/AIDS is crippling parts of Africa, with about 1 in 3 (or more) adults infected in Botswana, Lesotho, Swaziland, and Zimbabwe; 1 in 5 in Namibia, South Africa and Zambia, and more than 1 in 20 in 19 other countries of Africa, respecting neither rich or poor, including teachers, farmers, factory workers and civil servants. In 1998, Zimbabwe lost as many as 1,300 teachers and by 2020, the hardest-hit African countries could lose more than a quarter of their work-forces (UNDP, 2003:43). Though Sub-Saharan Africa accounts for nearly 70 per cent of HIV/AIDS cases, the epidemic is making unrelenting pressure on countries beyond Africa. Such as, nearly 0.5 million people are infected in the Caribbean, 1.2 million in East Asia, 1.2 million in Eastern Europe and the CIS, 1.5 million in Latin America and 6.0 million in South Asia. Already, about 7 million people are infected in India, China and the Russian Federation, and the figures are likely to rise in the coming years, projected to reach about 200 million by 2025 in these three regions as it has done in Sub-Saharan Africa, increasing from 7 million to 25 million within a space of a decade, reducing life expectancy rapidly by 35 years in Zimbabwe, 28 years in Botswana and Swaziland, and 24 years in Lesotho (Fig. 4.6). Table 4.8 gives an idea about the expected increase in the incidence of HIV/AIDS in China, India and Russia and its impact on life expectancy.

Table 4.8: Threats from HIV/AIDS and Reduction in Life Expectancy

Country	*Expected HIV/AIDS by 2005*	*Expected Reduction in Life Expectancy (in years)*
China	70 million	8
India	110 million	13
Russia	13 million	16

Source: UNDP 2003: Table 2.4

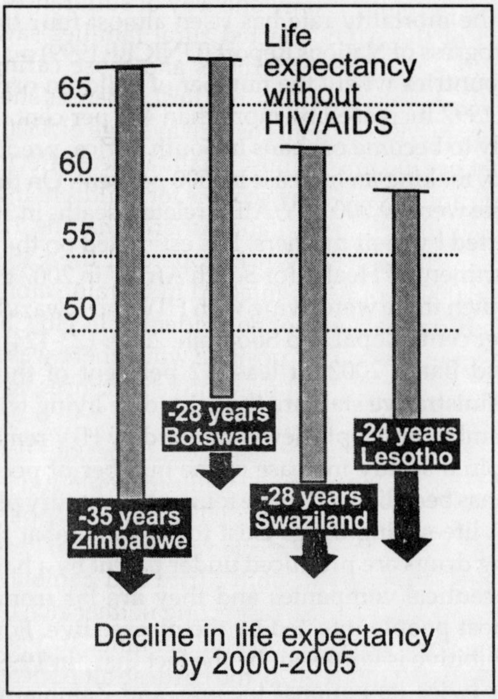

Fig. 4.6: Loss of Life expectancy due to HIV/AIDs in selected countries of South Africa

Source: UNDP, 2003

Gopal and Sookrajh (2009), quoting Taylor (1998) state that developing countries account for about 90 per cent of new infections of which Sub-Saharan Africa bears the largest burden, account for over two-thirds of HIV/AIDS patients, comprising nearly 21 million men, women and children, and 83 per cent of the world's deaths from the disease has taken place in Sub-Saharan Africa. Referring to the UN AIDS 2000 Report on the global HIV/AIDS epidemic, Sub-Saharan Africa is reported to have 4 million new infections in 1999, making it the leading cause of death. In 2000, 23.4 million Africans were estimated to have acquired this infection, 3.8 million of them infected during that year and 19 countries with the highest HIV prevalence in the world are in Africa. The UN AIDS 2000 further informs that in 7 out of 11 studies conducted, more than 1 woman in 5 in her early 20s was infected with the virus, and will die before reaching the age of 30 years.

According to Karim and Karim (2003), the record of South Asia is not less horrifying in this respect as about 5 million out of a total of 43 million people in South Asia are infected by this virus which is almost 15 per cent of the HIV positive in the world (Myeza and Francis, 2009: 111) with a prevalence rate of 24.8 per cent at the end of 2001, while the national incidence in antenatal clinic attendees has been about 6.8 per cent for the past five years. In young women, aged 20-30 years, the mortality rate has risen almost four times during this period. The 1999 Progress of Nations Report (UNICEF,1999) makes South Africa one of the seven countries where the number of children orphaned by AIDS between 1994 and 1997 increased by more than 400 per cent. In 2002, 300,000 children were likely to become orphans in South Africa, predicted to increase to 2 million by 2015, making an increase by 600 per cent. On the other hand, in Kwazulu-Natal there were 80,000 HIV/AIDS related deaths in 2001 when 40,000 children were infected by their mothers. It is estimated on the basis of a study made by the 'Department of Health for South Africa' in 2007 that 30.2 per cent of the pregnant women there were living with HIV and Kwazulu-Natal headed the list with 39.1 per cent (Gopal and Sookrajh, 2009: 123-124). According to a report of the World Bank, 2002, at least 12 per cent of the South African educators and administrative staff are thought to be living with HIV.

Notably, the number of people newly infected by HIV remains at 4 million per year, resulting in a steady increase of the number of people afflicted by HIV. The epidemic has been likened to the fourteenth century plague that swept Europe except that life-saving drugs exist today to combat the HIV impact. But these life-saving drugs are produced under patent by a handful of US and European pharmaceutical companies and they are far from affordable for countries where most people infected by the disease live. For example, how far the price is prohibitive is borne out by the fact that "in 1999, it would have cost Kenya at least twice its national income, and Zambia more than three times to provide treatment (Table 4.9). As a result, just 0.1 per cent of the 25 million people with HIV/AIDS in Sub-Saharan Africa have access to these life-saving drugs" (UNDP, 2001: 106).

However, generic versions of antiretrovirals are being produced far below global prices by manufacturers in Brazil, Cuba, India and Thailand. In February, 2001, the Indian company Cipla offered three-drug combination therapy at $600 a person per year to governments and $350 to Non-Governmental

Table 4.9: The Affordability Gap of Treating AIDS in 1999

	Switzerland	*Kenya*	*Uganda*	*Zambia*
Population	7 million	30 million	23 million	10 million
People with HIV	17,000 (0.24%)	2,100,000 (7%)	820,000 (3.6%)	870,000 (8.7%)
Cost of treating all infected people with antiretroviral drugs at global market prices, about $12,000 a person per year	204 million	25 billion	10 billion	10 billion
Cost of treatment as % of GDP	0.08	238	154	336
Public Healthcare spending as % of GDP, 1998	7.6	2.4	1.9	3.6
Total Healthcare spending as % of GDP, 1998	10.4	7.8	6.0	7.0

Source: UNDP 2001 (Taken from UN 2001; Hirschel 2000; World Bank 2001; UNAIDS 2006)

Organizations (NGOs), and the price is likely to fall to $200-250 with time, widening up the scope of treatment possibilities. In 1993, Brazil began producing the drugs and distributing them free, saving lives and money, reducing the number of deaths by half by 1996, while the government saved $422 million (UNDP, 2001: 106-107)

Young people infected by HIV accounted for 40 per cent of new infections in 2006, and 48 per cent of people living with HIV were women. Encouragingly, the number of patients in developing countries with access to antiretroviral HIV/AIDS treatment has increased in recent years though Uganda is the only Sub-Saharan country to have begun to reverse the epidemic once it reached crisis proportions. In Zambia, HIV proportions among women fell 4 percentage points between 1996 and 1999, offering hope of it becoming the second country to reverse the crisis. Senegal is another success story, having kept HIV/AIDS under control from the beginning through an immediate concerted response (UNDP, 2003: 43). In spite of the fact that the number of infected people in need of therapy is rising at a faster pace than treatment services, the proportion of young people having sex before age 15 has declined in Sub-Saharan Africa and the use of condoms has increased in 2006. "In addition some progress has been made in the development of new technologies for prevention. Critically, additional financial resources are being made available to fight the pandemic. Developing countries and their array of development partners in this area should spare no effort to achieve the goal of universal access to HIV prevention programmes, treatment, care and support by 2010 called for by the General Assembly in the 2005 World Summit Outcome" (United Nations, 2007: 252).

Very highly relevant to HIV/AIDS which cannot be ignored is physical violence and systematic harassment becoming a common experience with people living with the disease. In many societies some sort of stigma and prejudice work against these people, the company shunned by others in the society out of sheer ignorance about how far the disease could be infectious, and putting infected people to shame and disgrace, treating them almost as social outcasts. Some amount of social education is needed to dispel such wrong notions among people in general.

III. Malaria and Tuberculosis

One of the top killers among communicable diseases, malaria, is largely confined to the developing countries, infecting 500 million people every year, a little more than 10 per cent of the world's population, mostly children, one of them dying every 40 seconds 'resulting in a daily loss of more than 2,000 young lives worldwide'. Potentially, a fatal protozoan disease transmitted by a specific kind of mosquito, the genus anopheles, malaria is "utterly treatable", yet incredibly it still claims up to 3 million lives a year, mostly young children, about 90 per cent of whom live in Africa. The rest of the deaths occur in tropical regions of the Americas and Asia. Low-cost treatments exist but they do not reach the poor "causing great concern for the fact that according to current estimates, malaria causes up to 5 billion clinical cases per year. Virtually everybody in tropical Africa contracts the illness at least once a year. In some places, the entire population lives year-round with the malaria parasite in their bodies (although without clinical symptoms much of the time)" (Sachs, 2005: 196-197) The answer to the question – why malaria situation is so much worse in Africa, may be found in the fact that malaria and poverty are 'intertwined', and as such poverty certainly exacerbates malaria by impoverished households and governments without the financial means to fight the scourge. "Where malaria prospers most, human societies prosper least. The global distribution of Gross Domestic Product (GDP) in 1995, adjusted for purchasing power, shows a striking and unmistakable correlation between malaria and poverty. Poverty is concentrated in the tropical and subtropical zones, the same geographical boundaries that most closely frame malaria transmission. The extent of correlation suggests that malaria and poverty are intimately related. In fact a comparison of income in malarious and non-malarious countries indicates that average GDP (adjusted to give parity of purchasing power) in malarious countries in 1995 was US $1,526, compared with US $8,268 in countries without intensive malaria – more than a five-fold difference. Malaria-endemic countries are not only poorer than non-malarious countries, but they also have lower rates of economic growth. Between 1965 and 1990, countries in which a large proportion of the population lived with ... malaria experienced an average growth in per-capita GDP of 0.4 per cent per year, whereas average growth in other countries was 2-3 per cent per year" (Sachs and Malaney, 2002: 680-685).

According to Sachs (2005), Africa's vulnerability to malaria is greater than other regions because of its "unique disease environment", since "malaria is largely a tropical disease, and if warm weather is a pre-requisite of it, Africa has it". In this connection Sachs has made a very important observation on transmission of the disease. According to him, "the force of transmission of malaria in Africa is ... roughly nine times that of India because of difference in mosquito species". In explaining further the behavioural character of the mosquito he says, "some types of mosquitoes prefer to bite people whereas others feed off cattle. Transmitting malaria requires two consecutive human bites: the first for the mosquito to inject the parasite and the second for the mosquito to infect another person, roughly two weeks later. If the mosquito feeds frequently on cattle rather than on people, the odds are that at least one of the bites, if not both, will be taken from the cattle. In India, for example, the predominant type of anopheles tends to bite humans about one third of the time, and cattle the rest. Africa, sadly, has another predominating type which prefers human biting nearly 100 per cent of the time. Mathematically, the chance that an Indian mosquito will feed off two humans in a row is about one out of nine, whereas in Africa it's about one out of one, or nearly certain" (Sachs, 2005: 198-199).

Malaria thrives best in the tropics, reaching into the temperate zones, but the transmission becomes less likely with temperature falling below 18 C and malaria parasites cease development completely at temperatures below 16 C. The antimalarial efforts taken between 1946 and 1994, have been more successful in eradicating the disease in temperate regions of the world than in the tropics, the latter with increasing concentration of the disease, so that Africa alone accounts for 90 per cent malaria mortalities. During the first half of the twentieth century malaria had been terminated from the United States and from most of Europe. This could be achieved through changes in 'land use, agricultural practices, house construction and vector control. As a matter of fact, the great success of the global eradication programme carried on in the 1950s and 1960s achieved in India, Sri Lanka and former Soviet Union was largely possible due to widespread use of the insecticide DDT but the momentum could not be sustained and the eradication programme ultimately failed later on partly for the resistance in many countries to the repeated use of DDT in and around houses and also because of the loss of interest in pursuing it after the elimination of malaria from the United States and Europe. There was a period of about 25 years between the early 1970s and the late 1990s when even the industry lost interest in the development of insecticides to combat the menace as is clearly evident from the fact that only 3 out of 1,223 new drugs developed during the period 1975-1996 were antimalarial (Greenwood and Mutabingwa, 2002: 670-672). All this naturally led to a deteriorating situation with greater mortality and morbidity, especially in Sub-Saharan Africa. Nevertheless, India's efforts at eradicating malaria "brought down the case load from an estimated 75 million to a record low of 100,000 cases in the 1960s with negligible deaths " (Misra *et al.*, 2003: 58). This would be worth emulating. Very dismayingly, however, the stupendous success was very short-lived, the disease resurfacing as a result of utter complacency in the matter of keeping up the vigil. Cases of reappearances reported first from Assam in a very short space rose to 6.47 million in 1976 in the country. Concerted action in coping with the situation brought down the number of malaria patients to about 2.18 million in 1984 and the effect continued through the years that followed though several outbreaks of the disease were once again witnessed in 1994 'resulting in high mortality' partly due to the emergence of parasites resistant to available drugs. Further, the appearance of a more dangerous strain of malaria (Pf or Plasmodium falciparum) increasing incidence from 38.84% to 49.96 per cent between 1995 and 1999 came to be a cause of greater concern. Of the average number of 2 to 3 million malarial cases, 10 states share about 93 per cent of the total burden in India, of which two states – Madhya Pradesh and Orissa – together account for 50 per cent of mortality (Misra *et al.*, 2003: 58).

Inadequate drug supplies limit the access of people, particularly at the lower level of economy in even countries of medium economic development, and according to WHO, the range between countries with access to essential drugs is huge: in China 80-94 per cent of the population has access while in India it is 0-49 per cent, again with wide range. For example, in four southern states in India – Andhra Pradesh, Karanataka, Kerala and Tamil Nadu – the health facilities function better because drugs are distributed through the primary healthcare network (UNDP, 2003:103).

Though poverty accounts for 'some of the intense malaria transmission in poorer countries', the economic class of people makes little difference in its impact as is frequently evident from people belonging to wealthier classes in malarious region easily falling victim to this parasite. On the other hand, "personal expenditure on prevention methods, such as bednets or insecticides, increased funding for government control programmes, and general development such as increased urbanization can reduce malaria transmission... combined efforts at environmental management such as the draining of swampland acting as breeding ground of certain vector mosquitoes and indoor spraying of residual insecticides such as DDT" (Sachs and Malany, 2002: 680-685) have an effective role in the elimination process.

Tuberculosis takes an important place among the dreaded killers, infecting 60 million people every year. Even more than fifty years after the introduction of effective chemeotherapy, tuberculosis still kills almost 2 million people every year, making it along with AIDS, 'the leading infectious killer of adults worldwide with its toll rising. In fact, between 1997 and 1999, the number of new tuberculosis cases rose from 8.0 to 8.4 million and this trend continuing will make it one of the leading causes of adult mortality beyond 2015 (UNDP, 2003: 100). According to Sachs (2005), "ten thousand Africans die needlessly and tragically every single day – and have died every single day since 9/11, 2001 (when the World Trade Center was struck by extremists) of AIDS, TB and Malaria (Sachs, 2005: 215), deaths which are preventable".

Reversing these trends appears to be possible and for tuberculosis remarkable progress has been achieved in formulating a plan called DOTS or 'Directly Observed Therapy' a short treatment course which has proved to be effective in halting the spread of the disease. To keep up the effectiveness of DOTS and expand its service requires increasing political and financial support and health-system capacity for procuring a sustainable supply of quality drugs. There may also be instances of fighting drug-resistant types of tuberculosis as in Russia where the incidence of tuberculosis increased by 300 per cent between 1990 and 1996 with a substantial proportion of the cases turned out to be drug-resistant, which involved the implementation of "DOTS plus" to overcome the resistance.

Table 4.10 shows the occurrence of tuberculosis in countries according to their level of human development in 2005. Prevalence of this dreaded disease seems to maintain a low profile in the countries with high human development, with patients not exceeding 100 per 100,000 people in 65 out of a total of 70 countries while the number ranges between 101 and 200 per 100,000 people in only 5 countries, the highest of 150/100,000 occurring in the Russian Federation followed by 146/100,000 in Romania. The number of people suffering from tuberculosis, however, increases keeping some sort of parity with the level of human development. In countries of medium human development in only 43 out of a total of 85 countries, the number of tuberculosis cases remains under

Table 4.10: Prevalence of Tuberculosis (per 100,000 people) Worldwide, 2005

HDI Rank	*Total Number of Countries*	*Number of Patients*								
		0-5	*6-10*	*11-20*	*21-50*	*51-100*	*101-200*	*201-500*	*501-1000*	*above 1000*
1. High Human Development	70	10	12	13	18	12	5	0	0	0
2. Middle Human Development	85	0	3	1	11	18	11	28	11	2
3. Low Human Development	22	0	0	0	0	0	1	9	12	0

Source: Calculated from data provided by UNDP 2007/2008: Table 9

100/100,000, but exceeds 200/100,000 in 39 countries, and goes above 500/100,000 in 13 countries, recording the highest of 1,211/100.000 in Swaziland followed by 1,161 in Djibouti, over five to six times higher than the highest record of the HHD country. A similar parity is maintained in the prevalence of tuberculosis in countries with low human development where, barring one country (Benin with 144/100,000), in all other 21 countries the number of tuberculosis cases varies between 200 and 1,000 per 100,000 people, exceeding 500/100,000 in 12 or majority of the countries, the highest reaching 905/100,000 in Sierra Leone (which, incidentally, is much lower than the records of three middle-income countries, including Kenya (936/100,000)).

The close relevance between the levels of human development and prevalence of tuberculosis cannot be ignored, indicating some way of deliverance from the grip of this killer disease.

However, it is highly significant to note that the character and role of causes affecting human mortality is fast changing with change in life-styles and habits of people. This is very clearly mirrored in the historical changes of human mortality pattern as associated with the five epidemiological transitions in the past and future human societies. The three transitions that occurred in the past took place from— (i) external injuries as the major causes of death in the hunting and gathering society, to (ii) infectious diseases in the agricultural society, and from the former to the degenerative diseases like (iii) cardiovascular diseases in the industrial society. The future transitions are (iv) to cancers as the major causes of death in the high-technology society, and (v) to senescence or old-age frailty in the society of coming ages. But mortality from both cancer and senescence is likely to decline in near future and be, in all likelihood, replaced by three main causes of death or mortality resulting directly from changes in people's life-styles. They are smoking and consumption of alcohol and road accidents (U.N., 1999: 54-55).

IV. Tobacco, Alcohol and Injuries

As observed by Murray and Lopez (1997), 'tobacco was already a major cause of DALYs (disability-adjusted life year or the sum of life lost due to premature mortality and years lived with disability adjusted for severity) in established market and formerly socialist economies of Europe where "tobacco was estimated to have caused about 1.8 million deaths in 1990... The effect on mortality of the massive increase in tobacco consumption in developing countries, particularly during the1970s and 1980s, may well cause tobacco-related mortality to increase from 3 million deaths in 1990 to more than 8.4 million by 2020, 6 million of which would be in the developing world, making smoking by far the leading cause of death in the world in that year. Worldwide DALYs attributed to tobacco are predicted to rise from just under 40 million (1990, 2.6% of all DALYs) to 120 million (2020, just under 9% of all DALYs) (Murray and Lopez, 1997: 1440).

According to Lopez, 'in Western countries, smoking among males was associated with a 70 to 80 per cent excess mortality from all causes'. Increasing the relative risk of contracting lung-cancer cardiovascular disease, 70 to 80 per cent of the smokers dying of the latter before age 50, make smoking responsible for 'three-fold increase in mortality during middle age'. Further, smokers are '10 to 12 times more likely to die' of lung cancer. Lopez estimated that, in 1995, 1.9 million deaths due to smoking occurred in the developed countries, and it was 1.6 million in the developing countries, half of which occurred in China alone, though the distribution of smoking-related deaths has changed fast since then. WHO estimated that "by the mid-1990s there were about 1.1 billion daily smokers in the world or about one-third of the world's population (at that time) aged 15 years or over. Eight hundred million of those smokers were male and 700 million lived in developing countries" (UN, 1999: 27). Almost half (47%) of all men aged 15 and above smoked while only about 12 per cent women were smokers. In its geographical distribution, 37 per cent of men as regular smokers were concentrated in developed market-economy countries compared with 60 per cent in the countries with economies in transition and in China. The prevalence of smoking was ... high in other developing countries' for example, it was 40 per cent among men in India, Latin America, North Africa, Western Asia, and 54 per cent in other parts of Asia.

The impact of unhealthy life-styles, particularly in the wealthy societies with market economies is demonstrated clearly by the fact that in 1990, 24 per cent of all male deaths and 7 per cent of female deaths were due to smoking, as observed by Peto *et al.* (1994) and lung cancer, related to smoking, rose by 170 per cent on average for men between 1950-54 and 1980- 90 and by 230 per cent for women during this period. One needs to be cautious here that for populations of developing countries as a whole, nearly 40 per cent of adult mortality based on registered deaths giving causes of mortality does not reveal the real picture when almost 60 per cent of the deaths cannot be traced to the cause of death.

However, smoking is undoubtedly one of the causes of death in both developed and developing countries, though with marked regional variations, with 37 per cent of the regular male smokers of the world concentrated in the established market economies compared with about 60 per cent of men in the former socialist economies of Europe and China. As shown in Table 4.11, the Republic of Korea tops the list of countries with 68.2 per cent of its male population as regular smokers, and in altogether 8 countries, over 60 per cent of the males are regular smokers. Interestingly, only two of them are in Europe (the Russian Federation which also includes Asiatic regions, and Latvia) followed by another 14 countries where the male smokers range between 50 to 59.3 per cent of the total population. In comparison, the proportion of females as regular smokers falls far below that of the males in each of these countries, rising to the highest of 30.6 per cent in Fiji (against 59.3% for males) and 30 per cent in Russian Federation (against 67.0% for males). Smoking prevalence is considerably higher for males in many of the large developing countries, like China (61%), Indonesia (53%), India (40%), Brazil (39.9%), compared to some of the highly developed countries like the United Kingdom (29%), the United States (28.1%) and Germany (36.8%), though Japan with 59 per cent for the males was much ahead of many developing countries in this respect.

Among the 88 countries reported in Table 4.11, Benin has the lowest record of 8.9 per cent of the male adults addicted to smoking, while in Sweden females outrank males (with 24% and 22% respectively) as smokers and in Denmark males and females match each other with 37 per cent of each used to regular smoking. Further, the proportion of males and females involved in smoking is very close to each other in countries like Norway (36.4% males and 35.5% females), Canada (31% and 29% respectively), Iceland (31% and 28%), Ireland (29% and 28%), the United Kingdom (29.5% and 28%), New Zealand (24% and 22%) and the difference between the two sexes in this respect is quite negligible in some other countries, like Netherlands, Switzerland, Germany, Luxembourg, the United States and Finland. All of them with moderately high prevalence of smoking among both men and women are in the industrialized world, and taking the number of people belonging to both sexes together gives them a quite high percentage of the population involved in smoking. Apparently, however, in spite of prevalence of high percentage of male smokers in the developing countries, the presence of very low percentage of females as smokers in them (China with 7%; Sri Lanka with 0.8%; Indonesia with 4.0%; Thailand with 4%; India with 3%; Uzbekistan with 1%, Pakistan with 4.4%; Nigeria with 6.7%) keep the overall percentage of people engaged in smoking appreciably lower as compared with that of developed countries.

The estimate of tobacco-smoking mortality based on Peto-Lopez method (1991), as given in Tables 4.12 and Table 4.13, shows that, between 1950 and 2000, there had been about 62 million deaths in the developed countries (Europe, including the former USSR, Australia, Japan, North America and New Zealand) caused by tobacco smoking, 52 million of which were men. In the developing countries, smoking is causing 1.6 million deaths each year, predominately for men (1.4 million), and half of the total occurs in China alone (UN, 1999: 400-401). According to Murray and Lopez (1997), as mentioned earlier, the total number of deaths per year caused by tobacco smoking is likely to rise to 8.4 million by 2020, of which 6 million will occur in the developing world.

Table 4.11: Estimated smoking prevalence among men and women aged 15 and above, by selected countries ranked in order of male smoking prevalence

Rank	*Country (year of survey)*	*Men*	*Women*
1.	Republic of Korea (1989)	68.2	6.7
2.	Latvia (1993)	67.0	12.0
2.	Russian Federation (1993)	67.0	30.0
4.	Dominican Republic (1990)	66.3	13.6
5.	Tonga (1991)	65.0	14.0
6.	Turkey (1988)	63.0	24.0
7.	China (1984)	61.0	7.0
8.	Bangladesh (1990)	60.0	15.0
9.	Fiji (1998)	59.3	30.6
10.	Japan (1994)	59.0	14.8
11.	Sri Lanka (1988)	54.8	0.8
12.	Algeria (1980)	53.0	10.0
12.	Indonesia (1986)	53.0	4.0
12.	Samoa (1994)	53.0	18.6
15.	Saudi Arabia (1990)	52.7	N.A.
16.	Estonia (1994)	52.0	24.0
16.	Kuwait (1991)	52.0	12.0
16.	Lithuania (1992)	52.0	10.0
16.	South Africa (1995)	52.0	17.0
20.	Poland (1993)	51.0	29.0
21.	Seychelles (1989)	50.9	10.3
22.	Bolivia (1992)	50.0	21.4
23.	Albania (1990)	49.8	7.9
24.	Cuba (1990)	49.3	24.5
25.	Bulgaria (1989)	49.0	17.0
26.	Thailand (1995)	49.0	4.0
27.	Spain (1993)	48.0	25.0
28.	Mauritius (1992)	47.2	3.7
29.	Greece (1994)	46.0	28.0
29.	Papua New Guinea (1990	46.0	28.0
31.	Israel (1989)	45.0	30.0
32.	Cook Islands (1988)	44.0	26.0
33.	Jamaica (1990)	43.0	13.0
33.	Philippines (1987)	43.0	8.0
33.	Slovakia (1992)	43.0	26.0
37.	Cyprus (1990)	42.5	7.2
38.	Austria (1992)	42.0	27.0
39.	Malaysia (1986)	41.0	4.0
39.	Peru (1989)	41.0	13.0
41.	Uruguay (1990)	40.9	26.6
42.	Argentina (1992)	40.0	23.6
42.	France (1993)	40.0	27.0
42.	Hungary (1989)	40.0	27.0
42.	India (1980)	40.0	3.0
42.	Iraq (1990)	40.0	5.0

(Contd.)

Table 4.11 (*Contd.*)

Rank	*Country (year of survey)*	*Men*	*Women*
42.	Malta (1992)	40.0	18.0
42.	Mongolia (1990)	40.0	7.0
42.	Uzbekistan (1989)	40.0	1.0
50.	Brazil (1989)	39.9	25.4
51.	Egypt (2986)	39.8	1.0
52.	Morocco (1990)	39.6	9.1
53.	Lesotho (1989)	38.3	1.0
53.	Mexico (1990)	38.3	14.4
55.	El Salvador(1988)	38.0	12.0
55.	Italy (1994)	38.0	26.0
55.	Portugal (1994)	38.0	15.0
58.	Chile (1990)	37.9	25.1
59.	Guatemala (1989)	37.8	17.7
60.	Denmark (1993)	37.0	37.0
61.	Germany (1992)	36.8	21.5
62.	Norway (1994)	36.4	35.8
63.	Honduras (1988)	36.0	11.0
63.	Netherlands (1994)	36.0	29.0
63.	Switzerland (1992)	36.0	26.0
66.	Colombia (1992)	35.1	19.0
67.	Costa Rica (1988)	35.0	20.0
67.	Slovenia (1994)	35.0	23.0
69.	Swaziland (1989)	33.0	8.0
70.	Luxembourg (1993)	32.0	26.0
71.	Singapore (1995)	31.9	2.7
72.	Belgium (1993)	31.0	19.0
72.	Canada (1991)	31.0	29.0
72.	Iceland (1994)	31.0	28.0
75.	Australia (1993)	29.0	21.0
75.	Ireland (1993)	29.0	28.0
75.	United Kingdom (1992)	29.0	28.0
78.	U.S.A. (1991)	28.1	23.5
79.	Pakistan (1980)	27.4	4.4
80.	Finland (1994)	27.0	19.0
81.	Turkmenistan (1992)	26.6	0.5
82.	Nigeria (1990)	24.4	6.7
83.	Paraguay (1990)	24.1	5.5
84.	Bahrain (1991)	24.0	6.0
84.	New Zealand (1992)	24.0	22.0
86.	Sweden (1994)	22.0	24.0
87.	Bahamas (1989)	19.3	3.8
88.	Benin (1990)	8.9	6.6

Source: UN 1999 (Taken from WHO: Tobacco or Health: A Global Status Report, WHO, 1997a)

NA = Not Avilable.

Table 4.12: Estimated Death Caused by Smoking in the developed world, 1950-2000

Mid-Decade Year	*Men*			*Women*		
	Number of deaths in mid-decade year	*Percentage of all deaths*	*Percentage of deaths at ages 35-69*	*Number of deaths in mid-decade year*	*Percentage of all deaths*	*Percentage of deaths at ages 35-69*
1955	447,000	10	20	26,000	<1	2
1965	793,000	17	28	70,000	2	4
1975	1,119,000	21	31	165,000	3	7
1985	1,369,000	24	35	317,000	6	11
1995	1,442,000	25	36	476,000	9	13
Total of all deaths caused by smoking	52 million	20	30	10.5 million	4	7

Source: UN, 1999 (Taken from R. Peto, A.D. Lopez, J. Borehan, M. Thun and C. Health: 'Mortality from smoking in developed countries, 1950-2000,' 1994)

Table 4.13: Estimated Number of deaths from tobacco by Region, 1995

	Men	*Women*	*Total*
Developed Countries	1,440,000	475,000	1,915,000
Established Market Economies	840,000	375,000	1,215,000
Formerly Socialist Economies of Europe	600,000	100,000	700,000
Developing Countries	1,385,000	195,000	1,580,000
China	690,000	120,000	810,000
Other Asia and North Africa	500,000	40,000	540,000
Sub-Saharan Africa	90,000	5,000	95,000
Latin America and the Caribbean	105,000	30,000	135,000
World	2,825,000	670,000	3,495,000

Source: UN 1999 (taken from various sources)

Notably, the number of deaths from smoking has increased more than three times between 1955 and 1995 among males which, indeed, seems to be quite alarming, representing one-fourth of all deaths in 1995 (Table 4.12), increasing from one-tenth in 1955 with a steady rise all through the period.

On the other hand, the second human-induced cause of mortality is that of alcohol which, as claimed by Murray and Lopez (1997), has turned out to be 'a major cause of death and disability from injury'. According to WHO,1997, "recorded alcohol consumption is substantially higher in developed countries although the data do suggest that there has been a generalized decrease in consumption in many of the leading wine producing countries (France, Italy, Spain), as well as in Australia, Finland and North America in recent years. Overall, per capita consumption in the 15 European countries has declined by almost 20 per cent since 1980" (UN, 1999: 397). In comparison, alcohol consumption is rising rapidly in developing countries. For example, beer consumption has risen sharply since 1970s as reported in countries like Brazil, China, India, Mexico, Nigeria, South Africa and Uganda. While in Accra almost all alcohol is consumed by men, in Latin America and Africa women as well are significant consumers of alcohol. In general, women consume in less hazardous ways than men, 10-20 per cent of them in many developed countries drinking at hazardous levels, a level common among men in African countries

Excessive alcohol intake leads to the incidence of cirrhosis of the liver, certain cancers, certain types of cardiovascular and traumatic or violent deaths mostly due to accidents. On the other hand, the effect of alcohol consumption is not entirely negative. Consumed at low levels, in a restrained way, alcohol provides 'some protection against ischaemic heart disease', which cannot be ignored since "In established market economies, the protective effects of alcohol probably averted as many deaths as its harmful effects caused. However, most of the averted deaths would have been in older ages, since ischaemic heart disease, the disorder against which alcohol protects, is rare in those younger than 50" (Murray and Lopez, 1997:1440). As estimated by Murray and Lopez, 1.35 million deaths were caused annually by alcohol, of which 620,000 were because of disease and 625,000 out of injuries as affected by alcohol drinking. At the same time, alcohol intake prevented 470,000 deaths that would have taken place due to ischaemic heart disease. Therefore, the net annual global mortality from alcohol consumption is estimated at around 1.25 million of which 640,000 are estimated to occur in the developing countries and of the total number of deaths, 710,000 are males.

There is marked regional variations in the pattern of mortality related to the consumption of alcohol (Table 4.14). For example, the estimated impact of alcohol intake on injuries is particularly high in Sub-Saharan Africa (120,000 male deaths and 15,000 female deaths) per year. In Latin America, 85,000 males and 10,000 females die every year from injuries related to alcohol intake while almost another 100,000 die of injuries generally attributable to alcohol consumption in the EME (Established Market Economies) region, 70 per cent of them are males though in the same region alcohol is claimed to "save" as many lives from ischaemic heart disease as it claims from all other diseases combined (170,000 per year) (UN, 1997: 403). See Note D Table 4.14 gives a comprehensive idea about the number of deaths as well as life saved by region occurring in 1990 caused by alcohol. It can be observed that alcohol-induced diseases take a direct toll of 454,000 males and 165,000 females a year (1990) which is slightly less than the number of deaths caused by alcohol-induced injuries, much higher for males (524,000) than females (102,000). Surprisingly, the effect of alcohol in saving life of people subject to ischaemic heart appears to be quite substantial (See Note E).

Injuries, other than those induced by alcohol, is another leading cause of death particularly among young people, responsible for about half the deaths in the 10-24 year age group. There are seven causes of injuries: (i) RTAs (road

Table 4.14: Estimated Deaths caused (or prevented) by Alcohol use, by Region, 1990

(in thousands)

Region	*Men*			*Women*		
	Injuries	*Diseases*	*"Protected"*	*Injuries*	*Diseases*	*"Protected"*
Established Market Economies	70	115	-99	28	45	-76
Formerly Socialist Economies	84`	51	-62	15	19	-54
India	65	66	-37	16	24	-21
China	60	39	-5	10	17	-6
Other Asia and Islands	34	69	-20	6	21	-13
Sub-Saharan Africa	119	46	-13	16	17	-14
Latin America & Caribbean	85	62	-23	10	18	-15
Middle Eastern Crescent	7	5	-5	1	3	-6
World	524	454	-265	102	165	-206

Source: UN, 1997 (Taken from Murray and Lopez, 1996)

traffic accidents); (ii) war injuries; (iii) self-inflicted injuries; (iv) violence; (v) drownings; (vi) falls, and (vii) burns. Together, it is estimated that injuries account for 10-30 per cent of hospital admissions and about 78 million disabilities per year. For example, injuries are a leading cause of hospitalization in Indonesia and China. According to Murray and Lopez (1997a), more than 3.3 million deaths occurred in 1990 due to road traffic accidents (999,000), self-inflicted injuries (786,000), violence (563,000), drowning (504,000), and war injuries (502,000).

"With particular reference to persons aged 15 to 44, injuries in the developing regions led to 55 million DALYs lost among males in 1990, a third of all DALYs lost from all causes in this age group... For males the largest proportion of DALYs lost from all injuries was in SSA (Sub-Saharan Africa: 46%), followed by FSE (Former Socialist Economies: 40%), MEC (Middle Eastern Crescent:37%) and the smallest proportion was in EME (Established Market Economies: 28%) and India (25%). In SSA, violence (12%) and wars (11%) were the largest causes. Violence was a leading cause of DALYs lost in LAC (Latin America and the Caribbean: 11%) followed by RTAs. In the FSE and EME, RTAs were the leading cause (11% each) followed by self-inflicted injuries (6% and 5% respectively). Wars were the leading cause (14%) in the MEC followed by RTAs (7%)" (U.N, 1999: 419) In contrast, women are lesser victims of injuries because of comparatively less exposure and also because of their larger burden of disease from maternal conditions.

The estimates given in the "Global Burden of Disease" by Murray and Lopez (1997 c) show that deaths from injuries are expected to increase from 5.1 million in 1990 to 8.4 million in the year 2020, and among the deaths the RTAs and self-inflicted injuries would rank among the top causes of death in

2020, taking sixth and tenth position respectively (holding 9th and 12th position in 1990). Though RTA fatalities are gradually declining in the EME countries, African and Asian countries 'with relatively low vehicle densities are experiencing higher RTA fatality rates in proportion to population size than in Europe or the United States'. According to a study of the World Bank (1990), 70 per cent of 600,000 fatalities per year and 15 million injuries occur in lower- and middle-income countries. A more recent estimate made by the World Bank (1993) shows that there were 856,000 deaths from road accidents a year and 76 per cent of them did occur in the low- and middle-income countries where road-traffic comprises buses, trams, trucks, motorcycles, bicycles and other public service vehicles often using the same road-space. Almost 'eighty per cent of the road fatalities are males' and 'between 1968 and 1985, RTA fatalities increased by 150 per cent in Asian countries and by 300 per cent in African countries' in contrast to a decline of 25 per cent over the same period in European countries (*See* Note *F*)

Violence makes another important cause of mortality, particularly with high rates among males aged 15 to 44 years in Sub-Saharan Africa and Latin America and the Caribbean, contributing to 13 per cent and 11 per cent of DALYs lost in Sub-Saharan Africa and Middle Eastern Crescent (Murray and Lopez, 1996 a). Apart from the roles of alcohol, 'relative rather than absolute deprivation' has a very important role in the occurrence of violence. Fire arm-associated mortality happens to play an important role in violence among 15-19 year olds in the United States where 27 per cent of injury deaths result from fire arms (compared with about 1 per cent in the United Kingdom) linked in the US, particularly, to 'easy availability of fire arms, inner city deprivation and misuse of substances'. Violence against women pervading all societies across the barriers of economy, race, religion, education and so forth plays a very significant role in terms of human rights and dignity while war crimes contribute substantially to injury, mortality and morbidity.

Interestingly, apart from social costs, the burden of economic costs for premature suffering and death from all these types of injuries is becoming frightening in nature. A single case of the richest country of the world, the United States, may be a sufficient indication in this regard draining away a fabulous amount of financial resource which could have been much better utilized in making life more comfortable in many of the financially deprived countries. For example, in the United States, according to the National Safety Council (1993), there was a 'morbidity rate of 61 per 100,000, with 150,000 fatalities in all ages, being the leading cause of death to those aged 1to 44. All this led to a national economic cost estimated at US $400 billion. In Malaysia, increasing affluence and rapid industrialization' resulted in a surge of road and occupational injuries, 75 per of them involving people below 45 years of age at a national estimated cost at US $ 725 million, while in Thailand, "the death rate in 1987 from injuries was 33 per cent per 100,000 and 30 per cent hospital beds occupation was due to injuries. The economic costs to the nation was estimated at US $ 1.5 billion" (U.N., 1999: 436). One feels highly frustrated to consider the fact in this regard that while the 'lifetime costs of all injuries that occurred in 1985 in the United States were estimated at $ 158 billion, six times the losses for cardiovascular disease, yet a disproportionately small amount is spent on injury research, amounting only to '$ 160 million' in contrast to '$930 million for cardiovascular research". (*See* Note G)

On the whole, the expansion of healthcare systems and services becomes a more challenging issue, quite a daunting one for the 2 billion people living in the low-growth health economies of Africa and South East Asia as well as for more than 500 million people who live in fragile states. As a matter of fact, the low-growth health economies with 2.6 billion people and less than 5 per cent of world's health expenditure, "suffer from an absolute under-funding of their health sector, along with a disproportionately high disease burden. The persistence of high levels of maternal mortality in these countries – they have close to 90 per cent of all maternal deaths – is, perhaps, the clearest indication of the consequences of under-funding of health on the performance of their health systems... In fragile states, particularly in those located in Africa, health expenditure is not only low but barely growing at all, and 28 per cent of this little amount of growth in recent years is accounted for by external aid" (WHO, 2008: 105). However, health expenditure in other countries of this group is growing at 6 to 7 per cent per year. Health expenditure at this rate is likely to double in India by 2015 compared to 2005 (more than 80 per cent of which will be out of pocket), and increase by half elsewhere, except in fragile states (WHO, 2008:105).

C. WATER AND SANITATION

Water, the basic element of our life, has already become one of the most stressed resources, fast getting scarce in almost 60 per cent of the world, including large part of Africa, Asia and Latin America particularly affecting people with limited means. The Earth Summit in Rio de Janeiro adopted an agenda defining the overall goal of water policy as making certain that, 'adequate supplies of water of good quality are maintained for the entire population of this planet, while preserving the hydrological, biological and chemical functions of ecosystems, adapting human activities within the capacity limits of nature and combating vectors of water-related diseases'. The "Millennium Development Goals and Targets", adopted as a part of UN Millennium Declaration in 2000 set a target of reducing the proportion of people without sustainable access to safe drinking water by half by 2015 against the backdrop of more than one person in four in 9 out of 59 top priority countries in 1990s lacking access to safe water (UNDP, 2003: 2-3)

The UNDP (2006) quotes from the Holy Koran, saying that "by means of water we give life to everything". Nothing could be more than the absolute truth that 'people need water as they need oxygen' and life cannot exist without it, and stretching the truth further, clean water and sanitation are needed for sustaining health and human dignity, and water acts as one of the principal providers in supporting production systems and maintain livelihoods. As a basic ingredient of human life, water pervades all aspects of human development and freedom, and as such, its availability signifies security of human life, any disruption in that threatening health and livelihoods. Regrettably, however, today some 1.1 billion people in developing countries, or almost one-sixth of world population have inadequate access to water, and 2.6 billion people lack basic sanitation.

As reported by the UNDP, 2006, one can take comfort from the fact that "there is more than enough water in the world for domestic purposes, for agriculture and for industry. The problem is that some people – notably the poor – are systematically excluded from access by their poverty, by their limited legal rights or by public policies that limit access to the infrastructures that provide water for life and for livelihoods. In short, scarcity is manufactured through political processes and institutions that disadvantage the poor. When it comes to clean water, the pattern in many countries is that the poor gets less, pay more and bear the brunt of the human development costs associated with scarcity" (UNDP, 2006: 3). This is amply demonstrated by the fact that in some parts of India water is extracted by irrigation pumps from aquifers 24 hours a day for wealthy farmers, while neighbouring small farmers depend on uncertain rainfall. The underlying cause of this scarcity 'in large majority of the cases is institutional and political', rather than an actual deficiency of supplies and in many countries scarcity of water is more the result of public policies than the availability of this natural resource, encouraged further by overuse through 'subsidies and underpricing'.

However, going back three centuries, an historical assessment may enlighten us about the worsening situation and major causes leading to it in the availability of fresh water in the world. As mentioned by L'Vovich and White (1990), "from 1687 to 1987, the total amount of water withdrawn for all needs (agricultural, industrial-urban, and domestic uses) has increased probably 35-fold. At present, it is over 3,800 km^3. The rate of increase was greatest after 1950, and in volume equaled all gains prior to that period. This appears on all the continents. In Europe, North America, and Australia the acceleration in rate had begun to be felt in the nineteenth century. Although irrigation is the principal consumer, its share decreased during the past three centuries from 90 to 75 per cent. In industry and energy production, mainly in Europe, water use began to expand on the eve of the nineteenth century. Increasing at a very high rate, by 1890 it was 2.5 times as great as domestic-municipal withdrawal, and by 1989, it reached almost 20 per cent of total water withdrawal. The share of domestic-municipal withdrawal changed little and comprised about 5 per cent while water used by livestock decreased to 2 per cent. Reviewing the principal features of water use, two changes stand out: a very rapid growth of consumption use since the middle of this twentieth century, and rapid build up of huge amounts of waste water" (L'Vovich and White, 1990: 245).

Two principal causes affecting the distribution of water were urbanization and 'transformations in management practices in farms and forests'. According to L'Vovich and White, urbanization had two major impacts on water components of the environment, such as 'sewers and drains'. Domestic and industrial wastes instead of getting widely deposited on soils were channeled into rivers and other water bodies', and by increasing the quantity of impermeable surfaces through constructing roads and roofs of buildings, the surface runoff was increased, and the replenishment and underground water was decreased. In 1800, the population of the cities of the world was 29 million, or 3 per cent of the global total population when the volume of surface runoff was not great though pollution around the urban area was probably much higher than today. By mid-1986, urban population of the world had risen to about 2,200 million when industry and roads together occupied an area of about 1,200,000 to 1,400, 000 km^2. "Assuming that the impermeable surface within this territory comprises 25 per cent of the total land ... the surface runoff in the aggregated cities of the world would may have increased by as much as 163 km^3/year, and the total runoff by 137 km^3/year, while replenishment of the underground waters decreased by 26 km^3/year" (L'Vovich and White', 1990: 246)

Nevertheless, there have been big improvements in access to safe water and adequate sanitation in recent years, making fresh water available for nearly 2 billion people and 400 million having access to basic sanitation during recent years but regrettably, "these gains in many cases have passed by the poor. As a result nearly 30 per cent of the population of developing countries, more than 1.3 billion people, lack access to safe water – and nearly 60 per cent, or over 2.5 billion, to basic sanitation" (UNDP, 1998: 68) (Table 4.15) (Figs. 4.7 and 4.8).

But the link between water and food being strong, we have to look to keep the supply of fresh water undiminishing as necessary for human survival. Each of us drink "on average 4 litres of water per day in one form or another, while the water required to produce our food totals at least 2000 litres – 500 times as much. This explains why 70 per cent of all water use is for one purpose – irrigation. Another 20 per cent is used by industry, and 10 per cent for residential purposes. With the demand for water growing in all three categories, competition among sectors is intensifying, with farmers almost always losing" (Brown, 2006: 42).

Naturally, the supply of fresh water per person is declining with commensurate growth of population, and the supply is expected to fall by 73 per cent between 1950 and 2050 and, in other words, 'there will be scarcely one fourth as much fresh water per person in 2050 as there was in 1950' (Brown *et al.* 2000:37) unless very cautious efforts are made in the use of water and stopping its misuse or abuse. Sadly, we are not yet adequately conscious about the impending crisis in near future when growing scarcity of fresh water is likely to spell disaster threatening human survival. The dramatic fall in the availability of fresh water is singularly evident in rivers getting dry and water tables falling across the world.

Table 4.15: Lack of Access to Safe Water and Basic Sanitation — a regional profile, 1990-96

(*Percentage*)

Region	*People Without Access to Safe Water*	*People Without Access to Basic Sanitation*
Arab States	21	30
Sub-Saharan Africa	48	55
South-East Asia and the Pacific	35	45
Latin America and the Caribbean	23	29
East Asia	32	73
East Asia (excluding China)	13	
South Asia	18	64
Developing Countries	29	58
Least Developed Countries	43	64

Source: UNDP 1998 (Taken from Human Development Report Office)

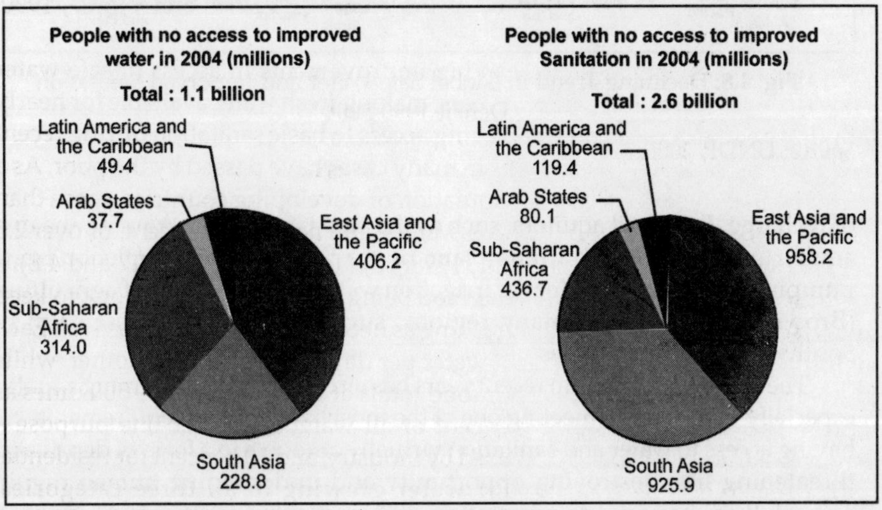

Fig. 4.7: People with No Accees to Safe Water and Improved Sanitation by Region, 2004

Source: UNDP, 2006

As Brown (2006) points out, scores of countries are overpumping water from aquifers causing constant fall in water table leaving hardly any chance of getting the aquifers recharged. This is happening in China, India and the United States specifically, the big grain producers facing growing need of water to irrigate fields. "There are two types of aquifers – replenishable and non-replenishable (fossil) aquifers. Most of the aquifers in India and the shallow aquifers under the North China Plain are replenishable. When these are depleted, the maximum rate of pumping is automatically reduced to the rate

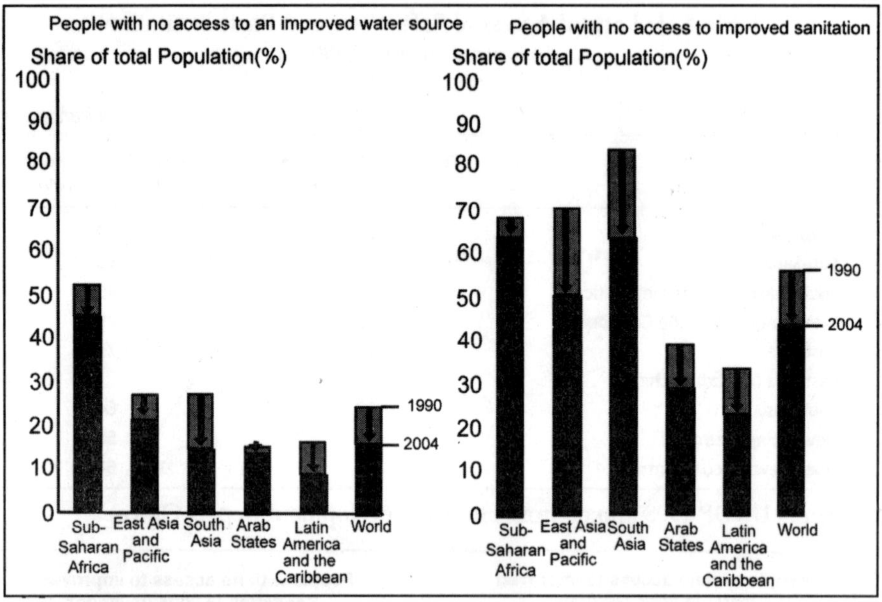

Fig. 4.8: Declining Trend in Global Safe Water and Imporved Sanitation Deficit 1990-2004

Source: UNDP, 2006.

of recharge. For fossil aquifers, such as the vast US Ogallala Aquifer, the deep aquifer under the North China Plain, or the Saudi aquifer, depletion brings pumping to an end ... the loss of irrigation water means the end of agriculture" (Brown, 2006: 42-43) in many regions, such as the Middle East or in the Southwestern United States.

The fall in water table in recent years has already made an alarming situation especially in the face of meeting one of the most basic human requirements. Not having access to water and sanitation virtually amounts to a form of deprivation threatening life, destroying opportunity and undermining human dignity. Though these basic needs vary widely with available opportunity of supply of fresh water, the minimum threshold is about 20 litres a day though" most of the 1.1 billion people categorized as lacking access to clean water use about 5 litres a day–one-tenth of the average daily amount used in rich countries to flush toilets. On average, people in Europe use more than 200 litres – in the United States more than 400 litres... Dripping taps in rich countries lose more water than is available each day to more than 1 billion people" (UNDP, 2006: 5-6).

Needless to say that access to water and sanitation is closely tied up with economic condition of the people, increasing wealth improving access though there are variations, such as Bangladesh and Thailand doing better in sanitation, Sri Lanka and Viet Nam in safe water than expected solely on the basis of income. In contrast, India and Mexico are in a worse position in sanitation

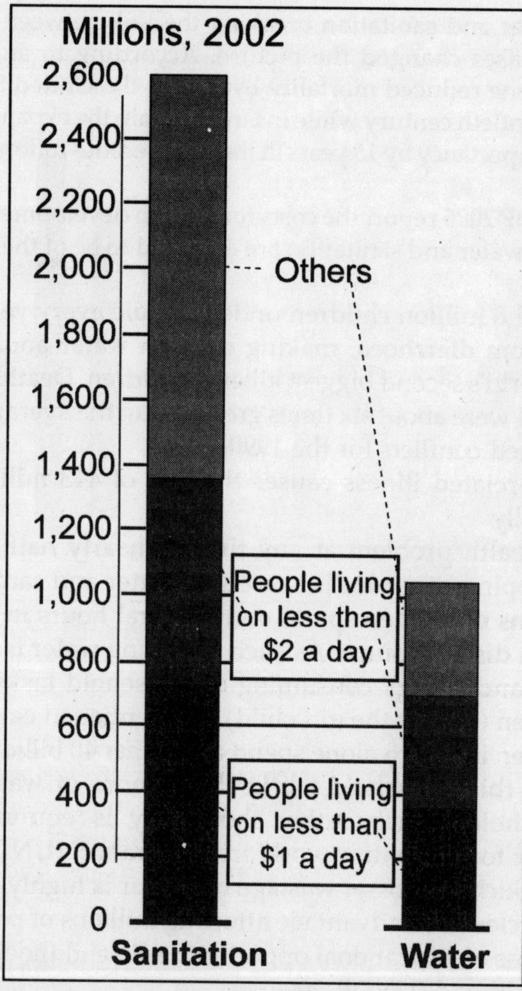

Fig. 4.9: Poor People Account for Most of the Water and Sanitation Deficit

Source: UNDP, 2006.

than expected since public policy in this regard is far from encouraging in improving the situation (Fig. 4.9).

Before turning to the situation affecting water availability, it is necessary to mention that, 'clean water and sanitation are among the most powerful drivers for human development contributing directly to extend opportunity, enhance dignity and help create a virtuous cycle of improving health and rising wealth'. People living in rich countries today are hardly conscious about the role of clean water in fostering social progress. For example, about a hundred years ago London, New York and Paris were centres of infectious diseases, with diarrhoea, dysentery and typhoid killing people at an early age. Sweeping reforms in water and sanitation breaking the link between dirty water and infectious diseases changed the picture. According to an estimate, water purification alone reduced mortality by half in the United States in the first third of the twentieth century while in Great Britain the expansion of sanitation increased life expectancy by 15 years in the four decades following 1880 (UNDP, 2006: 5).

In the UNDP 2006 report the costs for human development resulting from deprivation of water and sanitation are expected to be of the following order:

(a) Some 1.8 million children under five die every year, or 4,900 each day from diarrhoea, making unclean water and poor sanitation the world's second biggest killer of children. Deaths from diarrhoea in 2004 were about six times greater than the average annual deaths in armed conflicts for the 1990s
(b) Water-related illness causes the loss of 443 million school days annually.
(c) The health problem at any time of nearly half of all people in developing countries is caused by water and sanitation deficits.
(d) Millions of women spend daily several hours in collecting water from a distant source. As such collecting water is among the most time- and energy-consuming of household tasks of women and children (mostly the girl child). As mentioned earlier, women and children in Africa alone spend altogether 40 billion hours a year in doing this household job in the absence of water source in the household, which, rather shockingly, is 'equivalent to a year's labour for the entire workforce of France' (UNIFEM, 2008/2009: 130). Such a colossal wastage of labour is highly unjustifiable.
(e) Lifecycles of disadvantage affecting millions of people with illness and loss of educational opportunities in childhood lead to poverty in adulthood (UNDP, 2006: 6).

Undeniably, the crisis in water and sanitation is in reality a crisis for the poor since almost "two in three people lacking access to clean water survive on less than $ 2 a day, with one in three living on less than $ 1 a day. More than 660 million people without sanitation live on less than $ 2 a day, and more than 385 million on less than $ 1 a day" (UNDP, 2006: 7). There is, in fact, a close connection between the distribution of wealth and the distribution of piped water. For instance, 85 per cent of the households of the wealthiest 20 per cent of the population have access to piped water compared with 25 per cent of the poorest 20 per cent.

I. Water Availability

The constant fall in the water table is a grim reminder of the situation the world is going to face in near future irrespective of economic development or technological progress.

There is a large regional variation in water use, for example, in Europe where agriculture is mainly rain-fed, water withdrawals are dominated by industrial use. In Asia, in contrast, irrigation accounts for 85 per cent of all water use. There is a general opinion that the 'economics of water use does not favour agriculture' as there is a ratio of 70 to 1 between agriculture and industry, the latter always winning in competition where one thousand tons of water is needed to produce one ton of wheat worth $200 while using same amount of water expansion of industrial output worth $14,000 can be achieved (Brown *et al.*, 2000: 38).

Against the backdrop of water scarcity on a global scale aggravated by extensive over-pumping of underground water the prospect of irrigation is going to be a very crucial issue in future hitting hard on food production. Table 4.16 lists countries over-pumping the groundwater leading to fast depletion of this natural resource.

The 15 countries listed in the table experiencing over-pumping of underground water for use in agriculture, have a total population of about 3.2 billion, or about half of the global population, and offer a strong basis for an assumption that food production in the not very distant future is going to decline, making it more difficult to remove hunger and malnutrition particularly in developing countries dependent on food imports at times of lower production.

Table 4.16: Countries Overpumping Aquifers in 2005

Country	*Population (million)*
China	1,316
India	1,103
Iran	70
Israel	7
Jordan	6
Mexico	107
Morocco	31
Pakistan	158
Saudi Arabia	25
South Korea	48
Spain	43
Syria	19
Tunisia	10
United States	258
Yemen	21
TOTAL	3,262

Source: Brown, 2006

Undoubtedly, a situation of decline in food production is an unavoidable consequence of falling water tables adversely hitting harvests resulting from less irrigation, already happening in some countries, including China, the world's largest grain producer. A groundwater survey released in Beijing in August 2001, revealed that the water table under the North China Plain, producing 'half of the country's wheat and third of corn, is falling faster than earlier reported'.In another report, made by the Geological Environmental Monitoring Institute in Beijing, it was mentioned that the 'average level of the deep aquifer under Hebei Province in the heart of the North China Plain was dropping nearly 3 metres (10 feet) per year while it was falling twice as fast in some other areas. A World Bank report, apprehending 'catastrophic consequences', observed that "deep tube wells (drilled) around Beijing now have to reach 1,000 metres to tap fresh water" (Brown, 2006: 44). It is further reported that wheat farmers in some areas have to pump water from a depth of 300 metres, or nearly 1,000 feet, raising costs so high that farmers are sometimes forced to 'abandon irrigation and return to less productive dry farming'. Water shortages due to falling water tables have caused wheat harvest to decline in China from 123 million tonnes in 1997 to 95 million tonnes in 2005; and a similar decline has also affected the rice harvest, dropping from a peak of 140 million tonnes in 1997 to an estimated 127 million tonnes in 2005. Another World Bank study on China reveals a shortfall of 40 million tonnes of grain harvest due to complete depletion of the aquifer, enough to feed 120 million Chinese.

Constant fall in water table has also been experienced in India, falling by 6 metres (20 feet) per year in North Gujarat and wells are going dry almost everywhere in Tamil Nadu, with 95 per cent of those owned by small farmers having gone dry. With fall in water tables, well drillers have to go as deep as 1,000 metres in some locations and the availability of water for irrigation (as well as for domestic use and other requirements) is going to reduce the harvest severely in the near future. As reported by USDA, in parts of Texas, Oklahoma, and Kansas – three leading grain-producing states in the United States – 'the underground water table has dropped by more than 30 metres (100 feet)' making wells dry on thousands of farms in the southern Great Plains. However, the effect on grain harvest will not be as great as in India and China as only one-fifth of grain production takes place on irrigated land compared with 'three-fifths in India and four-fifths in China.

Pakistan faces a similar fate as India with overmining of underground water where the water table in wells near the twin cities of Islamabad and Rawalpindi has gone down by 1 to 2 metres a year between 1982 and 2000, and this is happening at a faster rate of 3.5 metres a year around Quetta in Baluchistan, which is predicted to go completely dry in another '15 years at the present rate of consumption'. In Iran, water from its aquifers is overpumped' by an average of 5 million tonnes per year, the equivalent of one-third of its annual grain harvest', and dry wells in eastern Iran are 'generating a flow of water refugees'. In the extensively developed irrigated agriculture dependent on water pumped from deep fossil aquifer, Saudi Arabia's farmers pump water from wells 4,000 feet deep (or four-fifths of a mile), and according to national experts, 'half' of the huge 'reserve of that fossil water, 462 billion tonnes, has probably disappeared by now', suggesting irrigated agriculture in Saudi Arabia may continue for another decade or so and vanish entirely after that making the farmers in Saudi Arabia completely dependent on shallow aquifers replenished by sparse rainfall. The water table 'under most of' Yemen, a neighbouring country, 'is falling roughly 2 metres a year', the extraction of 224 million tonnes exceeding the annual recharge of 42 million tonnes, dropping the water table by 6 metres per year in Sana'a Basin, which was going to be pumped dry by 2010 as predicted by the World Bank. Israel is depleting 'both of its principal aquifers' — the coastal and mountain aquifers - the last shared with Palestinians. In Mexico, '51 per cent of all the water extracted is from overpumped aquifers' (Brown, 2006: 44-48).

At the same time drying rivers and disappearing lakes all over the world are mute witness to a growing water crisis. Two rivers, the Colorado in the United States and the Yellow River in China provide major instances in this regard. As Brown (2006) puts it, "as the world's demand for water has tripled over the last half century and as the demand for hydroelectric power has grown faster, dams and diversions of river water have drained many rivers dry. As water tables have fallen, the springs that feed rivers have gone dry, reducing river flows" (Brown, 2006: 49).

The world's rivers are now obstructed by more than 45,000 large dams, those over 15 metres high (taller than a four-storey building) and all but 5,000 of them were built after 1950. China having only 8 large dams at the time of its revolution in 1949, built around 19,000 later. The United States has the second position with some 5,500 large dams followed by ex-USSR, Japan, and India, and among them the United States is estimated to have about 96,000 small dams (McCully, 1998: 3-4).

The combined effect of all the factors mentioned above, has made many of the large rivers dry, for example the Colorado hardly reaches the sea, the Yellow River has failed to reach the sea since 1985, the Nile in Egypt barely makes it to the sea following the construction of the Aswan Dam, the Sind in Pakistan has started running dry in its lower reaches; the dams being built in the upper reaches of the River Mekong by China, already have reduced its flow affecting countries like Cambodia, Laos, Thailand and Viet Nam through which it flows. Similarly, large dams built by Turkey and Iraq have reduced the flow of the river system of the Tigris and Euphrates, destroying 'more than 90 per cent of formerly vast wetlands that enriched the delta region'. A deteriorating system occurs in Central Asia with the flow of River Amu Darya cut off and thus depriving the Aral Sea from being fed while 'the diminished flow of the River Sys Darya keeps the Aral Sea from disappearing entirely'.

On the other hand, the increasing pressure on water crunch otherwise available for irrigation has further been accentuated due to the drying of lakes, best examples of which are provided by Lake Chad in Central Africa, the Aral

Sea and the Sea of Galilee in Jordon. Lakes in the United States may well meet the same fate as Owens Lake in California which dried up following the diversion of the Owens River. The water level in Mono Lake has dropped 35 feet since 1941 meeting the growing water requirement of Los Angeles. The diminished flow of the Jordon River has caused shrinking of the Sea of Galilee and has made the Dead Sea shrink' even faster than the Sea of Galilee', its water level dropping by some 25 metres (nearly 80 feet) over the past 40 years or so and with further diversion of water from the Jordon River along with fast-falling water table in Jordon, the Dead Sea is likely to 'disappear entirely by 2050'.

By far one of the greatest examples of damage was done to one of the world's greatest freashwater bodies, the Aral Sea, once the water of its two feeding rivers – the Syr Darya and Amu Darya — started to be diverted for cotton production from 1960 onwards leading to the shrinking of the sea and the consequent rise in salt concentration in its water until 'the fish died – having once produced 50,000 tonnes a year – and the centres of commerce around its shore were abandoned for ever.' The drop in annual flux of water from the two rivers from 65 billion cubic metres to 1.5 billion cubic metres is far from adequate to reverse the shrinkage and the shoreline now is 'up to 250 kilometers (165 miles) from the original parts has exposed a vast sandy bed contributing to wind-swept sand deposition on surrounding grassland and croplands' (Brown, 2006: 48-52).

China presents, perhaps, the worst case of disappearance of lakes where in its western province, Quinlai, the main course of the Yellow River was dammed, resulting in more than 2,000 out of 4,077 lakes disappearing over the last 20 years or so, and the situation in Heibi Province around Beijing is far worse, which lost 969 of its 1,052 lakes due to fast fall of the water tables. Similarly, lakes are disappearing in other Asiatic countries, such as India, Pakistan as well as in other continents for the same reason 'excessive diversion of rivers and overpumping of aquifers. In the Kashmir Valley of India many lakes have already disappeared while Lake Dal, which once covering 75 square kilometers, has shrunk to 12 square kilometers'. Lake Chapala, the largest lake in Mexico, the primary source of water to 5 million people in Guadalajara, the water volume has reduced by 80 per cent due to expanding irrigation (Brown, 2006: 52-53) (Fig. 4.10).

II. Changing Demands of Water and Water Scarcity

Population growth is one of the major factors affecting demands for water. As population increases, the demand for fresh water increases concurrently with a reduction in water availability. For example, per capita supply has decreased by a third between 1970 and 1990. Despite reduction in the growth rates of population, increase in the number of people "will be a major driver of water resource management ... (and) based on ... (population) projections the future for many parts of the world look bleak. The most alarming projection suggests

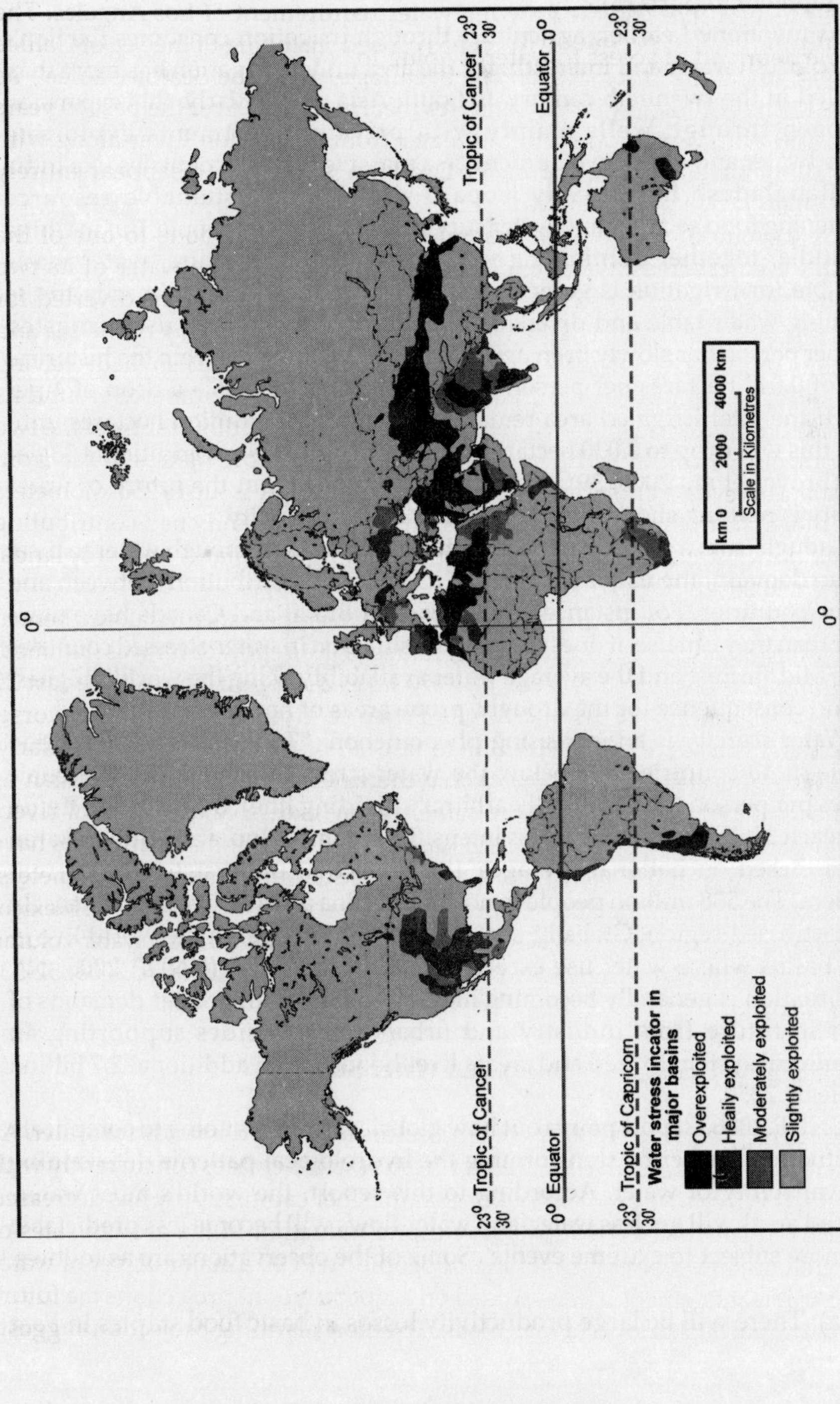

Fig. 4.10: Water Overuse around the World

Source: UNDP 2006.

that nearly 7 billion people in sixty countries will live water-scarce lives by 2050. Even under the lowest projection, just under 2 billion people in forty-eight countries will struggle against water scarcity in 2050 (UN, 2003: Water Development Report, 13)

As mentioned earlier, agriculture through irrigation consumes the lion's share of fresh water and interestingly, the area under irrigation has more than doubled in the twentieth century. In South Asia, particularly, this expansion has been through wells mainly with private investments exploiting groundwater and such development on a massive scale in countries like India and Bangladesh has already jeopardized this unsustainable resource threatening food security and water supply for millions of people. But in China and India, together dominating world irrigated agriculture, water made available for irrigation is going to get slowly reduced due primarily to the declining water table and diversion of water to non-agrarian uses. Irrigated area per person has slowly been declining since 1978, falling from the historical high of 0.047 hectares per person to 0.045 hectares in 1996 – a drop of 4 per cent. If the total irrigated area remains at roughly 263 million hectares until 2050, this will drop to 0.030 hectares per person by 2030 – an additional 33 per cent (Brown et al., 2000: 39), strengthening the idea that the future of water scarcity is nothing short of facing a future of food scarcity.

Though from a global perspective there is more than enough water to meet human demand, the major snag lies in its unequal distribution 'between and within' countries. For instance, even if each of Brazil and Canada have more water than they can use, it does not help the situation in water-stressed countries in the Middle East and the average water availability being the world's highest is of no consequence for the drought-prone areas of north-east Brazil.

Water scarcity is an increasing phenomenon. "Today, about 700 million people in 43 countries live below the water-stress threshold of 1,700 cubic metres per person (an admittedly arbitrary dividing line). By 2050, that figure will reach 3 billion, as water stress intensifies in China, India and Sub-Saharan Africa. Based on national averages, the projection understates the current problem. The 538 million people in northern China already live in an intensely water-stressed region. Globally, some "1.4 billion people in the world live in river basins where water use exceeds sustainable levels" (UNDP, 2006: 14). The situation is generally becoming more desperate with greater demands of water in future from industry and urban areas, besides supporting an agricultural system to feed and create livelihoods for an additional 2.7 billion people by 2050.

UNDP, 2006, Report points out how global warming is going to complicate the situation further by transforming the hydrological patterns determining the availability of water. According to this report, the world's most water-stressed areas will get less water, and water flows will become less predictable and more subject to extreme events'. Some of the observations are as follows:

(a) There will be large productivity losses in basic food staples in East

Africa, the Sahel and southern Africa due to marked reduction in water availability (causing up to 33 per cent decline in maize, more than 20 per cent for sorghum and 18 per cent for millet in East Africa).

(b) The decline in food production is likely to affect an additional 75-125 million people with a threat of hunger.

(c) The increasing rate of melting of glaciers will cause medium-term reductions in water availability in regions like East Asia., Latin America and South Asia.

(d) Disruptions to monsoon patterns in South Asia will cause more rain but with fewer rainy days and affect more people with drought. In 2005, more than 20 million people in the Horn of Africa were affected by drought and the growing sequence of periodical drought in many parts of India is a grim reminder of an erratic monsoon regime.

(e) Rising sea levels will cause fresh water losses in river delta systems in countries like Bangladesh, Egypt and Thailand (UNDP, 2006: 15).

III. Water and Sanitation as Human Rights

As observed in UNDP, 2006, Report, "exclusion from clean water and basic sanitation destroys more lives than any war or Terrorist Act. It also reinforces the deep inequalities in life chances that divide countries and people within countries on the basis of wealth, gender and other markers for deprivation... Whether viewed from the perspective of human rights, social justice or economic common sense, the damage inflicted by deprivation in water and sanitation is indefensible... The waste of human potential associated with unsafe water and poor sanitation ultimately hurts everybody" (UNDP., 2006: 27).

One of the major ambitious targets set by the United Nations in the 1970s and 1980s was "water and sanitation for all" within a decade which has, rather, sadly not yet been achieved although impressive progress in regard to access to safe water has been made in China and India since then and there are two regions – East Asia and Latin America – which are on the way to achieve the target in sanitation. An assessment of results obtained so far in two other regions shows that, 'on current trends Sub-Saharan Africa' will reach the water target of making it universally available' in 2040 and the sanitation target – a service for all – in 2076'. Discouragingly, however, South Africa is still off the track by 4 years for making water universally available while the Arab states will require another 27 years to do the same. As a result, on the whole, 234 million people are yet to get the benefit and altogether 53 countries are off the track. Similarly, 430 million people, with 74 countries are off the track for sanitation and in Sub-Saharan Africa the connection rates of water needs to increase from 10 million a year to 23 million a year in the next decade to make it available for all, and in South Asia the rate of sanitation provision has to increase from 25 million people a year to 43 million a year.

History tells us how safe water and improved sanitation have enriched human life with their direct impact on increasing life expectancy, decline in child mortality as reflected in the health of people in Great Britain where 'life expectancy increased in the four decades following the 1880s by an astounding 15 years with reduced child deaths' while in the United States 'half the mortality reduction in the first third of the twentieth century 'where at the beginning of the twentieth century 'infectious diseases accounted for 44 per cent of mortality in US cities'. It is claimed that, "by 1920 almost every big city in today's industrial world had purified water'. Contrarily, at the beginning of the twentieth century one in five people living in the developing world – about 1.1 billion in all – lacked access to clean water and about 2.6 billion people, nearly half of the total population of the developing countries did not have access to adequate sanitation.

The use and abuse of water on the one hand and the profound deprivation on the other are sharply reflected in the difference in them existing between rich and poor countries, highlighting the scale of deprivation and global inequalities. As mentioned earlier, both the WHO and UNICEF draw a dividing line at 20 litres of clean water per person a day from a source within 1 kilometre of the house as the minimum requirement for drinking and maintaining basic personal hygiene, and the personal threshold will rise to about 50 litres a day when bathing and laundry are included. The supply falling below that level would necessarily affect health and dignity of people. But the disparity in the use of water across the world is startling as earlier showcased for some rich countries. For example, in Europe average water use ranges between 200 and 300 litres per person a day which rises to 575 litres in the United States, reaching about 1,000 litres per person a day in some of the areas in the United States. By contrast, people in Mozambique in Africa use less than 10 litres a day per head, and 12-14 litres a day in Uganda. Besides, in dry season the use falls sharply as distance from where water is collected increases as happens in western India, the Sahel and East Africa, the water availability falling way below 5 litres a day and 8 litres a day in some areas of the city of Chennai in India (UNDP, 2006: 34-35).

Fresh water is not an inexhaustible resource considered in terms of its availability and its use by man under different economic and technological situations. An assessment of the freshwater resources in regional dimension becomes necessary to understand the situation. From Africa, for instance, the average annual flow for the 1921-1985 period was 4,000 km^3. Many of the rivers in this continent are exploited for power generation, irrigation making water table high in many areas with increased salinity. The mean annual runoff from Asia for the period 1921 to 1985 is estimated to be 13,500 km^3, about half of which originates in East Asia and only 7 km^3 from Arabian peninsula. There are large aquifers, many lakes of fresh water in Asia and irrigation is widely practiced in China, India, Pakistan, and in the Russian Federation. The mean annual runoff from Europe for the same period is estimated at 2,900 km^3 per year, most of which is from northern Europe. There are a large number of lakes and reservoirs as well as aquifers. "Over the last 200 years, industry, energy generation, agriculture and urban development have changed the pattern of runoff from the continent and altered its quality characteristics. Many rivers and lakes were badly polluted by discharges of untreated sewage, mine wastes and agricultural effluents" (United Nations, 2003: 1991). The average runoff from South America containing the world's largest river was calculated to be about 12,000 km^3 for the period 1921-1985. There are large aquifers, lakes and reservoirs in the continent but the water quality has deteriorated largely through discharge of agricultural effluents and mine wastes. The average annual runoff from North America over the period 1921 to 1985 was estimated at 7,900 km^3. But human activities throughout the nineteenth and the twentieth centuries has considerably changed the runoff pattern and the position of water table in many aquifers. Irrigation, inter-basin transfers of water for power generation caused problems and pollution is accelerated further by industry and mining. The rivers of Oceania are short and fast flowing, producing an average runoff of about 2,000 km^3. The amount of groundwater is considerable, but salinity induced by irrigation affects water quality (United Nations, 2003: 90-92). Notably, excessive use of large amount of water in the rich countries has no direct effect on reducing the use of water in poor or developing countries. The existing inequality or deprivation in the latter results from inadequate infrastructure for making safe water available for people living in these countries and the absence of that is defined by wealth as well as technology necessary for the development of water purification and supply systems. As observed, though most people without water live in South Asia, Sub-Saharan Africa has by far the lowest covering rates, only 55 per cent of its people having access to safe water while in South Asia the coverage is as low as that of Sub-Saharan Africa, with two of every three people in both regions lacking access. Half the people in East Asia and a quarter in Latin America lack access to even the most basic sanitation. Some 40 developing countries provide clean water for fewer than 70 per cent of their citizens, and 54 countries provide improved sanitation for fewer than half.

The close association between poverty and lack of access to safe water and improved sanitation cannot be ruled out, and is more marked for water since about 'a third of people' without access to a safe water source live on less than $ 1 a day. 'Twice this share live on $ 2 a day', implying that "660 million people lacking access to water have, at least, a limited capacity to pay more than a small amount for a connection to water service. Of this total some 385 million people fall below the $ 1 a day absolute poverty threshold. More than half of the 1.1 billion people without access are the poorest 40 per cent of the income distribution" (UNDP, 2006: 49), very strongly suggesting a 'two-way relationship between income poverty and deprivation in access to water'.

The strong association between poverty and access is displayed by sanitation as well when for the 'poorest two-fifths of households accounting for more than half the global deficit, the average rates being even far lower for sanitation than water when quarter of the richest 20 per cent of people in developing countries have no access to improved sanitation, available for half of the second richest 20 per cent. Almost 1.4 billion without access to improved sanitation live on less than $ 2 a day. Factually, the 660 million people living on less than $ 2 a day lacking access to water and the equally poor 1.4 billion people without access to improved sanitation are incapable of financing the water-utility services. A study of 17 developing countries carried out by Demographic and Health Survey for UNDP Report 2006, found that access to piped water was available for 85 per cent of the richest 20 per cent households and 25 per cent for the poorest 20 per cent, establishing further the link between wealth and access. The disparity in the availability of water and sanitation is more pronounced between the rural and urban areas as revealed in the fact that irrespective of economic status, some 92 per cent of urban people in the developing countries as a group have access to clean water against 72 per cent in rural areas while sanitation coverage for urban people is double that of rural.

The price of water seems to have an important role in providing access to clean water across the developing countries and it is 'inversely related'. For example, "some of the poorest people living in urban slums pay some of the world's highest prices for water. In Jakarta, Lima, Manila and Nairobi households living in slums and low-income settlements typically pay 5 to 10 times or more for their water than high-income residents of the same city. In Manila, an estimated four million people receive water resold through kiosks, pushcart vendors or tanker deliveries. Their average monthly water bills are $10-$20. By contrast, households directly connected to the utility pay an average of only $ 3 to $ 6 a month but consume five times more water... Some of the poorest people living in sprawling slum areas of Accra and Manila are paying more for their water than people living in London, New York or Rome" (UNDP, 2006: 52).

Set against a background of gross inequality profoundly harming human life in the developing countries in particular, provision of adequate amount of clean water and access to improved sanitation pose a great challenge. This is more so because of the growing population, likely to increase by 830 million over the next decade in the developing countries with 'Sub-Saharan Africa accounting for quarter of it and South Asia a third of it, which means that an additional 900 million need access to water' and '1.3 billion need access to sanitation by 2015'. This implies several hundred thousand new connections each day in some of the world's poorest countries. For achieving this goal it has been estimated that South Asia is required to provide sanitation coverage to 43 million people a year compared with 25 million people annually over the past decade. Sub-Saharan Africa faces a greater challenge where the coverage rates for clean water has reached an annual average of 10.5 million which needs to be increased to 23 million a year while for sanitation the coverage has to increase four times – from 7 million to almost 28 million people.

The prospects for the world achieving the MDG (Millennium Development Goals) set for water and sanitation, based on looking at the present trend of development in developing countries, including population giants like China and India, are that the share of people deprived of access to water is going to be halved, but achievement in providing adequate sanitation will fall far short of the targeted goal. Though there are regional variations, some regions will miss the water and sanitation targets and it is rather frightening to observe that on current trends "Sub-Saharan Africa will miss the water target by a full generation and the sanitation target by more than two generations. South Asia will miss the sanitation target by four years, and the Arab states will miss the water target by 27 years" (UNDP, 2006: 57). Table 4.17 shows the level of performance and the job yet to be done for attaining the goal of making access to clean water and improved sanitation for all people living in different regions

Table 4.17: The Millennium Development Goal Target: Past Performance and Future Targets for water and sanitation, by Region, 1990 and 2004

People with access to clean source of water (millions)				*Average Annual Number of People*	
Region	*1990*	*2004*	*Target 2015*	*Gaining Access 1990-2004*	*Needing Access to meet the target 2004-2015*
Sub-Saharan Africa	226.6	383.8	627.1	10.5	23.1
Arab States	180.1	231.8	335.8	4.7	6.5
East Asia and the Pacific	1,154.4	1,528.2	1,741.2	22.9	24.3
South Asia	840.6	1,296.4	1,538.1	32.5	22.1
Latin America and the Caribbean	334.3	499.0	527.8	9.0	6.1
World	2,767.7	4.266.4	5,029.5	79.5	82.4

People with access to improved sanitation (millions)				*Average Annual Number of People*	
Region	*1990*	*2004*	*Target 2015*	*Gaining Access 1990-2004*	*Needing Access to meet the target 2004-2015*
Sub-Saharan Africa	148.4	256.5	556.0	7.2	27.9
Arab States	120.6	196.0	267.2	4.9	6.9
East Asia and the Pacific	467.0	958.2	1,284.9	32.0	33.6
South Asia	242.9	543.8	1,083.3	24.7	42.5
Latin America and the Caribbean	279.6	423.2	492.2	8.6	8.4
World	1,456.9	2,663.9	3,994.0	77.5	120.4

Source: UNDP, 2006 (Taken from calculations made on the basis of data provided by WHO and UNICEF 2006 and UN 2005.)

of the developing world. As it shows, the current performance falls short of the level required, as a result of which 55 countries are estimated to be off the track in providing access to clean water, making 235 million people to miss the target and thus keeping a total of 800 million people without access to clean water even by 2015. On the other hand, there are 74 countries still off the track with 430 million people missing the target and leaving 2.1 billion people without access to improved sanitation.

These figures, however, do not give a full picture of the prevailing situation since they do not factor in the problems linked to quality and continuity of provisions, while reporting data for urban areas they mask the fact that the figures do not include the slums in them. Such is the case of the city of Mumbai, which it is claimed enjoys a safe water coverage of more than 90 per cent of its population but the figures do not include almost half of the city's 18 million people living in slums who do not figure in municipal data. Indonesia, claiming water coverage for 90 per cent of its urban population, shows that in Jakarta, its national capital, less than a quarter of its 12 million people living in slums enjoy access to clean water. Nairobi, claiming access to water for more than 90 per cent of its people, has more than 1 million – a third of city's population – living in slums, who 'depend on private water' as a secondary source. For Pakistan urban clean water coverage is reported to meet the requirements of over 90 per cent of urban population though in practice in the city of Lahore (with 5 million population) and Karachi (with 10 million), half the population living in slums have to depend on groundwater and canal water as sources 40 per cent of which is unfiltered. Across urban Pakistan, unclean water poses a constant threat to public health.

Further disturbing to note is the gap between water and sanitary provisions. "In all regions and in almost all countries sanitation provision lags behind access to water – and there is no evidence that the gap is narrowing. In South Asia access to improved sanitation is less than half that for water. Elsewhere, the gap in coverage ranges from 29 per cent in East Asia to 18 per cent in Sub-Saharan Africa" (UNDP, 2006: 36).As a result the full benefits of safe water cannot be derived since benefits derived from both clean water and improved sanitation are 'mutually reinforcing' – one having impact on the other. In this regard, the close relationship between sanitation and water pollution remains as a big problem. The age-old problem of keeping water and human waste (excrement) separate has been a great challenge in the absence of adequate infrastructure. For example, less than 14 per cent of human waste receives any form of treatment in Latin America, the remaining untreated waste getting dumped in rivers, lakes or allowed to seep through into ground water. Pollution from human waste is a great problem in China where nationally less than 20 per cent of municipal waste receives any treatment and 16 cities with more than half a million residents do not have any wastewater treatment facilities. According to the report of the Environmental Protection Administration in 2003, more than 70 per cent of the water in five of China's seven major river systems was too polluted for human use.

Discouragingly, the situation in this regard has not undergone significant improvement over the decades in India. As reported in the National Human Development Report 2001, "a majority of India's population does not have access to toilet facilities in their dwellings and lack sanitation facilities for the disposal of waste water. Apart from the availability of safe drinking water, lack of sanitation, particularly sewage and disposal of solid waste including 'night soil' has been observed as among the main reasons for prevailing ill health and morbidity levels in India. As per the 1991 census, less than one-fourth of the households in the country had a toilet facility within the premises of their residence, the proportion was less than 10 per cent for rural households and around 64 per cent for urban households" (Government of India, 2002: 40). By 2004, 91 per cent of urban population in India had access to clean water though "this access does not ensure adequacy and equitable distribution, and per capita availability is not as per norms in many countries. Average access to drinking water is highest in Class I towns (73%), followed by Class II towns (63%), Class III towns (61%), and other towns (58%). Poor people in slums and squatter settlements (sometimes representing more than one-third of the city –population) are generally deprived of these basic amenities" (Government of India, 2008: 162) (*See* Note *H*). As mentioned in that report, the quality of urban water supplied for people is also 'poor' and erratic, 'supplied only for few hours of the day', and in many cases it comes in a thin flow, besides much of it is also wasted as taps are inadvertently (or callously) kept open. Quite frustratingly, fewer than 10 per cent of households with piped water connections get water 24 hours a day and elsewhere two to three hours' supply is taken as the standard practice. In 2004, 63 per cent of the urban population had access to sewerage and sanitation facilities while sanitation coverage in the country, as on November 2007, was about 49 per cent (Government of India, 2008).

Nothing, perhaps, is more disappointing than allowing 1.8 million children die each year of diarrhoea, many for want of clean water and a toilet, as mentioned by the President of Brazil. According to another estimate, if universal access to water and sanitation were achieved, the world could avert 2 million deaths over the next decade. Much more financial investment in providing universal access to clean water and improved sanitation is justified by the cost-benefit ratio urging overwhelming support from every national government in the developing world (as well as donors). The likely benefits to be derived from greater investment and efforts are incalculable in saving human life and also economic benefits likely to accrue from such efforts as summarized in the UNDP Report, 2006, are as follows:

(a) If Millennium Development Goal Targets are achieved, there would be 203,000 fewer child deaths in 2015, of which 124,000 would be in Sub-Saharan Africa alone.

(b) "The economic rate of return in saved time, increased productivity and reduced health costs for each $1 invested in achieving the target is $8.", or eight times the money spent for achieving the target, while total economic benefits would amount to $ 38 billion of which $ 15 billion would be for Sub-Saharan Africa, $ 8 billion for Latin America and $ 5 billion for South Asia.

(c) Further, it is estimated that $1.7 billion a year in costs associated with the treatment of water-related infectious diseases could be saved by achieving the water and sanitation targets. In this regard, Sub-Saharan Africa would save about $2 per capita, equivalent to about 12 per cent of public health expenditure.

(d) Reduction in the rate of diarrhoea alone would allow people aged 15-59 to gain 3.2 billion working days, having a huge impact on production and the economy and, as mentioned earlier, another 20 billion working days would be available by releasing women from the arduous task of carrying water from a distant source, consuming time and energy to be used more fruitfully for productive work. (UNDP, 2006: 58-59). (*See* Notes I A and I B).

This is, nevertheless, a partial assessment of the benefits to be derived directly on achieving the target of making universal access to water and sanitation. It does not include the benefits for 'education, empowerment of women', for restoring human dignity or for the 'reduced anguish and suffering' associated with lower child death rates. Furthermore, if the world could achieve universal access to water and sanitation by 2015, as intended, 2 million deaths over the next decade could be averted. All that is needed to make this happen would be $10 billion annually to get the world on the path of making the target by 2015, which is a surprisingly small amount representing "about eight days of global military spending" (UNDP, 2006: 59). In view of the limited financial resources of the developing countries, where the current spending on water and sanitation per year averages between $14 and $ 16 billion, the aforesaid amount of $10 billion from donor countries would be a minimum financial threshold for low-cost technologies available in the developing countries (Fig. 4.11).

IV. Water Scarcity as the Scenario for Future Availability

Since clean water, with a provision of at least 20 litres per person a day considered as a part of a 'social minimum', is at short supply, the moot question is whether it is becoming a scarce commodity with increasing demand from a growing population with changing life-style intimately linked to improving economy.

Arguably, 'the world has far more water than the 1,700 cubic metres per person as the minimum threshold for meeting water requirements for agriculture, industry, energy and environment', but what is disturbing is, as

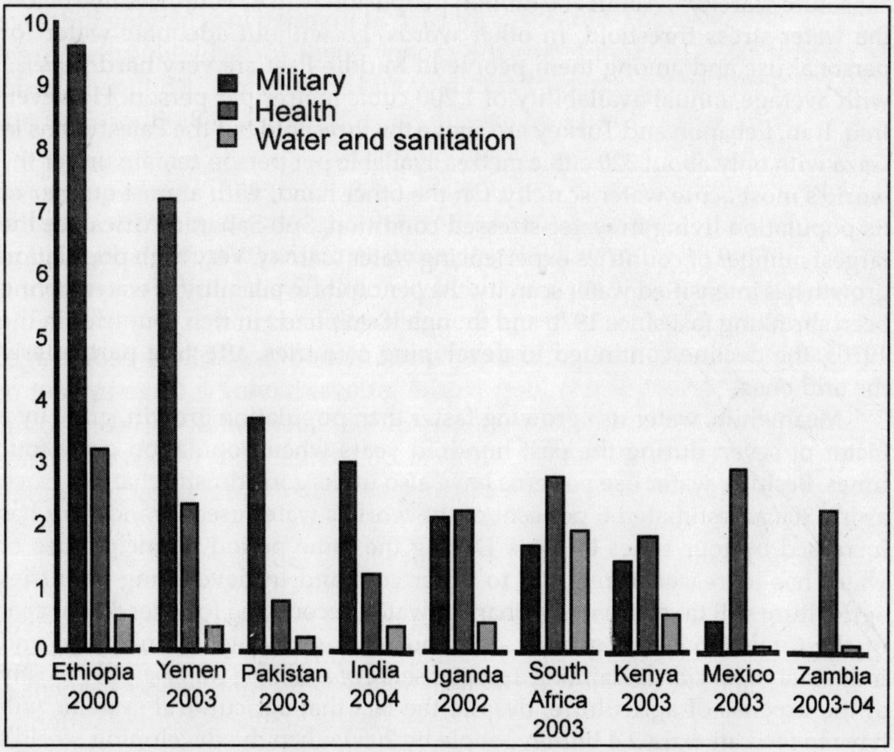

Fig. 4.11: Water: A Low Priority in National Budgets

Source: UNDP, 2006

mentioned before, the geographical distribution of fresh water when some countries have much more than others, making supply and demand irrelevant as water distribution does not follow the pattern of human distribution. For example, Lake Baikal situated in very sparsely populated Siberia locks up almost a quarter of the world's fresh water supply and Latin America with 31 per cent of global fresh water resources has 12 times more water per person than South Asia. Similarly, Iceland's water availability of more than 300 times the 1,700 cubic metres threshold is of no consequence to relieve water scarcity in regions short of water supply in China and India. Glaring contrasts are also presented by countries like the Republic of Congo having more than a quarter of Sub-Saharan Africa's water with 20,000 cubic metres or more for each of its citizens while Kenya, Malawi and South Africa are below the water-stress threshold. Further, Canada's overabundance of fresh water (90,000 cubic metres per person) is in sharp contrast with oil-rich Middle East remaining under great water-stress.

According to hydrologists while 1,700 cubic metres set the threshold in water requirement of a person, availability going down below 1,000 cubic metres is taken as "water scarcity", and below 500 cubic metres is taken as "absolute scarcity". About 700 million people today in 43 countries live below the water-stress threshold, in other words, go without adequate water for personal use and among them people in Middle East are very hard pressed with average annual availability of 1,200 cubic metres per person. However, Iraq, Iran, Lebanon and Turkey are above the threshold but the Palestinians in Gaza with only about 320 cubic metres available per person remain under the world's most acute water scarcity. On the other hand, with almost quarter of its population living in water-stressed condition, Sub-Saharan Africa has the largest number of countries experiencing water scarcity. Very high population growth has intensified water scarcity, the per capita availability of water having been shrinking fast since 1970 and though it stabilized in rich countries in the 1970s, the decline continued in developing countries, affecting particularly the arid ones.

Meanwhile, water use growing faster than population growth, grew by a factor of seven during the past hundred years when population grew four times. Besides, water use patterns have also undergone drastic changes, such as in 1900 an estimated 6 per cent of the world's water used by industry has increased by four times by now. During the same period municipal use of water has increased three-fold to 9 per cent and in developing countries agriculture still takes the major share of water, accounting for over 80 per cent of the total. However, greater demand from growing urbanization and industrial expansion is claiming a greater share of available water supply mostly at the expense of agriculture, despite the fact that agricultural systems will have to feed an extra 2.4 billion people by 2050 when the developing world's share of world population of 8 billion will mount to 82 per cent of the total.

In this context, the decline in available water is quite conspicuous in many countries, so much so that while by 2025 more than 3 billion people will be living in water-stressed countries, 14 of them will slip from a status of water-stressed to water-scarcity. Thus—

(a) The share of population in water-stressed countries in Sub-Saharan Africa will rise from above 30 to 85 per cent by 2025'.
(b) In the Middle East the average water availability falling by more than a quarter will aggravate the situation further, bringing it down to just over 500 cubic metres per person by 2025, when more than 90 per cent of the people will be living in water-scarce countries in this region.
(c) China and India, with the two largest populations of the world, are also turning into water-stressed countries. In India, for example, already 224 million people living in river basins with renewable water resources endowed with less than 1,000 cubic metres per person are approaching the water-scarcity threshold, primarily because more than two-thirds of the country's renewable water is concentrated in areas serving a third of the population.

In China, on the other hand, the national per capita availability of water being already low, a third of the global average, is accentuated further by unequal distribution of water within the country, with 42 per cent of its population – 538 million – living in the northern region have access to only 14 per cent of its water or 757 cubic metres per person comparable to that in some parts of North Africa, but lower than Morocco (UNDP, 2006: 135-138).

These are some of the instances showcasing an impending water crisis growing in intensity across the developing world causing serious concern, more so in a situation when more and more water is getting diverted to non-agricultural uses, making it increasingly difficult to feed the world population with a fresh addition of 2.4 billion of extra mouths by 2050 in a world containing 800 million malnourished people. The steady decline of the share of water for use in irrigated agriculture accompanied by a commensurate rise in the share of water used by industries and municipalities from less than 5 per cent today to over 25 per cent by 2050, as estimated, poses the most pertinent question of human development in its multi-facet dimensions, particularly keeping in mind that water is a finite resource in many situations.

Man's use of water exceeding the level required to 'maintain the ecological integrity' of river basins clearly demonstrates ecological stress, damaging the environment. This is obvious in many regions of South Asia, such as in the Indo-Gangetic Plain, the North China Plain and the High Plains in North America besides in areas where urbanization and industrial development are taking place at a fast pace. It is estimated that already many river basins covering more than 15 per cent of the world's land surface inhabited by 1.4 billion people are considered "closed", which means that in those areas water use exceeds the minimum recharge levels, or are near closure. For example, in northern China the recharge level for maintaining balance requires a quarter of the flow of Yellow River but only 10 per cent is left after human use, forcing the river to dry up at its lower reaches for a stretch of 600 kilometres inland for 226 days in 1977, reducing agricultural production and incurring a loss of about $1.7 billion that year. In Australia's Murray-Darling basin the environmental requirement for replenishment is estimated at 30 per cent but agriculture using almost 80 per cent of available water flows cause extensive environmental damage and drying up of the Murray River, which has failed to reach the sea in recent years.

Similarly, unrestrained overuse of groundwater causes the water table to fall faster than the groundwater could be replenished through natural processes. The most striking instances of this is provided by many countries today, particularly in those where water mined from underground sources is used primarily for agriculture, such as in India, Yemen and China where the water tables are falling at more than 1 metre a year. Apart from rivers drying up due to overuse of water causing severely reduced flows affecting rivers like the Nile, the Colorado, the Ganges, the Jordon, the Tigris-Euphrates, the Syr Darya – Amu Darya, lakes and inland water bodies provide another very important instance of asset depletion. The Aral Sea, as mentioned before, offers the most glaring case of injudicial water use for irrigation causing it to shrink from its original size (11,775 square kilometers in 1960) to one-fourth of that today. Overexploitation has been equally harmful in the case of many lakes, including the greatest of Africa's lakes, such as Lake Chad, Lake Nakivale, and Lake Nakran, of which Lake Chad has shrunk to 10 per cent of its former volume, the final phase of depletion of its water resource completed by enhanced irrigation demands, which quadrupled between 1983 and 1994.

There is another very important aspect directly related to water availability involving the risk, vulnerability and uncertainty or insecurity in its supply. The large global inequalities in earth's infrastructure have aggravated the situation. This is particularly felt in developing countries where the institutions and infrastructure like disaster management services for tackling the situations of unprecedented flood or drought are much weaker than they are in rich countries. Among the seasonal variables, droughts or failed rains cause enormous damage, depleting watersheds, farmlands, degrading land, destroying crops, and the poor people suffer most as a result of 'decreased production, loss of livestock, loss of soil fertility, and extreme shortages of drinking water'. In this regard, the worst affected region is Sub-Saharan Africa. 'In 2005, more than 20 million people were at risk from drought in the Horn of Africa'. Droughts are almost endemic, occurring every 3-5 years across much of Sahel, East Africa and Southern Africa. However, Sub-Saharan Africa is not the only region suffering from frequent droughts. About 15 per cent of people living in some parts of South Asia were affected by drought during the past two years. Middle East has the experience of more frequent and larger droughts. Agricultural output was down by 45 per cent in Morocco in the mid-19990s with a loss of 100 million days in agricultural employment. Ethiopia, on the other hand, covered with 12 river basins, has over 1,600 litres of water per person per year. But the vast majority of its population depending on rainfed agriculture, as in large parts of India, is faced with a great insecurity with variable rainfall, having pushed 12 million of its population today below the absolute poverty line in the second half of the 1990s. Rainfall variability in Ethiopia reduces its "potential for economic growth by a third ... (which) is estimated to increase poverty levels in 2015 by between a quarter and third, or some 11 million people" (UNDP, 2006: 157). This also happens to be the case in Mozambique, the weak infrastructure affecting the poor households very badly as living in low-lying areas along the river banks they mainly bear the brunt of flooding with 700 people dead in 2000 and a half million homeless, economic growth falling from 8 per cent in 1999 to 2 per cent in 2000. In contrast, Japan having been successful in mitigating the effects of floods the cost of which seldom rising above 0.5 per cent of GNI and loss of life are rare events makes a case for emulation.

At the same time water storage capacity of a country acts as a sure safeguard against checking floods or as a source of irrigation, mitigating water insecurity resulting from uncertain rainfall. Besides, water in storage for power generation is used for supplying as much as 22 per cent of electricity in Sub-Saharan Africa.

Though the 'world is not running out of water', many countries are already pressed for time to tackle the emerging problem of growing water stress, particularly unchecked population growth having aggravated the situation in the developing countries. The more important step to be taken for meeting this challenge is to treat water as a scarce resource at the national level and 'managing demand within frontiers of ecological sustainability', aligning 'wateruse patterns with the needs and demands of different users, including the environment' (UNDP, 2006: 169-170). All this brings us back to the most emerging issue almost threatening any solution to water scarcity unless use of water in agriculture is restricted in future. The link between water and livelihood is nowhere such a reality as in countries dependent on irrigated agriculture, playing a positive role in bringing down poverty levels often by 20 to 30 per cent under irrigation systems. But growing urbanization and industrialization gradually making greater demands on water is going to affect its supply for agriculture, eventually increasing the cost of irrigation which has doubled since 1989 in India, Indonesia and Pakistan. The growing competition between agriculture and non-agricultural sectors for water leads to overpumping underground resources already pushing the water table further down than presumed in China, South Asia and the Middle East, the overpumping rate already exceeding 25 per cent in China and 56 per cent in some parts of India. The ecologically sustainable solution of restoring the water table to its normal level would require cutting groundwater use from 817 million cubic metres to 753 million cubic metres which reduces irrigation to hit hard on food production in a large part of the developing world with severe effect on livelihood of agricultural workers. As is pointed out in UNDP 2006 Report, and elsewhere, there is likely to be a great reduction in agricultural output by 2050 when almost the entire increase of world population, increasing by about 80 million each year during the preceding three decades, will take place in the developing world, and by 2050, two-thirds of the world population will live in cities. The use of non-irrigation water will double, given the needs of industry, urban centres and livestock, as against the use of water increasing only 4 per cent leading to projected water withdrawals estimated to be 27 per cent higher in developing countries in 2025 than in 1995. Crop production in developing countries, at the same time, has to increase by 1.4 per cent per year, with 2.5 per cent in Sub-Saharan Africa to meet the rising requirements of food when irrigation slows down dramatically. However, to meet the growing demand for food, irrigated yields need to increase in developing countries by about one-third, as estimated, 'with production increasing by two-thirds' for which 'rainfed agriculture will have to account for 47cent of the total increase in cereal production'. Productivity of green water based on soil moisture and plant transpiration through water retention and improved tillage practices will take a very important role in enhancing yield rates of crops (UNDP, 2006: 176) (Fig. 4.12). This may be the only way to catch up with the dwindling source of water need for growing crops. Interestingly, among the developing countries rainfed agriculture takes a prominent role in agricultural production in all the

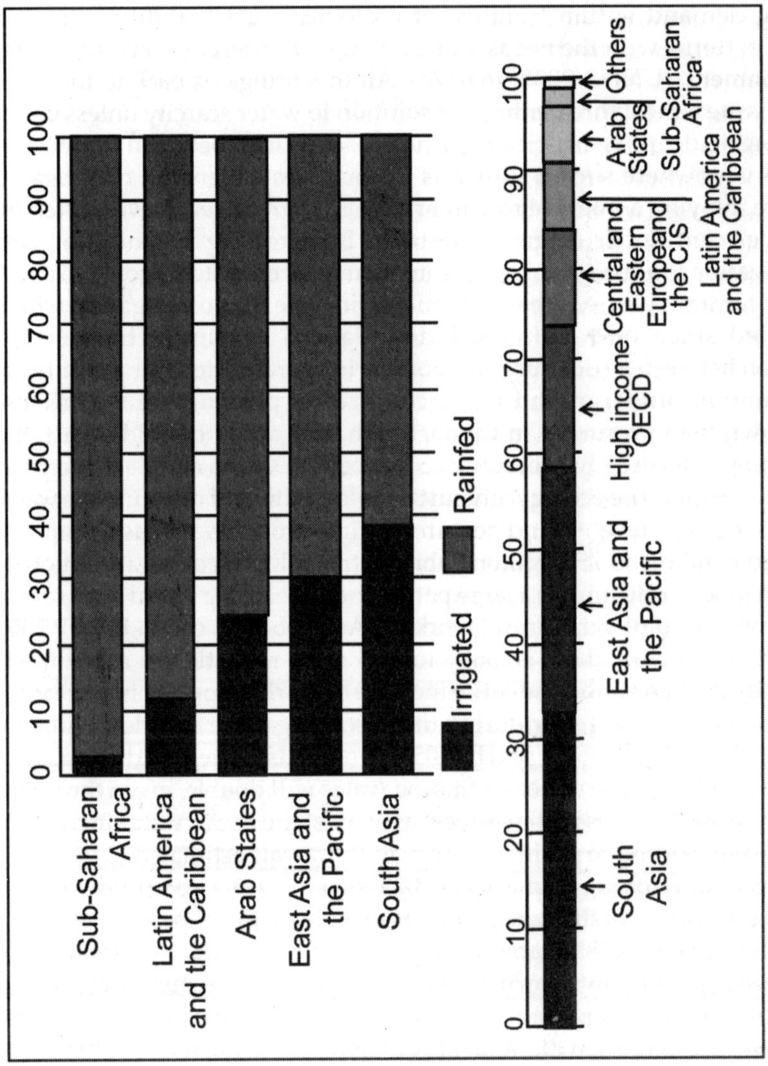

Fig. 4.12: Irrigated and Rainfed Agriculture (After UNDP, 2006)

Source: UNDP, 2006, FAC 2006

regions (exceeding irrigation agriculture overwhelmingly) — almost 98 per cent of the arable land in Sub-Saharan Africa, reaching nearly 52 per cent in South Asia. The challenging situation of declining water supply for agriculture requires greater emphasis on rainfed agriculture.

However, coming back to the overpowering problem of water and sanitation, it is the marginalized section of people who are the primary victims of being denied access to clean water and sanitation across countries mostly in the developing world and hence universal access should be the major aim for achievement by 2015. It is accepted as a fact that an effective political leadership has 'the potential to surpass targets and move rapidly towards universal provision' of these two most basic life-, and human-dignity saving services. Water considered as a basic human right for personal and domestic use in the United Nations Constitution, application of most basic technologies would make a huge difference, averting 600,000 child deaths in 2015. What is required to accomplish this cherished goal is the provision of piped water in the homes of the marginalized class of people in all the countries still lacking it, provide access to networks of sanitation built for this purpose and a water-sanitation infrastructure for processing waste water. The fact that "inadequate sanitation may kill large number of children, compromise public health, undermine human dignity and hold back economic growth" (UNDP, 2006: 61) is ignored by most of the countries giving short shrift to this basic problem, who should make it an important issue affecting life and economy of the country. The neglect of the issue is very strongly reflected in the national budgets, where it represents less than 0.5 per cent of GDP, falling to 0.1 per cent in Pakistan and Zambia, while expenditure on sanitation falls below that of water. "India spends more of its national wealth on military budgets than on water and sanitation. Pakistan spends 47 times more. In Sub-Saharan Africa low average incomes closely constraint public spending capacity. At the same time, Ethiopia, one of the poorest countries in the world with some of the lowest coverage rates (and some of the highest child death rates from diarrhoea), still manages to mobilize almost 10 times more for military spending than for water and sanitation" (UNDP, 2006: 62). Interestingly, the budget-outlay on water and sanitation in the Eleventh Five Year Plan (2007-2012) in India is 4.8 per cent of the total, with a little over 2.7 per cent for water and sanitation in the rural areas and about 23 per cent of the total budget allocation for the urban areas (or, altogether Rs. 175,000 crores, or approximately $ 3,420 million to be spent over a period of five years (Government of India, Eleventh Five Year Plan, 2007-2012, Vols I and II). Incidentally, diarrhoea claims some 450,000 lives annually in India, the highest for any country in the world, and 118,000 in Pakistan, underpinning the lack of access to clean water and improved sanitation as the main reason for this scale of human disaster. The success in increasing access to clean water is, however, minimal, depending largely not so much on the availability of funds but more on defined principles. For example, despite financial and technological constraints, as noted earlier, countries as diverse as Bangladesh, Brazil, Lesotho and Thailand have overcome those drawbacks in making exemplary progress through determined efforts. Bangladesh, having one of the lowest coverage of sanitation, took up an ambitious plan of making nationwide coverage in sanitation by 2010 with an annual target of increasing coverage for 2.4 million households. The total sanitation campaign in Bangladesh started by local NGOs in the late 1990s, involving 600 NGOs working in collaboration with district authorities to make the country 'a world leader in producing, marketing and maintaining low-cost latrines. At the end of 2000 there were 2,400 registered small-scale latrine production centres', the number rising to 3,000, bringing down the price of these cheap latrines further, putting the country on the path to attaining universal access by 2015. Thailand provides another success story in this regard, having increased national sanitation coverage rate since 1990 from 80 to 100 per cent, with more than 13 million people in its rural areas having gained access to sanitation in just two decades on the basis of development of technologies by the government affordable for and accessible to the poor, with provision for training and maintenance for which finance was provided to meet capital costs. Access to sanitation so far in both Colombia and Morocco looks promising, in Colombia about 86 per cent of people covered by 2005 which is 'far higher than its national income would permit' while in Morocco more than 70 per cent people were covered by 2003.

Further, it is highly promising in Lesotho, a very poor country, where a small pilot project for rural sanitation started twenty years ago with financial assistance from the United Nations Development Programme, has increased coverage from 15 to 32 per cent – higher for many countries with higher average incomes – and intends to make full coverage by 2010. In Pakistan, the appalling sanitary situation in the low-income informal settlement of Orangi in Karachi, home to a million people, is a success story achieved in collaboration with the local government in tackling the sanitary problem, making access almost universal. In China, the sanitary situation in rural areas was much less developed than in urban areas until the mid-1990s, but afterwards rural sanitation came to be an integral part of the health programme, and as a result rural sanitation coverage doubled in five years. Interestingly, the users meet 75 per cent of the cost while 15 per cent comes from the village council and the government provides the remaining 10 per cent towards the cost.

One of the most striking examples of success in providing sanitation facilities at small financial costs is offered by India. At the initiative of the NSDF (National Slum Dwellers Federation), SPARC (Society for the Promotion of Area Resource Centres) and Mahila Milan (a Mumbai-based NGO of women), more than 440 toilet blocks with more than 10,000 new toilets were constructed based on models developed jointly by them, financed by the government of Maharashtra, and the NGOs taking up the responsibility of maintenance. Another equally encouraging example of success in this regard is provided by West Bengal (India) where a drive for rural sanitation was initiated by the state government in Midnapore, the largest of its districts till 1990, where sanitation coverage was hardly 5 per cent but is now 100 per cent.

Across the state as a whole, 2 million toilets have been constructed and installed in the last five years bringing sanitation coverage from 12 per cent in 1991 to 40 per cent today. Similarly, a huge sanitation coverage programme was launched in Andhra Pradesh in 1994 based on relatively high-cost, heavily subsidized latrines compared with those in West Bengal ultimately with a poor result mainly because of mishandling of the campaign to motivate the poor.

The most daunting problem for the world as a whole is that 2.6 billion people – two and a half times the number of people without access to water – lacking sanitation facilities need to be rescued from this basic deprivation and if this can be accomplished, bringing it to 120 million every year between now and 2015, it will still leave 1.8 billion people without access to sanitation. As pointed out in UNDP 2006 Report, there are certain barriers working against fulfilling the target. These are 'national policy barrier, the behaviour barrier, the perception barrier, the gender barrier and the supply barrier'. As stated, the national policy in many countries is more favourable to improve conditions in making greater access to water than to sanitation. South Africa and Morocco are cited as cases in support of this view. Even from people's behavioural point of view greater priority is attached to water as a basic necessity than sanitation as 'lack of water is a more immediate threat to life than the absence of a toilet', since lack of the latter does not affect so much as lack of water when prevalent sanitation practices, such as defecating in fields or streams, may do away with the necessity of toilets. People's perception of improved sanitation is a very important factor in how they value keeping the environment clean and devoid of bad smell and the risk of contamination associated with defecating in the open. Besides, the cost of improved sanitation often acts as a deterrent for poor people when nearly 1.4 billion people without access to sanitation (mainly toilets) live on less than $1 a day and for many of them even a low-cost technology is beyond their financial capability. For example, in Cambodia where the daily wage of a rural labourer does not cover a family's basic nutritional requirements, leaving nothing for health, clothing and education, it would take 20 days' wages to purchase a simple latrine. "In Nairobi, constructing a pit latrine costs about $45, or two months' of income" (UNDP, 2006:120). In Viet Nam, already having achieved the Millennium Development Goal target in sanitation, the rural coverage has increased rapidly but the poorest households have been left far behind, with hardly 2 per cent of the poorest 20 per cent having access to sanitation. On the other hand, a gender barrier is very clearly expressed in women placing a higher priority on sanitation as a protection against loss of dignity in going to public places to relieve themselves. But this priority is ignored in a male-dominated society and will not get recognition till women are sufficiently empowered. Last of all comes the supply barrier where a 'mismatch between what people demand and what is offered by the government' as it happens when pour-flush latrines offered through government programmes are often discarded by people for the lack of 'secure water supply' or technologies become too expensive for people in general. As is evident from facts gathered from many countries, a planning frame of 10-15 years is required for making effective progress in sanitation which is much more than is needed for water but 'average donor cycles and national planning cycles operate over 2-3 year cycle'.

Overcoming these barriers is not a simple task but as instances show, they are not insuperable. However, one of the major hurdles towards implementing sanitation programmes is the source of finance. In the developing household resources make a critical source of investment but there are limits to what poor households can spare for this purpose while in many countries the governments are found to be taking interest in providing subsidies for household sanitation. Without subsidies it is not possible to make sanitation widely available and accessible to all as one of the most important steps towards health improvement pervading the life of people and ultimately contributing to build a strong and stable society. However remote the chance of making sanitation universal, the major section of people can be provided with this basic service through concerted efforts as demonstrated by many of the developing countries and made available for all in the long run.

Towards this goal the support of national policies and political leadership becomes an absolute necessity for taking up sanitation as a part of national development policy, as has been done in countries like Bangladesh, China, Lesotho and India. Second, 'public participation has to be a part of national planning at all levels' so that local committees involved in such programmes can identify low-cost appropriate technology as in Brazil, Bangladesh and in Pakistan's Orangi Pilot Project. This, further, requires a proper identification of people deprived of this service and reasons for that. Besides, the voices of women need to be given more importance helping better service provision. Finally, as financing has been identified as the matter of constraint, international partnership is likely to make a big difference based on 'participative national planning processes (which) could bring the Millennium Development Goal' on achieving universal success in sanitation 'within reach'.

5
'Making Poverty History'

An assessment of various forms of human deprivation pervading the life of people across the world, as made in the foregoing chapters, stresses the point beyond ambiguity that poverty remains as the primary source of basic negations in human life, people getting deprived of access to the basic rights to food, shelter, education, medical care, freedom of speech, freedom of movement, empowerment and participation in formulating policies for social development, or in simple words, what is basically good for them. This negative existence is a painful burden to men in poverty as well as it is equally burdensome to the society they belong to. Hence, eradicating poverty means releasing the poor from this position of negation or deprivation of basic rights which every human being should earn with birth. In other words, not being in poverty is an entitlement without which his right to basic requirements of life is not fulfilled. Regrettably, this does not happen and deprivation rules the lives of many people, keeping them deprived of their basis rights defined entirely by their social position, as governed by economic status, called 'poverty' and as long as they cannot get out of this 'trap', if one can call it so as it is a condition in which they live, where the condition is nothing but a creation of the society they are a part of. Arguably, entitlement is a quality to be earned or achieved, and not one to be offered to any one for making claim to any opportunity, like an employment enabling a person to acquire other requirements, by way of selling his labour. To explain the position further, it can be said that an uneducated person has only his physical labour to sell which does not take him far in having access to other rights as he is not left with enough income to 'buy' or access services available to a person selling his labour of higher quality, greatly improved through education and training, generally called skilled labour. On a comparative assessment, a man only capable of offering physical labour without skills remains stuck to a limited fulfilment of his rights to access other more important rights which would make life much more fruitful and less stressful.

The question arises here about earning this entitlement to basic rights where some people enjoy it while some others remain largely deprived of it, or in other words, one may ask why some people remain without much entitlement, that is, poor in achieving this entitlement while many people have this entitlement in greater measure enabling them to have much greater access to basic rights. How the degree of entitlement differs so widely and what could be the cause behind making such wide variations even among those with entitlements in the degree of access they enjoy, perhaps, demands an explanation. In the case of sustained poverty people are under a strong pressure to channel all their efforts to somehow meet the absolute need for food more than any other. To get out of this limited status ruled by poverty one needs the capability to compete with others in the race for improving one's status of entitlement, a competitive spirit associated with people not heavily burdened with poverty. The poor are generally bound to block any initiative when their skills are at a very low level. In contrast, people not suppressed by poverty have the opportunity to improve their capability, and as such their marketability, giving them greater entitlement, and making them capable of more competitiveness resulting in social mobility and economic difference. (*See* Note *A*)

"In the case of studying poverty, it is the failure to have the capability to achieve minimal levels of certain basic functionings that would occupy the centre of the stage. The capabilities to be adequately nourished, to be comfortably clothed, to avoid escapable morbidity and preventable mortality, and so on, become the appropriate focus variables" (Dreze and Sen, 1999: 8). Thus, poverty as capability failure needs to be considered "with the deprivation of living conditions, for example, lack of nourishment (rather than that of the income to buy nutrients), exposure to preventable diseases (rather than inability to buy medicine) and so on" (Dreze and Sen, 1999: 9) to be explained in terms of the 'effects of income on people's basic capabilities to lead minimally acceptable lives.' It, therefore, becomes obligatory for any policy strategy to put emphasis on the close relevance between income and poverty in analyzing the level of deprivation.

In confronting deprivation, Dreze and Sen advocate two approaches to enhance capabilities of people affected, one of which they call "growth-mediated security" aimed at promoting economic growth and for this "take the best possible advantage of the potentialities released by greater general influence, including not only an expansion of private incomes but also an improved basis for public support". The alternative, "support-led security", is "to resort directly to wide-ranging support in domains such as employment provision, income redistribution, health care, education, and social assistance in order to remove destitution without waiting for a transformation in the level of general affluence" (Dreze and Sen, 1999: 22).

In defining the policy strategies further one may view the opinion of other scholars commenting on the definition of poverty which is said to be 'largely concerned about material needs or it is about a much broader set of needs that permit well-being (or at least a reduction in ill-being)', the former position concentrating on the measurement of consumption determined by income, taking income-poverty relationship almost absolute in expressing the access of different people to their most pressing material needs (Hulme and Mosley, 1996: 105). However, according to Chambers (1983, 1995), there are many forms of deprivation that poor people suffer from not 'captured by income-poverty measures' including situations like "vulnerability to a sudden dramatic decrease in consumption levels, ill-health and physical weakness, social inferiority, powerlessness, humiliation and isolation. Such dimensions of poverty are significant in their own right and are also essential analytical components for the understanding of income poverty" (Hulme and Mosley, 1996: 105-106). Eastwood and Lipton (1999) apply the 'Food Energy Method' (FEM) to draw a poverty line where the FEM "identifies the level of consumption per person' which adjusted for Purchasing Power Parity (PPP), amounted to $ 30 per person per month (in 1985 prices) worth of goods and services consumed by an average citizen." (Eastwood and Lipton, 1999: 4).

Significantly, while the essence of the concept of poverty relates to deprivation, it has a multi-dimensional aspect "as a shortfall from some minimum acceptable standard of not merely consumption or income, but also elements of human capability as health, education, security and empowerment. The reason that the poor are deprived of these basic minimum indicators of well-being is that they lack resources" (Khan, 2006: 63). The World Bank in a recent report on attacking poverty has identified these resources as human assets, such as 'capacity for basic labour, good health and skills; natural assets, such as land; financial assets, such as savings and access to credit; and social assets, such as informal and formal social security and political power'.

However, according to Vandemoortele (2004), although poverty is broadly interpreted as multi-dimensional, its principal measure remains 'one-dimensional' and $ 1 per day per person (expressed in purchasing power parity of 1985) has become the 'international benchmark' for measuring the extent of poverty in developing countries based on studies in some 33 countries made in the 1980s. But this has been updated in terms of an expanded set of PPPs (values at 1993 prices) and the international poverty line has been set at $ 1.08 per day per person. As estimated by the World Bank (2002), the proportion of people in the developing countries living on less than $ 1 a day declined from 32 to 25 per cent between 1990 and 1995, decreasing by 1 per cent per year during that period, reducing the total number of poor people from 1.3 billion to 1.1 billion which is not very satisfactory as 'poverty trends for most regions showed little or no progress'. Actually, 'the incidence of income poverty remained largely unchanged in Sub-Saharan Africa, Latin America and the Caribbean and the Middle East and North Africa', the total number of income-poor people in these regions decreasing only by about seven million each year between 1990 and 1999. On the other hand, the achievement in reducing poverty in China and India had its impact felt largely in the reduction of global poverty but as it was not reciprocated in other regions of the developing world in the same measure so is unlikely to impact heavily on reducing global poverty which can not be halved by 2015, as targeted by the United Nations as one of its MDGs (Vandemoortele, 2004: 378) (*See* Note *B*).

The UN World Summit on Social Development held in 1995, which was participated by 117 countries including all EU governments, came to an agreement on two definitions of poverty: absolute and overall poverty of which absolute poverty was defined by the UN as "a condition characterized by severe deprivation of basic human needs, including food, safe drinking water, sanitation facilities, health, shelter, education and information. It depends not only on income but also on access to services". On the other hand, overall poverty was considered to take various forms, included "lack of income and productive resources to ensure sustainable livelihoods; hunger and malnutrition; ill health; limited or lack of access to education and other basic services; increased morbidity and mortality from illness; homelessness and inadequate housing; unsafe environments and social discrimination and exclusion. It is characterized by lack of participation in decision-making in civil, social and cultural life" It is an accepted view that income is important, but access to some of the basic requirements, such as safe water supply, sanitation, healthcare, education and roads takes an equal or more important role in eradicating poverty, particularly in the developing countries.

In the process of eradicating poverty and ensuring proper well-being of people, human development forms an integral part, as pointed out in the UNDP 1995 report, in defining the concept of poverty, and that human development has two sides one of which is the formation of human capabilities, such as improved health, knowledge and skills whereas the other is concerned about making use of the acquired capabilities of people for 'productive purposes, for leisure or for being active in cultural, social and political affairs', and 'the purpose of human development is to enlarge all human choices, not just income' (Gordon, 2004: 61-62).

The well-known poverty bench-mark set by the Millennium Development Goals of the United Nations at $ 1 a day per person is 'specially calculated' and 'adjusted' to 'the cost of living which varies between countries and means that a dollar commands a larger purchasing power in Delhi, Dhaka or Johannesburg than it does in New York'. "Adjustments are made by the UN using a set of conversion factors known as 'Purchasing Power Parity' (PPP) exchange rates. The PPP-adjusted dollars attempt to account for the greater purchasing power in the countries we study than market rates would imply" (Collins *et al.*, 2010: 5)

A. CHALLENGE TO ERADICATE POVERTY AND HUNGER

Between 1990 and 2004, the share of people living in extreme poverty fell from 33 per cent to less than 20 per cent, though the rates of progress vary geographically and are inconsistent. Africa presents a desperate situation in this regard, where despite several success stories of achieving food security, there 'has not been any sustainable reduction in the prevalence of poverty and hunger'. With agricultural productivity declining, 'latest FAO estimates indicate that, over the past 15 years, the number of undernourished people has actually increased to more than one-third of the continent's total population' (*See* Note C).

Poverty alleviation remains a daunting challenge today despite recent progress in economic growth improving the lives of millions of people in the developing world. There is sufficient indication that economic growth alone is not adequate unless it consists of pro-poor policies required for appropriate income distribution. As Islam (2006) observes, the pattern and sources of growth and the manner in which its benefits are distributed are extremely important for poverty reduction in which employment as a link between growth and poverty alleviation plays a very effective role. Hence, in the formulation of poverty reduction strategies employment needs to be integrated with them as a 'route out of poverty', though this is not very clearly specified as a means for achieving the Millennium Development Goal related to the eradication of poverty. In expounding further the close link between growth, employment and poverty reduction, Islam stresses that high 'rates of economic growth leading to increase in productive capacity, generating increasing employment opportunities which, in its turn, allows absorption of unemployed and the underemployed'. This has a very important effect on the economy and in moulding the life of the poor who with increase in income become capable of diverting a part of that to "education and skill formation of their children, thus raising the productive capacity of the future work force and creating necessary conditions for achieving higher levels of economic growth. The process would thus complete the virtuous circle of economic growth leading to poverty reduction via growth of employment with rising productivity, and reduced poverty creating the possibility of further increases in productivity and higher rates of economic growth" (Islam, 2006: 34) (Fig. 5.1). But this happens only when the policies of economic growth are basically pro-poor.

"Econometric exercises made by Islam (2006) based on household survey data in some countries like Bangladesh, Bolivia, Ethiopia, India and Viet Nam, provide stronger support to the hypothesis" that the 'impact of employment and labour market on poverty reduction remains strong showing different types of results: for example, "concentration of workers in agriculture causes poverty" while shift of workers to industry reduces poverty; further, "education of the workforce (or population as a whole) contributes to poverty reduction; and a higher dependency burden causes poverty" (Islam, 2006: 43). However, in many of the developing countries agriculture must have a prominent role for pro-poor growth. As is often the case, a structural shift of the economy towards higher productivity sectors generating higher incomes is more capable of reduction of poverty but this does not rule out the fact that agriculture can contribute to the pro-poor growth process. This has been very powerfully demonstrated by countries like Uganda and Viet Nam where the distribution of landholdings being egalitarian and crop production based on smallholders moved to higher-value products incorporating a greater degree of product diversification, raising productivity and the real wages of agricultural labourers proved to be very successful ways of making agriculture an equally effective

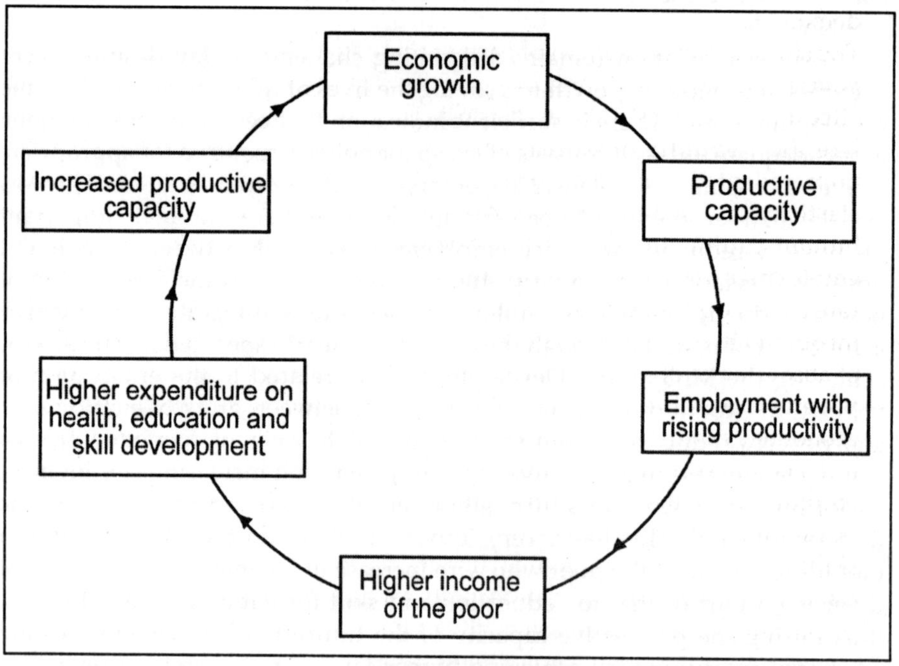

Fig. 5.1: Virtuous Circle of Links Among Growth employment, and Poverty Reduction

Source: Rizwanul Islam, 2006

instrument for improving the economic level of the poor boosted by pro-poor policies and thus fighting poverty effectively.

At the same time, the labour market may have significant influence on poverty reduction along with education and newly acquired skills through training and greater exposure. This is investment in 'human capital', as mentioned in an earlier chapter, playing a vital role in enhancing economic development benefiting the poor. Education and skills are the most important tools in enhancing the capability of the work-forces with integration into the growth process giving higher productivity and increasing incomes. A comparison of data between East Asia and South Asia confirms further that investment in 'human capital', in terms of both education and skill training proves to be 'an important ingredient in a country's strategy for pro-poor growth' with implications for development strategies and policies accelerating growth and poverty reduction (Islam, 2006: 59). The linkage between employment and poverty may be taken as a process in market economy in which economic growth inducing productive and remunerative employment leads to a reduction in poverty, 'the effectiveness of the process' depending very much on the 'strength of each link in the process'. "These links consist of the rate of growth of the economy, the output elasticity of demand for labour, and the ability of the poor members of the labour force to respond to increasing demand for labour, especially for more productive categories of employment. The simultaneous strength of each of these links promotes rapid reduction in poverty, while the weakness of any of these links weakens the poverty reduction effect of the process" (Khan, 2006: 64). This has been proved amply in countries like the Republic of Korea, Taiwan, Hong Kong and Singapore in their unprecedented economic growth since the 1960s, accompanied by a high output elasticity of demand for labour "endowed with steadily increasing levels of human capital in terms of better health and higher skills. As a result, employment increased rapidly and real wages increased roughly at the same rate as per capita output. Rapid economic growth resulted in low and stable inequality in the distribution of income and a dramatic decline in the incidence of absolute poverty" (Khan, 2006: 64-65). Entitlement to means of overcoming poverty is possible only if employment is 'productive, that is, accompanied by economic growth. At the same time, it may be noted that high economic growth not accompanied by a rapid growth of productive and remunerative employment will fail to bring about a commensurate reduction of the rate of poverty. Besides, growth in employment should be of a kind to suit the available skills of the labour force so that they are not deprived of access to employment opportunities. Similarly, growth might fail to reduce poverty if the distribution of productive resources, such as, land and physical capital remains highly concentrated. To make employment generation for the poor a more effective tool it is necessary to utilize the capability of the poor in 'an effective way to raise the rate of economic growth for which they need access to resources like land, other physical assets, credit and public services' ensuring complementarity between high growth and rapid poverty reduction'. "East Asia's success in eliminating much of poverty by growth-induced employment expansion, buttressed by redistribution of assets and egalitarian access to human capital, was a remarkable achievement. These countries should have reinforced this achievement by building a system of social protection for the workers, especially after they had reached high levels of income. Their failure to do so at the right time contributed to an unwarranted intensity of decline in the welfare of the workers, and to a descent into poverty for many of them, in the wake of the crisis of 1997" (Khan, 2006: 97-98).

The governments in many countries, such as South Korea, Singapore etc. have pursued a strategy of growth-mediated security by way of "promoting economic growth through skillful planning, facilitating wide participation of the population in the process of economic expansion, particularly through the promotion of skills and education and the maintenance of full employment, and utilizing a substantial part of the resources generated by rapid growth for extensive public provisioning of basic necessities" (Dreze and Sen, 1999: 24).

The UNDP (2003) report suggests sustained economic growth as the only means for poor countries to come out of the poverty trap. It points out that by directly increasing average household incomes with households sharing the rising national income, the extent of income- poverty is demonstrably reduced. The report further emphasizes ensuring poor people's share in overall growth, with a 'focus on extending their access to critical assets including land, making it easier to start small business enterprises, supporting labour-intensive exports and micro-finance. The report also stresses an increase in public investments in education, basic infrastructure, disease control and health as contributions to the reduction of non-income poverty. Greater investment in education and training skills of the labourers will enhance productivity and economic growth further raising poor people's incomes.

To escape poverty traps, countries require greatly expanded finance from donors to be invested more heavily in health, education, agriculture, water and sanitation as well as in other peripheral infrastructure to create conditions for sustained economic growth. In other words, to get out of poverty traps, the countries are urgently required to "reach certain critical thresholds – of health, education, infrastructure and governance – that will permit them to achieve take off to sustained growth" (UNDP, 2003: 18).

The process of emerging out of this situation, i.e. the poverty trap, is directly dependent on certain steps to be taken. They include – (a) a larger investment in human development, such as nutrition, health (including reproductive health), education, water and sanitation, gains in them contributing to raise the per capita income substantially, (b) More investments should be made for small farmers, especially in predominantly agricultural countries, and they need to be helped in increasing their productivity and thus 'break out of subsistence farming' which can be done with better technology, such as improved seeds, tillage and crop rotation systems, soil nutrient and pest management with fertilizers and pesticides besides investing in improving 'rural infrastructure creating irrigation facilities, storage and transport facilities, road connectivity between the rural settlements and outlying markets besides security in landholdings to protect the farmers' rights, (c) Investment in infrastructure such as power, roads, ports and communication supporting diversification into non-traditional areas of activities may open up prospects for some areas now lying dormant for lack of them, (d) Encouraging development of non-traditional activities in the private sector with particular emphasis on small-, to medium-sized enterprises, for which special industrial development policies need to be framed – such as in East Asia's "tiger" economies – "through tax holidays, export processing zones, special economic zones, science parks, investment tax credits, targeted funding for research and development and public grants of infrastructure and land.Without such special inducements it is difficult for small countries to gain a foothold in non-traditional areas of the world economy" (UNDP, 2003: 19). Many of them have already taken form and functioning in different parts of India with similar aim in view, (e) Greater emphasis needs to be laid on human rights and social equity, and political institutions must allow the poor section of the population, especially the women, to participate in decisions affecting their lives, ensuring 'women's right to education, reproductive health services, property ownership, labour force participation and secure land tenure', and (f) Environmental sustainability and urban management make a very important issue since biodiversity and ecosystems with life-support, such as clean water, air, soil nutrients, forests, fisheries and other key ecosystems have to be protected and sustained to ensure livelihoods, safe environments and livable cities. Further, many of the world's poorest places, such as Bangladesh, Maldives, etc. are in regions of enormous climatic 'variability and vulnerability, requiring sound ecological planning' (UNDP, 2003: 76-78).

One of the major thrusts or planning principles aimed at bolstering the economy of the poor countries in particular has been advocated as promotion of labour-intensive industries creating manufacturing exports which are likely to win 'half the battle of achieving sustained growth' especially for countries having a history of exporting primary products, or, in other words, in countries which for long have basically been agricultural economies. Policies promoting labour-intensive rather than capital-intensive industries, using available or imported simple technology suited to the situation, will increase employment opportunities much faster than agriculture can do, and raising productivity with access to a larger, slowly opening outer market will raise real wages. This has happened in East Asia's "tiger" economies since the 1960s supported by 'disciplined institutional capacity within governments'. The success of such pro-poor industrial development policies depends on the long-term growth of the economy in those poor countries which depends on exports (having a larger, as well as sustained, demand for products like processed food or parts of machineries involving high labour-cost in advanced countries, etc.) with moderate exchange rates making the products more export-oriented. Government protection to domestic markets along with export incentives in China and the Republic of Korea have been a great success, particularly so with Korea providing 'exporters with tax incentives and 'duty-free import of inputs' raising returns to capital invested.

A structural change in the economy in the manner mentioned above is very much to be desired in view of growing challenge of increasing number of unemployed. But "favourable recent global economic trends have not been fully reflected in commensurate growth in employment. Global unemployment reached a record of 195.2 million in 2006, although the rate of unemployment was unchanged from the previous year at 6.3 per cent. For the first time ever, the share of global employment in the service sector (40.0%) rose above that in agriculture (38.7%)... Employment opportunities in recent years have tended to favour skilled workers over the unskilled so that the poor, with their limited training, have benefited least. Nevertheless, some progress has been made as, among the employed, the number of those earning less than $ 1 per day fell by over 10 per cent between 2000 and 2006 (Table 5.1). Most of these workers moved into the group of those earning less than $ 2 a day, whose number rose from 768 million in 2000 to 846 million in 2006" (United Nations, 2007: 250).This is further indicated by increase in the growth of output from 5.6 per cent in 2000 to 6.9 per cent in 2006, employment of females increasing from 787 million in 2000 to 878 million in 2006, and increase in enrolment rate in primary education both for males and females, as shown in Table 5.1.

Table 5.1: Selected Indicators of Progress in Developing Countries 2000-2006

Economic Growth	*2000*	*2001*	*2002*	*2003*	*2004*	*2005*	*2006*
Growth of Output (percentage per year)	5.6	2.7	3.9	5.2	6.9	6.5	6.9
Poverty (millions)							
Population living on Less than $1 per day (millions)	1120.0		1067.0	—	986.0	—	—
Proportion of Population (%)	22.3	—	20.4	—	18.4	—	—
Working poor ($1 per day) (millions)	570.0	568.0	560.0	545.0	528.0	515.0	503.0
Employment (millions)							
Employment – Female	787.0	800.0	816.0	834.0	847.0	862.0	878.0
Employment – Male	1255.0	1276.0	1299.0	1325.0	1347.0	1371.0	878.0
Education (%)							
Primary Gross Enrollment rate – Female	78.9	79.5	79.2	81.3	83.7	—	—
Primary Gross Enrollment rate – Male	85.5	85.6	85.4	85.6	87.9	—	—

Source: UN, 2007: Strengthening Efforts to Eradicate Poverty and Hunger (based on World Economic Situation and Prospects 2007 & World Bank 2007)

For industries to provide greater openings for unemployed people capital-scarce economies require financing incentives such as "directed and subsidized credit, support to chosen sub-sectors, export subsidies, technology acquisition institutions and a host of other sector-specific interventions. Several South-East Asian countries have used export credits and fiscal incentives to raise returns to investment in exports" (UNDP, 2003: 79).

Osmani (1999) talking about social-security in South Asia, suggests the following measures as part of fighting poverty: (a) policies relating to land including land reforms, assistance for small and marginal farmers, crop insurance, and so on; (b) policies relating to employment putting emphasis on creating opportunities for self-employment as well as employment in non-agricultural activities, providing wage-employment in public works, etc.; (c) direct public provision of basic needs, such as food, housing, health, and so on. The results obtained through social security measures in control over land, self-employment, and wage employment, and public provision of basic requirements in countries like India, Bangladesh and Sri Lanka have been reviewed by the author (Osmani). According to his findings, 'by the mid-1980s about 1.5 per cent of the cultivated land in India had been acquired under ceiling laws passed in the 1950s, and less than 80 per cent of it was actually distributed', benefiting about 3.4 million persons receiving on an average a little over 1.3 acres of land, enhancing the security of livelihood of beneficiary families, more than half of whom belonged to the scheduled castes and tribes, forming the most disadvantaged group of the society. However, the total amount of land thus available for distribution among the landless people, those benefited, constituted a very small section of the total number of land-poor people in the country.

Social Security Measures in Some Countries/Regions of the Developing World

I. Bangladesh

Compared with India, Bangladesh presents even a poorer case of land reform, where the excess land amounts to about 1 per cent if the ceiling is applied strictly, and only 15 per cent of the potential surplus land has been acquired, more than a quarter of which has been distributed among landless peasants. In contrast, Sri Lanka holds a better record, with as much as 20 per cent of cultivated land acquired by 1972 and 1975, and 12 per cent of that was given to the landless and small peasants on the basis of the special nature of land reform law enacted with a special consideration for the plantation sector and a decision not to give away plantation land acquired by the government in excess of ceiling limit to the peasants 'on the ground of economies of scale' (Osmani, 1999: 306-07). As a result, the overall impact of land distribution as a policy for land reform giving security to people with the prospect of better livelihood was negligible, and as Osmani (1999) observes, the failure lies in the stipulation of an exceedingly high ceiling for land, such as 33 acres in Bangladesh and 25 acres (of paddy land) in Sri Lanka, leaving a very small surplus of land even if the ceiling is applied very strictly for distribution among the farmers. The high ceiling, fixed at 20 acres in Kerala and 25 acres in West Bengal, yields a surplus of less than 2.5 per cent of the cultivated land in Kerala and just over 1 per cent in West Bengal. At the same time, the presence of share-cropping as widely practiced in West Bengal and Bangladesh is an exploitative institution.

Osmani observes further that an alternative to land-based security is "to enable the land-poor to engage in those self-employed activities which are not crucially dependent on land. These so-called non-farm activities occupy a significant position in the occupational-structure of South Asia, accounting for 34 per cent of the rural labour-force in Bangladesh, 46 per cent in Sri Lanka, and 19 per cent in India" (Osmani, 1999: 318-319) In Bangladesh for example, of all the gainful activities – primary and secondary – by people having no more than 0.5 acres of land, over 50 per cent are engaged in non-farm activities when agricultural activities account for only a third of the total. However, all of those workers in the non-farm sector may not be self-employed, as many of them are wage labourers, the former depending largely on the availability of easy credit or subsidy as an 'essential prerequisite for ensuring security through self-employment.'

For an insight into the link between poverty and economic growth in Bangladesh, the trend in rural and urban areas needs to be disaggregated. Based on poverty line income the incidence of poverty in Bangladesh declined continuously from 1988-89 to 2000 on a 'Headcount ratio' (HCR) though according to the Cost of Basic Needs (CBN) method it increased from 1986 to 1992, increasing at a higher rate in rural areas, recording 8.1 percentage points increase during this period (1986-1992) and 2.0 percentage points in urban areas (Table 5.2). The poverty situation improved between 1992 and 1995 in both sectors, more in urban than in rural areas, being 9.9 and 4.5 percentage points respectively. In contrast, rural poverty declined by 3.7 percentage points in 1995 to 1999 while urban poverty increased by 1.6 percentage points. As observed by Rahman and Islam (2006), in their study of the relationship between poverty and different factors, the GDP growth rate shows the expected relationship with poverty reduction, a GDP growth rate higher than 6 per cent per annum leading to poverty reduction at a rate of 1.5 percentage points per annum (Rahman and Islam, 2006: 118). Poverty incidence is higher in rural households and the rural poor are mostly engaged in agriculture. Despite the share of agriculture in GDP having fallen from 50 per cent in the 1970s to about 25 to 30 per cent by the end of the twentieth century, it remains the highest sector for employment with more than 60 per cent of the labour force engaged in such activities. The decline of poverty incidence in the rural areas is plausibly linked not only to agriculture but also to other sectors since growth of industrial activities and urban service sectors attracts workers from the rural areas, increasing the average labour productivity of agricultural workers. As a matter of fact, with gradual withdrawal of the landless labourers and their migration to urban areas, rural wage rates will be subjected to an upward pressure. Poverty alleviation is linked with overall GDP growth and sectoral growth where increased food production will raise the real earnings of both the urban and rural wage labourers. As analysed by Rahman and Islam, "if total employment can be increased through high growth sectors where the elasticity of employment (high elasticity of employment implies an employment-intensive growth) has declined, with an associated increase in productivity and wage, this can be an effective step for poverty reduction. Leather and leather products, knitwear, readymade garments, food manufacturing, and printing and publishing are among such subsectors. Leather and leather products, and knitwear and readymade garments, are the

Table 5.2: Trends in Rural and Urban Poverty in Bangladesh 1985-86 - 2000

	HCR based on CBN method (%)			HCR based on DCI method (%)		
	Rural	**Urban**	**National**	**Rural**	**Urban**	**National**
1985-1986	53.1	42.9	51.7	54.7	62.6	55.7
1988-1989	59.2	43.9	59.1	47.8	47.6	47.8
1991-1992	61.2	44.9	58.8	47.6	46.7	44.5
1995-1996	56.7	35.0	53.1	47.1	49.7	47.5
2000	53.0	36.6	49.8	42.3	52.5	44.3

Source: Rushidan Islam Rahman and K.M. Nabiul Islam (2006) (based on Bangladesh Bureau of Statistics, various years and World Bank, 1998, 2002d)

Notes: HCR: Headcount Ratio; CBN: Cost of Basic Needs; DCI: Direct Calorie Intake

largest export sectors in Bangladesh. Thus the current sectoral growth pattern, if pursued vigorously, can help in both GDP growth and poverty reduction" (Rahman and Islam, 2006: 130).

But, differences in male and female wage rates have direct implications for gender equity and the overall functioning of the labour markets well as poverty levels The direct implications of the low female wage rates, much lower than the wage rate of male workers, becomes obvious from Table 5.3.A day's wage earned by a female worker even with year-round employment is not adequate to support a family of three members. The wage rate of women being 45 and 31 per cent lower than men in the rural and urban areas respectively, 'families depending on female labour are subject to more intense poverty. And the difference becomes more significant, especially in the rural areas as usually women get employment for fewer number of days than men do.

Besides, the difference in underemployment between males and females in Bangladesh is quite striking in view of existing wide disparity. For example, 'low male underemployment prevails in both rural and urban areas – 11.8 and 9.7 per cent respectively, implying that poor male workers are not actually without work; rather they are working for long hours'. In contrast, however, the unemployment and underemployment rates are much higher among women, 74 and 60.4 per cent remaining underemployed in the rural and urban areas respectively. As finally observed by Rahman and Islam (2006), ensuring participation of marginal and small farmers, high growth in agriculture can contribute to poverty reduction in Bangladesh with emphasis on crop intensity and crop diversification besides 'extensive extension service making findings of research centres effective in the efforts of farming. Growth of "non-crop agriculture" such as livestock, fisheries and horticultural production supported by public investment in infrastructure, availability of credit, quality control and development of marketing facilities and the like would be of great importance in such development activities. Lastly, non-farm activities can play

Table 5.3: Male and Female Wage Rates Relative to (Moderate) Poverty Line in Bangladesh 1989-1999

	Sector	*Wage Rate (taka per day)*		*Number of Persons who can be sustained above poverty level by a day's wage*	
		Male	*Female*	*Male*	*Female*
1989	Agriculture	31.6	22.7	2.1	1.5
	Non-agriculture	46.0	20.9	2.1	1.0
1995-1996	Rural	44.0	25.0	2.2	1.3
	Urban	60.0	36.0	2.5	1.5
1999-2000	Agriculture	63.0	35.0	2.5	1.5
	Non-agriculture	85.0	59.0	3.2	2.2

Source: Rahman and Islam, 2006 (Based on Bangladesh Bureau of Statistics, various years)

a very effective role taking people out of poverty provided the real wages offered to workers engaged in them is really attractive. Further, growth in employment and output in manufacture need to be higher so that the share of employment in higher-productivity sector can increase and the industries already identified for this purpose – food, textiles, leather and leather products, wood products, non-metallic minerals, etc., where 'output growth can be quite employment-intensive'. All with the necessary process of transformation there should be larger investments in education and skill-development opportunities to keep it going in reducing poverty' (Rahman and Islam, 2006: 155-156)

Most significantly, the poverty alleviation programme taken up by an Economics Professor, Muhammad Yunus, has a revolutionary impact on the Bangladesh economy. Introduced in 1976, it was a programme to provide credit to poor people prompted "by the idea of alleviating poverty in his war-torn (freedom movement in 1971) and desperately poor country, (which) looked as if he had developed an antipoverty device as much as a new form of banking" (Collins *et al.*, 2009: 155). Known as Grameen Bank (or Rural Bank) it focused on the poorest rural households – those owning less than half an acre of cultivated land. Members of these households formed groups of five members, representing a single sex in the beginning and came to be all women later, and a member of such groups met weekly with a Grameen worker, making a compulsory savings deposit into a jointly-owned fund to repay the loan each week at a fixed amount covering both principal and the interest taken from the Grameen Bank to use in a new or existing business. Successful on-time repayment of loans entitled the loanee to a bigger loan in future while availability of credit in this manner proved to be the 'most effective way to unleash the productive capacity of the villagers trapped by cycles of low incomes and low skills'. By 2000, the Grameen Bank had more than two million customers, or "members", as they were called, and the movement became increasingly popular as a pro-poor enterprise after 1998 flood doing away with rigidity and inflexibility in the lending system, bringing in a wide range of loan terms from three months to three years with scope for topping up their full value before they were fully repaid. Further, the bank introduced the 'Grameen Pension Savings' or GPS, which offered a good rate of interest in return for regular monthly deposits over a five-, or 10-year term.

Regarding use of the loan an insight into 237 loans based on dividing them into six categories, has been made as a case study, to ascertain the nature of the use of loans by members of the Grameen Bank. This is shown in Table 5.4 where 55 loans are shown against mixed categories, used for various purposes. Among the rest, the largest number of loans, representing 39 per cent by value of the total loans, were utilized for the purpose of retail or trading businesses and crafts, and 37 loans, accounting for 14 per cent of the total loan value went into asset acquisition and/or maintenance, in other words, for infrastructural development, and this along with the first one imply the use of 53 per cent of the loan is meant for productive purposes while a paltry 4 per cent of the total loan was used for consumption, which indirectly indicates the apparent better

Table 5.4: Number and Disbursed Value of Microfinance Loans from Grameen Banks, by Use Category

Category of use	*Number*	*%*	*Value (US$)*	*%*
Stock for Retail or Trading Business and Crafts	75	32	15,231	39
Asset Acquisition and/or Maintenance	37	16	5,583	14
On Lending to others Outside the Household	27	11	5,764	14
Paying Down Other Debt	25	11	3,413	9
Consumption	18	8	1,423	4
Mixed Uses	55	23	7,535	19
Total	237	100	38,951	100

Source: Collins et al, 2009. (US$ Converted from Bangladesh takas at $=60 takas)

financial position of the borrowers who are in a position to meet expenses on household consumption, including that on education and healthcare (with particular emphasis on girls' schooling) from their own earnings.

Highly interesting is the growing financial health of the Grameen Bank itself in which the "loans had grown rapidly to 27,970 million takas (US $ 466.16 million) by 2005, while savings had increased even faster, to 31,659 million takas (US $ 527.65 million) or 13 per cent bigger than the loan portfolio... Bangladesh's microfinance industry, one of the world's oldest and biggest, continues to develop at a rapid pace. The combination ... (of) open passbook savings, more flexible ways of lending, and commitment savings accounts – show how much has been achieved in Bangladesh to improve financial services for the poor. Muhammad Yunus's original vision for Grameen Bank helped the world see the power that access to simple loans can have in helping villagers build small businesses ... (and once these are in place) improved specialist services, loan, and insurance services can be developed. They would respond to demands for products for home improvement, medical and educational expenses (over and above which individual households are already doing within their means), and pensions, for example, as well as for micro-enterprises" (Collins *et al.*, 2009: 171 and 173).

The very purpose of micro-finance in Bangladesh having been the eradication of poverty, and remaining focused on the poor, the Grameen Bank has exhibited a surprising capacity to 'develop products and take them quickly to scale. This alone is very likely to enable them to absorb new ideas and technologies to improve further, thus make it a model of financial innovation for the rest of the world to emulate'. In full recognition of this novel idea translated into action with unprecedented success, Professor Muhammad Yunus was awarded the Nobel Prize.

II. Sub-Saharan Africa

Sub-Saharan Africa presents a case where social security, as observed by von Braun (1999) can only be achieved after economic growth which must result from agricultural growth 'with an increase in labour productivity', 'inseparable from improved production technology (capital) and nutritional well-being and health'."Improved social security is likely to make a substantial contribution ... by (a) reducing mortality rates; (b) reducing the number of births needed in a family to ensure a given number of surviving children; and (c) reducing the demand for surviving children" (von Braun, 1999:395). However, the ability of all people to acquire food for an active and healthy living is at the core of social security policy. According to von Braun (1999), the relevance of food security increases with poverty and increases further where environments have extremely deficient infrastructure, as in Sudan and in Ethiopia, where 'investment in rural infrastructure and policy reform for the stimulation of agricultural growth' should have high priority for ensuring social security. Unlike in Asian and many Latin American countries, such priorities have to be fixed to increase levels and reduce instability of domestic food production and develop market integration, and related delivery systems arising out of deficiency in infrastructure. Clearly, 'employment generation is a key requirement for the expansion of access to food for famine prevention' but at the same time price instability, due largely to production instability, hinders the process of social security. Hence, for increasing agricultural production and making it less shock-prone improved technology is one of the top priorities. Both trading scope and gain from specialization in production are enhanced by improved infrastructure. A conducive foreign trade policy contributes very effectively to basic social security by 'reducing the impacts of production fluctuation' (von Braun, 1999: 402-03).

In this context, one cannot forget the fact that Sub-Saharan Africa contains 16 of the 18 countries with the highest malnutrition in the world and about one-third of the total population in Sub-Saharan Africa suffers from hunger. The rural landscape in Africa is still dominated by smallholder subsistence farming, low technology and a weak-based agricultural production system resulting in "stagnating or declining agricultural productivity, weak backward and forward linkages between agricultural and other sectors" of production, "loss of competitiveness in world markets, increased food insecurity and natural resource and environmental degradation" (United Nations, 2007: 125).

According to Kanji (2004), countries in Sub-Saharan Africa continue to be dependent on primary commodities, 80 per cent of which are composed of oil and non-oil commodities, subject to price variations and declining terms of trade. The region's dependence on trade, measured as a share of GDP increased from 38 to 43 per cent between 1988-89 and 1999-00 but it became increasingly marginalized through a declining share of total world trade'. It is generally agreed that it is a region where the majority of the people are still largely dependent on small-scale agriculture, where the investments are low, markets are weak and disorganized, making rural livelihoods largely dominated by small-scale farming, agricultural labouring, petty trading and service provision, migration as well as remittances from the emigrants. In a situation like this, as

Kanji observes, it is inconceivable that in the foreseeable future the cities or commercial agriculture can offer employment to vast number of poor people in the countryside. As a way out of lasting poverty "small-scale agriculture is essential for progress on poverty reduction, including policies favouring small farms and related industries (the non-farm economy)" (Kanji, 2004: 246) In a somewhat different approach, however, it is very strongly advocated that to combat raging poverty in Africa as a whole, an agricultural transformation process becomes a priority in which the farmers are increasingly connected to markets through the agro-industrial and agro-business activities, and to make this effective there should be a response to the demands for agricultural inputs and to the demand and preferences of 'increasingly urbanized consumers through processing, transporting and delivering food and agricultural products'. Such an 'agriculture-industry-trade nexus' is of crucial importance contributing to agricultural transformation in Africa. (United Nations, 2007: 126-127).

III. Indonesia

In Indonesia, the expenditure distribution data are used to derive a "poverty line" below which people are classified as poor, not having adequate purchasing power to acquire the basic necessities of life. Consumption-poverty, as claimed by the Central Bureau of Statistics (CBS), declined from 70.1 per cent of the population in 1978 to 32.5 per cent in 1996 (UNDP, 2003) while alternative estimates put it at the highest of 40.1 per cent in 1976 and 11.3 per cent in 1996. Using the international poverty lines based on $ 1 a day per person and $ 2 a day per person, in terms of the former "the income/poverty in 2002 was a little lower than the 1996 benchmark (while) the higher poverty line (US $ 2 a day) suggests that it was still above (by 5.9 per cent) the pre-crisis (1997) level" (Islam, 2006: 314-315). (Fig. 5.2). A comparative assessment of data available from different countries in the developing world reveal that Indonesia is one of the few countries that are on track to attain the MDGs (Millennium Development Goals) in the perspective of reduction in deficient capabilities by 2015 (Table 5.5). However, as mentioned by Islam (2006), the estimates of the incidence of poverty have to take into account the fact that 'individuals and households may move in and out of poverty, thus making them vulnerable to' at least a temporary phase of poverty caused by any change in the economic functioning of the country where the diverse nature of the informal economy accounting for 67 per cent of the Indonesian workforce (1999) is characterized by "low levels of skill and productivity, low or irregular incomes, long working hours, small or undefined workplaces, unsafe and unhealthy working conditions, and lack of access to information, markets, finance, training and technology. A large proportion of those who suffer these conditions (making life highly vulnerable) are women and children" (Islam, 2006: 324). Considering the situation, creation of employment seems to be the 'core element of a national poverty reduction strategy' and specifically there is a need for 2.5 to 3 million new durable jobs to be created annually to absorb the growing workforce and to clear the 'backlog of the unemployed and the underemployed', and identification of sectors with high employment potential. The agricultural sector dominates the national employment scene, with 41 million workers, 80 per cent of them engaged in the informal economy and about 75 per cent of the total population living in rural areas, the latter providing the main livelihood. Hence, the strategy for poverty reduction needs to focus on the rural sector for development, including generating more employment by investment in infrastructural development and encouraging the development of small-, and medium-scale agribusiness activities, already targeted by the government. To quote Islam, "Improvements in land tenure security, enabling farmers to diversify into agribusiness through provision of greater information and capacity building, and nourishing linkages between small and large businesses" (Islam, 2006: 355) will be the most welcome way of generating more durable jobs and in arresting rural-urban migration in a very effective way. Since, the bulk of the Indonesian workers are employed in small-, and medium-sized enterprises (SMEs), further development of them 'represents a core plank of Indonesia's poverty reduction strategy'. Further, the significant lack in the provision of infrastructure with about 50 per cent of Indonesians having no access to electricity, with only 9.1 telephones per 100 people, and about 50 per cent of people lacking access to an asphalt road, and with 'between 6 and 11 million' people without any 'connection to any decent road or motorized transport network', are major handicaps keeping the country away from the path of development. Public investment in the development of infrastructure facilities in both rural and urban areas may be made into a 'tool of employment policy, and according to the studies of the ILO, the labour-intensive production

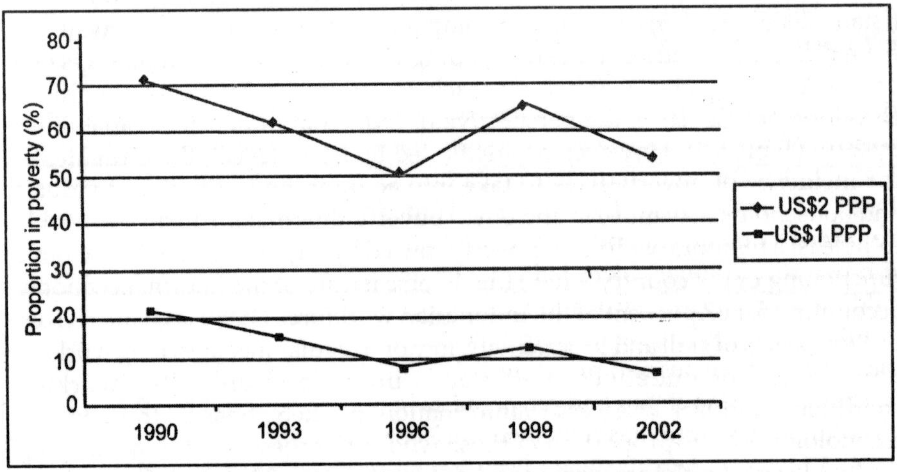

Fig. 5.2: Income Poverty, Indonesia, 1990-2002

Source: Rizwanul Islam, 2006

Table 5.5: Key Indicators of Deficient Capabilities in Indonesia relative to 2015 (MDG) targets

Indicator	2002	2015 Target
Extreme Poverty: percentage of population living below national poverty line	18.2	7.5
Child malnutrition:		
Percentage of children moderately and severely underweight	27.3	18.3
Under Five mortality rate (per 1000 live births)	46	32
Infant mortality rate (per 1000 live births)	33	17
Maternal mortality rate (per 100,000 live births)	307	124
Net Enrollment rate (boys and girls) (%):		
Primary	92.7	100
Secondary	61.7	100
Access to safe water		
Percentage of households with piped water	18.3	57.3
Percentage of households with access to improved water sources[a]	50	69.1

Source: Islam, 2006 (Based on UN Development Programme 2003)
Note: a: ten metres or further from excreta disposal

methods in Indonesia "can generate as many as 1.2 million durable jobs over four years without compromising standards of quality that one associates with equipment-intensive production techniques. This means that if general growth-induced employment creation is of the order of 2 million jobs annually, it has the potential to rise to 2.3 million annually with the adoption of labour-intensive production methods in public investment policy. This has been internationally confirmed, stressing the fact that "labour-intensive infrastructure programmes are 10-30 per cent less costly than more equipment-intensive techniques, reduce foreign exchange requirements by 50-60 per cent, and create up to five times as much employment for the same investment" (Islam, 2006: 356).

IV. China

Social security in China enjoys a sort of primacy in being the first of its kind taken up on a gigantic scale for a vast population and also some uniqueness in view of following some basic tenets of improving living conditions in general. Since, the formation of the new government in 1949, it guaranteed basic provisions to its citizens and made deprivation a concern of government policy, with two revolutionary steps being taken towards this end in the form of (a), building protective measures in land tenure and employment practices, and simultaneously, (b) implementing social-security programmes taking care of specific contingencies and protection of vulnerable groups. Notwithstanding the changes in leadership and major shifts in economic policies, the thrust to uprooting deprivation has remained unrelenting all through.

As a matter of fact, the major transformation in rural economy to stave off poverty – the most identifying feature of rural life – has been the disbanding of collective agriculture and shift to family farming since 1978, making this the central feature of the Chinese economy since 1958 (under this agricultural system the land remains under collective ownership but it was leased out to households according to 'Responsibility System'). Such economic reforms made the household economy flourish with dramatic effect on incomes and consumption of the rural population, the per capita income doubled between the nine years of reforms (1978-87) and the per capita consumption expenditure – a real index of the living standard -exceeded the total increase over the previous twenty-five years – 1952 to 1977'.

The low-income economy of China has been remarkably successful in reducing the threat of starvation and destitution of its citizens in a country in which the local calamities of 1959-61 combined with the upheavals of the Great Leap Forward took a toll of 10 to 30 million lives. Despite massive increase in farm output and rural incomes since 1978, the year when new 'reforms' started, there are still small 'pockets of transitory or permanent poverty' in China and as admitted by Deng Xiaoping (*Beijing Review*, 5 January 1985) some years ago "around 100 million people or about 10 per cent of the population still did not have enough to eat and wear" (Ahmad and Hussain, 1999: 250). Nevertheless, the achievement in providing social security to the people is astounding, surpassing that any country in the developing world has achieved so far. One of them is the implementation of a health charter in the 1950s, with priority to "preventive over curative healthcare" and the development of public hospitals and health centres, the coverage of over three-quarters of the rural population by health insurance being the most important event of the healthcare programme. The gain in life expectancy at birth from 40 years in 1953 to 68 years in 1981 is matched only by a few small countries in the developing world., such as Taiwan, Chile and Sri Lanka.

The two most important features of social security system in China are—(a) all rural households have access to agricultural land, either individually or collectively, and (b) recruitment and employment practices are geared to maintaining a high rate of employment in urban areas. A guaranteed access for all the rural households to agricultural land has enabled them to earn a living, making it the most effective issue in the transformation of rural economy, and the fundamental aim of land reform was to provide each rural household with 'at least enough land' and supports to meet basic needs. The reforms had a wide coverage, including 'middle' as well as poor peasants and they were also supplied with agricultural implements. The process eliminated landlessness as a cause of destitution. The most important concern in the distribution of land was to provide each household with the means to make a living and land has been assigned to households in proportion to their size. However, the entire system of agricultural activities remains subject to certain new conditions set by the government with the land reformation. "Households in receipt of collectively owned land have to pay land tax, local accumulation and welfare tax, and are obliged to sell to the government a portion of the produce. Taxes and the procurement quota obligation are divided prorata to the land area. In effect, households are subject to a lump-sum tax" (Ahmad and Hussain, 1999: 267). In comparison, it can be said that under collectivization, functioning as the central feature of Chinese economy between 1958 and 1978 and disbanded from 1978 onwards, 'a guaranteed income' was provided not only to shirkers but also to households lacking sufficient labour due to sickness, old age and deaths. Besides, 'financing social security programmes, rural health insurance and infrastructural work posed no special problem'. During the period of collective agriculture, there was a massive investment of resources and labour in infrastructure including a massive hydraulic construction and at the end of this period 45 per cent of the collective area in China was irrigated. The 'Responsibility System' has increased incentives though it has also increased the risks to be borne individually by rural households. Under this, the rural health insurance system undertaken by collectivization has been replaced with 'payment according to treatment thereby making households vulnerable with the risk of losing income'. Besides, mobilization of labour for infrastructural work has effectively diminished. On the other hand, the impact of the Responsibility System is quite remarkable considering rural incomes which have risen at record pace, increasing by 13 per cent annually between 1978 and 1987. The rural households have been provided with wider and more varied opportunities for earning an income and economic reforms have provided great stimulus to the growth of rural industries, a part of their profit used for subsidizing agriculture and financing rural social welfare which has been growing at 10 per cent per year during the period, 1978 and 1987 (Ahmad and Hussain, 1999: 268).

Apart from land tenure as the most important determinant of personal incomes in rural areas, the trade arrangements and pricing involving rural products take a determining role in the rural economy where the farmers sell a part of their produce (normally a quarter) to the government of which the latter resell about a third of its purchases back to rural areas, known as 'return sales', and in turn the farmers purchase agricultural inputs like fertilizers, manufactured consumer goods and grain to face shortfall in production. In practice, these proportions have increased since 1978 due to greater specializations in the cropping patterns, and in effect resale (return sales) of a larger proportion of agricultural produce to the rural areas was 'designed to alleviate nutritional contingency'. Further, 'poor areas or areas struck with natural disasters are supplied with grain either free or on loan', and 'the provision of grain is the most important component of social relief in rural areas'. At the same time, the government pricing policy has been revised with a substantial increase in procurement prices for agricultural produce since 1978 so that between 1978 and 1987, the prices of agricultural commodities rose by 77 per cent, contributing substantially to the increase in personal incomes in the rural areas.

The rural security arrangements in China are composed of two major components, which are — (a) 'measures to maintain income and promote the development of poor regions'; (b) 'public provisions for health, including cooperative rural insurance'. Of them, the first one comprises (i) social relief, including "wabao" ('a term dating from the 1950s, refers to the guarantee of food, healthcare, shelter, clothing and funeral costs for all citizens – belonging to households, or persons unable to earn a living and lacking relatives to depend on'): (Ahmad and Hussain, 1999,).; (ii) the second one is disaster relief; and (iii) the third represents development assistance to poor regions. In this context, social relief and disaster relief are meant to provide "a safety net to stop the rural population from slipping into the depths of deprivation", and social relief is primarily meant for the very poor, while disaster relief caters to a larger population against severe damage caused by natural calamities. According to Ahmad and Hussain (1999), "there has been an important difference between methods for relieving rural poverty in China and India. The Chinese government with its tight control of trade in agricultural produce and inputs, and with its relatively underdeveloped public finance system has tended to rely more on assistance through commerce than on assistance from the government budget. In contrast, the Indian government with its more developed public finance system and its looser hold over trade in agricultural produce and inputs has tended to rely more on budgetary assistance than on assistance through commerce" (Ahmad and Hussain, 1999: 273-274).

Besides land reform, giving the whole of the rural population access to land, the extension of healthcare to a vast majority of them has been another landmark achievement in social security with emphasis on public hygiene and preventive care followed by the provision of primary medical care at the grass-root level, the biggest of them being the 'Great Patriotic Health Campaign' of the 1950s targeting four pests: flies, mosquitoes, rats and sparrows (this last mistakenly). Subsequently, there had been many more campaigns, directed against opium addiction, diseases like small pox, venereal diseases, typhus, plague and schistosomiasis, emphasizing the promotion of hygiene and elimination of disease 'vectors'. As a result of all these movements the incidence of parasitic and infectious diseases could be dramatically reduced within two decades (1950-1970) though not entirely eliminated, but making an effective decline in the rates of mortality. Further, a major achievement during the period of Cultural Revolution (1966-1976) was the extension of primary medical care for the rural households, the establishment of rural health stations, training of a large number of paramedics – the 'barefoot doctors' as they were known (though they used to wear shoes, at least straw sandals as advised by Deng Xiaoping)— and sending medical personnel from cities/towns to villages when necessity arose. The idea of barefoot doctors (perhaps borrowed from the Soviet Union relying on paramedics for the same purpose) was unique and proved extremely useful as it was the only way to provide the vast rural population with sufficient trained medical persons who happened to be rural residents receiving an elementary medical training in a very short time, working part or full time dispensing basic primary healthcare and referring cases beyond their competence to commune or county hospitals. Around 1979, there were about two barefoot doctors per 1,000 inhabitants, and a proportion of these barefoot doctors acquired further medical training and upgraded to the status of rural doctors working full time. However, the term 'barefoot doctor' is no longer used. Another unparalleled achievement for China was its success in bringing about 85 per cent of its rural population under some form of health insurance scheme by 1979. (Ahmad and Hussain, 1999: 276-278).

On the other hand, as a counterpart of access to land in rural areas, a guarantee of employment for urban residents with urban registration represent the basic elements of social security in urban areas of China. They consist of 'occupational welfare' since 'urban residents receive their social security benefits and a wide range of social provisions through their work units' which vary with 'employment status and the status of the employment unit'. At present, the urban population derive benefit from three lines of defence against poverty, (i) Minimum Living Standard Insurance (MLSI); (ii) unemployment insurance, and (iii) living allowance for laid-off employees or an unemployment compensation scheme. In addition, there is an old-age pension scheme covering a part of the urban labour force. Formally, the MLSI extends its benefit to the bona fide urban population (those with urban registration, thus excluding migrants from rural areas not registered as residents with the urban authorities), covering 458 million or 36.2 per cent of the total population while the UI covers 212 million in the urban labour force, restricted to only employees and excludes 128 million people on the ground that they are classified as 'rural' despite the fact that they are urban residents without registration.

Entitlement to MLSI stipulated by the government is determined by the minimum living standard (or Benefit line) defined on the "basis of the cost of food, clothing and housing necessary for basic subsistence, giving also due consideration to the cost of schooling for children and public utility charges - water, electricity and fuel. Besides goods and services necessary for basic subsistence, two other considerations enter into the determination of local benefit lines. One is the public finance of the city and the other is the relation between the poverty line and other social security benefits" (Hussain, 2002: 315). The minimum wage in urban areas is fixed on the basis of the 'local labour market, the living standard and prices' is 'similar to a poverty line that sets an upper boundary to various social security benefits'.

But there is a section of urban population in China who, unlike other developing countries (and including some developed countries as well), remain completely outside the umbrella of urban social security. They are the floating population without urban registration, including all types of rural immigrant, amounting to as many as 50 million or about 9 per cent of the country's total urban population. As a result of a strict denial of any right to the floating population or the rural migrants to be absorbed in the urban force through government initiative, they do not feature in the statistics of the unemployed and as a result urban unemployment records have remained very low in China for any developing country, having been about 3 per cent during the period 1978-1987. There has been a massive influx of rural migrants into the cities creating problems of unemployment and destitution in urban areas. Dispatching the migrants back to the countryside had been pursued as a state policy in the late 1950s and early 1960s, which has become more difficult with time. Moreover, given such facilities as maternity benefits under labour insurance and widespread provision of child care facilities made available by work units (for those with urban registration), female labour force participation is exceptionally high in urban areas, though with a change in attitude on the part of the employers, reports of discrimination against women labour recruitment and low wage rates are becoming great hurdles for women in China, although the government is in favour of women's participation in labour force.

Urban registration is an important aspect of guaranteeing social security in urban areas, including conferring entitlement to obtain rations of selected necessities at low prices. Although most commodities have been de-rationed since 1978, items like grain, cooking oil and cotton cloth are rationed nationally. As observed by Ahmad and Hussain (1999), 'the urban social security system is largely anchored in the work units', providing housing to a large majority of the workers, continued even after retirement of the workers receiving pension and healthcare services from the work units including scare consumer goods. Private housing is extremely rare in urban areas. Heavily subsidized housing appears to be the most important benefit derived by urban dwellers under social security measures.

V. Latin America and the Caribbean

The development of the social security mechanism as a bulwark keeping poverty at bay has not been a uniform achievement for all the countries of Latin America and the Caribbean, nor has it been universal in most of them. The social insurance system emerging as early as in the 1920s in the most developed countries of Chile, Uruguay, Argentina, Cuba and Brazil evolved in a piecemeal fashion and resulted in a stratified social insurance system consisting of independent sub-systems with their own legislation, administration, financing and benefits, acquiring "a pyramidal structure, with relatively small groups of persons protected by privileged sub-systems at the apex and centre and the majority of the population with sub-systems providing less protection at the base" (Mesa-Lago, 1999, 359). Not only does this give rise to inequalities in the distribution of benefits but it also makes the costs prohibitive in giving universal coverage of providing generous benefits and liberal entitlement conditions, giving rise to financial imbalances to many sub-systems. To make the process more effective the 'unification and uniformity of the sub-systems and the elimination of the costly privileges' have been strongly advocated.

The second stage in the evolution of social insurance introduced in the 1940s took place in countries having a low level of industrialization with rural sector predominating over the urban sector. A managerial agency system was established in these countries which eventually covered the entire population by the end of the 1970s.

Lastly, the third group of countries, consisting of two sub-groups, such as the Least Developed Countries of Central America (except for Costa Rica and

Panama) and the Latin American Caribbean (except for Cuba) had the general managing agency covering virtually all the persons insured and in this sub-group, social insurance did not generally appear until the 1950s and 1960s. The other sub-group consisted of the non-Latin countries in the Caribbean achieving independence in the 1960s and 1970s. There was a national health system operating in the British colonies prior to independence and social insurance introduced there after independence had universal coverage by the 1980s, which came to be highly unified and financially solvent as in the first sub-group.

There are, however, some differences in the nature of social security systems prevailing in the three different groups of countries. For example, in the first or pioneer group of countries the first pension programmes were set up in the 1920s; more than 60 per cent of the total population came to be covered by the social security system and it became practically universal in the assistance given to the very poor. On the other hand, the typical features of the social security system developed in the intermediate groups of countries includes a pension programme introduced in the 1930s and1940s, covering between 18 and 52 per cent of the population, and most of the social security expenditure comprising about 3 per cent of the GDP and between 14 and 23 per cent of the national budget, goes into health programmes. In contrast, the first pension programmes were introduced still later in the countries of the third group in the 1950s and 1960s covering hardly 10 per cent of the total population, primarily concentrated in the capital and the largest cities. The social security expenditure amounts to only 2 per cent of the GDP and not more than 18 per cent of the national budget, some 80 per cent of which goes into health programmes and less than 20 per cent to pensions. Table 5.6 shows the proportion of expenditure under different heads through the social security systems in Latin America and the Caribbean countries.

Table 5.6: Number of Countries in Latin America and the Caribbean (LAC) providing Social Insurance and Family Allowance Coverage of Various Risks: 1922 — 1985

Programmes	*1922*	*1932*	*1942*	*1952*	*1962*	*1972*	*1982*	*1985*
Social Insurances								
Occupational Accidents and Diseases	12	18	23	27	27	29	30	30
Sickness and Maternity*	0	1	7	13	17	24	29	30
Old Age, Disability and Survivors	0	2	8	14	17	31	34	34
Unemployment	0	0	0	2	4	5	6	6
Family Allowances	0	0	0	1	5	6	7	7

Source: Mesa, Lago, 1999 (Based on US Social Security Administration, 1985)

Note: * In practically all the non-Latin Caribbean countries, social insurance/ sickness/maternity programmes cash awards are given, but healthcare is provided by public-health programmes.

Very strikingly, majority of the LAC countries do not offer any relief for the unemployeds and family allowances for the distressed made available only by about early 1950s, involving 6 to 7 out of a total of 34 countries in LAC by 1985. In comparison, old age, disability and security for survivors are the more common grounds for relief in which all the countries have taken interest and also most of the other countries have started providing relief since 1972 or thereabout.

However, the most important among social security measures in the LAC countries is the social insurance sickness and maternity programme, covering the entire employed labour force in half of the countries while in the other half, it covers only those in the public sector and in industry, mining, commerce and financial services. The countrywise differences in coverage are, however, interesting: for example, in Cuba all residents are covered but in Costa Rica and Chile the entire population except those with high incomes who are not working are covered. The major qualifying aspect of this coverage is that in most countries only urban wage-earners and their closest dependents are covered by social insurance while self-employed persons, agricultural workers, those in domestic service, the unemployeds and their dependents are not covered. Furthermore, in a quarter of the Latin American countries, the least developed ones, 'the coverage is limited to the capital and the larger cities' (Mesa-Lago, 1999:368-369), though it is reported that 'public healthcare is available for practically all the population'. However, apart from the low overall coverage, inequality in the degree of coverage among the different occupational groups, economic sectors and geographical areas is very strongly felt primarily caused by difference in income, labour skills and the power of pressure groups (*See* Note *D*). It is highly interesting to note that those below the poverty line 'are not normally protected by social insurance when proportion of people below the poverty line in LAC rose to 39 per cent in 1985. Nevertheless, the pioneer countries in respect of introducing welfare systems, tend to cover all the social risks, providing greater number of benefits as is observed in the case of Argentina, Brazil, Chile and Uruguay where all the social risks, including unemployment and family allowances, are covered, strengthening the fact that 'older the social security system, the greater the number of benefits provided'. Such social benefits also include 'housing plans, personal and mortgage plans, purchasing facilities, and other recreational, sporting and cultural services for insured persons and their dependents'. In comparison, in countries where social security programmes have been introduced much later, the coverage remained low with the majority of their population having access only to a small proportion of the public health and social assistance services available. Such inequality in benefits derived from social security services is much less in the non-Latin and Caribbean countries, enjoying a universal coverage combined with unified programmes though 'the civil servants enjoy better pensions than the rest of the insured'.

V. India

Poverty in India is an overwhelming feature keeping a little over 300 million, or more than a quarter of its total population, below the poverty line [poverty line is considered rupees 356 (approximately US $ 8)] monthly per capita consumption expenditure as the threshold for rural areas and rupees 539 [approx. a little over US $ 11] for urban areas in 2004-05), and the number has hardly changed over the last three decades despite economic development. But the redeeming feature is the reduction in its share of the total population of the country, reduced by 27.4 percentage points from 54.9 in 1973 to 27.5 in 2004. The fall in the share is no consolation, however, as on headcount ratio, the total number barely changed, remaining constant over two decades (321 million in 1973 and 322 million in 1983) before sliding down to 301 million in 2004 (Tables 5.7 and 5.8).

As reported by the Planning Commission, Government of India, the composition of the poor has been changing, with rural poverty becoming concentrated in agricultural labour while the urban poor are concentrated in casual labourers. As a result, agricultural labour accounted for 41 per cent of the rural poor in 1993-94 as well as in 2004-05, and the share of self-employed in agriculture among the rural poor has declined from 32 per cent to 21.6 per cent. During this period the casual labour households in urban areas accounted for 62.6 per cent in 1993-94, declining to 56.5 per cent in 2004-05.

A World Bank study (2000) found Indian poverty mostly rooted in rural areas with 'three of every four poor persons living in rural areas'. The wide disparity in poverty across Indian states and their uneven progress in poverty

Table 5.7: Percentage of People Below Poverty Line in India 1973-2004

Years	*Rural*	*Urban*	*Combined*
1973	56.4	49.0	54.9
1983	45.7	40.8	44.5
1993	37.3	32.3	36.0
2004	28.3	25.7	27.5

Source: Planning Commission, Eleventh Five Year Plan

Table 5.8: Number of Persons Below Poverty Line in India 1973-2004

(*in Lakh /00,000*)

Years	*Rural*	*Urban*	*Combined*
1973	2612.90	600.46	3213.36
1983	2519.57	709.40	3228.97
1993	2440.31	763.37	3203.68
2004	2209.24	807.96	3017.20

Source: Planning Commission, (Eleventh Five Year Plan, 2007-2012, Vol. III:80)

reduction through decades of development illustrate the evolution of poverty in India. The results obtained in poverty reduction vary widely, the better-off states having done better than many others. For example, the states which have succeeded in reducing the absolute number of the poor in rural areas despite overall increase in total population over three decades between 1973 and 2004-05, are Andhra Pradesh from 17.82 to 6.47 million; Karnataka from 12.84 to 7.5 million; Kerala from 11.14 to 3.24 million; Tamil Nadu from 17.26 to 7.65 million and West Bengal from 25.79 to 17.32 million, while in Assam and Gujarat the reduction was less impressive. Contrarily, the absolute number of the poor increased during the same period in Uttar Pradesh (including Uttaranchal), Rajasthan, Maharashtra, Nagaland, Madhya Pradesh (including Chhattishgarh) and Bihar, and the number remained roughly static in Haryana, Himachal Pradesh, Orissa and Mizoram (Planning Commision, 2008: 80). (See Note E)

According to the World Bank, the major factors taking an effective role in reducing poverty in India are— (i) faster growth, particularly agricultural growth raising agricultural wages and reducing the price of food; (ii) lower inflation; (iii) infrastructural development; and (iv) human resource development, particularly female literacy. Among them, rates of agricultural growth and simultaneous increase in rural wages were the major factors in bringing down the level of poverty across the Indian state. "Green Revolution technology, irrigation and infrastructure were associated with rising rural wages and increased rural farm employment such as in Punjab and Haryana which had the highest GDP per capita up to the early 1990s (and continued to be in the top five states in 1996-97) (The World Bank, 2000: 13). This finds corroboration in Fig. 5.3. In general, however, the incidence of income poverty among females is marginally higher in both rural and urban areas than that of males. Similarly, the other dimensions of poverty and deprivation, such as food insecurity, malnutrition, health sand education are associated more with females than males. Another very challenging aspect is poverty among children both in rural and urban areas, with 39 per cent of children below 15 years living in below poverty line households in rural and 41 per cent in urban areas in 1993-94 and 44 per cent in the rural and 32 per cent in urban areas in 2004-05. The dissimilarities are more pronounced in the headcount ratio compared between the socially disadvantaged groups, such as SCs (Scheduled Castes) and the STs (Scheduled Tribes) and OBC (Other Backward Communities) on the one hand and others (i.e. the higher castes) on the other. The former groups taken together accounted for 80 per cent of the rural poor in 2004-05, considerably more than their share in the total population.

In geographical distribution, a large proportion of the STs happen to be concentrated in forests of arid, semi-arid and sub-humid areas of central and eastern India with limited access to resources. The poor are also concentrated 'in areas where a significant proportion of agriculture is irrigated, and not rainfed'. Poverty from these regions cannot be reduced without effective land reforms and provision of agricultural services. Besides,"the poor have to be

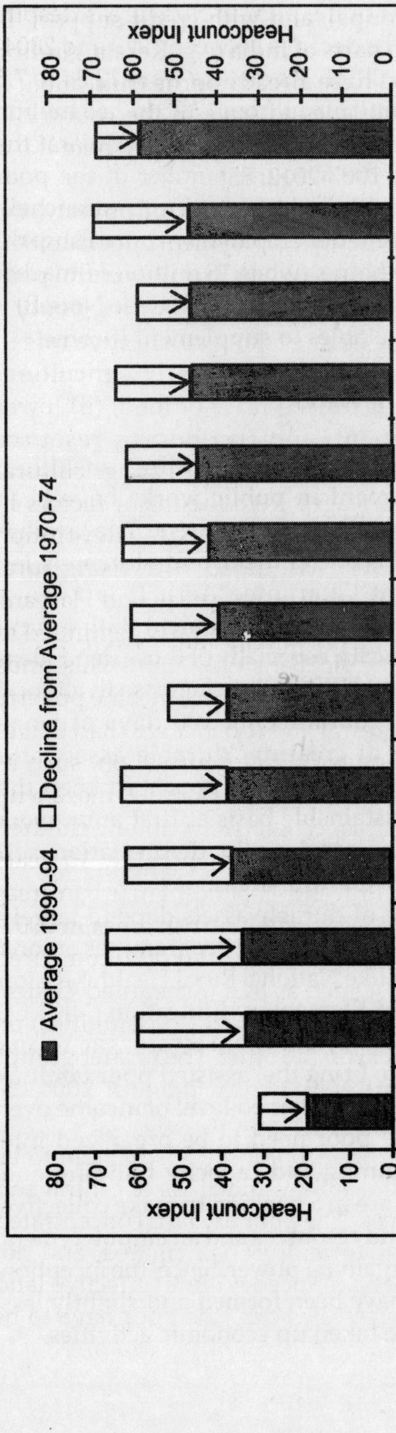

Poverty Reduction : Progress and Challenges

Fig. 5.3: Rural Poverty Levels by State

Source: The World Bank, 2000.

enabled to increase their income by diversifying away from agriculture, and relying on non-farm sources for at least subsidizing income. Casual labourers and artisans, especially women, can be brought together through social mobilization to form SHGs (Self-Help Groups), and with credit support, to diversify their sources of income. In certain parts of India (e.g. Kerala, Andhra Pradesh, Gujarat) the success of such efforts have already borne fruit, and it is essential to make sure that this programme goes to scale in the country in regions which need it most – in the eastern and northern parts – where it has hardly taken root" (Planning Commission, 2007-2012: 85)

In order to escape the net of inter-generational transfer of poverty, the poor need a safety net, at least among other benefits, employment opportunities, safe guard against medical care costs, and a home, (when 18 million rural poor in India do not have one) along with a small plot of land for rearing poultry, goats (and cattle, if possible) and grow vegetables to supplement income.

Rural Welfare Programmes for Poverty Alleviation

Workforce programmes already adopted in India, providing unskilled, unemployed workers short-term employment in public works have been effective in slowing down the severity of poverty to some extent. The Jawahar Rojagar Yojona (JRY) implemented in 1989 is one of them, and ultimately led to NREGA (National Rural Employment Guarantee Act). It is a wage employment programme implemented by the state government with central assistance aimed at providing 'enhanced security, especially of those dependent on casual manual labour'. The objective of the NREGA is to increase livelihood security of the people in rural areas by guaranteeing 100 days of wage employment in a financial year, aiming at creating 'durable assets and strengthening the livelihood resource base of the rural poor'. To keep the process of employment generation on a sustainable basis at first areas were chosen where the causes of chronic poverty were drought, deforestation, soil erosion etc. but it has now been extended to all rural areas.

Empowerment of workers and creation of durable assets depend largely on the linkages between NREGA and other development programmes among which are human development programmes like National Rural Health Mission (NRHM), National Mission for Literacy and Elementary Education (NMLE), Self-Employment Programmes, such as Swarnjayanti Gram Swarozgar Yojana (SGSY) launched in 1999 with an objective to bring the 'assisted poor families above the poverty line by ensuring appreciable sustained level of income over a period of time'. To achieve this the rural poor need to be organized into SHGs (Self-Help Groups) and through training and capacity building and provision of income generating assets, they are in a position to make collective decisions with prioritization of their needs and resources and an empowerment process like this improves their collective bargaining power. Since, the inception of SGSY in 1999, nearly 2.6 million SHGs have been formed and slightly less than five lakhs (half a million) of them have taken up economic activities.

As an integral part of poverty alleviation, India has already embarked on the programme of providing 'houses for all', and replacing the earliest housing programme initiated by the Government of India meant for the rehabilitation of refugees following the partition of India (1947). The Indira Awas Yojana (IAJ) was launched in 1996 by the Ministry of Rural Development, and has already been very popular with the rural population. The popularity stems from the fact that it is another 'self-help' kind of project enabling direct participation of the beneficiary in the construction of his/her home for which the state government facilitates the 'use of local, low cost, environment-friendly, and disaster-resistant technology and also encourages the construction of sanitary latrines and smokeless *chulhas* (oven)'. The houses are constructed without any assistance of a contractor as per the choice of design, technology and requirement of the beneficiary towards which he/she directly receives the fund allocated for the programme through the government in instalments

Interestingly, an amount of Rs. 25,000 (approximately.US $ 530) per dwelling in the plain areas and Rs. 27,000 (apprx. US $ 575) in the tribal and hilly areas is not adequate and this inadequacy means the beneficiary must give free labour for construction and this participation makes the scheme more meaningful. The estimated housing shortages in 2001 stood at 15.95 million requiring the Eleventh Five Year Plan (2007-2012) to fix its target of constructing about 15 million houses under the project. However, all is not well with this very important project. The primary hurdle is the selection of beneficiary, to be done by the Gram Sabhas (Village Councils/ Panchayats). This does not appear to be always carried out as "25 to 50 per cent" of the beneficiaries are not reported to be selected by the Gram Sabhas. Further, people having no land are not selected, thus depriving on the whole almost 8 million out of the 14 million houseless people who have no land. Moreover, the poorest among the BPL (Below the Poverty Line) households are left out while non-BPL households get selected. Besides, 'collection of illegal gratification' is another major problem affecting the process of selection.

On the other hand, IAY (Indira Awas Yojana) is also facing another serious problem as only about 50 per cent of the housing units built under this programme have sanitary latrines and even less than that have installed smokeless '*chulhas*' (which replace traditional ovens, which subject over a million women and children to respiratory problems and succumbing to inhaling smoke inside homes.). In addition, there is need for provision of minimum infrastructural facilities, such as internal roads, drainage, water supply stand posts — some of the immediate requirements to improve the quality of life, making it more secure.

Incidentally, HUDCO (Housing Uraban Development Corporation) started its rural housing activity from 1977-78 'by extending assistance for construction/ upgradation of rural houses' through the state governments and their agencies (like Housing Boards, Panchayat Development Boards, etc.), catering to the Economically Weaker Sections (EWS) with a monthly household income less than Rs. 2,500 (or approx. US $ 53) at a cost ceiling of Rs. 60,000 (approx. US $

1,275) per unit. Over the last 28 years (till March 2006), HUDCO has sanctioned 2,472 schemes with a loan amount of Rs. 58,070 million (approx. US $ 1,235 million) for the construction of 86.11 million housing units in various states of India (Planning Commission, 2007-2012: 94-98)

The incidence of poverty is very closely linked to the absence of social infrastructure. The close relationship between social and economic infrastructure is directly reflected in the incidence of poverty as shown in Table 5.9, the states with high infrastructure, such as Punjab (187.51), Kerala (178.68), Tamil Nadu (149.10), Haryana (137.54) and Gujarat (124.31) have much lower incidence of poverty with 9.15, 13.25, 22.85, 13.57 and 19.08 respectively in contrast to states like Orissa (81.00), Bihar (81.33), Madhya Pradesh (76.79) etc., which have much higher rates of poverty, 46.76 per cent in Orissa, 42.14 per cent in Bihar and 36.87 per cent in Madhya Pradesh while Rajasthan makes a contrasting case with quite low infrastructural development but having low incidence of poverty (18.71). However, creation of infrastructure, both physical and social, has been given much greater priority in the Eleventh Five Year Plan (2007-2012) (*See* Note *F*)

In recognition of the role played by infrastructure in reducing poverty, the Government of India took up a massive programme for the construction of rural infrastructure under Bharat Nirman in 2005, the first half of which was implemented during 2005-09, taken up in the Tenth Five Year Plan period to be followed by the second half during the Eleventh Five Year Plan period. The primary components of this plan are irrigation, drinking water, electrification,

Table 5.9: Index of Social and Economic Infrastructure and Incidence of Poverty (HCR 2004)

State	Index of Infrastructure	HCR (2004)
Andhra Pradesh	103.30	11.17
Assam	77.72	22.33
Bihar	81.33	42.14
Gujarat	124.31	19.08
Haryana	137.54	13.57
Karnataka	104.88	20.85
Kerala	178.68	13.25
Madhya Pradesh	76.79	36.87
Maharashtra	112.80	29.58
Orissa	81.00	46.76
Punjab	187.51	9.15
Rajasthan	75.86	18.71
Tamil Nadu	149.10	22.85
Uttar Pradesh	101.23	33.40
West Bengal	111.25	28.62

Source: Planning Commission. 11th FYP, Vol. III (Taken from Eleventh Finance Commission Report and Poverty Estimates by Planning Commission)

roads, housing, and rural telephones. The major targets under Bharat Nirman to be achieved by 2009 were as follows:

(a) *Irrigation*: To create 10 million hectares of additional irrigation capacity;
(b) *Roads*: To provide all-weather roads to every habitation over 1,000 population and above (500 in hilly and tribal areas);
(c) *Electricity*: To provide electricity to remaining 125,000 villages and to 23 million households;
(d) *Housing*: To construct 6 million houses;
(e) *Drinking Water*: To provide drinking water to 55,067 uncovered habitations by 2009. All habitations with failed sources and water quality problems were to be included; and
(f) *Telephone Connectivity*: To connect 66,822 villages with telephone by 2007 (Planning Commission, Eleventh Plan: 130)

The achievement in creating irrigation capacity against targeted potential of 4.30 million hectares in the first two years of the Bharat Nirman Programme was 2.587 million hectares, that is about half of the actual target while during the Eleventh Plan period the target is to create 16 million hectares potential. As pointed out in the report, 'restoration of the lost potential' by restoring tanks with external assistance and utilization of the groundwater potential in states with scope for that will form part of implementation of the project.

Improvement in rural connectivity as a key component of rural development and poverty alleviation has been taken up by 'Pradhan Mantri Gram Sadak Yojana' programme (PMGSY) launched in 2000 with a target of constructing 146,185 kilometres of rural roads and upgradation of nearly 194,000 kilometres of the existing rural roads. Actual progress under this programme during the period, 2005-07, is shown in Table 5.10.

Table 5.10: Progress in the construction of Rural Roads under Bharat Nirman during 2005-07

Status		*2005-06*		*2006-07*		*2005-07*		*Percentage Achieved*
		Target	*Achieved*	*Target*	*Achieved*	*Target*	*Achieved*	
New Connectivity	Length (in km)	15492.4	18053.7	35182.15	21422.85	50674.55	39476.55	77.90
	No. of inhabitations	7034	5552	16130	8279	23164	13831	59.71
Upgradation	Length (in km)	11394.4	3925.6	54669.26	46129.90	66063.66	50055.51	75.77

Source: Planning Commission, Eleventh Five Year Plan, Vol. III.

For providing safe drinking water, the target of 55,067 habitations lacking this facility were intended to be covered by the Bharat Nirman Programme as well as covering the 'slipped –back habitations'. Of homes lacking safe drinking water, 48 per cent got safe drinking water for the first time, and in the case of the slipped-back habitations, more impressively 63 per cent of them had been covered during the first two and a half years of the Plan. But only 5 per cent coverage of the water-quality-affected habitations makes a disappointing picture.

The twin objectives of electrification of 125,000 villages and electrification of 23 million BPL households make an ambitious effort under the Bharat Nirman Programme but progress has so far been far from satisfactory on both counts, achievement remaining at 36 and 6 per cent respectively.

In providing houses to the homeless rural population, 2.869 million houses have been constructed in the first two years of this programme out of a targeted total of 6 million houses to be constructed between 2005 and 2009.

Out of 66,822 revenue villages not having telephone services, 48,704 Village Public Telephones (VPTs) connectivity have been provided and another 18,118 VPTs were likely to be provided by 2008.

Programmes intending provision of improved livelihood opportunities are coupled with a number of programmes aimed at providing social services, called as flagship programmes, which include, (a) National Rural Health Mission (NRHM); (b) Integrated Child Development Services (ICDS); (c) Sarva Shiksha Abhiyan (SSA); (d) Mid-day Meal (MDM); (e) National Social Assistance Ptogramme (NSAP); (f) Total Sanitation Campaign (TSC); (g) Backward Regions Grant Fund (BRGF).

'The main aim of NRHM (National Rural Health Mission) is to provide assessable, affordable, accountable, effective, and reliable primary healthcare, especially to the poor and vulnerable sections of the population' (Planning Commission, Eleventh Plan; 133). The programme is going to be implemented in 18 states with poor health infrastructure but it has been noted in this connection that many of the primary health centres present in the states like Bihar, Chhattishgarh, Maharashtra, Orissa, Uttar Pradesh, West Bengal and other states are without regular water supply, electricity, medical provisions, and all-weather approachable roads.

The ICDS (Integrated Child Development Services) focuses on the interrelated needs of young children and pregnant and lactating mothers aiming at reducing the incidence of mortality, morbidity, malnutrition and school drop-outs and increase the capability of mother and family to take better care of health, nutrition and development needs of the child. Till March 2007, there were 6,284 sanctioned projects out of which 5,672 were operating through "*anganwadi*" centres established for this purpose helping 70.5 million beneficiaries 'comprising 58.1 million children below six years and 12.3 million pregnant and lactating mothers'.

As a multifaceted project it aims at 'raising the sex ratio of girls to boys in the age group of 0-6 years from 927 in 2001 to 935 by 2011-12 and to 950 by 2016'.

It will also reduce infant mortality rate to 28 and maternal mortality ratio to 1 per 1,000 live births. Further, malnutrition among children of 0-3 age groups will be reduced by half of its present level and anemia among women and girls is expected to fall by 50 per cent by the end of the Eleventh Plan (2012).

The principal objective of Sarva Shiksha Abhiyan (SSA) has been 'universalization of elementary education' in the age-group of 6-14 by 2010 and remove social, regional and gender gaps in primary education. As a result of the implementation of this project, the total enrolment in primary classes increased from 113.9 million to 130.8 million or 14.84 per cent between 2001-02 and 2004-05, and from 44.8 to 51.2 million or 14.29 per cent in upper primary classes. The most remarkable feature is increase in girls' involvement during this period, increasing from 50.3 to 61.1 million or 21.47 per cent in primary classes and from 18.7 to 22.7 million or 21.39 per cent in upper primary classes. (*See* Note G).

The Mid-day Meal (MDM) forms an integral part of encouraging efforts to make primary education universal, and launched in 1995, it focused on 'increasing enrolment, retention, and attendance and simultaneously impacting upon nutritional status of students in primary classes', meant specifically for educationally backward communities across the country. By 2006-07, the coverage of children in primary classes has risen to 120 million involving 950,000 primary schools, and is expected to reach out to 180 million children by 2008-09.

The National Social Assistance Programme (NSAP), launched in 1995, provides 'security to the poor and the destitute' with 'financial assistance to families with little or no regular means of subsistence, living below the poverty line'.

The objective of the Total Sanitation Campaign (TSC), launched in 1999, was to attain 100 per cent sanitation coverage in rural areas 'in terms of households, schools and *anganwadi* toilets' and providing education in hygiene to the people with a target of achieving 50 per cent coverage by 2015'. "The physical target of the Eleventh Plan (2007-2012) is to cover 69 million households, 25,769 sanitary complexes and 133,114 *anganwadis*, all the remaining schools and Rural Sanitation Marts/Production Centres" (Planning Commission, Eleventh Plan: 136).

The Backward Regions Grant Fund (BRGF) is another very important programme providing spatial coverage including 250 districts, based on 'well-conceived participatory district plans' for which the village panchayats as well have a direct role in planning and implementation of the programme. While this forms a part of the project, another part includes special plans for Bihar and Koraput, Bolangir and Kalahandi (KBK) districts of Orissa for bringing about improvement in 'such sectors as power, road connectivity, irrigation, forestry and watershed development', besides focusing on 'tackling the main problems of drought proofing, livelihood support, connectivity, health, education, etc.' as per local priorities in the KBK districts (Planning Commission, Eleventh Plan: 136)

According to official statistics the proportion of urban and rural poor coincided in 1983-84 but widened slightly in 1993-94, and there was little difference in the depth and severity of poverty between urban and rural areas in 1993 and 1994. The World Development Report 2000-01 recognises material poverty as measured by income and consumption but also encourages non-material considerations, such as low levels of education and health. It "also broadens the notion of poverty to include vulnerability and risk – and voicelessness and powerlessness" (Kumar, 2004: 276).

The percentage of people living below the poverty line in the urban areas of India has declined from 32.3 per cent in 1993-94 to 25.7 per cent in 2004-05. Interestingly, however, this is a percentage-wise decline since in absolute numbers it has increased during this period by 4.4 million. The incidence of urban poverty as well as the number of urban poor has increased in Orissa and Rajasthan while it declined in both respects in the States of Gujarat, Kerala, Punjab, Tamil Nadu and West Bengal during the same period. According to the estimates, the slum population in 2001 was 61.82 million, and of them 42.58 million lived in 640 towns and cities having a population of 50,000 and above. The slum population were in 8.2 million households. Every seventh person in urban India is a slum dweller (according to 2001 National Census).

The major objectives of the Eleventh Plan (2007-2012) for urban poverty alleviation concentrate on the following aspects:.

(a) "To provide them affordable shelter and decent living and working conditions;
(b) To make adequate provision of land for the poor in the Master Plan itself;
(c) To help in developing self-employment enterprises and job creation for the wage employment earners; and
(d) To protect the economic interest and safety of women and other vulnerable sectors of our society.".

For the development of slum areas, the major thrust has been on facilitating access to services for slum population, granting tenure security to slum dwellers, in situ upgradation and resettlement options for slum improvement, and using land as a resources for housing and shelter development for slum dwellers (Eleventh Plan, Vol II: 406). Interestingly, the housing stock in 2001 was 50.95 millions for 55.8 million urban households.

The Swarna Jayanti Shahri Rozgar Yojana (SJSRY), a centrally sponsored scheme, launched in 1977, was meant to provide gainful employment to the urban unemployed/underemployed living below the poverty line. Under the self-employment component of this programme (SJSRY), as reported, '891,679 beneficiaries and 934,649 persons were provided various types of skill training for people below 30 years of age. Development of Women and Children in Urban Areas (DWCUA) groups, numbering 52,701, have been formed benefiting 198,314 women, assisting them in setting up group enterprises who market their products themselves except in the eastern and North-Eastern states. 'A total of 177,508 Thrift and Credit Societies have also been formed under SJSRY project'.

There is an overall requirement of creation of at least 30 million additional jobs in the urban areas during the Eleventh Plan period (2007-12) to fill the ever-widening gap of unemployment and underemployment as a way for poverty alleviation. This has to be done, as indicated in the report of the Eleventh Plan, based on a target-oriented approach 'with specific focus on primary issues, such as skill upgradation, entrepreneurship development and wage employment. Skill development has to be correlated to the demands of industry and service sector in and around the city. The success in urban poverty alleviation will never be complete, as it is pointed out in this report, 'the improvement of slums and their upgradation, including the provisioning of urban basic amenities to the poor' forming key features of urban municipalities. Further, since "any urban poverty alleviation programme without adequate manpower equipped with the capability, capacity, and sensibility to effectively implement such social sector programme will not succeed" (Planning Commission, Eleventh Plan: 410), it becomes imperative that there are properly trained and dedicated manpower for doing that.

Notwithstanding the ambitious approach as laid out in all these programmes intended at poverty alleviation in rural and urban areas of the country, the overall performance has, so far, been far from encouraging in reaching their proposed targets since in almost every sphere the progress has largely been impeded by inept handling, lack of any sense of responsibility and coordination, absence of dedication and above all, by rampant corruption prevailing at various levels with financial involvement, leading quite often to diversion of funds, leaving funds unspent, not accounting for the expenses incurred and siphoning off often a large share of that to personal purses. As a result, much of what could be accomplished, has remained unaccomplished, slowing down the process of poverty alleviation in physical, economic as well as social perspective.

Nevertheless, according to Kaushik Basu (2004), economic reforms in the recent decades in India has helped in its economic growth, and decline in poverty. The annual per capita growth increasing from 1 per cent per year between 1960s and 1970s to about 3 per cent per year in the 1990s increased per capita consumption by about one-third over the last decade and a 5 to 10 percentage point reduction in national poverty rates. Despite this, Basu (2004) admits that 'millions of Indian citizens still live in poverty, reduction of which depends on' improving the investment climate to accelerate growth, and policies for empowering poor people to participate in and benefit from this growth. "Economic growth has been a leading determinant of poverty reduction ... Most of the reduction in poverty (in India) in the recent period has been a result of the increase in average consumption driven by economic growth. Clearly, policies that enhance overall economic growth will be key levers to improving incomes of poor people in India" (Basu, 2006: 154). Agricultural productivity with higher farm yields to small producers and providing higher real wages to agricultural labourers plays a significant role in alleviating poverty and non-farm growth, brought about by more conducive climate for small- and medium-scale industries which 'can continue to be the engine of growth and poverty reduction'. "The investment climate necessarily involves institutions, rules and governance as well as traditional questions of fiscal policy, public expenditure management, and taxation. It should be noted that the role of small- and medium-sized enterprises is critical. If the investment climate is improved for small and medium companies, it is likely that larger firms will benefit as well" (Basu, 2006: 157).

Empowerment means the ability of people to shape their own lives besides enabling them to contribute to, and participate in economic growth. The development of human capital, particularly education and health are included in this process with greater stress on female education and such demographic measures as reduction in infant mortality, maternal mortality and fertility rates, 'effective rights in property, especially land are of critical importance for women's economic and social empowerment in India' (Agarwal, 1995). Gender inequalities still persist in laws governing inheritance and holding of arable land – the most valuable form of property in rural areas in India. In spite of the fact that legal rights of the girls in the family to an equal share of the property with boys (brothers) have been accepted by the country's legal system (enforced through a dictat of the Supreme Court), in the absence of proper application of these laws the girls remain deprived as before. This disparity applies to various fields: for example, "among several religious groups women are entitled to smaller shares than men ... Women's rights in land are severely circumscribed and typically limited to usufruct....For Hindus in several states of north-west India, the rules of devolution specified in various tenancy and other land reform enactments give priority to male agnatic heirs, and supersede the rules of devolution spelt out in the Hindu Succession Act of 1956. In laws pertaining to the fixation of ceilings, many states allow additional land to a cultivator on account of adult sons, or allow adult sons to directly hold such additional land in their own right, but adult daughters are given no such consideration; and in several states, unmarried daughters do not figure either as part of the family unit, or as independent units" (Agarwal, 1995: 52), which is worst in northwest India. The question of women's empowerment gets illegally curtailed when they cannot enjoy their legal rights obstructed by many formidable obstacles in India.

According to Basu (2004), formerly the Chief Economic Advisor to the Government of India, overall economic growth depends on rapid export growth. A number of studies support the view that 'openness to trade and direct foreign investment would accelerate growth further and to encourage this, maximum reduction in import duties is very much welcome. The study made by a World Bank team along with the Confederation of Indian Industry (CII), covering about 1,400 industrial establishments from tradable goods

Table 5.11: Foreign Direct Investment: India and selected countries

(US $ Million)

	1995	1996	1997	1998	1999	2000
India	2,144	2,426	3,577	2,635	2,169	2,135
China	35,849	40,180	44,237	43,751	38,753	38,399
Malaysia	4,178	5,078	5,137	2,163	3,895	3,788
Philippines	1,478	1,517	1,222	2,287	573	2,029
Thailand	2,068	2,336	3,895	7,315	6,213	3,366
Brazil	4,859	11,200	19,650	31,193	28,576	32,779

Source: Basu, 2006 (Taken from IMF International Financial Assistance)

sectors in ten states to assess the investment climate found that "West Bengal, Kerala and Uttar Pradesh had poor investment climate compared to states such as Tamil Nadu, Maharashtra, Karnataka and Andhra Pradesh. There is little doubt that India's weak investment climate has prevented it fulfilling its vast potential. Its growth rate, while increasing in recent years, remains way below that of China. As the joint World Bank- CII study has established, India's low-cost advantage relative to other countries is whittled away by higher effective costs of interest rates, bureaucratic harassment, and regulatory hurdles. The low level of FDI into India (Table 5.11) is a signal of the importance of tackling problems in the investment climate" (Basu, 2004: 169-170).

Interestingly, China is the largest recipient of FDI among the six developing countries, receiving nearly 71 per cent of the total for all of them (in 1995) while India trails far behind with a paltry 4.2 per cent, which increased slightly through the following two years, but then dropped further in actual amount as well as in proportion to 2.6 per cent of the total available as FDI for the six countries in 2000. China's share as well fell sharply to 46.5 per cent of the total by 2000 when Brazil's share increased from 9.6 per cent in 1995 to 39.7 per cent in 2000, increasing by nearly seven times over a period of five years.

Nevertheless, the impact of India's economic growth on the poor of the country is reflected in the national poverty rate, and according to the Planning Commission (2001), it fell by about 10 percentage points between 1993-94 and in 1999-00, from 36 to 26 per cent. Using the Planning Commission's (2001) official poverty line, Deaton (2001) found the rural poverty rate during the same period 'fell from 37.2 per cent in 1993-94 to 30.2 per cent in 1999-00, while urban poverty fell from 32.6 to 24.7 per cent (Datta and Ravallion, 2002.) "It is notable that agriculture as a whole has lagged the non-agricultural sector in the 1990s; while India's aggregate GDP grew at a rate of 6.7 per cent per annum over the period 1993-1994 to 1999-2000, agriculture and allied services grew at only 3.2 per cent per annum. The importance of rural economic growth and agricultural growth in particular, to poverty reduction in India has long been recognized" (Datta and Ravallion, 2002: 97-98). As pointed out earlier, the proportion of people living below the poverty line has reduced from nearly 55 per cent in 1973 to about 27.5 per cent in 2004, but still over a quarter of the

Indian population were left poor even after 60 years of Independence, suffering from multiple deprivations such as income poverty, education, health, etc. (Planning Commission, Eleventh Plan, Vol III: 79-81). According to Datta and Ravallion (2002), 'higher farm yields and higher development spending reduce incidence of poverty and higher non-agricultural output per person lowers poverty in all the states of India. They conclude by saying that "the incidence of poverty has been falling at a little less than one percentage point per year over the main post-reform period ... States with relatively low levels of initial rural development and human capital development were not well suited to reduce poverty in response to economic growth ... Achieving higher aggregate economic growth is only one element of an effective strategy for poverty reduction in India. West Bengal, in this regard, provides a case of very high percentage increase in farm output accompanied by low poverty, lower than all the twelve states in the list of fifteen except one state, Punjab, while one does not provide any information. (Table 5.12). The sectoral and geographic composition of growth is also important, as is the need to redress existing inequalities in human resource development between rural and urban areas" (Datta and Ravallion, 2002: 106). Among the matters particularly important for pro-poor growth is expanding literacy as the most significant step for human resource development so that 'the capacity' of the poor 'to participate in the opportunities that come with economic growth' and women's literacy in particular 'is a slightly more significant predictor of elasticity of poverty with respect to economic growth' (Datta and Ravallion, 2002).

Among the 15 states listed in Table 5.12, Bihar has the highest record of the incidence of poverty in 1993-94 with 60.3 per cent of its population living below poverty line while Punjab has the lowest of 21.4 per cent, and in 7 other states (Assam: 44.5%; Madhya Pradesh: 44.0%; Rajasthan: 43.3%; Maharashtra: 43.2%; Orissa: 40.3%; Uttar Pradesh and Karnataka: 37.4%) it was higher than the national average of 36.0 per cent in 1993. Interestingly, the percentage increase in non-agricultural output per capita (29.3%) in Bihar though higher than among some other states between 1993-94 and 1999-00, the increase in farm output per hectare was very negligible (0.5%) during this period but Punjab with the lowest percentage of poverty scored much higher increase in non-farm output along with moderate increase in farm output. This indicates impact of economic growth on the incidence of poverty and its variation between the states of India, and this has relevance among economically better-off and poorer states. As observed by Deaton and Dreze (2002), poverty decline is closely related to the growth of per capita consumption expenditure (APCE), and they find a 'striking regional pattern': for example, barring Jammu and Kashmir, "the low-growth states form one contiguous region made up of the eastern states (Assam, Orissa and West Bengal), the so-called BIMARU (Bihar, Madhya Pradesh, Rajasthan, and Uttar Pradesh) and Andhra Pradesh. The high-growth states, for this part, consist of the southern states (except Andhra Pradesh), the western states (Gujarat and Maharashtra), and the north-western region (Punjab, Haryana and Himachal Pradesh). Poverty reduction is

Table 5.12: Relation between Economic Growth and Poverty in India

State	Headcount Index of Poverty for 1993-94	Share of National Poverty 1993-94 (Percentage)	Percentage Increase in Non-agricultural Output per Capita 1993-94 to 1999-00	Percentage Increase in Farm Output per Hectare 1993-94 to 1999-00
1. Andhra Pradesh	29.5	6.1	43.2	21.9
2. Assam	44.5	3.1	13.2	27.8
3. Bihar	60.3	16.7	29.3	0.5
4. Gujarat	33.7	4.4	38.1	29.9
5. Jammu & Kashmir	n.a.	n.a.	n.a.	n.a.
6. Karnataka	37.4	5.3	53.4	11.9
7. Kerala	28.8	2.6	73.4	64.0
8. Madhya Pradesh	44.0	9.2	18.1	26.8
9. Maharashtra	43.2	10.8	27.7	11.4
10. Orissa	40.3	4.0	26.3	36.7
11. Punjab	21.4	2.5	43.2	14.0
12. Rajasthan	43.3	6.1	44.3	64.5
13. Tamil Nadu	34.9	6.0	58.9	-6.4
14. Uttar Pradesh	40.1	17.8	37.3	26.1
15. West Bengal	25.9	5.5	49.4	83.0

Source: Datt and Ravallion, 2002
NA = Not available

overwhelmingly driven by the growth rate of APCE (Average Per Capita Consumption Expenditure), and in this respect, " (i) most of the western states (with the important exception of Andhra Pradesh) have done comparatively well; (ii) the eastern region has achieved very little poverty alleviation between 1993-94 and1999-00; (iii) and there is a strong overall pattern of 'divergence' (states that were poorer to start with had lower rates of poverty reduction)" (Deaton and Dreze, 2002: 3734-3735). Importantly, growth of real agricultural wages makes for substantial decline in poverty in rural areas, causing a 'sharp reduction of the rural headcount ratio with an increase of 2.5 per cent per year in real wages. However, it may be one of the reasons contributing to poverty reduction but there are other reasons equally important, if not more, such as educational levels, fertility rates, public health provisions (including clean water and vaccination), infant mortality etc. which are the other faces of human deprivation, playing a vital role in reducing poverty.

B. GLOBAL FOOD SITUATION AND ITS FUTURE

Talking of poverty alleviation, one cannot escape the major issue affecting people more than anything else as there is no suffering greater than suffering from hunger, an unavoidable requirement every few hours, when a human being can, perhaps, go without many other absolute necessities for the time being but never without food for a long time. Food as the absolute, immediate, unavoidable and unquestionably the most basic necessity for survival (quite apart from what a human body requires for healthy living) must take centre-stage in any consideration for the removal of poverty.

Comparing the situation of hunger and food as reported by Willy Brandt in 1980, with that of present-day one becomes more disillusioned by the fact that despite development over the intervening three decades in the developing countries, the improvement is not at all impressive with regard to hunger and destitution. For example, there were 800 million estimated to be destitute in the 'Third World' at his time (Willy Brandt) who could not "afford an adequate diet" even though "food production in all the developing countries rose by over two-and-a-half per cent annually between 1950 and 1975, but demand for food has grown by well over 3 per cent a year and populations and incomes have also gone up. As a result the developing countries have rapidly increased their imports of cereals, from relatively low levels in the 1950s to 20 million tonnes in 1960 and 1961, to over 50 million tonnes in the early 1970s, and nearly 80 million tonnes by 1978-79. On current trends the Third World could be importing 145 million tonnes of food by 1990, 80 million tonnes of which would be needed by the poorer countries of Africa and Asia. It is unlikely within the prevailing economic climate that these countries' own exports or even additional aid, can finance such massive food imports" (Brandt, 1980: 90-91). Comparison between the situation at the time of Brandt and today shows that his prediction still holds good on both counts – the number of people going hungry and countries dependent on food imports.

A clear picture of the present situation in this regard may be obtained from a report of the International Food Policy Research Institute, 2002, stating that "each day our world witnesses 800 million people go hungry and about 170 million children under 5 years of age suffer from malnutrition" (IFPRI, 2002). As a matter of fact, the number of hungry people in the world has increased from a "historical low of 820 million in 2000 to 852 million in 2002" (Brown, 2006: 163) and this is in spite of the landmark achievement of China in eradicating hunger with a "wholesale reduction in poverty associated with an eightfold expansion in its economy since the economic reforms of 1978 and the 50 per cent jump in its grain harvest between 1977 and 1986" (Brown, 2006: 163). However, the rise in the number of hungry people between 2000 and 2002 was largely due to increase in their number in Sub-Saharan Africa and India. (*See* Note *H*). As pointed out by Brown *et al.* (2000), 'the combination of a dramatic surge in production' matched by an equally dramatic reduction in population growth enabled China in effectively eliminating most of its hunger and malnutrition.

Both the growth of population, the number of mouths to be fed and the corresponding growth in the production of food will ultimately determine the global capacity to feed the people, providing enough nutrition to sustain the health of human beings.

I. How Many Mouths to Feed?

According to Meadow *et al.* (1972), in 1650, world population "numbered about

0.5 billion, and it was growing at a rate of approximately 0.3 per cent per year. That corresponds to a doubling time of nearly 250 years. In 1970 the population totaled 3.6 billion and the rate of growth was 2.1 per cent per year. The doubling time at this growth rate is 33 years... In 1650 the average lifetime of most populations in the world was only about 30 years. Since then, mankind has developed many practices that have profound effects on the population growth system, especially on mortality rates. With the spread of modern medicine, public health techniques, and new methods of growing and distributing foods, death rates have fallen around the world. World average life expectancy is currently about 53 years and still rising" (Meadows *et al.* 1972: 41 and 94). (The life expectancy today has crossed 80 years in some of the countries in the world with the lowest of about 64 years in some of LDCs). With slight decrease in fertility rates and greater decrease in mortality rates in subsequent years it was predicted that the world population would rise to 7 billion in another 30 years (i.e. around 2000) and with further lowering in mortality rate but with no better success in lowering fertility rate, in 60 years' there will be four people in the world for every one person living' at the time (around 1972) Meadows *et al* wrote their report. It is no surprise to note that the authors of "Limits to Growth" are not far short of the mark in their prediction and their accuracy is really very striking in assessing the number of world population today.

The United Nations estimated that, on 1 July 2007, the world population will reach 6.7 billion, with 5.4 billion (or 82 per cent of the world's total population) living in the less developed regions out of which 804 million reside in the 50 least developed countries, accounting for 12 per cent of the world population, while the share of the more developed regions will be 18 per cent of the total global population, or 1.2 billion inhabitants (UN: Global Population Prospects, 2006 Revision: Table 1) (Table 5.13). In the Demographic Year Book, 2007, the world population for 2007 is shown as 6,671 million which tallies with the figure presented by the United Nations: World Population Prospects (2006 Revision), besides giving interesting details of population by sex for all the major regions as well as the sex ratio (Table 5.14). According to this source, while in the world as a whole males outnumber females, there is a contrast between the developed world with female dominance maintained in each of the three regions (Europe, Northern America and Oceania), and the developing world where males are more numerous, thanks to Asia, though Africa and Latin America and the Caribbean follow the developed region pattern (*Demographic Year Book*, 2007).

According to the estimates (high variants) of the United Nations, global population is likely to reach 10 billion by 2050, and both the UN and the World Bank predict world population to rise to 11 billion by 2100. This is partly corroborated also in forecasts made by Lutz *et al.*, estimating that the world population will reach 6.6 to 11.4 billion in 2050 and as calculated by the US National Research Council using a different method, the 2050 world population will be between 7.9 billion and 10.9 billion (Keilman, 2001: 490).

Table 5.13: Population of the World, Major Development Groups and Major Regions 1950, 1975, 2007 and 2050, according to different variants

Major Area	*Population in millions*			*Population in 2050 (millions)*			
	1950	*1975*	*2007*	*Low*	*Medium*	*High*	*Constant*
World	2535	4076	6671	7792	9191	10756	11858
More Developed Regions	814	1048	1223	1065	1245	1451	1218
Less Developed Regions	1722	3028	5448	6727	7946	9306	10639
Least Developed Regions	200	358	804	1496	1742	2002	2794
Other Less Developed Countries	1521	2670	4644	5231	6204	7304	7845
Africa	224	416	965	1718	1998	2302	3251
Asia	1411	2394	4030	4444	5266	6189	6525
Europe	548	676	731	566	664	777	626
Latin America and the Caribbean	168	325	572	641	769	914	939
Northern America	172	243	339	382	445	517	460
Oceania	13	21	34	42	49	56	57

Source: UN: World Population Prospects: The 2006 Revision, Vol. II, Table 1.

Table 5.14: Population of the World, Major Development Group, and Major Regions by Sex and Sex Ratio, 2007

Major Area	**Population (in millions)**			
	Total Population	**Males**	**Females**	**Sex Ratio**
World	6,67	3,363	3,308	101.7
Developed Regions	1,107.49	538.23	569.26	96.6
Developing Regions	5,564.	2,826	2,738	100.6
Africa	965	481	484	99.4
Latin America and the Caribbean	540	282	288	97.7
Asia	4,029	2,063	1,966	104.9
Europe	731	352	379	92.8
Northern America	342	169	173	97.3
Oceania	34.49	17.23	17.26	99.9

Source: Demographic Year Book 2007

Interestingly, all sources more or less agree on the number of world population hovering around 11 billion by 2050. Looking at the trend, one cannot miss incongruity in the pace of growth in that world population was increasing at a faster rate till the close of the twentieth century, then peaked at that time and has been gradually slowing down later. From an annual rate of increase averaging about 0.5 per cent annually in the first two decades of the twentieth century, it rose to around 1.0 per cent per year during the 1920s to 1940s, rising further to 1.8 per cent per year in the 1950s, reaching 2.0 per cent in the 1960s. The rate of growth despite 'sticking to 1.7 per cent in the 1970s' fell back afterwards, coming down to 1.2 per cent per annum at present, large enough to offset the effect of the growing population base, so that the size of annual increments have been falling to an estimated 77 million from almost 83 million during the 1980s, compared with 10 million at the opening of the twentieth century, 20 million a year in the 1920s and 50 million in the 1950s (Fig. 5.4). As mentioned earlier, the world population reached the mark of its first billion not until the beginning of the nineteenth century – to be precise, around 1804 – and it took another 123 years to attain the second billion in 1927. While the third billion took only another 33 years, the fourth only 15 years, and both the fifth and the sixth just 12 years. Looking from another angle, there was virtually no net increase in population during the first millennium, remaining static at about 300 million from the time of Christ's birth, whereas the second millennium witnessed a 20-fold growth of population, and while the "first 1900 years from Christ's birth added 1.3 billion, the last 100 years have added 4.4 billion to the world population " (Daniels *et al.*, 2005: 89).

Obviously, with a declining rate of fertility, the latest trend of population growth is likely to continue despite the declining rate of mortality, which is expected to fall below 1 per cent during 2015-20., as projected by the United Nations (2003) and will not be much above 0.3 per cent by the mid-twenty-first century (the same rate of increase as existed around 1750) and at this rate of growth the annual increment of world population is likely to be less than half by 2045-50 of what it will be around 2020, that is half of 70 million (Daniels *et al.*, 2005: 89).

Nevertheless, what is worrying against this background is the great unevenness of population growth across the world, the highly disproportionate growth getting more and more concentrated in the countries of the developing world, especially in the Less Developed Regions (LDRs), which account for nearly 90 per cent of the world's 3.8 billion extra population during the last fifty years or so. As per UN projections, between 2003 and 2050, the population of the LDRs will increase by 2.6 billion (the world population in 2050 is shown as 8.9 billion), leaving only 0.6 billion increase of the total world population for the More Developed Regions (MDRs). Further, within the LDRs, the 49 least developed countries are now witnessing 'the greatest upward surge in population', increasing their share of world population growth from 13.7 per cent in 1950-2003 to 36.5 per cent in 2003-2050, containing 18.8 per cent of the total world population by 2050 compared with 8 per cent in 1950. This unevenness in the geographical distribution of world population comes into sharper focus comparing the situation between Europe and Africa, the former having much more than double of Africa's population in 1950, is going to have about one-third of the latter's population in 2050. Africa will increase its share of world population from 8.8 per cent in 1950 to 20.2 per cent in 2050 while for Europe its share of world population is expected to go down from 21.7 to 7.1 per cent over the same period. This declining share is also predicted for another big partner of the developed world, Northern America, its share of world population falling from 6.8 per cent in 1950 to 5.0 per cent in 2050. In contrast, both Asia and Latin America and the Caribbean are going to increase their share: for Asia it is from 55.5 per cent of the world population in 1950 to 58.6 per cent in 2050 while for Latin America and the Caribbean the share of world

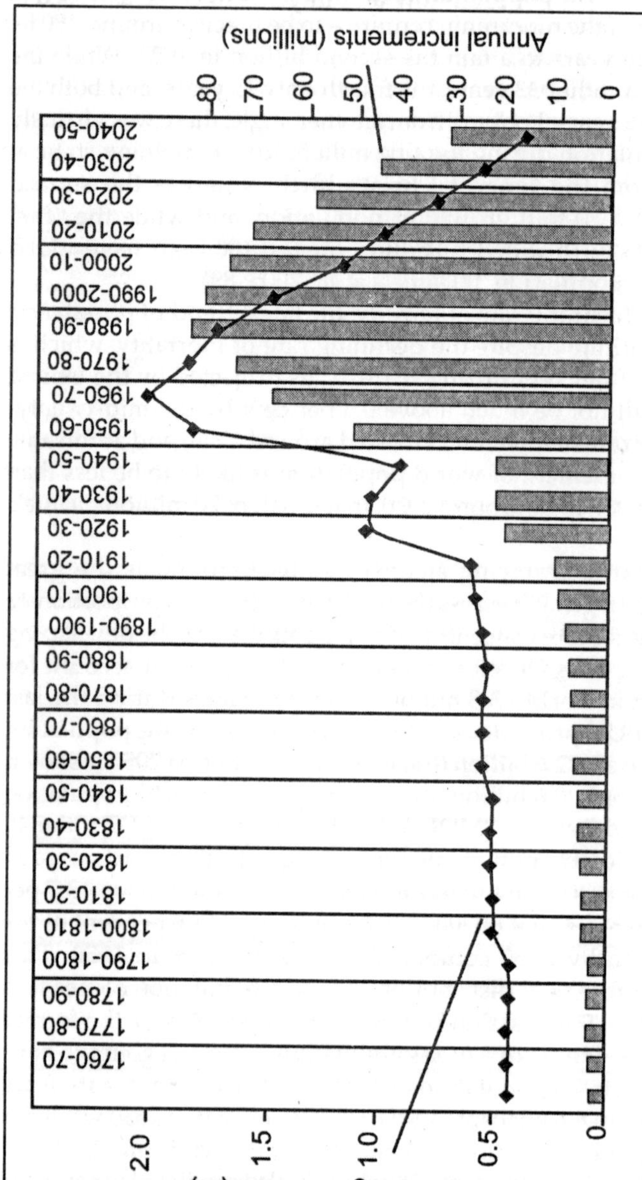

Fig. 5.4: World Population Growth, 1750–2050

Source: Daniels *et at.*, 2005;

population increases from 6.6 per cent in 1950 to 8.6 per cent in 2050 (see Table 5.15 and Fig. 5.5)

All this speaks of a great and growing imbalance in the concentration of world population, the overwhelming burden of which, accounting for almost 71 per cent in 1950 and about 87.4 per cent in 2050, lies with the developing world, leaving hardly 12.6 per cent of the world population in 2050 for the developed world, decreasing from 28.5 per cent in 1950. In the context of individual countries, China and India continue to dominate the scene in having the two largest populations of the world, together accounting for over one-third of the total global population. Interestingly, the number of countries with large populations is increasing, only 8 of which (on the basis of 2003 boundaries) contained 50 million each in 1950, the number having grown to 22 in 2003 and is expected to be 34 by 2050 (Fig. 5.6). However "one conspicuous feature of the recent past is the rapid growth of countries with a significant Muslim presence, with some of the highest growth rates occurring in the Arabian sub-continent and with the world's main zone of high population growth rates dominated by the swathe of Islamic countries that stretches from Pakistan westwards to Africa's Atlantic coast (Fig.5.5). This is a trend that seems likely to become even more marked over the next few decades" (Daniels *et al.*, 2005: 92) (*See* Note *I*).

In considering the number of mouths to be fed, those already in the grip of poverty, therefore suffering from hunger get the priority, though people lying outside this group, constituting the majority of the humanity, have an unquestionable claim on the food available and, in making any estimate of food the need of the entire human community has to be taken into account (without forgetting the fact we are also responsible for feeding the domestic animals sometimes with food we produce as our own feed). However, the policy of food security in this sense is needed for both the sections of the humanity – (a) those in hunger, and (b) the rest of the population.

It is claimed that the number of the undernourished people in the developing world – where they are primarily concentrated – has declined by 3 million between 1990-92 and 2001-03, from 823 million to 820 million (FAO 2006 a). Among the regions making the 'highest gains is East Asia while Sub-Saharan Africa had the most severe prevalence of undernourishment' with the highest increase of undernourished (37 million) during this period, and South Asia had the largest concentration (about 300 million) of hungry people. Overall, the number of hungry people increased by 50 million in the developing world apart from China during the period (1990-92 to 2001-03) (United Nations, 2007: 96-97).

However, according to FAO's estimates, "worldwide 848 million people suffered from chronic hunger in 2003-05 (the most recent period for which individual country data are available), the figure being slightly higher than the 842 million people who were undernourished in 1990-92, the WFS and MDG baseline period. The vast majority of the world's undernourished people live in developing countries, which were home to 832 million chronically

Table 5.15: Population Growth, 1950-2050, for Development Categories and Major Areas

Category and Major Area	*Population (millions)*			*Share of World Population*			*Change (millions)*		*Share of World change (%)*	
	1950	*2003*	*2050*	*1950*	*2003*	*2050*	*1950-2003*	*2003-2050*	*1950-2003*	*2003-2050*
World	2519	6302	8919	100.0	100.0	100.0	3783	2617	100.0	100.0
MDR	813	1203	1220	32.3	19.1	13.7	391	16	10.3	0.6
LDR	1706	5098	7699	67.7	80.9	86.3	3392	2601	89.7	99.4
LDC (within LDR)	200	718	1675	8.01	11.4	18.8	518	956	13.7	36.5
Africa	221	851	1803	8.8	13.5	20.2	629	953	16.6	36.4
Asia	1399	3823	5222	55.5	60.7	58.7	2425	1399	64.1	53.4
Europe	547	728	632	21.7	11.5	7.1	180	-96	4.8	-3.7
L. America	167	528	768	6.6	8.4	8.6	361	240	9.5	9.2
N. America	172	319	448	6.8	5.1	5.0	148	129	3.9	4.9
Oceania	13	31	46	0.5	0.5	0.5	19	14	0.5	0.6

Source: Daniels *et al*., 2005 (Based on data from UN 2003)

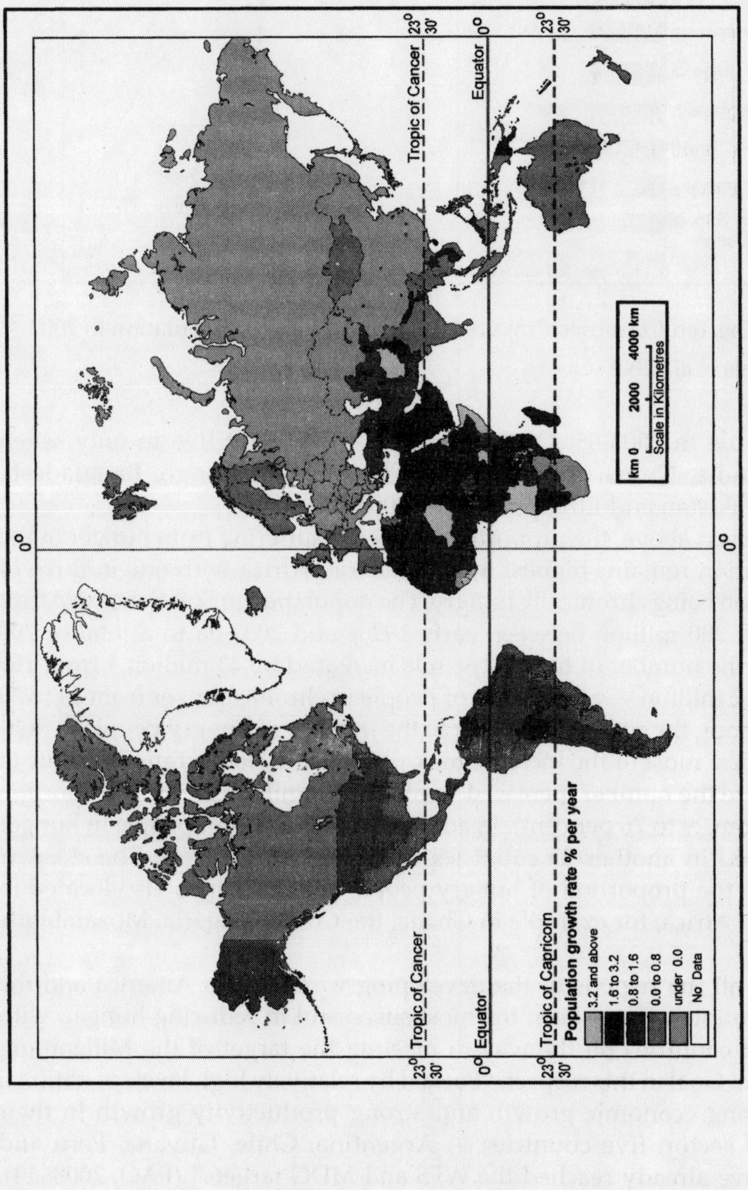

Fig. 5.5: Annual Average Rate of Population Change (%) 2000-2005

Source: Daniels *et at.*, 2005.

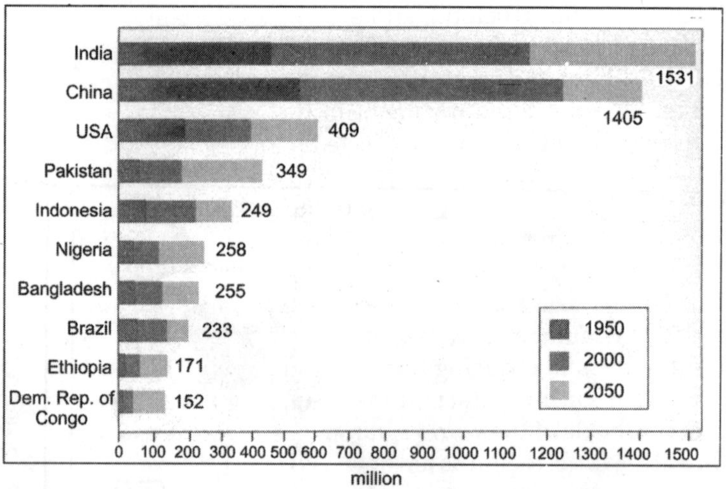

Fig 5.6: The Ten Countries Projected to have the Largest Population in 2000
Source: Daniels *et al*., 2005

hungry people in 2003-05. Of these people, 65 pr cent live in only seven countries: India, China, the Democratic Republic of Congo, Bangladesh, Indonesia, Pakistan and Ethiopia" (FAO, 2008: 12)

As observed above, the proportion of people suffering from hunger in the total population remains highest in Sub-Saharan Africa with one in three of the population being chronically hungry. The population in Sub-Saharan Africa increased by 200 million between early 1990s and 2003-05 to a total of 700 million and the number of hungry people increased by 43 million – from 169 million to 212 million — a reduction of people in chronic hunger from 34 to 30 per cent. Among the countries adding to the number of hungry people in Sub-Saharan Africa, most of the increase took place in the Democratic Republic of Congo (where the number increased from 11 to 43 million and the proportion increased from 29 to 76 per cent). In addition, the number of people in hunger also increased in another 25 countries of Sub-Saharan Africa. The steepest reduction in the proportion of hungry people is in countries also located in Sub-Saharan Africa, for example in Ghana, the Congo, Nigeria, Mozambique and Malawi.

Among all the regions of the developing world, Latin America and the Caribbean countries have been the most successful in reducing hunger, with 10 out of 12 countries on the way to meeting the target of the Millennium Development Goal in this respect. "Backed by relatively high levels of national income, strong economic growth and strong productivity growth in their agricultural sector, five countries — Argentina, Chile, Guyana, Peru and Uruguay have already reached the WFS and MDG targets" (FAO, 2008: 14). Similarly, Costa Rica, Jamaica, Mexico and Cuba have reached the targets for hunger reduction in 2003-05, but countries like El Salvador, Guatemala, Panama and Haiti are trailing behind, and among them Haiti with 58 per cent of its population steeped in hunger has the highest record in the world.

On the other hand, countries in the Near East and North Africa have the lowest levels of undernourishment in the developing world. However, the conflicts in Afghanistan and Iraq have an important impact, nearly doubling the number of people in hunger, from 15 million in 1990-92 to 28 million in 2003-05. In North Africa, according to the estimates of the FAO, 3 per cent of the total population were chronically hungry in 2003-05 (4.6 million against 4 million in 1990-92).

In contrast, Asia and the Pacific present a mixed picture of reduction and increase in the number of hungry people. On the whole, Asia has recorded a modest progress in reducing hunger, bringing down the proportion from 20 to 17 per cent, or the reduction in number from 582 to 542 million, though almost two-thirds of the world's hungry people are still concentrated in Asia. Southeast Asia sub-region as a whole has done very well in hunger reduction, and Viet Nam has already reached the MDG target by 2003-05 and Thailand also has made very good progress. In comparison, China and India as the "main driving forces behind hunger trends" present an almost contrasting case, together accounting for 42 per cent of the chronically hungry people in the world. China, however, has experienced spectacular success in eliminating hunger, reducing the number of chronically hungry people from 178 million in 1990-92 to 123 million in 2003-05 while India after registering impressive gains between 1990-92 and the mid-1990s, the progress has stalled since about 1995-97, and there has been an increase of 24 million more undernourished people by 2003-05 resulting from "the continuation of the declining per capita growth rate in total dietary energy supply and higher per capita dietary energy requirements" (FAO, 2008: 15-16)

According to FAO, in 2007 altogether 47 countries in the world were faced with food crisis and required immediate assistance, 27 of these countries were located in Africa, 10 in Asia, and the remaining 10 in other parts of the world. The same source observes that higher food prices have "triggered" an increase in global hunger, adding 75 million more to the estimated 848 million malnourished people in 2003-04, and bringing it to 923 million people in 2007, which was more than 80 million higher than in 1990-92. By region, the rising prices of food made a further addition of 41 million hungry people in Asia and the Pacific to their previous total, 24 million in Sub-Saharan Africa, 6 million in Latin America and the Caribbean, and 4 million in Near East and North Africa. But, using a different methodology, the United States Department of Agriculture (USDA) came to the conclusion that the 'impact of high food prices resulted in an increase in the number of undernourished people of 133 million in 70 countries' (FAO, 2008: 8) (*See* Note *J*).

However one interprets it, the 923million people estimated in 2007 as suffering from acute hunger deserve immediate attention in making adequate food available for them in addition to food consumed by the remaining section of the world population. Taking together the actual requirement of the two sections of the world population, i.e. people without adequate food and people with adequate food, the FAO has come to an agreement on the total requirement of food now and in years to come assessed on the principle of making food available of similar nutritional value.

The answer to the question – how much food we need to fulfil the requirements of entire humanity today – rests on the individual requirement which has been calculated as "per capita value of 2,200 kcal/day as an adequate average of global food energy needs. The global food supply available in 1990 could make this diet adequate not only in terms of total energy intake, but also in terms of its quality... But food consumption surveys have repeatedly shown actual intakes below 2,200 kcal/day. The two most representative US investigations resulted in means of less than 2,400 kcal for males and less than 1,600 kcal for females... Prevalence of such low intakes in many affluent societies – between 10 and 25 per cent below standard recommendations – clearly indicates that there is no single minimum of food energy supply applicable to all populations. Further, energy balance studies of traditional societies confirm that not only particular groups, but whole populations can live with food energy intakes well below the standard, western-set, metabolic expectations" (Smil, 1994: 263).

Consequently, as reported by Smil (1994), 'The World Hunger Programme prepared such estimates for more than 130 countries, containing 92 per cent of global population' with daily per capita averages (of consumption) ranging from a little less than 1,800 kcal in Afghanistan or Bangladesh to about 2,300 kcal in a number of rich countries, with the global mean at about 2,000 kcal, came to the conclusion that the "current global consumption would have to increase by less than 2 per cent if all people were to get enough food energy intake for their height" (Smil, 1994: 262). Taking FAO's food availability figures into account one can come to a firm conclusion that the "current global food production is sufficient to provide adequate food for the world's population. Even if all of the poor world's population would raise their health status and activity level to those prevailing in rich countries, the average 1990 global food energy consumption would have to increase by only 10 per cent, and it would still remain well below the available mean supply... This conclusion "starkly confirms that the current pattern of malnutrition and hunger is unrelated to food availability, but is instead a function of global entitlements to food" (Smil, 1994: 264).

Needless to say that the author (Smil, 1994) makes here a highly relevant point, strongly indicating the ground truth that it is not inadequate production responsible for food shortages in many countries coupled with rising prices curtailing entitlement of people to food but it is mostly due to severe lack of transportation, storage and distribution systems and the entire marketing system further crippled by manipulation at various levels of administration with a habit of looking away from those creating artificial deficits (crisis) through clandestine hoardings of food (as has often been the case in many states of India, where the farmers time and again do not get the cost-price for their products, such as potatoes or tomatoes, and so left them unsold) and not having a fair principle for complying with the need of providing adequate storage space available with the local administration. On the other hand, civil wars in Afghanistan, Angola, Ethiopia, Mozambique, Somalia, Sudan and Zambia had an upsetting effect on food production, or natural disasters like heavy floods in Bangladesh have been the reason for crop failures, finally leading to food deficits. Such deficits often have to be covered by huge food imports from other countries with surplus production (balancing the shortages and keeping world production at a level adequate for meeting the global requirements),

In addition, the total harvests meant primarily for human consumption get depleted through their use as a feed for domestic animals in many countries, amounting to about 700 million tonnes a year (though most of it reappears through consumption of animal products at the rate of 400 kcal/day), apart from the loss of about 600 kcal/day as post-harvest wastes and losses through distribution amounting to another 800 kcal/day.

C. SCOPE FOR FURTHER EXPANSION OF GLOBAL GRAIN PRODUCTION

There however, cannot be any complacency in the present situation of food availability thinking that improvement in the distributional system and removal of other hindrances will be adequate in providing universal access to food. Even if impediments hampering equity in the distribution of food are rightly handled, the available food is not going to be adequate to meet the requirements for a long time for a growing population, particularly because of change in food habits according to tastes/demands with rising income and consequent increase in entitlements, which are likely to create more stress without enhanced production. All this leads to the fundamental question –is there any further scope for expansion of production of grain crops in the world?

First of all, the scope for an expansion of the area under crops seems to be rather limited. Total world supply of arable land is about 3.2 billion hectares, and about 0.4 hectare of arable land were needed per person at the productivity level around 1972 (Meadows *et al.*, 1972: 60). About 45 per cent of the world's present cropland lies in Europe, the former Soviet Union, North America and Oceania with high per capita production of food. Nearly all the potential arable land in the Middle East and South Asia are under cultivation and the same is true through most of the Far East. There are major reserves of unexploited land to be found in sparsely populated parts of Indonesia and Cambodia. In Sub-Saharan Africa and Latin America 'under one-fifth of potential arable land is currently cultivated but it seems highly likely' that a significant part of the increasing demand for food in Sub-Saharan Africa will be met by expanding the cropland involving long-distance migration of the rural population to more remote and less exploited areas. Similarly, exploitation of the main tracts of unexploited land in the Congo and Amazon basins is restricted by distances from main areas of population concentration and poor communications. Table 5.16 gives a clear idea about the scope for further expansion of cropland in the

Table 5.16: Distribution of Arable Land in the World, by Regions and Cropland Irrigated, 1991

(*Percentage*)

Region	*Potential Arable Land (in hectares)*	*Cropland 1991 (in hectares)*	*% of Potential Arable Land that is cropped*	*% of Cropland that is irrigated*
Sub-Saharan Africa	815.7	142.0	17.4	2.9
Middle East	72.3	70.2	97.1	16.7
South Asia	230.2	227.6	98.9	33.2
Far East	235.9	185.2	77.9	38.1
Latin America	889.6	153.3	17.2	10.9
Europe/Former Soviet Union	N.A.	366.9	—	10.1
North America/Oceania	N.A.	282.2	—	7.7
World	—	1441.6	—	16.8

Source: Tim Dyson, 1996.

NA = Not available.

future in view of growing demand for food which, by any stretch of imagination, 'will not be met by the simple conversion of new land to crops, more so, as the available unexploited land is most probably' of lower agricultural potential. "Instead, the overwhelming bulk of future food demand must be met through more productive use of existing cropland' (Dyson, 1996: 117), keeping the fact in mind that a little over 1.4 billion hectares of land remaining under crops in around 1991 represent the availability of land suitable for cropping.

Apart from this visible limitation, "losses and degradation of arable land are worldwide phenomena caused by growing populations, industrialization, and environmental degradation" the effect of which are already felt in many poor and populous countries. China provides the most important case "whose potential expansion of arable land ... limited to about one-tenth of the currently cultivated area," has had a net loss of almost one-fifth of its arable land since the mid-1950s. (Smil,1994: 275) The UN Food and Agricultural Organisation's (1981) "detailed appraisal of long-term agricultural prospects in 90 poor countries (excluding China) found that 51 of them had abundant or moderately abundant reserves of arable land – but the population of these countries was just one-third of the studied total. In contrast, 18 countries with extreme land scarcity, already cultivating more than nine-tenths of the potentially arable land, supported one-half of all the population in the studied set" (Smil, 1994: 275).

Detailed assessment based on evaluation of soil and climate finds the potential for farmland expansion would bring the total amount of cropland close to 2 billion hectares, larger than the cropland in 1990, 'but the total addition to new farmland is likely to be much less than during the past 60 years when cultivated area expanded by about 50 per cent'. Importantly, expansion of "the cultivated area by no more than 20 per cent by 2050 (the slowest growth rate since 1800) would add about 300 MHa of crop fields, enough to produce an equivalent of at least 400 MT of grain even with average global yields locked at the 1990 level. Obviously, expanding the cultivated area will not suffice to feed an additional 2-2.5 billion people. Intensification of farming rather than extension of cultivated land will have to provide the bulk of new harvests" (Smil, 1994: 275).

Specifically, from the middle of the twentieth century the global population has outgrown the pace of expansion of the cropland area and a new era has begun stressing the need for raising land productivity, ending the 10,000/- year era of steady expansion of land under crops. While the 'grain area has increased by about 9 per cent since mid-twentieth century, global population has grown 132 per cent, seven times faster' (Brown *et al.*, 2000). This resulted in the sharp fall of grain area available for each person, making it half of what it was in 1950, from 0.24 to 0.12 hectares which is likely to fall further to 0.08 hectares by 2050. The per capita availability of land shrank by 38-56 per cent between 1950 and 1998 in Pakistan, Nigeria, and Ethiopia, and we can expect a further fall in per capita share by another 55-63 per cent by 2050 which will bring down the available amount of land to 300-700 square metres in these three countries, making the area available per head a third or less than that in 1950. In East Asia, the per capita land for producing cereals stands at 200 to 600 square metres. In Asia alone, grain area per person is 800 square metres, and 11 countries are going to cross this threshold by 2050 compelling them and many other countries to depend on food imports. Apart from North Korea which imports less than 20 per cent of its grain requirements from abroad, countries like Sri Lanka, Japan, Taiwan, South Korea, and those in Middle East and North Africa depend heavily on imports (Brown *et al.*, 2000: 62-64). In India, 5.5 million tonnes of wheat were imported in 2006-07 which, however, declined to 1.8 million tonnes in 2007-08 (Government of India, Economic Survey 2008-09: 194) with production of 78.57 million tonnes and 77.63 million tonnes respectively.

Obviously, per capita land for cropping is shrinking at an alarming rate in countries like Bangladesh, India and in African countries as a direct consequence of population growth. In Bangladesh, for example, average farm size has already fallen below 1 hectare and the: "strong tradition (in Bangladesh) of bequeathing land in fixed proportions to all male and female heirs has led to increasing landlessness and extreme fragmentation of agricultural holdings" (Brown, 2004: 148), making them so small that they are economically inoperative. In India, the number of farms with less than 2 hectares (5 acres) of land increased from 49 million to 82 million between 1970 and 1990, and with a similar trend continuing, the country has now more than 90 million farms of this size making them economically inoperative for a family of 6 members. A similar trend of reduction in farm-size prevails in Africa, where the world's fastest population growth is taking place. In Nigeria, for example, the per capita grainland will shrink from 0.15 hectares now to 0.06 hectares by 2050.

I. Efforts to Raise Land Productivity

According to Brown (2004), during the first half of the twentieth century the world's grain production increased nearly three times from 631 million tonnes in 1950 to 1,835 million tonnes in 2000, the gain during this period, as claimed, being "nearly double that from the beginning of agriculture some 11,000 years ago until 1950" (Brown, 2004: 146). The tripling of grain harvest between 1950 and 2000 was made possible partly by multiple cropping, with such combinations of crops as wheat and corn in northern China, wheat and rice in northern India, double or triple cropping of rice in southern China, southern India, and in nearly all the rice-growing countries in South-east Asia. For example, very impressive results were obtained from double cropping of winter wheat and corn in the North China Plain, raising the production to the US level, with a yield of winter wheat in the North China Plain of nearly 4 tonnes per hectare and corn of 5 tonnes per hectare, and these two crops, grown in rotation, give 9 tonnes of cereals per hectare. Besides, China's double-cropped rice yields 8 tonnes per hectare. In India, wheat followed by rice using the same land in successive order as wheat is harvested in time to plant rice, throughout the states of Punjab, Haryana and Uttar Pradesh, gives 3 tonnes of wheat and 2 tonnes of rice per hectare. In contrast, however, the harvested area in Japan has dropped from 5 million hectares in 1960 to 2 million hectares in 2005 mostly due to decline in the practice of double cropping as farmers are being drawn away by industries offering higher wages. Decline in multiple cropping has also led to the shrinkage of grain area in South Korea after 1965 and the same thing happened in Taiwan. Industrialization is likely to affect agriculture in a similar way in China and India in future. The double cropping combination of winter wheat with soybeans in the US, having been practiced since 1996 boosted grain production quite substantially (Brown, 2006: 164-165).

There are at least three ways of raising land productivity: raise the yield per crop; increase the number of crops per year grown on the same piece of land through multiple cropping; and, getting more out of the existing harvest by processing 'crop residues through ruminants to produce meat and milk'.

II. Role of Water in Increasing Food Productivity

One of the major ways of raising yield is irrigation and at the global level "average crop yields on irrigated land are roughly three times yields on unirrigated land, and it has been estimated that 36 per cent of the current world harvest comes from the irrigated area" (Dyson, 1996: 117). As Table 5.16 shows, by 1991, 17 per cent of the world's total cropland was irrigated. The largest fractions of cropland being irrigated are in the Far East and South Asia, China and India having respectively 48 million and 46 million hectares of cropland irrigated, from which they produce more than half of their domestic output of food crops. Among the other major irrigators are the former Soviet Union with

21 million hectares of crop land irrigated, the United States where 19 million hectares are irrigated, Pakistan with 17 million hectares irrigated and 5 million hectares in Mexico – all of which are estimates of net irrigated area and do not include repeat irrigation for multiple cropping.

According to experts, further scope of increasing irrigation facilities is very limited as countries like China and India have already exhausted irrigated potential. This is also applicable to the Middle East while in Sub-Saharan Africa irrigation is practiced only in a few countries like South Africa, Nigeria and Madagascar and that too for export crops. In the Sahel sub-region large areas in the eastern and western parts are already under severe strain of water scarcity. Despite this, some expansion of harvested area has taken place in recent years in some countries of the Middle East, such as in Syria, Morocco, Sudan, Iraq, Saudi Arabia and Egypt partly through increased multiple cropping and 'by reclaiming and irrigating significant tracts of steppe and desert land'. Even in South Asia and Far East there has been some minor expansion (Dyson, 1996: 117-118 and 122).

However, any future increase on a larger scale in the production of food crops will have to be realized from greater efficiency of water use. In this connection an idea may be obtained from Table 5.17 about yield rates of cereals and their trend during the second half of the twentieth century. As the table shows, the average rate of yield has increased by about 42 kilograms per hectare at the global level since 1981 which had been about 41 kilograms since 1951. At the regional level, Sub-Saharan Africa, Middle East, and South Asia with an average rate of annual increment lower than that at the global level, registered considerable increase except in Sub-Saharan Africa which had very little of it. In comparison, the Far East, North America, Oceania, Europe and the former Soviet Union experienced rates of increment higher than that of the global average which became higher since 1971 in the Far East, from 1981 in Europe/FSU but declined from 1971 in North America/Oceania. Latin America shows a contrasting case with increment rate lower than global average, remaining at that level all through the period.

Looking at the average yield of cereals per hectare in each of these regions, it was the highest (over 3.8 metric tonnes p/ha) in the Far East, in around 1990 followed by North America/Oceania (over 3.7 metric tonnes per/ha) and the lowest was in Sub-Saharan Africa (with little over 1.1 metric tonnes per/ha). The corresponding expected increment by 2020 shows nearly the same pattern, the Far East maintaining its lead with nearly 6 metric tonnes per/ha, followed by North America/Oceania with 5 metric tonnes and Europe/FSU with more than 4.2 metric tonnes, ranking third in order as in 1990, while Sub-Saharan Africa is going to have the lowest rate of yield (about 1.6 metric tonnes) as it was in 1990. Interestingly, the expected rate of yield of cereals per hectare in 2020 is going to be lower than the world rate (nearly 4 metric tonnes) in Middle East (over 2.5 metric tonnes), South Asia (over 3.4 metric tonnes), Latin America (3.1 metric tonnes), similar to the situation in 1990 (Dyson, 1996: 124-125).

Worldwide 70 per cent or even a little more of the water taken from

Table 5.17: Average Cereal Yields, Levels, Trends and Projections, World Regions, 1951-2020

Regions	*Average Yield around 1990 (metric tons per hectare)*	*Average Annual Yield Increase (kg. per hectare per year)*				*Projected Yield in 2020 based on 1981-93 Increment (metric tons per hectare)*
		1951-93	*1961-93*	*1971-93*	*1981-93*	
Sub-Saharan Africa	1.165	10.2	10.8	11.7	14.4	1.597
Middle East	1.642	19.1	21.0	23.4	31.3	2.581
South Asia	1.919	27.4	32.9	40.4	50.6	3.437
Far East	3.817	67.0	82.7	92.3	72.7	5.998
Latin America	2.119	30.2	34.2	38.9	32.7	3.100
Europe/Former Soviet Union	2.816	45.7	47.4	43.3	47.7(59.1)	4.247
North America/ Oceania	3.734	55.5	50.0	42.5	42.4	5.006
World	2.711	40.9	45.2	45.5	42.6	3.989

Source: Tim Dyson, 1996: Table 4.6

Note: (a) "1990 yields are averages for sample countries for years 1989-91.

(b) The 1981-93 average cereal yield increase for Europe/FSU is given in parenthesis. This figure of 59.1 kg is biased upwards by an exceptionally high regional yield recoded for 1993. This was largely due to a reduction in the area planted in the former Soviet Union which means that Europe (where yields are higher) is overrepresented in the 1993 regional composition. Therefore for present purposes the lower yield increment of 47.7 kg for the period 1981-92 has been preferred." 1996 : 124.

underground sources or aquifers, or from rivers and other water bodies are used for irrigation and this rises to 85 per cent in Asia though in Europe the water-intake for agriculture, which is largely rainfed, is minimal. As a matter of fact, industry always takes the upper hand in a competition for the use of water since the ratio of water use between the two (70 to 1 as 1,000 tonnes of water is necessary to produce 1 tonne of wheat worth, say, $ 200 when a similar amount of water helps in industrial expansion giving an output of $ 14,000, as mentioned previously quoting Brown, 2000) is hardly competitive.

Nevertheless, irrigation has so far dominated the scene in the use of water in almost all the countries of the world, the irrigated areas having increased three times during the last half of the twentieth century, climbing from 90 million hectares in 1950 to nearly 270 million hectares in 2000. Irrigation outstripped growth in population between 1950 to 1978, increasing irrigated land per person from 0.037 hectares to 0.047 hectares, 'an increase of one-fourth', after which the growth, as mentioned elsewhere, slumped, falling behind the growth of population since 1978, shrinking by 8 per cent per person (Brown, 2004: 154).

In order to achieve more efficient use of water as the major way of raising crop productivity, one has to consider the fact that 'crop usage of irrigation water never reaches 100 per cent' as part of it is lost through evaporation, percolation downward and some of it runs off. Furthermore, soil type, temperature and humidity also affect the amount actually available for irrigation. "Water Policy analysts Sandra Postel and Amy Vickers found that surface water irrigation efficiency ranges between 25 and 40 per cent in India, Mexico, Pakistan, the Philippines and Thailand; between 40 and 45 per cent in Malaysia and Morocco; and between 50 and 60 per cent in Israel, Japan and Taiwan" (Brown, 2006: 167)

The situation has become worse with growing water scarcity as a worldwide phenomenon. As the major rivers run dry, the overpumping of aquifers - a new development mostly during the last half of the twentieth century – done beyond their sustainable capacity has further aggravated the situation, a widespread practice in China, India and the United States (and also Saudi Arabia and North Africa) – the first three countries accounting for nearly half of the world's grain harvest. Overpumping of aquifers is depleting water-sources faster, triggering a fall in water tables. According to Sandra Postel, the author of 'Pillar of Sand', the annual overpumping of aquifers at 160 billion cubic metres or 160 billion tonnes if it goes unchecked would amount to a loss or deficit of 160 million tonnes of grain on the basis of 1,000 tonnes of water used to produce 1 tonne of wheat – or half of the US grain harvest. A similar amount of grain can feed 480 million people (at the rate of consumption of just over 300 kilograms or one-third of a tonne per person per year) and this is going to result from unsustainable use of water. (*See* Note *K*).

In Punjab, the bread-basket of India, producing a surplus crop through double cropping, the water table is falling 0.6 metres per year forcing farmers to dig shallow wells deeper. However, the Government of India, keen to increasing irrigation through major and minor irrigation projects, has increased irrigation from 81.1 million hectares in 1991-92 to 102.8 million hectares by 2006-07, though only 87.2 million hectares of the total potential is utilized (Government of India, Economic Survey 2008-09: 177). The need for irrigation water to keep production at present level has already outstripped the supply.

Acute water scarcity is also felt by the United States where, as mentioned earlier, the irrigated agriculture in the southern Great Plains based largely on water pumped from Ogallala Aquifer is slowly getting depleted, with a fall in the water table in the absence of proper recharge. The shrinkage of irrigated area in the United States expanding further will cause grain production to decline by 17 per cent, as pointed out by Brown (2004).

Augmentation of the supply of water for irrigation purpose depends on adopting more irrigation-efficient technologies and raising the price of water, as intended in China, which may be one of the measures urging users to more efficient uses, stopping overuse and wastage. The key to efficient use of water lies in other ways to irrigate crops, which include flood or furrow irrigation, overhead sprinkler, and drip irrigation. Among them "furrow irrigation, probably the earliest form, is used with row crops, with a small trench being cut near each row of plants. Flood irrigation traditionally used, is now being reconsidered since recent research indicates that at least in some situations periodic flooding will produce the same yield as continuous flooding, but use much less water" (Brown, 2004: 157). Overhead sprinkler irrigation is widely practiced in the southern Great Plains of US utilizing ground water and compared to furrow or flood irrigation, it reduces water use by 30 per cent. Switching from 'a high-pressure to a low-pressure overhead sprinkler system can boost irrigation efficiency from 65 to 80 per cent while shifting to a low-energy precision application system it can further raise it to 90 per cent or more. On the other hand, 'Drip irrigation enjoys better advantage as it is capable of reducing water use by half of that in furrow irrigation. Moreover, drip irrigation raises yields because of steady supply of water carefully geared to crop needs with minimal loss to evaporation and as it is labour-intensive and water-efficient, it suits 'countries with underemployment and water shortage'. However, till now only a few countries depend heavily on drip irrigation, such as Cyprus, Israel and Jordon, while "among the three big agricultural producers this more-efficient technology is used on less than 1per cent of irrigated land as in China and India and roughly 4 per cent in the US'. According to the findings of Sandra Postel and her colleagues, drip irrigation reduces water use by 30-70 per cent and raises yields by 20-90 per cent. This 'combination of reduced water use and higher yields can double water productivity', presenting an attractive prospect (Brown, 2004: 157-158). Postel reports that 'India has about 10 million hectares of agricultural land that can profitably be irrigated with drip irrigation system (already practised widely in tea gardens of India). China has similar potential to tackle growing scarcity of water. As pointed out by Brown (2004), "a rice crop yielding four tonnes per hectare uses little more [water] than one that yields two tonnes per hectare simply because so much of the water used to produce rice is lost through evaporation from the water surface. Simply put, raising land productivity also raises water productivity" (Brown, 2004: 158) (*See* Note *L*).

At the same time, revival of traditional water harvesting practices, as prevalent in India since pre-historic times, well-tested and very efficient as well as eco-friendly in water management answering well to drinking and irrigation requirements, may be tried to meet the challenge of growing water demand in a situation of increasing water scarcity. As reported by the Centre for Science and Development (India), India receives 400 million hectare metres (mha m) of rainfall annually over a geographical area of 329 million hectares (mha), which, if evenly spread would uniformly submerge the entire land surface to a depth of 1.28 metre. The distribution of rainfall varies widely – both spatially and temporaly – some areas (e.g. the Thar desert) receiving 200 mm annually while Cherrapunji in the North East receives 11,400 mm. annually. According to Pisharoty, one of the leading Indian meteorologists, almost no area in the country receives less than 100 mm a year which is sufficient to meet local drinking water requirements provided the rain is harvested properly and where it falls. The rainfall is largely concentrated during four months of the year does not occur daily and in most parts of the country it takes place during not more than 50 days and does not fall over a period of 24 hours. Heavy showers are more common and most of the country receives rain for just about 100 hours each year and Pisharoty stresses the fact that 'half the annual rainfall is precipitated in just one-fifth of the total hours of rain in a year. Thus if a town receives 80 cm of rain, half of it, that is 40 cm falls in just 16 hours. In the country as a whole, half of the total annual rainfall is received in about 20 hours'.

To harvest this torrential rainwater, Pisharoty recommends a system of 30,000 tanks, each measuring 100m x 100m x 120m, 10 cubic hectare metres in volume in each of the districts, numbering more than 300 in the country, receiving more than 500 mm of rain in a year. In areas with 800 mm, "assuming a collection efficiency of 50 per cent over a catchment area 30 times the tank's size, and an annual evaporation loss of 2 m, each tank will store water to a depth of 10 m. In areas averaging only 200 mm of rain annually, a catchment area 50 times the size of the tank, with 60 per cent collection efficiency, and 4 m evaporation loss, the tank will store water to a depth of 8 m. The cumulative harvest of water would total 90 mha m, almost equal to a quarter of the total annual rainfall of the country and more than all the surface water considered usable today. The surface area of the tanks would cover 3 per cent of India's land surface" (Agarwal and Narain (Ed), 1999: 314-315).

However "rough estimates" these are, the scheme seems to hold high prospects if treated seriously which is admirably suited to the technological background of the rural areas in India. In this connection it may be said that tank irrigation, as a practice, is fifteen to sixteen centuries old in this country, revival of which is likely to be an inexpensive and dependable source of irrigation as well as for supply of drinking water for the local community. Admirably, the Planning Commission of India, in realizing the importance and efficiency of traditional water harvesting systems, recommended "other sources", consisting of 'anicuts' in the south, 'kuhls' in the hilly areas, 'bund haras' in Gujarat, Maharashtra and Mysore, 'ahars' and 'pynes' in Bihar and various types of lift irrigation. To take lift irrigation first, this involves water being lifted by manually operated wooden devices, for example in West Bengal, from rivers or tanks or natural water bodies into the crop-fields through dug-out small channels. The 'bund haras' refer to bunds which are very efficient structures in utilizing every drop of water in the stream. Rooftop water harvesting structures, dugwells and stepwells are also the traditional harvesting structures used as sources of drinking water in India over the centuries. "Traditional water harvesting systems like tankas or kundis, used in arid regions of Rajasthan" for generations have fallen into disrepair in the modern days. "However, rooftop harvesting structures have been promoted actively in Mizoram and have proved to be successful... especially where the supply of piped water is very expensive, (and) has now been successfully accepted in Meghalaya as well...However, the quality of stored rainwater in Mizoram is found to be too pure for human consumption and lacking in the necessary minerals requiring some treatment... Traditional water harvesting systems definitely have relevance in areas where water scarcity is acute or where groundwater is either brackish or too deep to obtain cheaply. However, in some areas, a supplementary source may well be necessary. Traditionally in all these areas people have developed cultural practices which encourage judicious use of water, but now these practices are dying. Water conservation education should be encouraged across the country" (Centre for Science and Environment, 1999: 331 and 335) (*See* Note *M*).

III. Fertilizers as Productive Incentives

While at the global level potential supplies of cropland and water needed for agriculture are still relatively large, "natural availability of the three macronutrients – nitrogen, phosphorus and potassium - has been below the level needed to produce the global harvest for the past two decades... and no new natural sources of macronutrients are waiting to be tapped: in order to increase their supply we have to synthesize more ammonia and mine more phosphorus and potash. The relative scarcity of nitrogen in the biosphere has made it the most common limiting nutrient in plant growth" (Smil,1994: 273). But phosphorus is even rarer in the biosphere than nitrogen.

However, about half of the nutrients applied to the world harvest is represented by synthetic nitrogen fertilizer and with sufficient water available, it is the 'most important determinant of crop yields', and regional differences in fertilizer use closely correspond to agricultural intensification existing at regional levels. Use of fertilizers is particularly high in the Far East, China doubling its average applications from around 144 to 284 kg per hectare of cropland between 1980 and 1990, resulting in a sharp rise in food output. Of the 67 developing countries studied, only Egypt (360 kg), North Korea (410 kg) and South Korea (440 kg) used more fertilizer per hectare around 1990 than China. It is also relatively high in Europe/FSU, with a maximum reached in Germany (520 kg) and the Netherlands (614 kg) (Dyson, 1996). However, the yield gains from fertilizer use are higher with the availability of irrigation facilities, and also the type of fertilizer used, the placement and timing of application affect yield rates which vary widely between countries. Nonetheless, considerable scope exists to raise yields through the application of fertilizers, especially in countries in Sub-Saharan Africa, the Middle East, South Asia and Latin America. The recent fall in the international price of urea, the principal nitrogen fertilizer used in most developing countries primarily for the reason of the easier handling of this solid material, is another factor for its popularity while in the US ammonia, another variant of nitrogen, is widely used (Dyson, 1996: 119).

As the use of world fertilizer increased from 14 million tonnes in 1950 to 141 million tonnes in 2000, "in some countries it began to press against the psychological limits of plants to absorb nutrients. In response, fertilizer use has leveled off in the United States, Western Europe, Japan, and now possibly in China. In these countries, applying additional nutrients has little effect on production. Some parts of the world, such as the Indian subcontinent and Latin

America, can still profitably use additional fertilizer. But for the world as a whole, the rapid growth in fertilizer use – the engine that helped triple the world grain harvest since 1950 – is now history" (Brown, 2004: 151).

More unfortunate is the role of fertilizer washed out from land where applied and carried by running water into rivers and oceans leading to 'algae blooms' using up all available oxygen in the water, creating dead zones for fish ultimately hitting food output and throwing people out of livelihoods dependent on fishing and constructing river crafts. 'The accumulation of nitrates in underground water supplies in Western Europe led to European Union regulations to restrict fertilizer use.' As a matter of fact, in many countries, including the United States and Japan, fertilizer use has already reached a level where it no longer affects crop yields. A declining response to the application of fertilizers is inevitable in the most intensively farmed regions, but still now a large majority of the world's farmland is short of the needed macronutrients, tackling which by fertilizers should be guided by every caution in their excessive use so that harm to land or water, as mentioned above, can be avoided.

Incidentally, Africa has shown a brilliant example by simultaneous planting of grain and leguminous trees (in India there is a long practice of combining leguminous and non-leguminous crops in the cropping system in many places to make up for deficiency of nitrogen in the soil). "The trees start to grow slowly, permitting the grain crop to mature and to be harvested. Then the trees grow quickly to several feet in height, dropping leaves that provide nitrogen and organic matter – both sorely needed in African soils. The wood is then cut and used for fuel. This simple, locally adopted technology, developed by scientists at the International Centre for research in Agroforestry in Nairobi, has enabled farmers to double their grain yields within a matter of years as soil fertility builds" (Brown, 2006:166). Similarly, a combination of trees and shrubs yielding continuous or cyclical harvests of food (fruit, nuts, fodder (leaves), fuel or timber have proved to be highly beneficial agricultural practices in some countries remaining under the stress of water scarcity.

Nevertheless, despite efforts like this and others at local levels, the unmistakable evidence of loss of momentum in raising food production cannot be missed. What is needed to confront the situation facing limitations set by water scarcity and use of fertilizers in higher doses is a new mindset with a new way of thinking, as observed by Brown (2006). According to him, 'shifting to more water-efficient crops wherever possible ... boosts water productivity'. For example, rice production around Beijing is being 'phased out' because rice is a thirsty crop. 'Similarly, Egypt restricts rice production in favour of wheat'. Any measure like this raising crop yields on irrigated land also raises the productivity of irrigation water. Keeping the use of water at a level that can be sustained by aquifers and rivers worldwide involves measures being taken not only in agriculture but also 'through the economy'. For example, apart from adopting more water-efficient irrigation practices and water-efficient crops, it is necessary to use more water-efficient industrial processes and more water-efficient household appliances. (Brown, 2006: 170).

While land productivity in the world as a whole tripled since 1950, some countries have done even better than that. For example, China, France, the United Kingdom and Mexico quadrupled their wheat yield per acre between 1950 and 2000. India did nearly the same and the corn yield increased four times in the United States. For the world as a whole the yield of grain per hectare rose by 2.1 per cent a year between 1950 and 1990, but between 1990 and 2000, the annual increase fell to 1.1 per cent (Brown, 2004) (Table 5.18). The role of biotechnology in obtaining higher yields through the development of crop varieties reducing insecticide use, more drought- and salt-tolerant crops alleviating water shortages may be of great help in future years to come.

As observed by Smil (1994), 'some of the poor world's traditional soil and water conservation practices' hold 'considerable promise for more sustainable farming' and greater dependence on improved local staple, such as traditionally grown roots and tubers in the tropics, have great advantages over cereals, fitting better the high-humidity climates and giving high yields with more flexible planting and harvesting. Even keeping the planted area at 1990 level, a 50 per cent increase in root crop yields over the next 60 years will provide food energy equivalent to 50 million tonnes of grain. Cultivation of abandoned fertile land, almost a worldwide phenomenon associated with low-intensity use of farm land, opens up another very strong possibility of increasing the availability of cropland. In China, for example, such lands are estimated to be as high as 5 per cent of all cultivated land. Idle land of this type also occurs in countries in West Africa, the Caribbean Islands, Southeast Asia and even in densely populated areas of central Java. Further, "extension of good agronomic practices and the most suitable cultivars to many poor countries offers a far from exhausted opportunity for higher yields... Even relatively slow worldwide advances would have the desired effect. Average cereal yields increased about 2.5 times between 1950 and 1990 (from 1.1 to 2.8 t/ha), and by 55 per cent in the 20 years between 1970 and 1990. If global intensification (new cultivars, multicropping and higher fertilizer) were to raise the mean yield just by one-third by the year 2050, the global harvest would increase by an equivalent of at least 600 million tonnes of unmilled grain" (Smil, 1994:-277).

Land availability is a lesser problem today than the challenge thrown by water becoming more scarce with increasing withdrawals, averaging about 90

Table 5.18: Increase in World Grain Yield per Hectare, 1950-2000

Year (tons)	*Yield Per Hectare[1] (per cent)*	*Annual Increase (per cent)*
1950	1.06	
1990	2.47	2.1
2000	2.75	1.1

Source: Brown, Lester R., (2004) (Taken from USDA, Production, Supply and Distribution, updated May 2001).

Note: 1. Yield for 1990 is three-year average.

per cent of total water use in Asia and Africa, 60 per cent in South America and 50 per cent in North and Central America. In order to lessen worldwide demand on water for agriculture, there are some practical suggestions for increasing the production of three leading, relatively salt-tolerant grain crops – barley, sorghum and wheat — using 'properly managed salt-water' preferably in arid countries which has a potential of exceeding the average 30 per cent gain in production. Immediate benefit from the introduction of new food crops is not possible within the space of one generation. But following Plotkin (1986), Smil observes that the commercial potential of some of the new tropical crops is quite promising. "The most likely candidates for diffusion and breeding include amaranths (grain of Andean origin with high protein content), fruit of pupunha and buriti palms for eating, and those of pataua and babassu palms for edible oils. Naturally, such introductions, as well as the continuing improvement of leading crops and maintenance of viable agro-ecosystems, will require the maintenance of sufficient biodiversity – but this need is by far the most difficult production factor to evaluate" (Smil, 1994: 278)

IV. Role of Protein in Food Deficiency

'The second most important way to raise both land and water productivity is to produce animal protein more efficiently', when already 38 per cent or about 730 million tonnes of world grain harvest is used as animal feed which may go up or down depending on future grain prices.

The demand for animal food - meat (beef, pork, poultry and mutton) as well as eggs and milk – rises with income in every society, its consumption having increased from 47 million tonnes in 1950 to 260 million tonnes in 2005, more than doubling the per capita consumption from 17 to 40 kilograms (perhaps, as an innate habit bequeathed by man's hunting-gatherer forefathers). Though growth in meat production was primarily concentrated in Western industrial countries and Japan, later, over the last two decades it has increased rapidly in East Asia, especially in China, the Middle East and Latin America. Among the four different types of meat, beef followed by pork and poultry is consumed more than mutton.

Land too arid for crop production, covering a large part of the globe, accounting for double of the cropland, is used as grazing land for animals setting up different types of livelihood for people engaged in pastoral activities. For example, millions of people in Africa, the Middle East, Central Asia, western China and parts of Indian subcontinent thrive on the herds of cattle and flocks of sheep and goats providing meat, milk, hides, etc., for their livelihood. The lands on which the animals are grazed have no other better use as they are not suitable for crops and with little unused grazing land available now, any future increase in meat production has to depend largely on feeding animals with grain. This has a lot of influence on grain production in the world and is, at present, quite an effective force shaping production trends. In its turn the consumption rate of grain by each animal acts as a deciding factor about which animal is more preferred than the other. For example, the cattle in feedlots require about 7 kilograms of feed concentrate per additional kilogram of live-weight compared to ratios of 4 to 1 for pigs and 2 to 1 for chicken so that tightening of grain supplies shifts emphasis (or preference) from beef cattle to pork and more to poultry, which explains 'why world poultry production overtook that of beef in 1995' (Brown, 2000 and 2004). Between 1990 and 2005, world beef production increased only 1 per cent a year compared to 2.5 per cent for pork. The most rapidly growing source of meat during this period was poultry, expanding nearly 5 per cent annually. "The rapid growth in poultry production, going from 41 million tonnes in 1990 to 80 million tonnes in 2005, enabled poultry to eclipse beef in 1995, moving it into second place behind pork. World pork production, half of it in China, overtook beef production in 1979 and its lead has continued to widen since then. World beef production is constrained by both grazing limits and the inefficient feedlot conversion by cattle, though it is continuing to expand, but slowly. Indeed, within the next decade or so, fast-growing, highly grain-efficient aquacultural output may overtake beef production" (Brown, 2006: 171) (Table 5.19).

Table 5.19: World Growth in Animal Protein Production, by Source, 1990-2000

Source	*Annual Rate of Growth (percent)*
Aquaculture[1]	11.4
Poultry	4.9
Pork	2.5
Beef	0.5
Oceanic Fish Catch 1	0.1

Source: Brown, Lester R. (2004) (Taken from FAQ Year Book of Fishery Statistics).
Note: 1 = 1990-98 only.

However, two other very important sources of protein are eggs and milk. The consumption of eggs is growing three times as fast as the consumption of meat. The growing production of eggs and milk is indicative of their popularity as both sources of protein food as well as an important provides of livelihood for millions of people -especially dealing with milk in its various forms giving rise to chains of small — to medium-scale dairy industries in many countries. The agricultural residues like rice-straw, wheat-straw, and corn-stalks have a great unrealized potential for feeding ruminants, such as cattle, sheep and goats instead of feeding them with grain saved for human consumption, which means that it gives 'grain crops yield a second harvest' – the meat or milk produced with straw and corn-stalks.

The cattle and water buffalo as the major sources of milk in India are particularly good at converting coarse roughage into milk, the cattle figuring prominently in the dairy industry, the production of milk in the country increasing from 20 million tonnes in 1961 to 79 million tonnes in 2000 – achieving almost a fourfold increase. Quite appreciably, milk became the leading farm product in 1994 and India overtook the United States as the world's leading producer of milk and milk products in 1997, accomplishing this 'entirely by using farm byproducts and crop residues, avoiding the diversion of grain from human consumption (Fig. 5.7) (*See* Note *N*). Milk production per capita rose from 0.9 litres per week to 1.5 litres in India between 1961 and 2000, produced 'almost entirely by small farmers possessing one to three cows. Integrated with agriculture, milk production in India involves an estimated 70 million farmers, making a supplemental income, and a large number of milk corporations have come up utilizing milk produced by farmers in the rural areas, some of them functioning as quite large-scale enterprises engaged in the production of a wide and increasing range of milk products enjoying country-wide market. In its turn, ownership of a few cows/buffaloes provide a means of supplying dung used both as fuel (in the form of sun-dried cakes) and manure and if any new, affordable source of energy is introduced (as is done in the production of gas from cow dung), that will release an unaccountable amount of animal manure to fertilize soil for boosting the future of organic agriculture.

Similarly, China also has a huge potential to feed ruminants with rice and wheat straw as the world's leading producer of rice and wheat, and coming second in corn production, with an estimated production of 500 million tonnes of straw and corn stalks, most of which is either burned, or used as fuel. At the same time, ruminants produce soil-enriching manures returning nutrients to soil and in the face of grain becoming scarce, besides China, some other countries like the United States, Canada and France 'may follow India's example of using ruminants to convert crop residues into food'. (Brown, 2004).

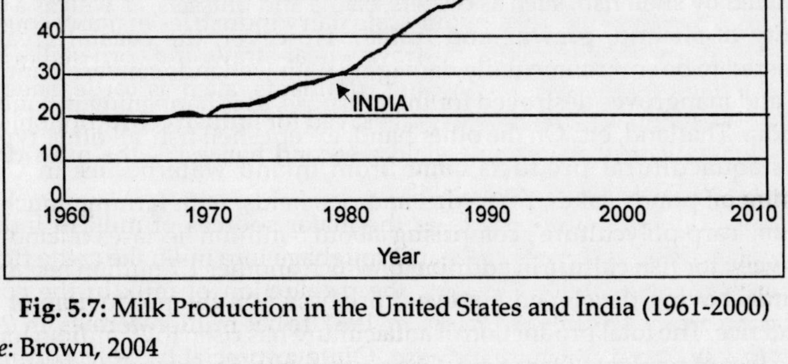

Fig. 5.7: Milk Production in the United States and India (1961-2000)
Source: Brown, 2004

The supply of animal protein can be increased in three ways without consuming more grain: improvement in the efficiency of grain conversion into protein, 'shift from the less efficient forms of conversion, such as beef or pork, to the more efficient ones, like poultry' or farmed fish; and greater reliance on 'ruminants to convert more roughage into either milk or meat'. (Brown, 2004)

Another very important source of protein is fish, available mainly from two sources, the oceans and through aquaculture in different types of water bodies. The oceanic fish catch increased from 19 million tonnes to 88 million tonnes between 1950 and 1988, but slumped afterwards, production falling far behind the growing world demand. While the per capita catch increased from less than 8 kilograms in 1950 to the 'historical peak' of 17 kilograms in 1988 (more than doubling over a space of less than four decades), but it then dropped by 4 per cent, bringing down per capita availability to 16 kilograms between 1988 and 1997. Undoubtedly, the five-fold growth in human appetite for sea food pushed the catch of oceanic fish in general to 'their sustainable limits or beyond', so much so that of the 15 oceanic fisheries, 11 are in decline, and "the next half-century is likely to be marked by the disappearance of some species from markets, a decline in the quality of sea food caught, higher prices, and more conflicts among countries over access to fisheries" (Brown, 2000: 50).

With the oceans pushed to their limits, future demand for sea food can only be met by fish farms engaged in fish culture, rearing a wide variety of both wild and locally available types for an ever-expanding market, more so because of greater preference for fish protein with people scared about cholesterol limits. The insatiable appetite for fish with more people moving up the food chain has given aquaculture a leading place in supplementing, or rather compensating for, the decline in oceanic catch. Thus, while oceanic catch has not increased appreciably since 1990, falling far behind the growing demand for sea food, aquacultural output 'expanded from 13 million tonnes of fish in 1990 to 31 million tonnes in 1998, growing more than 11 per cent a year', and is likely to 'overtake the production of beef by 2010'.

China takes the leading position as producer of aquacultural fish, accounting for 21 million tonnes of the global output in 1998, the catch divided evenly between the coastal and inland areas of which the coastal areas are dominated by shell fish, such as oysters, clams and mussels, as well as a small quantity of shrimp, prawns and finfish. However, the coastal areas are considered to be environmentally damaging with wetlands converted into fish farms and mangroves destroyed for that purpose, as is happening in countries like India, Thailand, etc. On the other hand, except shell fish, all other types of fish as aquacultural products come from inland waterbodies in China, consisting of ponds, lakes, reservoirs and rice fields, with farming much of it based on 'carp polyculture', comprising about 5 million hectares of land used exclusively for fish culture in addition to which another 1.7 million hectares of rice lands having developed a polyculture based on the production of both fish and rice. The total production of aquaculture has risen to 29 million tonnes, double that of poultry. "Over time, China has evolved a fish polyculture using four types of carp that feed at different levels of the food chain, in effect emulating the natural aquatic ecosystem. Silver carp and bighead carp are filter feeders, eating phytoplankton and zooplankton respectively. The grass carp, as its name implies, feeds largely on vegetation, while the common carp is a bottom feeder, living on detritus that settles to the bottom. Most of China's aquaculture is integrated with agriculture, enabling farmers to use agricultural wastes, such as pig manure, to fertilize ponds, thus stimulating the growth of plankton. Fish polyculture, which typically boosts pond productivity over that of monoculture by at least half, also dominates fish farming in India" (Brown, 2004: 160). However, the Chinese fish farmers are feeding more grain concentrates to raise pond productivity, increasing the yield from 2.4 tonnes in 1990 to 4.1 tonnes in1996.

On the other hand, the United States is the world's leading producer of catfish with a total production of 270,000 tonnes cultured, concentrated in four states of the country, Mississippi, Louisiana, Alabama and Arkansas, of which Mississippi alone has 45,000 hectares for its culture. With meat consumption growing twice as fast as the population and global demand for fish available from both sources – oceans and aquaculture – outpacing that of population, examples set by China and India in expanding fish culture inland may be adopted in other countries to meet this challenge.

D. HOW MANY PEOPLE CAN THE EARTH FEED ?

Having an idea about the number of people inhabiting this planet now expected to live in around 2050, the question arises if the earth can feed them, providing the right amount of nutrition. As has been observed earlier, not only does food energy intake vary widely but the amount of food consumed by an individual also differs widely depending on economic and other factors determining entitlement. While the daily per capita food energy may vary from an average of 1,800 kcal to over 2,600 kcal/day for those in rich countries with a global mean at about 2,000 kcal/day (with a little over 1,800 kcal/day in countries like Bangladesh and Afghanistan while according to *The Economist*, 13 July, 1991, the Irish consume 3,700 calories a day, more than Americans averaging 3,600 calories), in actual amount the variation seems to be much wider as the average at the US level is 800 kilograms per person per year for food and feed, whereas it is 200 kilograms for an Indian per year. Hence, any final estimate is very difficult to make unless a standard for average consumption per person is taken as a base keeping in mind wide variations in food habits as well the fact that many people are strict vegetarians.

Even a conservative estimate shows that with 'more efficient use of agricultural resources – as shown in Table 5.20 – more careful treatment and storage of harvested crops, and more healthy ways of eating worldwide, agriculture was capable of recovering inefficiencies equivalent of about 60 per cent of actual food consumption in 1990, which would have been enough to feed an additional 3.1 billion people or a total global population of 8.4 billion

Table 5.20: Estimates of Efficiency Gains in Global Food Production Achievable by 2050

Changes Compared to 1990 practices	*Gains Equivalent to global 1990 food energy consumption (per cent)*
1. Improved Field Efficiencies	
(a) Better Agronomic Practices (raise average yield 20 per cent)	22
(b) Higher Fertilizer Uptake (raise nutrient use efficiency 30 per cent)	7
(c) Reduced Irrigation Waste (rainwater use efficiency 30 per cent)	7
2. Reduced Waste	
(a) Post-harvest Losses (lower by 20 per cent)	6
(b) End-Use Waste (lower by 20 per cent)	8
3. Healthier Diet (limit Fat intakes to 30 per cent of total energy)	10
Total Gain	60

Source: Smil, 1994

"without cultivating more land and without increasing fertilizer and water inputs. If everybody were to receive 2,200 kcal/day, the total would be reduced to 7.7 billion people, 2.4 billion above the 1990 total (5.3 billion)." (Smil, 1994: 272)

The standard recommendation for food consumption for 25-year-old adults is 2000 kcal for housewives and 2500 for male office workers. This has to be viewed against the expected production of 3 billion tonnes of food required by 2020. Experiences of increases in average cereal yields are quite reassuring about the capacity of the earth to meet the challenge of the growing population. For example, yield rate 'increased 2.5 times between 1950 and 1990 (from 1.1 to 2.8/ha), and by 55 per cent in the 20 years between 1970 and 1990, and the global harvest is likely to increase by at least 600 million tonnes of unmilled grain by 2050 with the help of further intensification of agriculture based on new cultivars, multiple cropping and higher fertilizer application during the intervening years (1990-2050) (Smil, 1994) (Table 5.20). Assuredly, without waiting for epoch-making bio-engineering breakthroughs, enough additional food needed for human consumption could be achieved with "relatively small extension of cultivated land, gradual (and very slow in a long-term perspective) intensification of cropping, cultivation of idle lands, high-efficiency irrigation in the driest regions, and some smaller, but locally critical contributions (reduced beef production, agri-silviculture, aquaculture, new crops, saline irrigation) could combine to produce an additional food equivalent of about 1,500 (1,400 – 1,600) MT of unmilled grain (Table 5.21) ... more than enough to feed an additional 2-2.5 billion people even with rather high postharvest losses and with fairly inefficient animal feeding... It would seem realistic to conclude that the Earth could support a population of 10-11 billion people during the next century (21st century)" (Smil,1994: 279-280).

Table 5.21: Estimates of Additional Harvests Achievable by 2050

Agronomic Measures	*Harvest Equivalents (Million tons from unmilled Grain/Year)*
Intensification of Existing Cropping Practices (raise yields 35 per cent)	700
Extension of Cultivated Land (20 per cent increase compared to 1990)	500
Cultivation of Idle Land	75
High-Efficiency Irrigation	100
No beef production with intensively grown feed	50
Irrigation with saline waters	25
Agrisilviculture, Aquaculture and New Crops	50
Total	**1500**

Source: Smil, 1994.

However, hunger, defined as a 'chronic shortage of nutrients needed for growth and good health', was even affecting about 12 million children and 8 million adults in the United States in the late 1980s (Brown, 1987) (Hunger in the U.S., Scientific American 256, No.2), so one needs to be cautious about formulating proper strategies to be adopted for eradicating hunger.

Since 'water scarcity translates into food scarcity', multiple cropping is virtually dependent on the availability of irrigation water. The harvested area can be expanded through multiple cropping in a country like India, with water stored during the monsoon season for which farmers need to be properly trained in water harvesting techniques to help in water storage. Incidentally, there is no disagreement among experts about the role of irrigation and as such the supply of water is a 'much more significant constraint to raising food production than do any likely foreseeable difficulties relating to soil or land' (Dyson, 1996). In fact, most of the future production of food 'will have to come from' irrigation in the Far East and South Asia where more than half of the present production comes from the irrigated areas. While major agricultural areas like northern China and western India face acute seasonal water shortages, the whole of the Middle East is under great strain and all areas are increasing their pressure on pumping underground water almost to its depletion, unless restrained.

Interestingly, given appropriate technical support, management and institutional changes, there is scope for growing a lot more food from existing water supplies in all the regions of the world. Besides, the prospects for further expansion of irrigation capacity of the world cannot be ignored which expanded by about 3.6 million hectares in the early 1990s, most of it happening in the Far East and South Asia. The ratio of irrigated area per thousand people in the world was 45 hectares in 1991, being more or less the same as in 1961, and with increases since early 1990s it is likely to cover about 350 million hectares, though the ratio per thousand will slightly be lower – 44 ha/1000. According to World Bank estimates, 'the remaining potential irrigable area worldwide is about 137 million hectares, half of which lies in Asia. The major part of this expansion is likely to be accomplished through building of much smaller dams, introduction of low cost water pumps and drip irrigation' (Dyson, 1996: 152-153). However, a major concern in this respect arises from the loss of surface water used for irrigation through a network of dams channeling water through a network of canals, a large percentage of which never reaches the cropland as much of it is lost due to evaporation. It has already been mentioned earlier in this chapter how water efficiency varies widely, grading down from 50 to 60 per cent in Israel, Japan and Taiwan to 25 and 40 per cent in countries like India, Mexico, Pakistan, the Philippines and Thailand. Irrigation water efficiency is affected not only by the type and condition of irrigation systems but also by soil type, temperature and humidity, high temperature in arid regions making evaporation loss much higher than in humid regions. As a way of increasing irrigation efficiency, China, as declared in 2004, was going to raise water by incentives for adopting more irrigation-efficient technology with an aim for raising it from '43 per cent in 2000 to 51 per cent in 2010 and to 55 per cent in 2030' (Brown, 2006).

Higher fertilizer input is considered as another very important single factor contributing to increasing global food output. The production of fertilizer increased ten-fold between 1950 and 1990. In 1990, the global consumption of nitrogen, the major ingredient of chemical fertilizer, was about 80 million tonnes, or about 15 kg per person, and the requirement by 2020 will be around 120 million tonnes (taking per capita figure constant) but a 'more realistic estimate' of nitrogen requirement required for the production of nearly 3 billion tonnes of cereals' by 2020 would amount to about 155 million tonnes to 170 million tons, or a doubling of world synthetic nitrogen production. This is considered quite feasible in view of the limitless supply of atmospheric nitrogen while there is no shortage of either methane or coal used in the production process. Gradual upgrading and expansion of nitrogen processing capacity of the developing countries during the intervening years will make all of them, except Sub-Saharan Africa, capable of production of the required amount of nitrogen. At the same time, presence of ample mineral reserves of two other major types of chemical fertilizers – phosphates and potash – though disproportionately located (about 50 per cent in the Middle East and about 25 per cent in South Africa) will not be a problem for their increased production.

The use of pesticides will go on increasing worldwide but will be of much less importance than chemical fertilizers and is likely to be replaced by 'alternatives' going under the title of 'integrated pest control'. On the other hand, biotechnology with 'its advances in conventional technology will remain the primary source of growth in crop and animal production'. For quite some time among some of the promising developments are the new varieties of rice currently being developed in China and at the International Rice Research Institute in the Philippines. The new inbred rice has characteristics like 'very sturdy stems, vigorous roots, few unproductive tillers and three or four panicles (each with 200-250 grains) giving yields 30 per cent higher than semi-dwarf rice varieties.' Hybrid rice for the tropics is another very important development and with the new inbreds subject to hybrid technology, the 'combined effect may raise rice yield by up to 50 per cent, thus make a vary significant contribution to increasing the average cereal yield in the Far East to 6 tonnes per hectare.' India and Egypt may also derive similar benefit from these developments (Dyson, 1996: 158).

Another highly significant development of new varieties is '*triticale*' – a cross between wheat and rye – being developed in Mexico, which is a 'fast-growing high-protein cereal' giving higher yields than wheat with 'considerable tolerance to both salt and drought'. It is being increasingly grown in Australia, France, Poland, Mexico and parts of the former Soviet Union. One should also take note of new cultivars of pearl millet developed at the International Crop Research Institute for the Semi-Arid Tropics in India, which have great resistance to mildew disease and higher heat tolerance, giving twice the yield of other varieties during the severe drought in Sub-Saharan Africa in 1992. (Dyson, 1996: 159).

Interestingly, the leading centres engaged in research in biotechnology are mostly in the hands of private companies, mainly centred in North America and Europe, largely interested in developing commercial food crops meant for developed country markets. Besides, the inevitable time-lag, often 10 to 15 years, necessary for the development of staple food crops, a priority for the developing countries, becomes a major handicap for them to enjoy the real benefit of biotechnological achievements.

Significantly, moving from animal protein to high-quality protein, available from plant sources provided by beans or tofu made from soybeans and being more land-efficient, is going to take a very important role as sources of protein. As observed by Brown (2006), "most of the world's fast-growing soybean harvest is consumed indirectly in the beef, pork, poultry, milk, eggs, and farmed fish that we eat. Although not a visible part of our diets, the incorporation of soybean meal into feed rations has revolutionized the world feed industry, greatly increasing the efficiency with which grain is converted into animal protein" (Brown, 2006: 173). The farmers of the world produced 220 million tonnes of soybean in 2005, 15 million tonnes of which was consumed directly as tofu or meat substitutes while the bulk of the remaining 205 million tonnes, leaving some aside saved for seed, was crushed 'to extract 33 million tonnes of soybean oil, separating it from the highly valued, high protein meal'. By 2006, perhaps 2 million tonnes (7%) of these 33 million tonnes will be heading to service stations as biodiesel... (significantly) "the 144 million tonnes of soybean meal that remain after the oil is extracted is fed to cattle, pigs, chicken, and fish, enriching their diets with high-quality protein... The world's three largest meat producers – China, the United States, and Brazil – now rely heavily on soybean as a protein supplement in feed rations" (Brown, 2006: 173-174).

With cropland becoming scarce all over the world with continuous encroachment from urbanization, industrialization, road-building, etc.

Table 5.22: Sample Countries with Declining Cereal Yields between 1981-86 and 1987-92

Region	*Countries*
Sub-Saharan Africa	Mozambique (-39), Angola (-19), Malawi (-15), Somalia (-9), Zaire (-8), Zimbabwe (-7), Niger(-4), Rwanda (-3), Ivory Coast (-2)
Middle East	Iraq (-6), Syria (-4)
South Asia	Afghanistan (-8)
Far East	Myanmar (-4), Malaysia (-1)
Latin America	Mexico (-12), Cuba (-12), Haiti (-8) Colombia (-2), Ecuador (-1)
Europe/FSU	Romania (-21), Yugoslavia (-7), Bulgaria (-4)

Source: Dyson, 1996.

Note: Figures in parenthesis are percentage declines. All other sample countries not appearing in the list but affected by land degradation had increasing yields.

protecting farmland assumes special significance and in this regard Japan sets a brilliant example 'successfully protecting rice fields within the boundaries of the city of Tokyo, thus enabling Japan to remain self-sufficient in its staple food – rice' (Brown, 2004) (Table 5.22).

E. MEASURES AIDING AGRICULTURAL GROWTH AND A WORD OF CAUTION

The formidable challenge of meeting the global requirement for food in the near future in the face of growing population as well as the growing demand in response to improvement in lifestyles of people mainly in the developing countries, appears to be less reassuring under the great threat of monopolization of food production by big corporate houses. These corporations, as David Ransom says, are in the process of "gobbling up the world market in seeds" in spite of the fact that "peasant farmers" across the world "who save their own seeds, can, and do, still feed the world". (Ransom, 2010:5). According to Ransom, "one by one the chemical giants rebranded themselves as 'life science corporations', taking a giant step in having full control of the world's seed market." The top 10 seed corporations (Appendix A) already control 67 per cent of the global proprietory seed market of which the top three alone control half of all patented seeds, accounting for 82 per cent of the global commercial seed market. Among them, "Monsanto's seeds occupy 87 per cent of the total area devoted to genetically engineered seeds worldwide".

The most important fact about the seeds is that there are almost half a million known plant species, and more still to be discovered. Yet we now rely on just 15 species to provide 90 per cent of our food crops. Interestingly, "the world's top six agrochemical manufacturers are also seed-industry giants; sales of one promote sales of the other" (*New Internationslist*, 2010: 10).

However, in the backdrop of ever-tightening grip of the corporate houses the only redeeming factor is presented by peasant farmers, saving agriculture from a catastrophic future. According to Ransom, "a billion small farmers still save their own seeds and produce most of the world's food. The largest study of its kind, involving nearly 9 million farmers, covering 200 projects on 28 million hectares in 52 countries, indicates that crop yields increase on average by 73 per cent under small-scale, sustainable methods, and that 'sustainable practices can lead to substantial increases in per hectare food production" (Ransom, 2010:6). Here, 'substantial' in Ransom's opinion, "means something quite specific: not following industrial agriculture, which consumes a third of the world's depleted reserves of fossil fuels" and "not relying on monocultures of a few staple crops, which increases genetic vulnerability – when only genetic and cultural diversity stand any chance whatever of adapting to runaway climatic change". According to a report of the FAO in 2009, "Through the continuing shift to commercial agriculture, much of the diversity that still exists remains under threat...With the disappearance of lifestyles and languages across the globe, a large amount of knowledge about crops and varieties is probably being lost, and with it much of the value of the genetic resources themselves. Cultural diversity matters too" (Ransom, 2010: 7).

Ransom, in his keynote article, 'Seed Savers', referred to above, puts emphasis on 'listening' to stories about peasant cultures in countries like India and Africa "with renewed respect and appreciation of the subtle wisdom that grows only at the pace of nature itself". Land reform, according to Ransom, makes the first step since "diversity cannot flourish (and seeds will never be saved from extinction) unless peasant farmers finally come to control the land they work...Second, vital knowledge about seeds, their cultivation and preservation, is held largely by women. This says less about the stereotypical nature of women than about the culture of agriculture, where women play a leading role almost everywhere...Third, if half the resources (including government subsidies) currently lavished on corporate monoculture were used to promote sustainable food production instead, then human ingenuity might finally make rural life less laborious, more rewarding than it currently is. Finally, we can learn from the natural history of seeds themselves" (Ransom, 2010:7)

The case of Andhra Pradesh in India, cited by Sue Branford, branding the corporate giants as "merchants of death", demonstrably brings out the hollow claim when farming comes under the control of the corporate houses. For example, as mentioned by her, in 1980, the Government of Andhra Pradesh promised local farmers untold wealth if they produced cash crops, particularly cotton, buying on credit a 'package' of high-yielding varieties of hybrid cotton, fertilizers and pesticides supplied by the corporate houses. It went on well for a few years to be followed thereafter by soil losing its fertility built up over generations using traditional methods, now requiring larger application of fertilizers and pesticides, and ultimately the entire system was wrecked all too soon by a 'freak weather event'. "By the turn of the century, almost every rural household in the region was forced to sell cattle and land in a desperate bid to stave off bankruptcy. Overwhelmed by the ignominy of reducing their families to penury, farmers began to kill themselves, often swallowing the herbicide they had bought on credit. Some 150,000 farmers committed suicide in India between 1997 and 2005" (Sue Branford, 2010: 19).

Though not directly linked to the death of farmers in India, the corporations trying hard to push hybrid or GM crops everywhere for maximizing sales and profits, are "leading to the extinction of a huge number of local crops developed over thousands of years. It is a frightening prospect: with climate change, farmers will need to have access to as wide a diversity of crops as possible" (Sue Branford, 2010: 19). Referring to crops grown by farmers in another region of Andhra Pradesh, Jaideep Hardikar mentions that the farmers "grow between 6 and 25 crops from up to 80 seed varieties. This rich diversity, they say, is an insurance against aberrations of nature or market fluctuations. Some seeds survive untimely rain. Others resist drought. They never fail" (Jaideep Hardikar, 2010: 9). Five thousand women of the region, 'representing the poorest of the poor of the peasant world' have joined through village-level voluntary groups for seed collection and preservation. "Apart from their seed collections, these women manage some 60 community seed banks with over 80 retrieved land races (wild harvests) that have been obliterated by modern agricultural practices – vibrant reminders of what once dotted India's agricultural panorama" (Jaideep Hardikar, 2010: 9).

Interestingly, according to the findings of a recent UN study, (*New Internationalist*, 2010: 11) 'genetic uniformity increases crop vulnerability to pests, diseases or climate change.' According to this report, "in about 60 countries genetic vulnerability is considered a potential threat to agricultural production...China reports cases in which rice and maize varieties have become more uniform. In Lebanon the decrease in the production of almonds has been attributed to genetic vulnerability of the few varieties grown. Madagascar reports that the rice variety *Rojomena*, appreciated for its taste, is now rare. The *Botojingo* and Java varieties of the north-eastern coastal area have disappeared. Wild yam species are also considered likely to disappear soon. In India, a large number of rice varieties in Orissa, rice varieties with medicinal properties in Kerala, and a range of millet species in Tamil Nadu, are no longer cultivated in their habitats. Yemen reports that finger millet varieties as well as oilseed rape varieties, important traditional crops, are no longer grown, or only grown in specific areas. In Albania, all indigenous wheat and many maize varieties have reportedly been lost. It is highly significant to note that "since records began, some 600 plant species may have become extinct" (*New Internationalist*, 2010)

One may, however, take some comfort in the fact that commercial seeds do not as yet dominate the world. As it appears, (a) "three-quarters of the world's farmers still grow locally bred seed varieties and/or save their seeds, (b) 85 per cent of global food production is consumed close to where it is grown, (c) ten years after the first Genetically Modified (GM) seeds were planted, only four crops (maize, soybean, canola and cotton) and two traits (pesticide tolerance and insect resistance) have been commercialized; just four countries

(US, Canada, Argentina and Brazil) account for over 90 per cent of the total GM crop area" (*New Internationalist*, 2010: 11)

Since GM seeds aim at replacing local seeds developed through generations of hard labour, great care and wisdom, farmers, especially in the developing world must preserve local plants that have been developed for centuries so that what remains of food-crop varieties may yet be saved. The recently concluded convention on Biodiversity in Nairobi urged everybody to start investing in indigenous knowledge as the only sensible option for the development of sustainable agriculture capable of providing food for all.

6
Strengthening Food Production System

A. LAND POTENTIAL

Interestingly, changes in land use over the past three centuries - 1700 – 2000 – show that the world's forests and woodlands diminished by 1.2 billion hectares, or 19 per cent, and similarly grasslands and pastures declined by 560 million hectares, or 8 per cent of the estimate. Contrastingly, croplands brought into cultivation show a net increase of 1.2 billion hectares – a 466 per cent increase in less than three centuries. In absolute terms, the accelerating pace in agricultural expansion as a contrast to depletion of forests and grasslands is shown by their being more increase in cropland over the 30 years between 1950 and 1980 than in the 150 years between 1700 and 1850. According to Richards (1993), world cropland increased by 331 million hectares after Second World War in comparison to its growth by 272 million hectares in the eighteenth and nineteenth centuries, but the rate of change was far from uniform across the world. The processes involved in changing the character of land use were largely determined by European political and economic control which is clearly reflected in agricultural expansion, forest clearing, wetland drainage, irrigation of grasslands, expansion of human settlements, etc., taking place all over the world, hastening the process of land transformation in an uneven manner (Richards, 1993: 163).

Since 1950 total grain area in the world has grown 19 per cent, but global population has grown 132 per cent, seven times faster, reducing grain area per person by half, falling from 0.24 to 0.12 hectares (Brown, 2000). Some 90 per cent of the land surface of the earth that is not desert or occupied by permanent ice, is occupied by man, 17 per cent of which is productive and used for crops, while the rest remains under other uses including urbanization, etc., or forests, pastures or lying practically unused. The uncultivated land that could be planted with crops is marginal, contributing only a small addition to existing cropland. In 1950, 593 million hectares 'were used for grain production which, by 1990, increased by 21 per cent to 720 million hectares, 'but there was a 139 per cent increase in grain production resulting from 'more than a doubling of average yield – increasing from 1 tonne per hectare per year to about 2.3 tonnes per hectare per year (Ehrlich *et al.*, 1993: 7).

The tropical rainforests, apparently holding the largest prospect of land converting into agricultural use, have an important role in supporting human life as rainforests, by storing carbon and controlling the hydrologic cycle, providing fresh water. Besides, even if reclaimed, these lands are not very suitable for conventional farming as such lands quickly degrade into wasteland if put to plough.

On the other hand, much of the land which is very suitable as cropland is often sacrificed to meet the growing demands of urbanization, industrialization, construction of roads and for meeting other infrastructural requirements, amounting to some 5,000 square kilometres lost annually to urbanization in East Asia alone (Ehrlick *et al.*, 1993). Loss of fertile land will intensify further in the future, using '15 per cent of land now under forests and 40 per cent of that in croplands' for the production of biomass energy crops (*See* Note *A*).

Our most precious natural capital is getting fast depleted, with "erosion of topsoil up to centimetres per decade," amounting to a global loss of "some 24 billion tonnes" annually, 'in excess of the natural rate of soil regeneration", while the remaining top soil on earth's cropland is getting lost at an average rate of 7 per cent per decade (Brown and Wolfe, 1984) making agricultural practices unsustainable in the near future. It needs to be borne in mind that "the billions of tiny organisms per gram of rich agricultural soils" involved in maintaining soil quality and transferring nutrients from soil to crops" are "threatened by many aspects of modern industrial agriculture" (Ehrlick *et al.*, 1993: 8). Soil erosion leading to land degradation, taking a measurable toll on food production in many countries, suggest a need to adopt farming practices that will protect soil from erosion, paying handsome dividends in the long run. Since 1945, according to a study made by the United Nations Environment Programme (UNEP), 'nearly 11 per cent of the world's vegetated land has suffered moderate to extreme degradation, and another 6 per cent has been highly damaged.' (*Ehrlick et al.*, 1993).

Although farmers cannot control the conversion of cropland into non-farm uses, it is within their competence to check the losses of soil and land from severe erosion. For example, terracing, a time-tested method for dealing with water erosion, has been practiced in rice fields throughout the mountainous regions of Asia for a long time giving rich dividends. On less steeply sloping land, contour strip farming, as done in the United States Mid-West, works well. Another soil conservation technique introduced recently is conservation tillage which includes both no-till and minimum-tillage. Besides reducing both wind and water erosion, this practice 'helps retain water, raises soil carbon content, and reduces the energy needed for crop cultivation'. In this practice farmers avoid ploughing by simply 'drilling seeds directly through crop residues (of the previous crop grown) into undisturbed soil, controlling weeds with herbicides'. The undisturbed soil beside the narrow slit in which seeds are inserted, covered by crop residues, resist both water and wind erosion.

Already the no-till area in the United States, introduced in the 1990s, has gone up from 7 million hectares in 1990 to 25 million hectares in 2004, used widely for the production of corn and soybeans. The no-till practice is spreading fast, as mentioned earlier, in the western hemisphere where it covers 24 million hectares in Brazil, 18 million hectares in Argentina, and 13 million hectares in Canada while in Australia 9 million hectares practice no-till culture (Brown, 2006: 150). Recent FAO reports refer to the growing popularity of no-till farming in Europe, Africa and Asia.

In addition, many countries are seriously engaged in combating another menace, advance of the deserts into settled areas. This is happening in quite a number of countries destroying cultivable lands. Algeria provides one such example where trying to halt the northward advance of the Sahara Desert the country announced in 2000 that it is 'concentrating its orchards and vineyards in the southern part of the country', hoping that these perennial plantings will halt the desertification of its cropland. Responding to the severe drought, Morocco, in 2005, decided to convert cereal-planted lands into less vulnerable olive and fruit orchards. Similar concern about the march of Sahara made Nigeria propose to plant a Great Green Wall of trees, 5 kilometres wide and stretching for 7,000 kilometres across Africa. Senegal located at the western end of this proposed wall, 'strongly supports the idea'.

Similar steps are contemplated by China trying to halt the advance of deserts and 'it is paying farmers in the threatened provinces to plant their croplands in trees'. China is going to create a Great Green Wall, planting 10 million hectares of grain land with trees, which is one tenth of China's present cropland. The situation is equally serious in Inner Mongolia where the advancing front of the desert makes it necessary to plant desert shrubs for stabilizing sand dunes while sheep and goats have been banned in many areas. Already 7,000 hectares of land have been reclaimed by planting desert shrubs on abandoned cropland. One of the major policy changes in agricultural practice in Inner Mongolia is to replace a large number of sheep and goats with cattle, having already increased the number of cattle from 30,000 in 2002 to 150,000 by 2007. The cattle are kept in enclosed areas and fed on 'corn stalks, wheat straw, and the harvest from a drought-tolerant forage crop resembling alfalfa, grown on reclaimed land, and which 'will double the income within the country' (Brown, 2006: 151).

As Brown observes (2006), the only way to eliminate overgrazing on the two-fifths of the earth's land surface classified as rangelands is to reduce land used for grazing animals. India has shown a very good example in this regard, keeping animals within enclosures and feeding them inside and already having developed a thriving dairy industry, the largest in the world.

B. WATER FOR SURVIVAL

Taking into account all three major sources, the contemporary world's fresh-water supply (1965) amounts to approximately:

River run-off, including anthropogenic transformations	38,830 km³/year
Water and Ice run-off from polar regions	2,900 km³/year
Direct Underground run-off into Oceans	2,400 km³/year
Total	**44,130 Km3/Yr.**

(L'Vovich *et al.* 1993: 237) (*See* Note *B*)

From 1687 to 1987, the total amount of water withdrawal for all needs has increased about 35-fold, to over 3,600 km³. The rate of increase was the greatest after 1950. Although irrigation is the main consumer, its share has markedly declined during the past three centuries, from 90 to 75 per cent, but use for industry and energy production expanded on the eve of the nineteenth century, 2.5 times, water-use for domestic-municipal purposes expanded to 20 per cent by 1980 from 5 per cent earlier, and use for livestock decreased meanwhile to 2 per cent.

With population growing, the supply of fresh water declines, and its supply is predicted to fall by 73 per cent between 1950 and 2050, reducing its availability to scarcely 'one-fourth as much per person in 2050 as there was in 1950' (Brown, 2000) It is, indeed, a very dismal scenario. As a matter of fact, the per capita supply of fresh water has diminished by a third between 1970 and 1990 and there is little doubt that population increase (along with changing habits of using as well as wasting more water with growing affluence) is the main driving force in intensifying the water crisis in the world today. "The most alarming projections suggest that nearly 7 billion people in sixty countries will live water-scare lives by 2050. Even under the lowest projection, just under 2 billion people in forty-eight countries will struggle against water scarcity in 2050" (United Nations, 2003: 12-13).

Sandra Postel and Brian Richter cite (in their book entitled, '*Rivers for life: Managing Water for people and Nature*) South Africa's National water Act 1998 as a model for other countries. "The act focus on two broad needs", the first of them is "for satisfying basic water needs of everyone for drinking, cooking, sanitation, and other essential purposes, treated as nonnegotiable allocation" while "the second is the water needed to support river ecosystem functioning in order to conserve biodiversity and secure the valuable ecosystem services they provide to society" (Brown, 2006: 152).

As mentioned earlier, meeting the target of halving the proportion of the world population without sustainable access to drinking water means providing services for an additional 100 million people each year (at the rate of 274,000 people/day) from 2000 to 2015. In effect, an estimated total of only 901 million people gained access to water supply during the decade of 1990s. The challenge for sanitation is far more daunting. The pace at which about 1 billion people gained access to improved sanitation during the 1990s needs to be accelerated to allow provision for sanitation to increase by 125 million people each year (at the rate of 342,000 people/day) until 2015. Taking the overall situation, Asia's needs outstrip those of Africa, Latin America and the Caribbean taken together in absolute terms (United Nations, 2003: 110-112).

What is of immediate need is restoration of the supply of fresh water which is getting fast diminished through withering processes induced by human action. Maintaining minimum river flows especially needed for satisfying the specific requirements of downstream aquatic ecosystems, such as flood plains, river deltas and wetlands, is far from an easy task.This automatically leads to the role of irrigation in boosting food production. But according to the opinion of many experts, expansion of global irrigation capacity is now slowing and it is unlikely to increase in the next few decades at the same pace as in the past. Against this backdrop, both the Far East and South Asia face the greatest challenge where most of the required increase in future food production will depend on irrigation, which produces more than half of their present harvest. Acute seasonal water shortage is already felt in major producing countries like northern China and western India, and the situation is more precarious in Middle East countries. All these countries depend heavily on underground water used mainly for irrigation but overpumping accelerating the rate of fall in the level of water in the aquifers, as mentioned earlier, is going to be a serious problem with devastating consequences. Similarly, eastern and southern Africa with almost all their crops grown rainfed face an acute shortage of water. In contrast, however, in the predominantly temperate regions of North America/ Oceania and Europe/FSU where bulk of the crops are grown on rainfed lands, water supply will not be much of a problem in increasing agricultural production in the future (Dyson, 1996: 150).

Loss of irrigation water due to inefficient management in its supply is a greater problem than the problem of water lost through seepage, waterlogging or salinization of fields, making 8-12 per cent of the world's irrigated area salinized. Postel (1994) estimated that "salinization affects 28 per cent of the irrigated area in the United States, 23 per cent in China, 21 per cent in Pakistan, 11 per cent in India and 10 per cent in Mexico. Although salinization tends to lower yields, it should be noted that some cereal crops like barley, wheat and sorghum are comparatively salt-resistant" (Dyson, 1996: 151)

Raising the price of water (for all users, but especially for irrigation) seems to have a great restraining effect on its misuse/overuse as well as wastage, mitigating problems of waterlogging/salinization. At present, hardly 40 per cent of the water diverted for irrigation (constituting 69 per cent of total water used for human consumption in various areas) reaches the cropland, the larger proportion getting wasted in transit or otherwise. According to many water experts even modest efficiency gains will effectively increase the returns in the form of larger food production. Expansion of the world's irrigation capacity is still a possibility, for it was expanding at about 3.6 million hectares per year in the early 1990s, mostly in the Far East and South Asia. The ratio of irrigated area to population in 1991 was about the same as it was in 1961, as mentioned earlier, - 45 hectares irrigated land for every thousand people and is likely to change, very little, to 44 hectares/1000 by 2020 due to population growth, while the predicted area under irrigation by then will be about 350 million hectares. According to World Bank estimates 'the remaining potential area worldwide is about 137 million hectares', roughly half of it lying in Asia, and by 2020 "there will certainly be considerable further expansion of irrigation capacity, particularly in the Far East and South Asia" (Dyson, 1996: 154).

Increase in irrigation capacity undoubtedly depends largely on replenishment of supply available from rivers/underground sources. Australia and the United States provide some of the commendable instances in this regard. In Australia, the permanent closure of the Mowamba aqueduct after 100 years of use raised the flow of the Snow river to the natural level of 6 per cent, increasing it to 28 per cent in the long run. In the Murray-Darling basin release from a storage facility in the basin enhanced flow, helping in restoring the natural wildlife population. Most encouraging and the 'largest example of returning water to restore and support marine habitat occurred in California' in the Sacramento-San Joaquin river system. The increased flow of the two rivers helped 'protect the San Fransisco Bay's rich aquatic river system (Brown, 2006: 153).

According to Postel (1990), about 33 per cent of the crops harvested come from 17 per cent of land, which is irrigated. But the growing demand for water by non-farm users and rising cost of installing irrigation systems have caused diminishing rate of land brought under irrigation during the last decade or so. In addition, the rising rate of irrigated land losing fertility or gone out of production due to waterlogging and salinization resulting from irrigation have been an alarming signal as 'about a quarter of the world's irrigated land has already been affected by salinization'. Postel also refers to the dangers of increasing pumping of underground water almost in all countries by far out-stripping the rates of recharge for irrigation purpose, estimated to be between 6.5 to 8 billon gallons annually in the United States, making further extraction highly expensive and uneconomic as a consequence of the rapid fall of the water table. The rapid expansion of agricultural area under irrigation in the 1960s and 1970s is now facing a decline in irrigated area compelling farmers in the southern Great Plains of the United States to turn to dryland farming. India is facing a similar situation, especially in the northern parts 'where green revolution success was built on overdrafts of groundwater' (Ehrlick *et al.* 1993: 13).

Looking from another angle, irrigation holds prospect for better agricultural returns based on more efficient practices, such as making a 'combination of widespread adoption of proper planting (earlier, denser), better crop rotations, and careful choice of cultivars' which would 'increase average crop yields by 20 per cent during the next 50-60 years', provided with more nutrients and water. Further, accepting the fact that the total distribution, seepage and evaporation losses in traditional ridge-and-furrow irrigation amount commonly to 50 to 60 per cent of carried water, and loss from evapo-transpiration by crops is about one-third in surface irrigation in the United States, it is possible to make gains without further investment by way of irrigating every other furrow saving about one-third water with slight decline in crop yields. "Better matching of crops with natural moisture supply is another highly effective alternative. For example, replacing corn with sorghum can lower water needs by 10 to 15 per cent, and planting sunflowers instead of soybeans as an oil crop can save easily 20 to 25 per cent of water. If the current water use efficiency in field cropping averages 30 to 40 per cent for the poor world, and 40 to 50 per cent in rich countries, there is no reason why these rates could not be increased by at least half during the next two to three generations. Even if the average water use efficiency were to go up by no more than 30 per cent, and if no less than 1,200 kg of water were needed to produce a kg of crops, the global annual gain would be equivalent to about 250 million tonnes of additional grain. This amount must be reduced by water needed for increased productivity owing to agronomic advances: assuming that at least one-fourth of expanded harvests will come from irrigated land, the net gain would be 100-150 million tonnes of grain" (Smil, 1994: 270)

However, agricultural production and food security across the world is likely to be heavily impacted by climate change, in all likelihood it will increase the risks of a reduction of agricultural productivity in developing countries. As reported by the United Nations (2007/2008), "Climate change will be superimposed on wider pressures on water systems. Many river basins and other water sources are already being unsustainably mined. Today, around 1.4 billion people live on 'closed' river basins where water use exceeds discharge levels, creating severe ecological damage. Symptoms of water stress include the collapse of water systems in northern China, rapidly falling groundwater levels in South Asia and the Middle East, and mounting conflicts over access to water... In a world that is already facing mounting pressure on water resources, climate change could add around 1.8 billion people to the population living in a water scarce environment – defined in terms of a threshold of 1,000 cubic metres per capita per annum – by 2080" (UNDP, 2007/2008: 95).

In the Middle East., already the world's most water-stressed region, nine out of fourteen countries have an average per capita water availability below the water-scarcity threshold, mentioned above, while decreased precipitation projected for Egypt, Israel, Jordon, Lebanon and Palestine, rising temperatures and changes in run-off patterns will adversely influence the flow of rivers making much less water available for human consumption, with a 59 per cent decline in renewable water availability in Syria by 2050. (*See* Note C).

The most adverse effect of climate change with consequent water shortage is gong to hit heavily on agriculture, and as predicted, "by the 2080s, agricultural potential could increase by 8 per cent in developed countries, primarily as a result of longer growing seasons, while in the developing world it could fall by 9 per cent, with Sub-Saharan Africa and Latin America projected

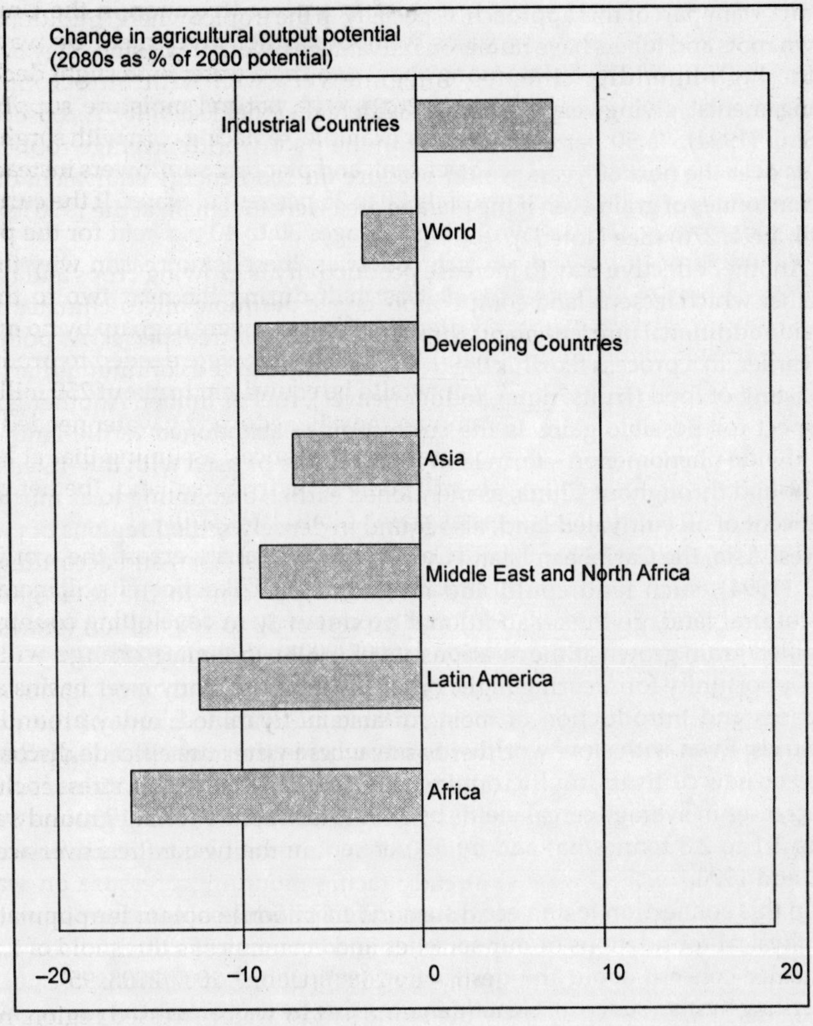

Fig. 6.1: Climate Change will Hurt Developing countries Agriculture

Source: UNDP, 2007/2008

to experience greater loss" (UNDP, 2007/2008: 91) (Fig. 6.1)., making developing countries more dependent on food imports from the rich world with their farmers losing in their share in world trade.

C. CHANGE IN CROPPING PRACTICES

One way of deviating from the traditional way of practicing agriculture is to replace it by new methods yielding encouraging results. Conservation tillage practice is one of them which also includes reliance on improved local staples as an important part of this approach, especially in the tropics where 'traditionally grown roots and tubers have numerous advantages over cereals since they better fit the high-humidity climates, integrate very well with intercropping arrangements', giving very high yields with higher yield stability. As observed by Smil (1994), "a 50 per cent increase in the poor world's average root crop yields over the next 60 years would produce the food energy equivalent of 50 million tonnes of grain even if the planted area were to remain at the 1990 level" (Smil, 1994: 276) (See Note *D*).

Another effective way to increase productivity is growing crops and trees together which lessens land competition, create desirable micro-climates and provide additional nutrient plants through leguminous tree species. As pointed out earlier, in a process like this the trees and shrubs give a continuous harvest consisting of food (fruits, nuts), fodder (leaves), fuel or timber. Another good prospect for possible gains is the cultivation of abandoned fertile land – a worldwide phenomenon – farm land which is idle or used with low-intensity use (found throughout China, as mentioned earlier, amounting to as much as 5 per cent of all cultivated land, also found in densely settled regions of parts of West Asia, the Caribbean Islands and in South East Asia) and according to Smil (1994), such land could add another 3 per cent to the total global agricultural land, giving an additional harvest of 50 to 100 million tonnes of unmilled grain grown at the average rate of productivity in 1990.

Opportunity for securing higher yields is enhanced with good agronomic practices and introduction of most suitable cultivars to many of the poor countries. Even with slow worldwide advances in the intensification process based on new cultivars, multicropping along with higher fertilization resulted in increases in average cereal yields by about 2.5 times between 1950 and 1990 (from 1.1 to 2.8 tonnes/ha) and by 55 per cent in the twenty years, between 1970 and 1990.

In this connection losing ecosystem diversity while not undermining the bio-physical foundations of our societies and economies is often raised with particular concern about the destruction (extinction) of millions of species imperiling human survival. Genetic engineering by way of transferring foreign genes into crop plants to develop new strains more resistant to pests and diseases, plants 'that are more productive under wider ranges of climatic conditions, or that are tolerant of soil salts or aridity are gaining ground'. Further, 'grain varieties might be developed with nitrogen-fixing capability', requiring less fertilizer, but such development takes time and it is unlikely that genetic engineering can 'induce yield increases in key crops comparable to those achieved by Green Revolution' (*Ehrlick* et al. 1993: 15-16). But as has been expressed by Smil (1994), mentioned earlier, a more effective way of raising the overall production of food, particularly for human consumption, is to rely more on 'rational use of existing agricultural inputs' and reducing post-harvest losses, estimated to be as high as 40 per cent of the global harvest which could be lowered by 'controlling rodents, insects, fungi and other organisms that attack food storage'.

On the other hand, the practice of providing farm subsidies prevalent among rich nations poses a very important hurdle impacting heavily on the agricultural production of the developing countries. "In 1997, the Earth Council published a study entitled '*Subsidizing Unsustainable Development*'. Its purpose was to identify and tabulate environmentally destructive government subsidies. It found an astonishing number of examples – at least $700 billion worth per year ... to encourage the use of water, the burning of fossil fuel, the use of pesticides, fishing and driving ...If these subsidies of $ 700 billion per year were diverted into funding environmentally constructive activities, such as investing in renewable energy, tree planting, family planning, and the education of young women in the developing countries, our future could be far brighter" (Brown, 2004: 240).

What is, however, really threatening is that the rich countries spend $ 279 billion of this annually as farm subsidies and as the Financial Times pointed out, "the cash subsidy to a dairy cow in the EU exceeds the EU development assistance per person in Sub-Saharan Africa. The principal motive behind spending such stupendous amounts of money annually as subsidy to farmers is not as innocuous as it looks since the main intention of these affluent countries is to keep world prices of certain commodities — which they can produce – low enough to prevent any developing country to compete with their own products in the world market. In other words, it is nothing but a deliberate attempt to keep the developing countries away from selling their products. For example, "the United States spent $ 46 billion on farm subsidies. These encourage overproduction of farm commodities, which are then sent abroad with another boost from export subsidies. The result is depressed world market prices, particularly for sugar and cotton, the two commodities where developing countries have the most to lose" (Brown, 2006: 137). The EU, in 2004, accounted for $ 133 billion spent on farm subsidies much of which went to European farmers growing sugarbeet at a cost much higher than sugarcane produced in the developing countries. Nevertheless, the interest of European farmers fortified by subsidies received encouraging their production, enabled EU to dump about 6 million tonnes of sugar in the world market but regrettably this is one of the farm produces in which developing countries have a strong comparative advantage because of which they should have been allowed an opportunity to market their product. Fortunately, however, the EU announced in 2005 to reduce its 'sugar support prices by 40 per cent' which might act as a positive step to discourage European farmers in making excess production to depress the world market price of sugar exported, leaving the market partially open for the sugar producers of the developing markets (Brown, 2006).

However, as Brown (2006) observes, the price of sugar might increase with it being used for the production of ethanol to compete with rise in oil prices, and thus providing developing countries further scope for their product in the world market.

Another crop enjoying the privilege of getting subsidies is cotton in the United States and according to Brown (2006), "these subsidies to just 25,000 cotton farmers exceed US financial aid to all of Sub-Saharan Africa's 250 million people. And since the United States is the world's leading cotton exporter, its subsidies depress world prices for all cotton producers" (Brown, 2006: 138). However, strong protests from cotton-producing countries in Central Africa, like Benin, Burkino Faso, Chad and Mali supported by Brazil succeeded in convincing the WTO about the serious implication of this practice and in response the WTO panel 'ruled that the United States had to eliminate subsidies'.

At the same time, debt forgiveness assumes a greater significance as an essential step towards the eradication of poverty. 'For example, with Sub-Saharan Africa spending four times as much on debt servicing as it spends on healthcare, debt forgiveness can help boost living standards in this last bastion of poverty' (Brown, 2006: 138). Rightly, the heads of the G-8 Group of Industrial countries meeting in July, 2005, came to an agreement to the 'cancellation of the multilateral debt of the poorest countries owing to the World Bank, the IMF, and the African Development Fund, which immediately benefited 18 of the poorest countries (14 in Africa and 4 in Latin America), offering 'a new lease of life', while another 20 debtor countries are also to get similar benefit on fulfilling necessary conditions. In support of this Oxfam International views are that it is inhuman to expect those countries with incomes of hardly a dollar a day to spend a part of that to service debt.

Continuing with agricultural subsidies, it may be pointed out that state subsidization does not stop at particular crops; it rather covers a wider area, such as governments in many countries, including some of them in the developing world, subsidizing the use of irrigation water, fertilizers and pesticides, and the consumption of food itself. For example, irrigation water is subsidized for keeping down the price of food crop in many countries. India is one of them where the state of Punjab, the country's bread-basket, goes a step further in providing electricity free for operating irrigation pumps, leading to over-pumping and consequent fast depletion of underground sources. In contrast, China acts much more wisely introducing a price increase on water used by farmers in steps over the next five years, reducing subsidization of water use.

Important instances of subsidization of food consumption are found in Iran where $46 billion a year is spent on the subsidization of bread consumption, amounting to $ 63 per person/year with the government buying wheat from farmers at roughly 70c per kilogram, making it into flour and selling it to bakeries at 2 c for a kilogram. This can be considered as an indirect subsidy to irrigation water which is so very scarce in the country.

Ocean fishing is another part of the world food economy which is heavily subsidized, ensuring that each country 'maximized its share of the ocean fish catch with unintended consequences of overfishing and thereby destruction of the fisheries themselves' (Brown, 2004). Similar destructive ways were prevalent in Australia and the United States where tree cutting used to be subsidized by the government, allowing the logging companies in the State of

Victoria, Australia, to receive subsidies for logging timber while in the United States construction of new roads in the natural forests used to be subsidized to 'facilitate clear-cutting by timber companies' till it was stopped in 1999 under pressure. Subsidies in these environmentally destructive activities need to be eliminated and shifted to environmentally constructive ones and one of the earliest attempts for this was initiated in the United States in 1934 when a nationwide Soil Conservation Service was created, taking the responsibility of protecting the 'agricultural resource base for future generations'. "Farmers were paid to plant wind-breaks, to strip-crop, and to adopt other cropping practices that would protect their soils from wind erosion. This reduced soil erosion, helping to bring the disastrous Dust Bowl era to an end" (Brown 2004:243). Another example of subsidization of environmentally constructive effort in the United States is provided by tax credits for investment in the generation of power from wind though, unfortunately, this was later discontinued but given high priority since then in Denmark, making it the leading country in the world in wind power generation per person and in manufacturing wind turbines on the basis of a modest subsidy

D. RESTORATION OF FORESTS AND CONSERVATION OF SOILS IN AID OF AGRICULTURE

Having discussed in Chapter 4 the impact fertilizers and pesticides can make on food production, particularly in the developing countries, we can turn to two other important issues maintaining a direct link with agricultural development as an instrument for fighting hunger and poverty.

(i) Forest Depletion: The Vanishing Green

Forests hold a very important place in the eco-economy of countries as engines of growth and sustaining developmental trend.

As one of the great inherited resources of the earth, the clearing of forests has been the most important factor that has altered the face of the earth to accommodate growing human demand through centuries and consequently happens to be the greatest factor in transforming natural landscape. The clearance of forests, getting momentum in the emergence of a an integrated world economy through European expansion from the later fifteenth century onward, brought in sweeping changes in the character of landscape as well as growing agricultural prosperity. As reported by Williams (1993), "before the great outward movement from the continent, Europeans had colonized vigorously and cleared their own forests during the Middle Ages... The surge of clearing had more or less spent itself by 1300, when the Plague, possibly deteriorating climate, and certainly, economic recession resulted in a period of stagnation and even abandonment of clearings. Yet, during the early sixteenth century, economic life experienced a resurgence, population rose from approximately 71 million in 1600 to approximately 200 million by 1800, and the forests were attacked again" (Williams, 1993: 181). The pace of forest exploitation (required for wood as fuel, needed for growing industrial activities in soap, glass, tin, lead, copper and salt-making, as well as charcoal-making, iron-, and metal-smelting, etc.) quickened from the early decades of the nineteenth century. Besides, a growing and spreading world population, in its search for food, caused more land to be brought into cultivation, so that the world's arable land increased to become three times the extent it had been in 1860, effecting a huge transformation of forest land to agriculture. Thus, in the space of only 118 years, forests and woodlands were eliminated on an average of 34,000 km^2 per annum. However, according to a modified estimate, a total of 42,000 km^2 of forest areas were converted into agricultural uses and 8,000 km^2 were modified per annum between 1860 and 1978. It is possible to calculate roughly the areas of the different major natural ecosystems that were affected by the transformation of land to agriculture (Table 6.1). As a consequence of the rapid disappearance of forests, the loss (of species) amounted to about 20 per cent of existing species. Soils in the tropical areas 'once opened up to the sunlight and direct impact of rain, become degraded irreversibly through laterization, oxidization and erosion'. Further, 'the beneficial effects of forests in regulating flows and preventing soil erosion' are largely curtailed. As a result of the fast rate of felling, it is difficult to make a correct estimate of how much of the (original) forest survived this prolonged human onslaught.

According to FAO (1986) "At present, forest and woodland stands at 40.9 × 10^6 km^2, which is still nearly three times the area of land in cultivation... the forests of the developed countries of the temperate world (including China) show a slow but steady net increase due to forest reversion, as marginal land

Table 6.1: Global Areas of Different Ecosystems Converted to Cropping, 1860-1978

(x 000 km^2)

	Forests	Woodlands	Savannas	Grasslands	Swamps	Deserts	Total
Developed Regions							
Canada/US	511	130	107	849	—	?	1,597
Europe	78	03	18	84	34	—	217
USSR	439	136	17	884	—	32	1,509
Oceania	239	123	133	51	—	05	551
	1,267	392	275	1,868	34	37	3,874
Developing Regions							
Africa	180	289	239	325	—	31	1,064
Asia	592	628	500	374	177	73	2,344
Latin America	384	253	161	401	01	31	1,233
	1,156	1,170	980	1,100	178	135	4,641
World Total	2,423	1,562	1,175	2,968	212	172	8,515

Source: Williams, 1993 (Taken from Revell 1984, based on Richards, Olson and Rotty, 1983)

is abandoned and agricultural production is concentrated in more favourable areas and as active replanting programmes are initiated"... (There is, however,) "decline of forested areas, most of which is occurring in the developing tropical world. About 56,000 km^2 per annum is cleared in tropical America, 37,000 km^2 in tropical Africa, and 20,000 km^2 in Asia (FAO, 1982)... Nevertheless, some of the biggest surprises are in those countries"... (such as)" the significant increase of 14,000 km^2 in India (+0.21%) or 19,700 km^2 in China (+1.71%) and the remarkable decrease of 21,400 km^2 in Australia (-1.64%) and 254,000 km^2 in the United States (-1.64%)... as new farmland areas have been cleared (in the United States) in response to high agricultural prices and as large tracts of forests have been taken over for urban and industrial uses" (Williams, 1993: 195-196).

However, considering the continuous loss of forest species, especially in the tropical world through unabated deforestation, it is predicted that 21 to 24 per cent of Asia's forest species, and 16 to 35 per cent of Africa's are likely to become extinct in this process under the growing pressure of "countries like China ... buying and importing massive quantities of central African timber. Much of this then becomes furniture that's sold to North America and Europe." According to Alexander Belokurov, WWF International's forest protection programme officer, "Such major promoters of tropical forest destruction as the increasing demand for timber, charcoal and food for urban populations, and such large-scale industrial drivers as cattle ranching, soy production, oil-palm plantations, timber-exploitation and infrastructure development have all played a much higher role in the disappearance of tropical forests in recent years" (*New Scientist*, 12 May 2007: 44).

The forested areas of the earth was estimated at 5 billion hectares at the beginning of the twentieth century, but have shrunk to 2.9 billion hectares (Brown, 2004), almost evenly divided between the two worlds, the developing (tropical and sub-tropical) and the developed (temperate and Boreal). Most of the loss is concentrated in the developing countries, averaging about 13 million hectares annually or 6.5 per cent of its forests per decade, in contrast to the industrial countries gaining up to an estimated 3.6 million hectares of forest land each year constituting mainly the abandoned cropland besides the spread of commercial forestry plantations, as in Russia. According to a report of World Resources Institute's Forest Frontiers Initiative made in 1997, "only 40 per cent of the world's remaining forest cover can be classified as frontier forests, which they define as 'large, intact, natural forest systems relatively undisturbed and big enough to maintain all of their biodiversity including viable populations of the wide-ranging species associated with each type" (Brown 2004: 56).

Protecting the remaining 2.9 billion hectares of forests and replanting the areas already cleared of forest cover is of paramount importance to save humanity from more disastrous consequences ultimately of most severe effects on world food production. As mentioned earlier, deforestation not only accelerates the flow of water back to the ocean, it also reduces the water carried inland by forests acting as 'conduits'. Besides, "if the rain forests are burned off and planted to grass for cattle raising, then the cycle of rainfall is dramatically altered – three-fourths of the rainfall runs off and returns to the sea the first time it falls, leaving little to be carried inland" which adversely affects agriculture (Brown, 2004: 57).

Equally alarming is overgrazing causing fast deterioration of rangelands, which is twice the size of cropland, comprising one-tenth of the earth's land surface, mostly semi-arid, supporting the world's 3.3 billion cattle, sheep and goats and providing livelihood for 180 million people worldwide surviving as pastoralists."An estimated 180 million people worldwide make their living as pastoralists tending cattle, sheep, and goats. Many countries in Africa depend heavily on their livestock economies for food and employment. The same is true for large populations in the Middle East, Central Asia (including Mongolia), northwest China, and much of India...Australia, whose land mass is dominated by rangeland, has one of the world's largest sheep flocks of 117 million sheep...Grass-based livestock economies also predominate in Argentina, Brazil, Mexico, and Uruguay. And in the Great Plains of North America, lands that are not suited to growing wheat are devoted to grazing cattle" (Brown, 2004: 58-59) The rangelands are exploited by large-scale commercial ranching, production of global beef and mutton while the feedlots play a minor role in this respect (Table 6.2) though the latter hold a much greater prospect for returning degraded land from overgrazing to regeneration of vegetation with gradual increase in stall-feeding based primarily on crop-residues, as is done in India.

Bearing in mind the fact that meat protein makes an important part of diet (beef consumption alone per person being 69 kilograms in Argentina, 45 in the

Table 6.2: Domesticated Ruminants by Country, 2000

Country	*Cattle and Buffalo (in million heads)*	*Sheep and Goats (in million heads)*
Argentina	55	17
Australia	26	117
Bangladesh	24	35
Brazil	169	31
China	127	279
Ethiopia	35	39
France	20	11
India	313	181
Mexico	30	16
Nigeria	20	45
Pakistan	45	72
Russia	28	16
United Kingdom	11	45
United States	98	9
Other	509	868
World	11,510	1,780

Source: Brown, Lester R., 2004 (Taken from FAO, FAO STAT, 2001).

United States, 39 in Brazil, 36 in Australia), half of all grasslands in the world are already lightly or moderately degraded and 5 per cent severely degraded, which is going to increase pressure to use food crops as animal-feed, as practiced in many countries, gradually leaving less for human consumption. The situation seems to be more serious in view of the fact that the fodder requirement to increase the share of stall-fed animals is still not adequate in almost all the developing countries.For example, the demand for fodder in India was estimated at 700 million tonnes as compared with the available amount of 540 million tonnes (Brown, 2004). Africa faces greater difficulty where even the sustaining capacity of the grasslands is slowly diminishing in supporting the huge number of livestock (230 million cattle, 241 million sheep and 209 million goats; Brown, 2004) with extremely limited prospects for turning to stall-fed practice as grain is scarce. However, turning slowly more to feeding crop residues is the only major alternative to overcome the annual loss of rangeland productivity estimated to be $ 7 billion in Africa and $ 8 billion in Asia, the two continents together accounting for two-thirds of the total global loss from rangeland production due to degradation.

As has already been mentioned for 'restoring the earth's health', it is highly imperative to protect the remaining 2.9 billion hectares of forests and thus 'arrest reducing rainfall run-off, associated flooding and soil erosion, recycling rainfall inland, restoring aquifer recharge' enabling humanity ultimately to buildup a new, stronger economy. Brown (2006) observes that there is vast unrealized potential in all countries to lessen the demand pressure shrinking the forest cover. "In industrial nations the greatest opportunity lies in reducing the quantity of wood used to make paper, and in developing countries it depends on reducing fuelwood use" (Brown, 2006: 143). Paper-recycling at the rate it is taking place in Germany (72% in Germany; 66% in South Korea and 48% in the United States by 2003) is likely to reduce production of paper using wood pulp by one-third in the world. In addition, there is an enormous possibility in reducing the use of paper for 'facial tissues, paper napkins, disposable diapers and paper shopping bags' which could be made out of reusable cloth. However, fuel makes the largest demand on wood an alternative to which may be found in 'solar thermal cookers, with electric hot-plates fed by wind-generated electricity, or with some other source of energy'. Further, forest plantations are likely to reduce pressure on existing forests remarkably. By 2000, 187 million hectares of land has been planted with trees worldwide, comprising less than 5 per cent of the total 2.9 billion hectares of remaining forest lands and equal to nearly one-fourth of the 700 million hectares of cropland. Such tree plantations provide wood for paper mills, for the lumber market, meeting about 12 per cent of world wood production, while 88 per cent still comes from natural forest stands (Brown, 2006). According to FAO, with expansion of plantation area and yield rises, the total production doubling in the next three decades may 'one day satisfy most of the world's demand.' Despite this prospective future it is greatly disheartening to notice the apparent lackadaisical approach to the protection of natural forests, making it highly vulnerable since only "about 290 million (out of 2.9 billion) hectares of global forest area are legally protected from logging and an additional 1.4 billion hectares are economically unavailable for harvesting because of geographical inaccessibility or low-value wood. Of the remaining area available for exploitation, 665 million hectares are undisturbed by humans and nearly 900 million hectares are semi-natural and not in plantation" (Brown, 2006: 144).

However, the potential of the vast deforested land lie in their use for growing 'trillions of genetically improved trees for food', consisting mainly of nuts as a source of high protein to supplement meat especially in the diet of people in developing countries, as well as trees grown for use as fuel or for conversion into ethanol, as observed by Reed Funk, professor of plant biology at Rutgers University. This offers a promising situation for the eradication of hunger and poverty.

(ii) Soil Conservation in Aid of Future Agricultural Practices

Loss of soil, which takes centuries to form, has the most drastic effect on agricultural productivity particularly in view of shrinking cropland area and reducing total production of crops. The United States learnt the lesson the hard way in the formation of the 'Dust Bowl' in its southern Great Plains, resulting from extensive wind erosion following successive periods of over-ploughing coupled with droughts in the early 1930s, and had to adopt planting of wind breaks and strip-cropping which ultimately checked soil erosion. However, because of faulty policies of "the farmers planting everything in sight", the USA was subjected to the loss of about 3.08 billion tonnes of top soil by 1982, when loss of one inch of top soil reduces the yield of corn and wheat by 6 per cent. As a consequence, a programme on soil conservation by way of retiring most of the erodible land was launched in the United States in 1985 (the US Conservation Research Programme), 'initiated and supported by environmental groups', which succeeded in converting nearly 15 million hectares of cropland, roughly 10 per cent of the national total, to grassland and thus reduced excessive soil erosion nationwide by some 40 per cent, giving a big boost to global food security. A similar case happened in the Soviet Union between 1954 and 1960 following 'overexpanding ploughing' of the grasslands in Central Asia, primarily concentrated in Kazakhstan in an attempt to enlarge the wheat area to be equal to the wheat area of Canada and Australia combined, reaching 25 million hectares by 1960, which, however, started shrinking after 1984 or so, dropping to 12 million hectares by 2001 due to loss of top soil resulting in declining yields, with a great apprehension that the cropland wrested from original grassland may have to be abandoned with further loss of top soil, given the yield of wheat per hectare in the croplands has already come down to less than 1 tonne as compared with 7 tonnes per hectare in France, the leading wheat producer in Western Europe (Brown, 2004).

Among the developing countries China is a big victim of soil erosion triggered by extension of ploughing into much less productive land, highly vulnerable to wind erosion. This is largely the result of cropland being devoured fast by expanding cities, industries, roads, etc., with growth of population as well as economy since 1950, tripling the demand for food. The process of desertification as a consequence of pushing the agricultural frontier further into the semi-arid north-west resulted in the loss of a million tonnes of top soil, 'a depletion of its natural capital', and has already had disastrous effects in the form of sand storms affecting several major cities (Brown, 2004).

The 'collision course of population growth and the degradation of cropland', as Brown (2004) calls it, has a similar disastrous effect on Africa where, according to Rattan Lal, an internationally noted agronomist at Ohio University's School of Natural Resources, "soil erosion and other forms of land degradation have reduced Africa's grain harvest by 8 million tonnes or roughly 8 per cent." Further, he expects the loss to climb to 16 million tonnes by 2020 if soil erosion continues unabated. Among the countries experiencing unusually heavy soil losses are Nigeria, Rwanda and Zimbabwe. Nigeria, Africa's most populous nation, is suffering from extreme gully erosion. In January 2001 Nigeria's Minister of Environment, announced that the country was "losing some 500 square kilometres of cropland to desertification each year (threatening) severe food shortages" (Brown, 2004: 66-67).

Many other countries in Africa, such as Algeria, Ethiopia, Kenya and Somalia are facing a similar situation of land degradation due to soil erosion and abandonment of cropland, meaning they face the great responsibility of feeding a projected 2 billion people by 2025 (instead of today's 800 million) with far less top soil. According to US analysts, 'Mexico is forced to abandon 1,036 square kilometres of farmland to desertification each year' due to soil erosion. Soil, 'the most precious natural capital is, as estimated, getting lost at the rate of about 24 million tonnes annually and the remaining top soil on earth's cropland is disappearing at an average rate of 7 per cent per decade globally' (Ehrlick *et al.*, 1993: 18). resulting from human activities like agriculture, overgrazing and deforestation.

E. ENVIRONMENTAL CONSTRAINTS LIMITING FOOD PRODUCTION

As predicted by Ehrlick *et al.*, (1993), some environmental constraints assuming global proportions are going to threaten food production required for sustaining the growing population during the present century.

(a) Greenhouse Warming, Ozone Depletion and Pollution

Rapid climate change with greenhouse warming represents a great threat to food production and it is predicted that "climatic zones will shift as much as 50 times faster than they have in the 10,000 years since the dawn of agriculture" (Ehrlick *et al.*, 1993: 18). According to the International Panel of Climate Change (IPCC), 'the average world surface air temperature has risen by between about 0.3 and 0.6 degrees during the last hundred years', believed to be due to human-induced release of greenhouse gases from activities like burning of fossil-fuels, cement production and deforestation, which are also responsible for ozone depletion. At the same time rice cultivation on an extensive scale is the source of emissions of methane gas-originating further from the guts of livestock - and nitrus oxides are also created by nitrogen fertilizers as well as tropical grasslands. Importantly, agricultural activities are at the root of generating one-sixth of the total human-induced warming (Dyson, 1994: 137-138).

"Global atmospheric concentrations of carbon dioxide and nitrus oxide have increased markedly as a result of human activities since 1750 and now far exceed pre-industrial values determined from ice cores spanning many thousands of years. The global increases in carbon dioxide concentration are due primarily to fossil fuel use and land-use change, while those of methane and nitrus oxide are primarily due to agriculture" (IPCC, 2007: 2). Concentration of the most anthropogenic greenhouse gas, carbon dioxide, has increased from about 280 ppm during the pre-industrial period to 379 ppm in 2005, exceeding the atmospheric concentration 'by far the natural range over the last 650,000 years (180 to 300 ppm).' The concentration growth-rate has been larger during the last ten years – 1995 to 2005 – an average of 1.9 ppm per year - than it was in 1960-2005 - 1.4 ppm. By far the largest contribution was from the use of fossil fuels. The net effect of human activities since 1750 has been one of warming, and eleven of the last twelve years (1995-2006) 'rank among the 12 hottest years in global surface temperature (since 1850).The total temperature increase from 1850-1899 to 2001-2005 is 0.76 degrees C (0.57 degrees to 0.95 degrees)' (IPCC, 2007: 3-5). Against this, however, one has to consider the fact that human-induced processes also have a partially 'offsetting (cooling) effect', most importantly – "humanity is constantly stirring up tiny particles into the atmosphere like smoke, dust and sulphur oxides and these, acting as nuclei for cloud droplets to form, have a significant cooling influence by reflecting solar radiation back into space, thus lowering surface temperature."

As mentioned in Chapter 5, global warming might be partly beneficial to some group of crops, such as crops like wheat, barley, rice and potatoes due to higher levels of carbon dioxide in atmosphere having a "fertilizing effect" on crops, and temperature rises will benefit crop production at higher latitudes with longer growing seasons especially in northern Europe, Asia and North America though reductions in soil moisture and soil fertility hastening soil erosion might be a cause of reduction in potential cereal yields in northern mid-latitudes of the United States, southern Europe and parts of Ukraine and Kaszakhstan.

But of greater serious concern is the rise of sea level in the low lying coastal areas of many countries in the world due to global warming, hitting hard on croplands concentrated mostly on the flat plains. On the basis of high population growth projections, IPCC estimated that, coastal flooding would affect 134-332 million people with a rise of sea level for 3 degrees – 4 degrees rise in temperature. "Observations since 1961 show that the average temperature of the global ocean has increased to depths of at least 3,000 m and that the ocean has been absorbing more than 80 per cent of the heat added to the climate system" (IPCC, 2007: 7). Global average sea level rose at a rate of 1.8 (1.3 to 2.3) mm per year over 1961 to 2003. The rate was faster over 1993 to 2003, about 3.1 (2.4 to 3.8) mm per year. The rate of observed sea level rise increased from the 19th to the 20th century. The total 20th century rise is estimated to be 0.17 (0,12 to 0.22) mm. per annum (IPCC)) (Table 6.3)

Table 6.3: Observed Rate of Sea Level Rise and Estimated Contributions from Different Sources

Source of Sea Level Rise	*Rate of Sea Level Rise (mm per year)*	
	1961-2003	*1993-2003*
Thermal Expansion	0.42 ± 0.12	1.6 ± 0.5
Glaciers and Ice Caps	0.50 ± 0.18	0.77 ± 0.22
Greenland Ice Sheet	0.05 ± 0.12	0.21 ± 0.07
Antarctic Ice Sheet	0.14 ± 0.41	0.21 ± 0.35
Sum of Individual Climate Contributions to Sea Level Rise	1.1 ± 0.5	2.8 ± 0.7
Observed Total Sea Level Rise	1.8 ± 0.5	3.1 ± 0.7
Difference (observed minus sum of estimate climate contributions)	0.7 ± 0.7	0.3 + 1.0

Source: IPCC, 2007.

According to the UNDP 2007/2008 Report, the rise of sea level by 1-metre would displace 6 million people in Egypt, besides flooding 4,500 km^2 of farmland where 17 per cent of the population live, about 4 million of whom live below the poverty line. In Viet Nam, 22 million people would be displaced and the sea level rise 'would expose' about 45 per cent of Mekong Delta's land area by 2030 to 'extreme salinization and crop damage through flooding., inundating much of the delta. In Bangladesh, 1 metre rise in sea level would inundate 18 per cent of land area, directly threatening 11 million people and the impact on river levels from sea rise could affect over 70 million of the country's population. Perhaps, Maldives will be worst hit with 80 per cent of its land area lying hardly 1 metre above sea level and any change is likely to inundate the entire group of islands. The Caribbean Sea Islands are similarly vulnerable to change in sea level where even a 50 centimetre increase would submerge over one-third of the Caribbean beaches devastating their tourist industry, while a rise of 1 metre would permanently submerge about 11 cent of the land area in the Bahamas (UNDP, 2007/2008: 100-101) (Table 6.4).

Further, climate change with rising CO^2 levels will lead to unprecedented losses of biodiversity and damage ecosystems during the 21st century. A rise in temperature above 2 degrees C will enhance the rates of extinction and the pace of environmental degradation with 'coral, wetland and forest systems suffering rapid losses'. In 2005, the Millennium Ecosystem Assessment found that 60 per cent of all ecosystem services were either degraded or being used

Table 6.4: Rising Sea Levels: Social and Economic Impacts (Impact % of global total)

Magnitude of Sea Level Rise (m)	*Land Area*	*Population*	*GDP*	*Urban Area*	*Agricultural Area*	*Wetland Area*
1	0.3	1.3	1.3	1.0	0.4	1.9
2	0.5	2.0	2.1	1.6	0.7	3.0
3	0.7	3.0	3.2	2.5	1.1	4.3
4	1.0	4.2	4.7	3.5	1.6	6.0
5	1.2	5.6	6.1	4.7	2.1	7.3

Source: UNDP 2007/2008 (Taken from Dasgupta, Susmita, B. Laplante, C. Meisner, D. Wheeler and J. Yan, 2007: 'The Impact of Sea Level Rise on Developing Countries: A Comparative Analysis,' Policy Research Working Paper 4136, World Bank.)

unsustainably. The loss of mangrove swamps, coral reef systems, forests and wetlands was highlighted as a major concern, with agriculture, population growth and industrial development acting together to degrade the environmental resource base. "Nearly one in four mammal species is in serious decline" (UNDP 2007/2008: 102)., making the world's poor who depend most heavily on these species having to bear the brunt of the actual cost.

People become very vulnerable in the face of such changes in environmental resources, the most glaring instance of which is provided by the destruction of wetlands which 'harbour biodiversity, provide agricultural, timber and medicinal products, sustain fish stocks' and 'buffer coastal and riverside areas from storms and floods, protecting human settlements from sea surges'. "During the 20th century, the world lost half of its wetlands through drainage, conversion to agriculture and pollution" (UNDP, 2007/2008: 102) while the destruction still continues unabated, threatening more intense storms and sea surges in the future, with greater exposure of mangrove areas, such as the Sundarbans in both Bangladesh and West Bengal (India) to these natural elements portending incalculable loss of human life and property.

Even more threatening are the changes in ecological maps, almost getting 'redrawn' over the past three decades with isotherms moving towards North and South Poles "at a rate of about 56 miles per decade", making species follow the climate zones. Already "changes in flowering seasons, migratory patterns and distribution of flora and fauna have been detected in the world" (UNDP, 2007/2008: 102). There has already been a great loss of species due to climate change. But the situation will be far worse, according to the IPCC, with 20-30 per cent of plant and animal species, including polar bears and fish species feeding on coral reefs, facing increased risk of extinction if global average temperature increases exceed 1.5 – 2.5°C. Climate models suggest that some of the most serious disruptions of agriculture will result from drying of the central parts of northern continents, regions that now constitute the world's principal breadbaskets (UNDP, 2007/2008:102). This cannot be compensated by prospective increase in agricultural yields more climatically suited, such as, 'Canadian Shield to which Iowa's present climate may migrate' because of thin nutrient-poor soils.

Another growing danger with highly disastrous consequences of human-induced pollution is "earth's toxic burden", which impacts heavily on human and all other forms of life with great reduction in global food-producing capabilities. Most important of these polluting sources are chemicals, manufactured and synthetic chemicals, generally organic in nature, which are dangerous in many ways as being 'highly persistent and found in remote corners of the globe far from the place where they are manufactured (or even used). For example, the polar bears living within the Arctic Circle have 'high concentrations of Persistent Organic Pollutants (POPs) in their fatty tissues', resulting in the formation of 'deformed female organs' among 1.5 per cent of all female bears (Brown, 2004: 131).

In 1999, some 7.8 billion pounds of toxic chemicals – 28 pounds for each person – were released into the U.S. environment. "Exposures to varying levels of toxic chemicals and various combinations can lead to birth defects, impaired immune systems, damage to the central nervous system (including mental retardation), respiratory illnesses, a disruption of endocrine systems and hormonal balances, and cancer of almost every kind", effects liable to reduce life expectancy. (Brown, 2004: 133). Further, the environment can be damaged by pollutants. For example, acid rains from sulfur dioxide emissions have damaged forests in industrial region in Europe, North America and China. According to a report in 2000, one-quarter of forests in Europe were damaged by acid rain while it killed all the trees in a 3,500 square kilometre area in Siberia. Developing countries are equally bearing the brunt of chemical pollution, such as in India the underground water system has become contaminated with pollutants like 'cyanide, cadmium and lead' coming from industries ranging from textiles to metal electro-plating in Ludhiana in Punjab. This is not a lone case as pollution of underground aquifers is quite rampant in other parts of the country. Even Japan, a highly industrial country, is not exempted from it as Tokyo is considered the world's capital of dioxins (a product of burning plastic), coming from the incineration of municipal waste.

Perhaps, the major way of getting rid of toxic pollution is recycling, 'an economical alternative to rising land-fill costs, it also greatly reduces ecosystem damage', and in this some countries like the United States, Brazil, Japan and India have already made remarkable progress not only in reducing environmental pollution but also in generating employment (in Brazil an estimated 150,000 people make a living by collecting used beverage cans and taking them to the recycling centres while in many of the big Indian cities thousands of people make a living out of collecting from garbage dumps and selling the articles collected for reuse).

Undoubtedly, "a great deal of the environmental destruction caused by Homo Sapiens (in such forms as deforestation, desertification, wetland destruction, toxic pollution of air, water and land, and releases of greenhouse gases into the atmosphere) is a direct consequence of the struggle to feed a rapidly expanding population. At the same time, environmental damage constitutes an increasingly important constraint on the future expansion of food harvests" (Ehrlick *et al.*, 1993: 20). To overcome the hurdles to increase food production and meet the challenge of growing needs in response to a growing population as well as growing purchasing power (provided by economic growth in the developing counties, in particular), the first and foremost task is conservation of soil and water, followed by 'restoration ecology' as a central discipline and, finally, establishing more integrated pest management systems to reduce the harmful effects of pesticide uses. Further, though politically challenging, the entire success of making all efforts at eradicating hunger and poverty will come to nothing unless the existing world trade system (by WTO) for food is effectively changed, making it much more friendly for the developing economies and the most counterproductive practice of rich nations giving their farmers agricultural subsidies is abolished once for all, creating an atmosphere of fair competition.

At the same time conservation of wild nature has to be taken more seriously because apart from deriving benefit from wild nature through 'ecological services like climate regulation, soil fixation and nutrient cycling', the direct harvest of wild species for food, fuel, fibres, and pharmaceuticals is of immense value for human society. According to the estimate made by Constanza team, the aggregate annual value of nature's services (updated to 2000 US $) amounts to' $ 18 trillion to 61 trillion, around a rough average of $ 38 trillion, similar in size to global Gross National Product (GNP) (Balmford *et al.*, 2002: 950). In contrast, conversion of biomes give short-term benefits, such as, the total economic value of intact mangroves compared with benefits obtained from conversion of the same into aquaculture, as in Thailand, (shrimp farming) exceeded by around 70 per cent from the latter ($ 60,000 compared with $ 16,000 per ha). Canada provides an example where total economic values were higher when the wetlands remained intact (arising from sustainable hunting, angling, and trapping), exceeding returns for conversion by about 60 per cent ($ 8,800 compared with $ 3,700 per ha). It is certainly true that conversion has been economically beneficial in many instances, particularly in cases in the past when forest and wetlands were converted into prime agricultural land. But "conversion of remaining habitat for agriculture, aquaculture or forestry often does not make sense from the perspective of global sustainability" (Balmford *et al.*, 2002: 951).

F. STABILIZING POPULATION AND UNIVERSALIZATION OF EDUCATION

Providing all humanity with food security depends on two other achievements, the stabilization of population and making education universal. Each of these is an extremely difficult objective to fulfil as hurdles on the path to solution are incredibly large in number and governed by circumstances often beyond our means to circumvent. Nevertheless, they are not insurmountable with a judicious approach and carefully prepared plans. For example, stabilization of the world population means halting further population growth and making it stable (not declining) so that we could have a firm idea about the amount of food needed to feed mouths not only today but also tomorrow, that is for another four decades or so from now. To guarantee food security for a stable population requires firm, calculated steps towards combating the factors responsible for unchecked population growth in some parts of the globe while in other parts it no longer remain as a problem. "Although rapid population growth continues in scores of countries, the world is beginning to divide into two parts: one, where population growth is slowing as fertility falls, and another, where population growth is slowing as mortality rises. One way or the other, population growth will slow. Rising death rates from AIDS have already reduced the projected population for 2050 by more than 150 million, which represents a failure of our political institutions unmatched since the outbreak of Second World War (Brown, 2004: 213).

There is, however, no point in remaining satisfied with the idea that world population is growing slowly, since even a slow growth adds up to quite a large number on top of what it is today. In reviewing the trend of population growth as extant today it shows that world population growth averaged barely 0.5 per cent a year during the first two decades of the twentieth century, but rose to about 1.0 per cent a year during the 1920s, 1930s and 1940s (as mentioned earlier) after which the rate increased to around 1.8 per cent in the 1950s, and reached 2.0 per cent in the 1960s. In absolute number of fresh additions, the annual increment was barely 10 million at the beginning of the century and rose to 20 million annually in the 1920s, 50 million in the 1950s and nearly 83 million during the 1980s, the annual rate of increase averaging just under 2 per cent between 1960 and 1999.

There is, nevertheless, enough evidence to justify the fact that the rate of world population growth 'has peaked and is now slowing, though the absolute volume of increase remains very high', having fallen in recent years to an estimated 77 million a year. The decline in the rate of increase started during the 1970s, falling to about 1.7 per cent per year and is now reckoned to be around 1.2 per cent per year and these latest trends are likely to continue (Daniels *et al.*, 2005: 88-89) (Fig. 6.2).

As projected by the United Nations in 2002, the rate of growth of world population is expected to fall further, going down below 1 per cent during 2015-20 (it exceeded that rate almost exactly a hundred years ago), and is expected not to be much above 0.3 per cent by mid-century (21st century), a level not experienced 'since around 1750'. Despite this, the annual increments are likely to stay above 70 million a year till about 2020 (because of the growing population base), but are expected to "more than halve by 2045-50". On the basis of this, the total world population is likely to reach 8.9 billion by 2050, nearly 3 billion larger than it had been in 2000, "but with majority of this addition occurring in the first quarter of the century and bringing up the 8 billion mark by 2028" (Daniels *et al.*, 2005: 89).

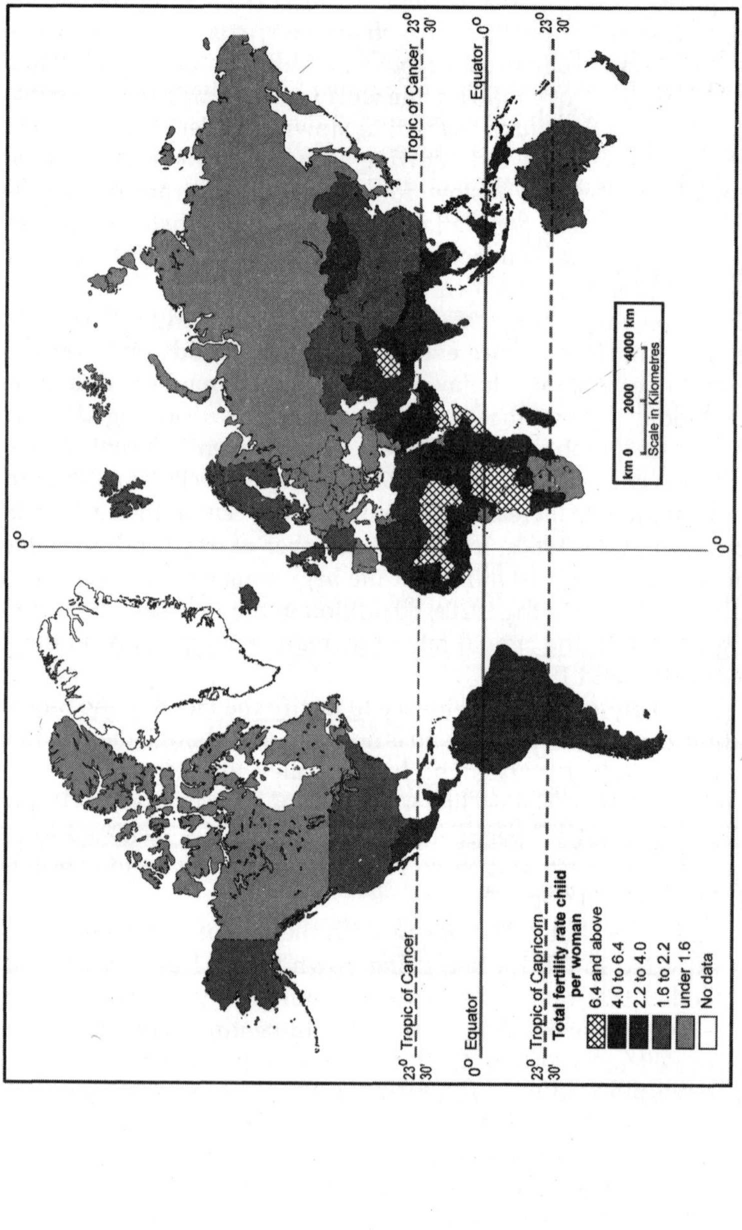

Fig. 6.2: Geographical Distribution of Total Fertility Rate, 2000-2005

Source: Daniels *et al.*, 2005

What is really of serious concern about the trend of world population directly associated with our major issue about stabilizing population, is the unevenness of population growth with disproportionate concentration in the Less Developed Countries (LDCs) of the world. Almost 90 per cent of the world's 3.8 billion increment took place in these countries, their contribution rising to about 95 per cent of the global total by 2000-05, and according to U.N. projections, the LDCs will be home for another 2.6 billion people between 2003 and 2050, incorporating almost the entire growing population, leaving hardly 0.6 per cent for the Developed Regions. This is happening in the 49 countries of the LDCs, which are experiencing the 'greatest upward surge in the population'. As a result of this unevenness in growth rate there has been substantial shift in the geographical distribution of population in recent decades, establishing itself as a growing trend. For example, the proportion of world population concentrated in the Developed Regions (MDRs) gradually declined from 32.3 per cent in 1950 (almost a third of the world) to 19.1 per cent (or, one-fifth) by 2003 and is likely to slide down to 13.7 per cent by 2050. In sheer contrast, the share of the 49 countries though rising slowly between 1950 and 2003, is expected to increase from 10.4 per cent in 2003 to 18.8 per cent in 2050. (Daniels *et al.*, 2005: 91)

Reduction in mortality and increasing life expectancy have to be factored in to understand the trend of population growth and its future indications. The crude death rates (deaths per 1000 people) across the world have fallen by more than half during 1950-55 to 2000-05 from 19.8 to 9.1, happening almost the same way in the LDRs, declining from 27.9 to 15.1 per 1000 during this period. The rate rose further in the MDRs later but it declined by about two-thirds, dropping to 8.8 per 1000 which is clearly reflected in life expectancy. The earlier increase (from 1950-55 onwards for 25 years) in life expectancy in the LDRs, increasing from 41 to about 57 years, is attributed to the use of improved techniques bringing down the number of deaths from diseases like malaria, smallpox, tuberculosis and cholera. But those countries, mainly in Africa, facing two challenges — the HIV/AIDS epidemic and high infant and childhood mortality — have the lowest life expectancy affecting population growth. The impact of HIV/AIDS on the growth of population may be ascertained from the UN projection about 38 highly affected African countries, that by 2015, they are going to have a population of 91 million or 10 per cent lower than 'it would have been without 'AIDs'. In seven countries (Botswana, Lesotho, Namibia, South Africa, Swaziland, Zambia and Zimbabwe) with HIV/AIDS prevalence rates at or above 20 per cent there is going to be a larger impact with a decline in population by 19 per cent. The other factor, infant mortality, determining the rate of population growth, presents similar sharp contrasts between the two worlds – the MDRs and the LDRs – when four-fifths of deaths in the MDRs are now accounted for by people aged 60 or over against two-fifths in the LDRs, while those aged between 0-5 years represent less than 2 per cent of all deaths in the MDRs but over a quarter in the LDRs (two-fifths in Africa).

Arguably, early, premature deaths in very high proportions of the total population in the LDRs resulting from HIV/AIDS and infant mortality act as significant brakes in slowing population growth, finally lowering the total global population which, otherwise would have increased faster than it is doing under present circumstances. Obviously, these are social inadequacies in relief provided to the people and steps towards improving the situation, through making better medical facilities available for countries suffering from HIV/AIDS and high rates of infant mortality, are bound to check the heavy human casualties, thereby arresting once for all the decline in population, especially in countries of the Sub-Saharan Africa as well as India (with 3.7 million adults now HIV-positive, India is home to more infected people than any other country except South Africa, as reported by Brown, 2004). According to UN estimates, based on 2002 projections, 'the world population in 2050 is expected to be nearly half a billion (479 million) lower than it would have been without the deadliest epidemic of our time, HIV/AIDS' (Daniels *et al.*, 2005: 98-99)

Infant mortality, as mentioned in an earlier chapter, is intimately connected with the rate of fertility, and significantly fertility behaviour is the ultimate key to determine the size of future world population. Any attempt at making world population stable as far as practicable cannot be fulfilled without changing the existing behavioural pattern of fertility across the world which is extremely variable now. As observed by Daniels *et al* (2005), even minor changes in fertility rate make large differences in population size over a long time.

An important issue in this respect is whether in the least developed countries, particularly in Africa, reduction in fertility rates already initiated over the last couple of decades can be continued and intensified further to bring down the family-size as well, offering double benefits, first, reduction in population and improvement in family economy with lesser number of mouths to be fed and cared for, and second, bringing below-replacement rate fertility (number of children below 2.1 per woman) now recorded by many European countries. Important to note in this regard is the record of the LDRs where the overall TFR (Total Fertility Rate) stayed at 6.2 (children per woman) for 15 years from 1950-55 onwards, falling to half afterwards, while in Asia and Latin America the TFR fell from close to 6 to 2.5 (per woman) during this period though it was far from even in all parts. For example, China made a special case bringing it down to 1.8 when in several countries of Western Asia the rate of fertility was above 4 children per woman. Variation in fertility rates, however, was less in Latin America where they range between 1.5 for some Caribbean countries to 4.4 in Guatemala. In many of the developed countries the fertility rates have gone below replacement rate, such as in Austria, Belarus, Bulgaria, Czech Republic, Estonia, Greece, Hungary, Italy, Latvia, Lithuania, Poland, Slovakia, Slovenia, Spain and Ukraine where the TFR is 1.3 children per woman or even lower than that. (Fig. 6.2). On the whole, achievements so far made in this direction amounts to nothing less than a 'reproductive revolution', as observed by Daniels *et al.* (2005).

However, unless fertility rates fall substantially, preferably to replacement-fertility level, especially in Africa, South Central and Western Asia, the ultimate aim of halting population growth is not possible. The slow realization of the benefits of a small family is one of the great inducements, no doubt, in reducing overall population size but to make the transition really effective, family planning programmes adopted and executed with proper zeal and purpose become an absolute necessity. At the same time, its effectiveness, again, depends very much on how common people accept it as the most important tool for restraining family-size which, in turn, is largely dependent on education of the people who are to be convinced of the benefits to be derived from a small family resulting from lowering fertility rate.

One way of doing this is by ensuring universal education reaches the 115 million children who do not attend school (Brown, 2006). As Amartya Sen observes, "illiteracy and innumeracy are a greater threat to humanity than terrorism." The central role of education, especially female education, in human progress is very forecefully demonstrated by the fact that, "the achievement level of children correlates closely with the educational level of their mothers. Children of educated mothers are better nourished not necessarily because the family income is higher but because their mother's better understanding of nutrition leads to a better choice of foods and healthier method of preparation. Educating women is the key to breaking the poverty cycle" (Brown, 2006: 125).

Fertility falls with a rise in female education in every society for which data are available and 'mothers with at least five years of schooling lose fewer infants during childbirth or due to early illnesses than happens with less educated females.' According to economist Gene Sparling, convinced from his study of 72 countries in 2001, "the expansion of female secondary education may be the single best lever for achieving substantial reductions in fertility" (Brown, 2006: 125), and for making education universal, Gene Sparkling believes that every plan for spreading education should aim at 'getting to the hardest–to-reach segments of society, especially poor girls in rural area'. Girls benefit specially because 'drawn to school by lunch' (which is provided as an educational inducement programme in many countries for attracting students from very distressed families unable to feed their children properly, as in India), they stay longer, marry later and have fewer children. Even agriculture gains from education as farmers with education can follow the instructions on a bag of fertilizer, and people with education are capable of avoiding risks of infection, and learn about viruses causing HIV/AIDs and other killer diseases. "The World Bank estimates that funding of about $ 12 billion a year would be needed to achieve universal primary education in more than 80 countries that are unlikely to reach this goal by 2015". (Brown, 2006: 126-127).

In taking a stock of the present situation leading to the stabilization of world population, it is reassuring to see that there are about 42 countries which have already stabilized their population, or have a declining population. The countries with the lowest fertility rates are Japan, Russia, Germany and Italy where population is likely to decline over the next half-a century. There is another large group of countries having fertility rates reduced to replacement level, including China and the United States. The family-size in China, at 1.8 children per couple, is now below the level in the United States. But by restricting to the one-child per couple principle, China is going to face a severe crisis of labour in the not very distant future coupled with a growing burden of old-aged people. On the other hand, a third group of countries, including Ethiopia, the Democratic Republic of Congo and Sudan are projected to increase their populations by more than two times by 2050. UN projections have made three assumptions about the growth of world population: the highest one of 10.6 billion, the medium one of 9.1 billion by 2050 while the lowest one is 7.8 billion in 2041, assuming that the world will quickly move below replacement-level fertility to 1.6 children per woman followed by a decline, the last one needs to be the target for our success in eradicating poverty, hunger and illiteracy (Table 6.5) (*See* Note *E*).

Table 6.5: Fertility Levels in Selected Countries in 2001

Country	*Average Number of Children Per Woman*[1]	*Population Mid-2001 (million)*
Countries with Fertility at or Below Replacement Level[2]		
Russia	1.2	144
Italy	1.2	58
Japan	1.3	127
Germany	1.3	82
Poland	1.4	39
Australia	1.7	19
United Kingdom	1.7	60
China	1.8	1,273
France	1.8	59
United States	2.1	285
Countries with Fertility above Replacement Level[2]		
Brazil	2.4	172
Indonesia	2.7	206
India	3.2	1,033
Pakistan	5.6	145
Tanzania	5.6	36
Saudi Arabia	5.7	21
Nigeria	5.8	127
Ethiopia	5.9	65
Democratic Republic of Congo	7.0	54
Yemen	7.2	18

Source: Brown, Lester R., 2004 (Taken from Population Reference Bureau, 2001 World Population Data Sheet)

Notes: 1. The average number of children born to a woman in her lifetime is also known as the TFR

2. Replacement Level Fertility is an average of 2.1 children per woman

Brown (2006) observes that slowing world population will depend on all women having access to family planning which is not available at present for 201million couples, and meeting this should have top priority. According to the estimate of the United Nations, 201 million women at present deprived of contraceptive facilities if provided with access will "prevent 52 million unwanted pregnancies, 22 million induced abortions, 1.4 million infant deaths and prevent about 142,000 pregnancy-related deaths every year" (Brown, 2006: 130). In this regard a very encouraging picture is presented by Iran which introduced its national family planning programme in May 1993, in which all religious leaders were included, introducing all contraceptive measures including male sterilization (a first among Muslim countries), contraceptives, use of the pill, etc, all free of charge and thus "Iran also became a pioneer requiring couples to take a class on modern contraceptive before receiving marriage license" (Brown, 2006: 129). In contrast India introduced a family planning programme at the national level in 1951 which has not been very effective.

In order to remove any confusion it is important to note that rapid economic growth is not a prerequisite for reduction in fertility rates. Bangladesh providing a successful case in this regard, reducing a fertility rate of nearly 7 children per woman in the early 1970s to 3.3 children in the first decade of the 21st century in spite of the average income ($200/year) being among the lowest in the world. Further, according to a study, the Bangladesh Government "spends $ 62 to prevent a birth, but saves $ 615 on social services expenditure for each birth averted – a 10-fold difference in cost. Based on the study's estimate, the programme prevents 890,000 births annually. The net savings to the government total $ 547 million each year, leaving more to invest in education and health care" (Brown, 2004: 224). Incidentally, the United Nations following the results of the International Conference on Population and Development held in Cairo in 1994, estimated that 'there were an additional 122 million unwanted pregnancies by 2000, one-third of them were aborted while an estimated 65,000 women who did not wish to be pregnant died in childbirth and another 844,000 women suffered chronic or permanent injury from their pregnancies.' (Brown, 2004).

Despite all these reassuring facts one needs to be doubly cautious about the fact that 'the threat of continuing population growth in more than hundred developing countries is all too real' and world population cannot be stabilized unless the world moves quickly below replacement level fertility to no more than 1.7 children per woman. This is much more imperative as it will bring world population to a manageable level of food security provided the food production system and the supporting systems remain undisturbed.

G. INTERNATIONAL MIGRATION AND POPULATION CHANGE

"Migration is an increasingly important component of population change" (Daniels *et al.*, 2005: 103), affecting countries with outward or inward flow of

Table 6.6: Remittances from Emigrants as share of GDP (Gross Domestic Product) 2004

(*Percentage*)

Country	*Share of GDP*
Albania	11.7
Bosnia and Herzegovina	22.5
Dominican Republic	13.2
El Salvador	16.2
Haiti	24.8
Honduras	15.5
Jamaica	17.4
Jordan	20.4
Kiribati	11.3
Lebanon	12.4
Lesotho	25.8
Moldova	27.1
Nepal	11.7
Nicaragua	11.9
Philippines	13.5
Samoa	12.4
Serbia and Montenegro	17.2
Tajikistan	12.1
Tonga	31.1

Source: Puri, 2008 (Taken from World Bank, 2006, Global Economic Prospects)

people in large numbers, mainly for economic rather than social reasons, though the latter at certain instances has acted as a great 'push factor'. Migration has no upsetting effect on world population as such, but it disturbs the distribution pattern and influences the economy of both the receiving and the sending country. While in the twentieth century, leading up to the First World War, about 55 million people moved from Europe to the 'New World', relieving population pressure in Europe, it had some effect on the population trends, one of them being reduction in natural increase in the countries from where people moved away, such as the MDRs. "Migration exchanges are now constituting a larger proportion of their rates of overall population growth than used to be the case and are increasingly determining whether their populations grow or decline" (Daniels *et al.*,2005: 103). Besides, the volume and nature of migration is changing. According to the estimates of the United Nations (2002), about 175 million people live outside their own country which is more than double the number in 1975. As regards the nature of movement, currently over 20 million people move from the LDRs to the MDRs every year, which accounts for over seven-tenths of the overall increase of population in the MDRs while the impact has been much less on the source regions, the LDRs, wherefrom the net emigration is removing less than 3 per cent of the natural increase (the Caribbean and Central America are the only regions from where more than one-twentieth of natural increase have been removed).

International migration is expected to remain high in the next few decades to 2050, the MDRs continually gaining 2 million people every year (UN,2003). Table 6.6 brings out the geographical distribution of international migrants based on their remittances.

More interesting in this regard is the current migration pattern as influenced primarily by changing economic scenarios. For example, the pull exerted by more attractive destinations in economic (as well as social) terms is leading to outward migration from urban centres in Bangladesh, Cambodia and Sub-Saharan Africa. Even the larger developing countries with high economic growth, such as China, Brazil and India, "have to meet formidable development challenges of integrating and supporting their own low-income populaces" (Puri, 2008: 9).

Significantly, migration at the international level is encouraged by declining rates of population growth and ageing, with concomitant declining labour supply and increased dependency ratio, in developed countries along with potential surpluses in developing countries, with fast growing young populations generating a flow of technical labour from the latter to meet the labour gap in the former. Currently, the United States with a shortage of 10 million skill-jobs by 2010 provides the example of an ever-widening vacuum to be filled in by skilled workers from developing countries. In addition, there has also been a steady increase in the demand for low-skill, labour-intensive services such as in construction, agriculture and caring services. Most of these skill-shortages are being met by developing countries, and countries like the Philippines, several countries in Africa and South Asia are sending out young people for tapping these opportunities. Philippines provides an interesting example, responding to the growing global demand for nurses by producing more nurses. The country produces "20,000 graduates per year and of 177,000 Philippine nurses, 85 per cent are estimated to be working overseas… (on the other hand) Indian migrants, particularly those in the United States, have made significant contributions abroad to the recent growth of the Indian software industry, not only by transferring knowledge and technology but also by opening up new markets for Indian products and services " (Puri, 2008: 12). Migration from developing countries is further encouraged by the growing demand for skilled, semi-skilled or even non-skilled workers for hard-core non-tradables, that is, services which cannot be outsourced. In response to increased productivity and ageing populations (it is estimated that by 2025 between one-fifth and one-quarter of the population in Eastern Europe will be 65 and older) there will be a larger need for fast food workers, home-health aids (specially for physically handicapped old people), cashiers, and janitors, on top of computer engineers, doctors and trained nurses. Migration also takes place because of wage differentials between sending and receiving countries. At present the United States attracts 55 per cent of the skilled migrants while Europe's share is only 5 per cent. Interestingly, there is also an inverse flow of migration, taking place with the shift of manufacturing operations from developed to developing countries displacing local workers, especially in textiles, electronics and auto manufacturing to exploit the advantages of cheap labour, cheap land and many inducements provided by the developing countries. Similarly, the scope of outward flow of migrants from the developing countries has partly been limited by growing outsourcing of particular services, such as information-technology-enabled services, through providing opportunities for the generation of employment in countries like India, China, the Philippines, Ghana, Bangladesh, Uruguay, Costa Rica and some Caribbean countries (Puri, 2008: 25-26).

The real impact of movements of workers (skilled, semi-skilled; low-skilled labour) assumes greater significance from the fact that according to the United Nations estimates based on 2005 data, some 200 million people live outside their country of birth, an increase from 175 million in 2000, with the global labour force amounting to 2.8 billion people in 2005 of which 2.25 billion were from developing countries. 'Persistent demographic and economic imbalances' remaining in force will aggravate the situation further and according to demographic projections, by 2025, North America, Europe, the Russian Federation, the high-income countries of East Asia, the Pacific, and China will lose 29 million people by 2025 (involved in migration and partly also due to low birth rate,) a figure expected to rise to 244 million by 2050, enhancing demographic imbalance further with great impact on labour-scarcity in developed countries. This is in sharp contrast with increase in labour force in developing countries, rising to about 1.55 billion by 2050 (Puri, 2008: 32-33). At the same time shortage of skilled labour due to migration is felt in many of these developing countries which cannot be removed as long as there are such sharp wage differentials.

Another very important feature of emigration from developing countries are the direct benefits derived by developing countries in the form of the flow of remittances from emigrants as part of their foreign earnings making a stable source of development finance in home country. According to Puri, in 2001, the remittances received by the home countries were "equal to 42 per cent of total Foreign Direct Investment (FDI) inflows to those countries and double that of official development assistance flows" which were at $180 billion in 2005 (Puri, 2008: 44)). According to UN 2007, emigrants' annual remittances to countries of their origin increased over the period 2000-2006 in the following order in $ billions: 2000: $ 85.6 billion; 2001: $ 96.5 billion; 2002: $ 113 billion; 2003: $ 142.1 billion; 2004: $ 160.4 billion; 2005: $ 188.0 billion; 2006: $200.0 billion. As Puri (2008) observes, actual remittances by the emigrants to their home country 'could be twice [as much] if the unrecorded flows could be captured'. Of the recorded remittances, 70 per cent was sent to Asia and Latin America while Sub-Saharan Africa's share was only 5 per cent of the total remittances. They played a key role in the economy of some countries (Table 6.6)., having a great impact on poverty reduction. Though China and India are the largest recipient countries, for some small countries the remittances received form a substantial part of the GDP. "In actual terms, remittances – which are essentially private flows – have a tremendous value and are more stable than export earnings, impacting on poverty alleviation by flowing directly to often low-income recipients. Those remittances are channelled into household consumption expenditures, education, health – and increasingly, investments in homes, farms and businesses. In addition, remittances produce spillover effects for the economy as a whole in terms of domestic savings, advanced education and higher consumption. With regard to their impact on poverty reduction, a Millennium Development Goal, an analysis of data taken from household surveys indicates that remittances have been associated with declines in poverty-headcount ratio in some countries: 11 percentage points in Uganda, 6 in Bangladesh, and 5 in Ghana. Another study showed that a 10 per cent increase in remittances reduces poverty by 3.5 per cent" (Puri, 2008: 47-48),

The growth of remittances in recent years (having doubled over the past five years between 2000-2005) and constituting a significant portion of the GDP in some low-income countries, efforts have been made to maximize the returns from these remittances with particular reference to development in general along with the well-being of the migrants and their families. For this purpose Brazil, Mexico, Panama and Turkey have already taken measures for productive uses of these remittances while the Indian government have 'floated specialized bonds, raising nearly $ 10 billion to be used for development purposes' (Puri, 2008: 49).

I. INTERNATIONAL TRADE

Inequality in income between rich and poor countries becomes a controlling factor in the long run in the sharing of world trade. The gap in personal income is getting wider between the citizens of the richest and poorest countries. The average American was 38 times richer than the average Tanzanian in 1990. Incomes have been increasing more rapidly in China and (less rapidly) in India than in the high-income countries over the past two decades, reversing a trend towards 'global inequality that started in the 1820s and continued till 1992', but based on 2000-05 growth trends 'it will take India until 2106 to catch up with high-income countries, while for Latin America until 2177 and Africa until 2236 to catch up with high-income countries even if growth stops in the latter today'. (UNDP, 2005).

However, inequality in income in the poor countries may have a very strong effect on poverty reduction if even a small share of the income of the top 20 per cent of the people in a developing country could be transferred to the bottom 20 per cent of the income-group. For example, a transfer of 5 per cent of the income of the richest 20 per cent in Brazil would lift 26 million people out of poverty, cutting it (poverty rate) from 22 to 7 per cent, and in Mexico about 12 million people would escape poverty, reducing the rate from 16 per cent to 4 per cent (UNDP, 2005). Economic growth, therefore, becomes central to all efforts for eradication of poverty and "trade is a driver of global inequality as well as prosperity" (UNDP, 2005: 116).

Entry into the markets for higher-value added manufactured goods brings success in world trade and it is only a small number of countries in East Asia in the developing world that could increase their share in world market in recent years, doubling their share to 18 per cent of the total world manufactures exports since 1980, with China among them doubling its share every five years. China alone supplies one-fifth of the world's clothing, one-third of the world's mobile phones, and "it is the world's largest exporter of domestic appliances, toys and computer electronics" (UNDP, 2005: 116). Between 1980 and 2000, the value in manufacturing world trade grew by more than 5 per cent a year in the developing countries, mostly from East Asia at double the rate in industrial countries but the latter controlled more than 70 per cent of manufacturing value added worldwide. In contrast, much of the rest of the developing world, excluding Mexico, has little presence in world manufacturing trade.

Undeniably, the key to success in global markets depends mostly on the development of industrial capabilities, without which it is difficult to compete with countries industrially advanced in making wares for the world market. In addition, participation in world trade can increase inequality in which small producers quite often take a part from which only big producers make gains as demonstrated by countries like Kenya, Zambia and Brazil. For example, "the increased exports of high value-added fruits and vegetables from countries like Kenya and Zambia have been concentrated in large, capital-intensive farms with weak links to the rest of the economy. Similarly, in Brazil, the world's fourth largest agricultural exporter, large commercial farms and agribusiness firms dominate the $ 20 billion export market; just four or fewer firms account for more than 40 per cent of exports of soy, orange juice, poultry and beef. The other face of Brazilian agriculture is scarred by mass poverty. More than 10 million people in rural areas live below the poverty line, most of them small holder farmers or landless labourers" (UNDP, 2005: 123)

This aside, while international markets create opportunities, they also involve risks so that many countries depending heavily on trade face high market risks generally linked to price fluctuations and policy change in importing countries, such as problems faced in exporting bananas and sugar to the European Union by some countries. Even exporters of some manufactured goods also face a similar fate from competition with other countries, as in the case of Bangladesh and Nepal, where millions of jobs in the garment industry are threatened with destruction by competition from China.

In such situations women suffer disproportionately from weak labour rights in many developing countries where they are not protected by labour laws and have no access to social security. This is the case with women workers in Chile, and in 2003 in at least 16 countries, including Bangladesh and Malaysia, with 'weak labour rights and discrimination' the women workers were not in a position to 'negotiate reasonable wages and conditions'.

On top of all this, sadly the system of tariffs in the international market is very unfair, being "lower a country's income, the higher the tariffs against their exports. Although industrial countries apply very low average tariffs in their trade with each other, they reserve some of their highest import barriers for the world's poorest countries. On average, low-income developing countries exporting to high-income countries face tariffs three to four times higher than the barriers applied in trade between high-income countries. ...The effective US import duty for countries like Viet Nam and Bangladesh is some 10 times higher than for most countries in the European Union... (As a result) the tariff structure prevents developing countries from adding value to their exports" (UNDP, 2005: 126-127). The escalation of tariffs is quite intentional, intended to compel the agricultural producers in the developing countries to sell their products unprocessed, thus transferring the value of the product to processors and retailers in rich countries. For example, while countries like Cote d'Ivoire and Ghana are compelled to sell unprocessed cocoa beans because of high tariff barriers, Germany becomes the largest exporter of processed cocoa and European companies swallow the bulk of the final value of Africa's cocoa production.

On the other hand, the developing countries impose high tariffs on trade between themselves, which are sometimes higher than those imposed by industrial countries, for example, more than 20 per cent on exports from low-, and middle-income countries to South Asia. This should be discouraged on the basis of talks at the highest government level of affected countries so that lower rates determined through such talks may speed up economic cooperation as well as economic development of these countries.

Global partnership in international trade forms a very important vehicle for fulfilling the interests of developing countries with a foothold in the competitive market through exports and imports. It is disappointing to note that there has been an acute lack of progress in trade negotiations under the auspices of the World Trade Organization's (WTO) Doha-round of talks in 2003, aimed at giving the developing countries the rightful access to the international markets by withdrawing restrictions put and exercised so far by rich nations. The WTO 'promises that international trade can play a major role in economic development and the alleviation of poverty'. "The September 1999 Annual Meetings (of the World Bank and IMF) resulted in a clear mandate for the IMF to integrate the objectives of poverty reduction and growth more fully into operations... Oxfam have long argued the trade regime has been insufficiently focused on the conditions of the poor" (Kapstein, 2006: 177). It is observed by liberal internationalists that increasing foreign aid is not enough but building up their capacity to join the world economy with a focus on helping poor countries to escape poverty is more urgent. Kapstein (2006) is very categorical about the absence of some degree of honesty among the rich nations where poor nations are concerned. In his words, "The liberal internationalists would also emphasize that granting foreign aid to developing countries without opening industrial world markets to their exports is like giving a child a scholarship to Harvard or Oxford, but then denying the student a job upon graduation because of his or her race or gender. It follows that the emphasis on 'pro-poor' aid in the presence of industrial world protectionism provides a shining example of great-power hypocrisy. It also severely undermines aid's effectiveness, and in so doing reduces domestic political support for that particular use of tax revenues. Reform of the trade regime in a way that promotes market access, therefore, would be a priority of the liberal internationalist, and pointing out the shameful nature of industrial-world protectionism would motivate many a jeremiad" (Kapstein, 2006: 183). It is clearly suggested by Kapstein that trade negotiations under the patronage of the WTO should not leave developing countries, particularly the small ones, 'on the margins with little influence over the proceedings'. "Negotiations should be modified to incorporate developing world concern while still holding them responsible for meeting certain trade liberalization measures" (Kapstein, 2006: 187).

Besides, for developing the capacity of the developing countries to supply export markets, they are in urgent need of external assistance in building additional infrastructure. There are, however, some encouraging signs indicating a slight reduction in average tariffs imposed by developed countries on exports from developing countries over the period between 2000 and 2005 which is likely to boost up trade and reduce trade imbalances to some degree (Table 6.7), provided the trend continues to bring further reductions in tariff (which depends more on political compulsions of the countries involved than their economy).

Table 6.7: Tariffs on Imports from Developing Countries 2000-2006

(Percentage)

Nature of Imports	*2000*	*2001*	*2002*	*2003*	*2004*	*2005*	*2006*
Agricultural Goods	9.4	9.3	9.5	9.4	9.2	8.9	—
Textiles	6.6	6.6	6	5.8	5.2	5.3	—
Clothing	10.8	11.3	10.7	10.4	9.2	8.9	—

Source: United Nations 2007 (Taken from World Bank and other sources)

Interestingly, the value of the world's total GDP has increased by a factor of approximately nine since 1950 and during the same period world trade has increased more than 18 times. According to the recent trend, trade in services has grown rapidly, now accounting for over 23 per cent of the world exports. While trade is always considered beneficial to economic development of a country involved in it, the export side of trade is more important as it helps 'to protect existing jobs as well as to create new ones' while imports have 'the potential to cause job displacement'. The TNCs (Transnational Corporations) taking control of manufacturing and service industries have an increasing tendency to seek new locations for their industries based on lower wages and less restrictive regulations offered by host countries. This has a significant impact on employment generation helping the economy of the countries providing location for these new industries, and this happened mostly in the developing countries which have a large pool of unemployed labour available at low wages. However, quite often factors like "lower productivity, higher training costs and less efficient transportation and telecommunications infrastructure outweigh the wage cost advantages of the developing countries". This is not to say that footloose production will not be relocated. "A good example is the recent emergence of the Indian subcontinent as a favoured location for outsourcing soft-ware development, call centres, technical support and related work from the US and the UK. Other examples include the relocation by US firms of ticket processing and other high volume, labour-intensive, routine functions to Jamaica and insurance claims processing to Ireland" (Daniels *et al.*, 2005: 302).

The relocation of activities by TNCs besides contributing to changes in geographical distribution of population also helps the national economy in fighting poverty and hunger, though its overall impact cannot be that large to bring significant change to the situation of prevailing deprivation. On the other hand, Foreign Direct Investment (FDI) aimed at developing infrastructure or removal of lacuna in some of the vital services for people living lives of abject poverty play a more important role in equipping people to face challenging situations. FDIs from the world bodies like the World Bank, the International Monetary Fund (IMF) or some of the rich countries mostly programme-bound with specific agenda for improving essential services like water, sanitation, health, education, transport, energy etc., are more welcome as they not only directly address the root-cause of human suffering but also ultimately take an effective role in building human capital, or get involved in an asset-building exercise extremely beneficial as an instrument for enhancing the entitlements of the people, or, in other words, provide direct access of the so long deprived people to unbounded benefits (in various forms, such as employment leading to other desired objects making life worth living, life becoming less vulnerable to any external force, entitled to freedom of expression and participation in social activities on a par with all others, and having no fear of being marginalized or secluded from the main stream of the community).

Contrastingly, FDI made by TNCs do not have an all-competitive character as in most of the cases such investments are primarily aimed at the proliferation of the particular interests generally fulfilled through fresh locations of their activities in a more congenial place to cut the cost of production, and quite often taking advantage of the local markets where activities are relocated with significant consumer potential. However, it is true that FDI, in such cases, can be greatly beneficial to secure access to 'capital, technologies and organization expertise with the net effect (of) modernization of infrastructure, to increase industrial capability and to improve the quality and breadth of much-needed financial, business and professional services'. Undoubtedly, the importance of non-governmental organizations, such as "TNCs, and regional trading blocks like the EU and NAFTA for shaping the economic geography" of the world is significant (Daniels *et al.*, 2005: 307). But what is lacking in their efforts is that the programme launched by them is entirely guided by the interests of building their own business empire and not tailored to the requirements of people of the countries where they set up their enterprises. Making more profits through such relocated, expanded enterprises is the primary motive rather than taking any effective role in eradicating poverty, hunger or improving the economy befitting the local requirements are never within their purview (See Note *F*).

7
Conclusion

Not long ago, the world leaders, meeting at the largest-ever gathering of heads of state, both rich and poor, in 2001, made the Millennium Declaration to do away with poverty once for all, promote human dignity and equality and achieve peace, democracy and environmental sustainability. The Millennium Development Goals thereby targeted to "attack inadequate incomes, widespread hunger, gender inequality, environmental degradation and lack of education, healthcare and clean water" (UNDP, 2003: 1). The primary thrust was towards reducing human deprivation where it prevails in various forms through efforts at a global level. Altogether 8 goals with targets for each were clearly specified in the Declaration, as described here for keeping our aim correctly focused:

Goal 1: Reduction of poverty and hunger by half between 1990 and 2015 of the proportion of people with income below $ 1 a day.

Goal 2: Achievement of Universal Primary Education, ensuring that by 2015, children everywhere both boys and girls will be able to complete primary education.

Goal 3: Promoting Gender Equality and empowering women with elimination of gender disparity in primary and secondary education, preferably by 2015 and in all levels of education not later than 2015.

Goal 4: Reducing Child Mortality by two-thirds between 1990 and 2015, the under-five mortality rate.

Goal 5: Improving Maternal Health and reduce maternal mortality rate by three-quarters between 1990 and 2015.

Goal 6: To combat HIV/AIDS, Malaria and other major diseases keeping two targets in view: (a) to reduce HIV/AIDS by half by 2015 and reverse the spread; (b) Malaria and other major diseases incidences to be halved by 2015 and reverse the trend.

Goal 7: It consisted of three targets: (a) to ensure environmental stability, integration of the principles of sustainable development into country policies and programmes and reverse the loss of environmental resources; (b) Number of people without sustainable access to safe drinking water to be halved by 2015; (c) To achieve significant improvement in the lives of 100 million slum dwellers by 2020.

Goal 8: To develop a global partnership for development consisting of seven targets: (a) To develop 'an open, rule-based, predictable, no-discriminatory trading and financial system, nationally and internationally; (b) To address the special needs of the least developed countries especially in tariff and quota-free access for exports and more generous official development assistance for countries committed to poverty reduction; (c) To address the special needs of landlocked countries and small Island States with programmes of sustainable development; (d) To deal with the debt problems of developing countries in order to make debt sustainable in the long run; (e) To develop and implement strategies for decent and productive work for youth; (f) To provide access to affordable essential drugs in developing countries; and (g) To make available the benefits of new technologies, especially information and communication technologies (UNDP, 2003).

Rescuing people from deprivation is a very complicated issue involving action in several areas primarily aimed at eradication of poverty and hunger. As observed earlier, poverty and hunger go together consistent with lack of entitlements, or, in other words, absence of capability on the part of the people to acquire an adequate amount of food for sustaining body and health. Food may be considered as a basic right of any human being but its supply is always conditional, depending on some kind of exchange rather than to be obtained free or forcibly. People living in a society need to pay the price for food either through labour or through money earned, which means to have access to food one needs to have the income which gives one entitlement to food. Such entitlement is not freely given to anybody but everybody has to earn it with their own efforts, to use as the medium of exchange.

This means incomes are crucial, and in a world where more than 1.2 billion people or one in every five survives on less than $ 1 a day, it obviously means income may be the only hurdle to attain entitlement and hence, one may think, the best way to guarantee food security is raising the level of income or one's earning by way of selling his/her labour. But, while raising income seems to be the only way to create entitlement, which is directly dependent on economic growth of a nation, there are many instances to show that economic growth is not enough to guarantee food security. For example, if economic growth is more of a concentrated nature, benefiting mostly the rich or well-to-do section of the society leaving the poor largely to their fate (as has happened in many developing countries including India during the decolonized period), the purchasing power of the poor does not improve much to increase their access to food and other basics, thus keeping their entitlement mostly unfulfilled. Besides, "poverty has increased even in some countries that have achieved overall economic growth, and over the past two decades income inequality worsened in 33 of 66 developing countries with data" (UNDP, 2003: 5) in the absence of 'policies that strengthen the links between economic growth and poverty reduction'.

We must remember in this context that eradication of poverty and hunger should not be treated as isolated issues, to be taken separately as a problem to be solved in an independent manner. Poverty and huger affecting a section of the society is the result of malfunction of the entire system on which the society depends. Some weakness in the system keeps a section of population deprived of rights to basic supports for sustaining body and health, and such deprivation results from a lacuna in their capabilities. Now, the question is, who is responsible for this lacuna, and how can it be overcome? The simple explanation is that the lacuna refers to purchasing power of the person, which may be equated with labour he/she is capable of selling to obtain desired objects (in this case, say, food and other basics). The removal of such lacuna, indeed a time-taking process, depends on a broad-based economic growth rather than a narrowly-based one, taking care of human development with greater investment in basic requirements like education, health, nutrition, water and sanitation services so that these are available for people long deprived of them, thus contributing greatly to improving the quality of labour people can sell (having education and training giving skill, better health and much less vulnerability to upsetting situations like loss of livelihood, etc.) for personal gain. This leads finally to an increase in human capital, which will act as a vastly enhanced asset for national gain in all fields of activities accompanied by a huge cut on expenditure for keeping people alive. Undeniably, inactive, unemployed/unhealthy people are a drag on any nation where the greatest resource, the human capital, goes unutilized apart from being a drain on the society in many ways. Once this unwanted drain is stemmed it turns into a national asset to enhance national development.

1. LAND IN LIMITED SUPPLY

However, increasing entitlements through improvement in human capital is not sufficient to make food available to eradicate hunger for all, which also depends on factors such as food production, and its availability, dependent on transport, storage, communication and a host of factors ultimately controlling food distribution. Among the most important factors responsible for food production is land, perhaps the most valuable of natural resources (capital) which is far too limited in supply (not renewable while reducible through degradation). Demand for land for food production is irreversible and growing need for cereals in the face of a growing population (until population is stabilized) will make a greater demand on land in future.

According to Pfaunder (1902), the total cropland in the world amounted to 2,174 million hectares at his time and Meadows *et al.* (1972) put the amount of arable land, potentially suitable for agriculture on earth at 3.2 billion hectares, approximately half of which was under cultivation while the remaining half "will require immense capital inputs to reach, clear, irrigate, or fertilize before it is ready to produce food" (Meadows *et al.*, 1972: 58). Land available per person at that time was 0.4 hectares, sufficient for meeting requirement of food crops for a person (though to feed world population at US standards, 0.9 hectares per person would be needed). The supply of land remaining static and population growing unchecked reduced per-capita availability of land later, and as estimated by Brown (2000), it came down to an average of 200 to 600 square metres in East Asia, going below the threshold of 600-700 square metres per person, besides giving rise to landlessness. This has already happened in Bangladesh where landlessness among rural households rose from 35 per cent in 1960 to 53 per cent in the early 1990s despite the success achieved by Bangladesh in slowing population expansion, with the population growth rate declining from 2.8 per cent in the late 1970s to 1.5 per cent in the early 1990s (Brown, 2000: 64). The fixed base of land, as it appears, could not take the pressure of further growth of population however small it was, with land suitable for cropping being lost to urban-industrial expansion eating into the existing resource, the amount of land available for cultivation is getting further dwindled.

However, expansion of agricultural land, though limited by land degradation due mainly to human-induced causes, discussed earlier, is not ruled out even in the semi-desert regions of Sub-Saharan Africa where 'much of the increased food production will come from area expansion and shorter fallow' (Dyson, 1996). According to the same source, the resource-poor farmers of Latin America and South-East Asia will 'certainly continue to expand into tropical rain forests', though because of the cycle of erosion setting in following few years of cropping and deteriorating yields, they will move on to clear new areas. Proper rehabilitation or restoration of degraded land holds considerable scope for expansion of agricultural land On the other hand, the presence of considerable supplies of unexploited fertile land, even in some of the world's most densely settled areas, as mentioned earlier, suggests that there is little real shortage of land for agriculture, when significant amounts of croplands remaining unused in North America and Europe, and in China alone about 5 per cent of its total cropland remains idle (Dyson, 1996).

2. IS FOOD ADEQUATE FOR FEEDING WORLD POPULATION?

The average annual production of cereals in the world between 1989 and 1991 was a little over 1,923 million metric tonnes (Dyson, 1996). According to FAO (2006) following few years of stagnation, world cereal production increased sharply in 2004-05, reaching 2,065 million tonnes, a 9 per cent increase over the previous year. Interestingly, global utilization of cereals continued its upward trend, but has not exceeded production. However, production in 2005-06 indicated a decline primarily due to unfavourable weather conditions in some developed countries but there had been a significant increase of 4.4 per cent from the previous year in low-income food-deficit countries. Excluding China and India, production in the remaining low-income food-deficit countries increased by 8 per cent which means the position of cereal production was good in almost all sub-regions of the world except in some countries, such as in South Africa, Morocco and Somalia affected by drought (FAO, 2006: 96).

But production of cereals alone does not explain the whole food situation relative to the demand for food and its movement (export and import). It is guided by food price as well as the availability/scarcity of food other than cereals acting as essential substitutes (supplements) among people in all countries at all times. Examining first of all the prices of food one may come to a conclusion about food situation. The prices of cereals increased by 21 per cent between 2000 and 2006 and continued rising in the first half of 2006. The prospect for a lower wheat production, a limited stock and expectation of a greater demand caused a rise in prices. Lower stocks and prospects for reduced production have also affected coarse grain markets. Similarly, the 2006 prospects for rice production pointed to a modest growth, leading to rising costs of production and a fall in profitability. However, expectations of a limited production increase was to cause a decline in per capita availability of rice in 2006-07. On the other hand, growing trade restrictions in 2006 (and in the following years as well) to limit meat output, trade and prices as safeguards against animal diseases including avian influenza happening on a large scale across the world, had the most dampening effect on this industry as one of the main sources of food. Further, sugar prices reached their highest level in 25 years in 2006, due to higher energy prices and a continuing supply deficit for the third consecutive year. One of the major steps towards eradicating poverty is encouraging farmers in developing countries to produce more without any fear from unethical or unfair competition from farmers in the industrial countries, quite often protected by their governments with subsidies for keeping down the price of their production in the international markets. The industrial countries spending $ 279 billion a year as subsidies should be stopped to give farmers of the developing counties a fair deal to sell their products as export items and thus strengthen their economy.

Undeniably, the prices of agricultural commodities have been rising sharply in 2006 and 2007, continuing even in early 2008, and factors responsible for that range from supply as well as demand-side making a complicated picture, and it is worthwhile to note that nominal food prices in the world doubled between 2002 and 2008, which were 64 per cent higher by mid-2008 than their mid-2002 levels (the only comparison to this is for the period in the early 1970s in the wake of the first international oil crisis). However, there are enough indications of food prices stabilizing in 2008-09 and "subsequently fall though will remain above their pre-2004 trend level for the foreseeable future" (FAO, 2008: 9). (*See* Note *A*) (Unfortunately, the prediction does not appear to be dependable as world wheat prices, as reported in the press – *The Guardian*, the U.K., 25 September 2010 – "climbed steadily until the beginning of August 2010").

Besides raising crop yields with various measures like greater efficiency in water use, switching over to more useful practices like crop rotation, crop combination, use of resistant varieties of seeds, increasing inputs of synthetic fertilizers where needed most and trying out new varieties of cultivars (such as triticale and pearl millet), greater emphasis should be laid on more efficient production of protein available from sources like meat, fish, poultry and milk. Among them meat and poultry still take centre-stage in meeting the world's growing appetite for protein. Because of this world's meat production increased from 47 million tonnes in 1950 to 260 million tonnes in 2005, more than a five-fold increase, 'roughly double that of population, raising per capita intake worldwide from 17 to 38 kilograms. This has prompted the shift to grain-based production of animal protein, making great demand on grain as animal feed (of the grain harvest of 1.85 billion tonnes in 1998, 36 per cent, or about 700 million tonnes were going to be fed to livestock and poultry), taking roughly 7 kilograms of grain to produce a 1- kilogram gain in live weight, 4 kilograms for pork per kilogram of weight gain, over 2 for poultry and less than 2 for herbivorous species of farmed fish like carp, tilapia and catfish (Brown, 2006:171). For this purpose, soybean plays a big role, 144 million tonnes of it after oil is extracted is used as soybean meal to feed cattle, pigs, chicken and fish, and the world's 'three largest meat producers – China, the United States and Brazil – now rely heavily on soybean meal as protein supplement as stall-feed. This explains why the production of soybeans increased 13 times between 1950 and 2005, jumping from 16 million tonnes to 220 million tonnes.

However, global beef production has not increased much in recent years, growing less than 1 per cent a year between 1990 and 2005, in comparison to pork production which grew by 2.5 per cent annually, and poultry by nearly 5 per cent. The production of poultry increased from 41 million tonnes in 1990 to 80 million tonnes in 2005, which meant it outstripped beef, taking second place next to pork. The oceanic fish catch, on the other hand, has not increased much since 1990, having increased much faster than population growth between 1950 until 1988, dropping after that by 4 per cent between 1988 and 1997. According to marine biologists, an annual catch of about 97 million tonnes may be taken as the threshold production and going beyond would make the catch unsustainable (Brown, 2000).

Of the 17 major ocean fisheries, 9 were in a state of decline largely due to over -fishing and pollution. Besides this threat to sea-food supply, the spawning grounds and nurseries of many aquatic creatures are fast disappearing as 'coastal wetlands, mangrove forests and coral reefs are destroyed', the damage of the coral reef as breeding ground for fish in tropical and sub-tropical waters expanding worldwide from 10 to 27 per cent between 1992 and 2000. Further, "the damming of rivers is depriving many species of their spawning ground. Other rivers are drained dry, with the same effect. Still others are simply too polluted for fish to survive". It is highly important to note that "some 90 per cent of oceanic fish rely on coastal wetlands, mangrove swamps and rivers as spawning areas. Well over half the original area of mangrove forests ...has been lost. The disappearance of wetland in industrial countries is even greater. In Italy, whose coastal wetlands are the nurseries for many Mediterranean fisheries, the loss is a staggering 90 per cent" (Brown, 2004: 52-53). Alarmingly, inland fisheries too are suffering very badly with serious impact of environmental mismanagement in the form of water diversion, acidification (acid rain) and pollution. The most glaring examples are provided by the Aral Sea (as mentioned earlier) the shrinking of which has made it 'biologically dead' for fishing because of exceptionally high salt content which yielded 60,000 tonnes of fish per year as recently as 1960. Similarly commercial fishing has dropped by 97 per cent in the Azov Sea in Russia over the last quarter century because it has dried up increasing the 'levels of salt, petroleum wastes, heavy metal pollution, and radioactive materials'. Inland waters in the United States and Canada provide two other important examples of environmental damage with severe impact on the fishing industry. In Canada alone there are 14,000 dead lakes rendered dead for fishing for pollution making fish unsafe for human consumption. The mercury level in fish largely coming from the smokestacks of coal-fired power plants in the United States make fish similarly unsafe for eating from about 50,000 fresh water lakes (Brown, 2004). According to the same source, fisheries off the coast of Goa in India, are facing a similar fate of collapse unless over-fishing is stopped forthwith. Brown (2004) warns about the future that lies ahead for humanity in the great loss of ocean fish as one of the most important sources of food and livelihood in various forms for a long time. "If the oceans cannot sustain a catch of more than 93 million tonnes (which it cannot with fast decline of fish in major oceanic fisheries) and if world population continues to grow as projected, the oceanic fish catch per person – which has already declined by 9 per cent since it peaked in 1988 – is likely to drop to 10 kilograms per person in 2050 (having been 17 kilograms in 1988). The generation that came of age during World War Second saw the fish catch per person double during their life times. Their grandchildren, the children of today, may witness a one-third reduction" (Brown, 2004: 55).

In comparison, against the background of dwindling supply of oceanic fish, aqua cultural output is growing fast, increasing by more than 10 per cent a year, the output expanding from 13 million tonnes of fish in 1990 to 42 million tonnes in 2003, and is likely to overtake the production of beef by 2010 (Brown, 2004). As discussed earlier, China is taking a leading place in the world's aqua-culture, where it has been integrated with agriculture so that farmers now may use agricultural wastes (pig or duck manure) for fertilizing ponds stimulating the growth of plankton as fish feed. 'Fish polyculture boosting pond productivity over that of monocultures by half, is widely practiced both in China and India'. The herbivorous species, such as carp, dominates aquaculture in China and India, while it is catfish in the United States, and tilapia in several countries, and shellfish, consisting mostly of oysters, clams and mussels which are produced in the coastal regions of some other countries taking aquaculture seriously including Thailand, India and Viet Nam. Viet Nam took up a plan to develop 700,000 hectares of land in the Mekong Delta for developing aquaculture, producing 1.7 million tonnes of fish and shrimp by 2005 which is going to increase further in the future. (Brown, 2006). However, aquacultural operations involving farming of salmon and shrimp, accounting for 3.6 million tonnes of output, comprising less than 9 per cent of the total global farmed fish, growing fast, are considered 'inefficient or disruptive', as salmon is fed other fish as fish meal made mostly from low-value fish caught for this purpose, while shrimp farmed mostly in coastal mangroves leads to their destruction upsetting ecosystems (Brown, 2006: 173).

Two very promising sources of animal protein gaining more importance than ever are milk and eggs, both of which are taking an important place in competition with other protein sources, such as meat and poultry, the production of the latter growing nearly three times as fast, and 'growth in the demand for fish is also outpacing that of population'.

Smil (1994) observed that with an average intake of 2200 kilo calories per head, the current food production is adequate to feed the world population even if the health status of all poor people in the world increases. According to him, the existing problem of hunger has no relevance to present food production or any deficit in it but is, instead, a function of global entitlements to food, a result of utter inefficiency in the function of national governments in the achievement of people's entitlements. As aptly said, "Good governance – the rule of law, transparency, the elimination of corruption, sound public administration, and respect and protection for human rights – supports efforts to achieve food security for all " (International Food Policy Research Institute, 2002: 4).

3. UNDERNOURISHED OR HUNGRY PEOPLE ACROSS THE WORLD

The most pertinent question is why, then, so many people mostly concentrated in the developing countries, have no easy access to food considered to be adequate to feed the entire world population ? How could it happen when "since the early 1970s food production in the developing countries has tripled," more than keeping up with population growth ? In addition, the real prices of the main cereal crops have dropped 76 per cent. Between 1980 and 1995, per capita food production increased 27 per cent in Asia and 12 per cent in Latin America. But in Sub-Saharan Africa (till now the greatest victim of food insecurity) it fell by 8 per cent. Although hunger is most prevalent in South Asia, it is declining – while in Africa about one-third of the population is undernourished, and the number is increasing. Interestingly, if all the food produced worldwide were distributed equally, every person would be able to consume 2,760 calories a day (hunger is defined as consuming fewer than 1,960 calories a day) " (UNDP, 2003: 87)

According to UNDP 2003 Report, every day 799 million people in developing countries, representing about 18 per cent of the world population, go hungry. In South Asia 1 person in 4, and in Sub-Saharan Africa 1 in 3 remain hungry. India was the home of the largest number of hungry people in the world, containing 233 million, in Sub-Saharan Africa they numbered 183 million, in China 119 million, the rest of East Asia and the Pacific had 74 million, Latin America had 55 million and the Arab states had 32 million. However, the number of hungry people gradually started declining in the developing countries, coming down from 21 to 18 per cent between 1990-92 and 1998-00. The largest reduction took place in China, where 80 million people could escape hunger, in contrast to its increase in 25 developing countries, while worldwide the number fell by 20 million between 1991 and 1999 though it was masked by population increase in most of the countries.

Nevertheless, the overall situation since then does not show much improvement and according to FAO estimates, 'the number of chronically hungry people' in the world increased by 75 million by 2007 over and above FAO's estimates of 848 million undernourished or hungry people in 2003-05, bringing it to 923 million in 2007 in the world, much of the increase attributed to rising food prices. The task of bringing down the number of hungry people by half by 2015, the target under the Millennium Development Goals, becomes more difficult to achieve in view of rising food prices and the uncertain global situation beset further with problems resulting from climate change. The impact of rising food prices is very pronounced, putting into reverse the trend in the reduction of hungry people achieved in bringing the world ratio down from 20 per cent in 1990-92 to less than 18 per cent in 1995-97 and just above 16 per cent in 2003-05. High price rises for food are already moving it back towards 17 per cent worldwide. The largest increase in the volume of hungry people as affected by price rise occurred in Asia and the Pacific and in Sub-Saharan Africa in 2007, together sharing 750 million (or 89%) of the world's hungry people in 2003-05. The rising prices have added another hungry 41 million to Asia and the Pacific and 24 million to Sub-Saharan Africa's total hungry people. In their turn, Asia and Africa together contain more than three-fourths of the developing world's low-income food-deficit countries, and in Africa prevalence of hungry people exceeded 35 per cent of the total population in 15 out of 16 countries. Rising food prices have been relatively less serious in Latin America and the Caribbean, the Near East and North Africa (FAO, 2008: 6-8).

The reasons leading to continuing rising food prices across the world are explained by FAO (2008) as the change in agricultural policies in recent years in some of the world's major cereal producers, such as China, the European Union, India and the United States resulting in 'significantly lower levels of cereal stocks compared with earlier years'. The "lower stock levels contribute to higher price volatility in world markets because of uncertainties about the adequacy of supplies in times of production shortfalls" (FAO, 2008:10). Second, climate change with extreme weather conditions accompanied by droughts and floods happening in 2005-07 severely affected major cereal producing countries, bringing down world cereal production by 3.5 per cent in 2005 and 6.9 per cent in 2006. The lower crop yields in two successive years aggravated the fears of the cereal producing countries about available stocks. This was further bolstered by rapid rises in petroleum prices until mid-2008, more than tripling since 2003, exercising a two-pronged effect, as increases in fertilizer prices escalated costs of production raising the price of food crops nearly three times, while the transport costs doubled in 2006-08.

Pressure on food prices also came from another side. The increasing focus on biofuel made large demands on some food crops as its raw material, such as sugar, maize, cassava, oilseeds and palm oil which raised their prices in the world market. An estimated 100 million tonnes of cereals or 4.7 per cent of global cereal production in 2007-08 went into biofuel.Another important factor in pushing up the price of food is the increasing demand for food with economic growth in some of the developing countries making greater demand for human consumption.

Despite increases in cereal production, by 4.7 per cent in 2007 'and a projected 2.8 per cent in 2008', food prices are likely to remain above those of previous ten years (FAO, 2008), due to increase in purchasing power and 'changing dietary preferences' of hundreds of millions of consumers in the developing world particularly the larger countries with economic growth continuing at about 6 per cent per year. Further, the growing demand for biofuels, to counteract the spiralling prices of oil, with much larger demand on arable land on the one hand, and increasing cost of cereal production because of higher prices of fertilizer and rising transportation costs on the other, are likely to have a paralyzing effect on food production, making global security unsustainable.

In their geographical distribution, 65 per cent of the global number of hungry people in the world are concentrated in only seven countries, India, China, the Democratic Republic of Congo, Bangladesh, Indonesia, Pakistan and Ethiopia, while by virtue of their size China and India together account for 42 per cent of the chronically hungry people in the world. With the number of chronically hungry people rising steeply with soaring food prices, it is going to be very difficult to keep to the target of halving the number by 2015. According to FAO's estimates, hardly one-third of the developing countries have succeeded in reducing the number of chronically hungry at all since 1990-92, and, though 25 countries were on track in 2003-05 to achieve the target, that was before food prices started rising. Food prices going still higher will make this challenge far more difficult than it is now.

Looking at the trends in food crisis, an average of 15 African countries faced food crisis annually during 1993-2000, requiring emergency assistance in the form of aid, food imports or both, but the number of such countries facing food crisis increased to 47 in 2007, 27 of them in Africa and 10 in Asia, and the remaining 10 in other parts of the world. Human-induced crisis, sometimes combining with natural disasters (in the form of drought or prolonged dry spells, floods, cyclones, hurricanes, earthquakes and volcanic eruptions) have aggravated the situation with fall in the production of food crops in these countries, jeopardizing food security. Significantly, this has a tremendous pressure on meeting bills for food imports which, according to FAO's estimate, 82 LIFDC (Low Income Food Deficit Countries) were expected to spend nearly US $ 169 billion in 2008 compared with US $ 121 billion in 2007, a 40 per cent increase and, as observed, it could increase four times by the end of 2008 compared with 2000 (FAO, 2008). Countries facing the extremely worrying prospect of it being beyond their means to feed their own people are shown in Table 7.1. Interestingly, of the 34 countries listed, 17 in food crisis and another 17 at high risk, most of them are in Africa (27 out of 34) though some other countries not included in the table are approaching a similar fate in a short while, such as Bangladesh owing to any natural disaster, like floods (quite a frequent event), since it is already listed as experiencing 'severe localized food insecurity'. Similarly, rising wheat prices in Pakistan may bring it down to the level of food crisis in near future.

Table 7.1: Countries at Risk of Deteriorating Food Security Due to High Food Prices

	In Food Crisis		*At High Risk*
1.	Central African Republic	1.	Cameroon
2.	Democratic Republic of Congo	2.	Comoros
3.	Cote d' Ivoire	3.	Djibouti
4.	Eritrea	4.	Gambia
5.	Ethiopia	5.	Magagascar
6.	Guinea	6.	Mongolia
7.	Guinea-Bissau	7.	Mozambique
8.	Haiti	8.	Nicaragua
9.	Kenya	9.	Niger
10.	Lesotho	10.	Occupied Palestinian Territory
11.	Liberia	11.	Rwanda
12.	Sierra Leone	12.	Senegal
13.	Somalia	13.	Solomon Islands
14.	Swaziland	14.	Togo
15.	Tajikistan	15.	United Republic of Tanzania
16.	Timor-Leste	16.	Yemen
17.	Zimbabwe	17.	Zambia

Source: FAO, 2008.

4. MEANS OF IMPROVING ACHIEVEMENT OF ENTITLEMENTS

As has been pointed out earlier, hunger is not entirely confined to lack of food, as a manifestation of deprivations in life but, instead, it means deficiencies in food entitlements, and at the same time it also includes deprivations in related essential services, such as education, healthcare, safe drinking water and sanitation among some others equally important as basics of life. "Food entitlement differs from food availability in that it indicates what a person can command with income and thus consume, rather than what is available in the market" (UNDP, 2003: 87). Viewed from another angle, 'income' of a person empowers one with entitlement providing access to food (and other requirements of life, obviously the quantity determined by the level of income) which, in other words, is the purchasing power earned with income and that is only possible if one has his/her labour to sell in the market. The quantity of purchase naturally depends on the quality (as well as nature) of labour one has to offer to earn an income, and the latter, as imperative, varies accordingly with demand for the particular type of labour supplied by the person involved. The quality or nature of labour one specializes in has to be earned through education/training in skill which on its own, depends on what one can afford to have as well as on the capability of a child receiving it largely determined by good health at tender age. Both education and health acting as correlated conditions in achieving entitlement, the condition of health is again determined by two very important basic services in life, and they are access to safe drinking water and sanitation, both essential for averting diseases. On the other hand, the huge impact of education (especially of girls and women) and good health in lowering the rates of infant mortality, maternal mortality and fertility rates have far reaching consequences in raising the level of entitlements and thus making deprivation non-existent.

In order to attain the goal of doing away with deprivations, the national governments, particularly in the developing countries, should get the priorities right. Rapid economic growth is essential for attaining greater degree of entitlements which, acting as a means for the removal of deficiencies, is in urgent need of healthy, well-nourished, literate citizens, which are essential preconditions for successful pro-poor economic growth. "Universal access to primary and preventive healthcare and access to clean water and safe sanitation are critical for people's good health and good nutrition. Like good health, education has a tremendous impact on economic growth and on the material well-being of individuals. Educating girls, especially, has beneficial effects on family size, spacing of births, childcare practices, child nutrition, and household income." (IFPRI, 2002: 3).

Basic education and basic healthcare are the 'cornerstones of both human capital development and population stabilization'. Universal education is the only way to narrow the gap between rich and poor apart from contributing to the achievement of personal entitlements. The central role of universal education in human progress recognized by the United Nations has made it a goal to be achieved by 2015, which involves enrolling 115 million children out of a total of 680 million of primary school age in the developing countries not going to school. Enrolment does not mean fulfilment of the requirement since according to experience, just over half of the enrolled children in school finish their course, and this is only one in three in Sub-Saharan Africa. Besides, of the world's 879 million illiterate adults, two-thirds are women. Further, high household costs per pupil, ranging between 10 to 20 per cent of the household budget in many countries, like Bhutan, Burkina Faso and Uganda (UNDP, 2003) discourages primary education among children (while lower costs in Myanmar and Viet Nam contribute to higher enrolments). As long as primary education is not made free, combined with further incentives like mid-day meal, reading materials and school uniform given free of cost, primary education cannot be made universal. Furthermore, the existing gender differences in enrolments and drop-outs make a very strong case for action in many countries, such as in South Asia and Sub-Saharan Africa. In order to eliminate these disparities steps need to be taken including ensuring schools are not far from their homes, lowering household costs on education, 'scheduling lessons flexibly' enabling girls to help with household chores (sharing their mother's responsibilities) and care for siblings, having women teachers, assuring parents about security of their girls, and having separate toilet facilities for girls (where there is coeducation). (UNDP, 2003).

Giving very high significance to women's education is amply justified by the benefits society derives in many ways with far reaching consequences. For example, children of educated mothers are better nourished as mothers' better understanding of nutrition leads to better choice of food to make children healthier, reducing the rate of infant mortality. Women's education leads to smaller families as they are conscious of as well as capable of adopting family planning measures when lower rates of infant mortality act as a security for cutting down pregnancies, itself making impressive reductions in fertility rates leading to stabilizing the population of the countries involved. Girls' education also acts as a great deterrent to early marriage of girls so that the reproductive period is effectively shortened and widening the space between births keeps mothers in much better health with a much longer life expectancy. Ethiopia has pioneered approaching poor families pleading with them to stop girls marrying off at an early age and instead sending them to school, while some countries, Brazil and Bangladesh among them, offer small scholarships for girls where needed, thus encouraging their parents to send girls to school.

The school lunch programme providing at least one good meal a day, especially needed by poor children, introduced in 1946 in the United States, has proved the best among all incentives so far for increasing enrolment and the attention span of children, improving their academic performance and attendance as well as continuity in their studies in schools. As has been mentioned earlier, girls benefit more since 'drawn to school by the lunch, they stay in school longer, marry later, and have fewer children' (Brown, 2006), and thus ultimately contribute to stabilizing population and take a vital role in removing poverty and hunger in the long run. According to Brown (2006), a school lunch programme adopted and implemented in 'the 44 lowest-income countries would cost an estimated $ 6 billion a year beyond what the United Nations is spending in its efforts to reduce hunger'. Further, efforts are also needed to improve nutrition of these children in poorer countries before they go to school for which the WIC (Women, Infants and Children) Programme implemented in the United States about 25 years ago if introduced, may bring enormous changes in their lives in the 44 lowest-income countries, offering nutritious food to needy, pregnant and nursing mothers, and pre-school children.

Health poses another great challenge to achieving entitlements when 30,000 of the world's children die everyday from preventable causes, like dehydration, hunger and diseases, as well as more than 500,000 women dying every year, or one every minute a day, in pregnancy and childbirth. Most distressing, "a pregnant woman is 100 times more likely to die in pregnancy and childbirth in Sub-Saharan Africa than in a high-income OECD country" (UNDP, 2003: 97). The most shameful failure of development is further revealed in the fact provided by the same source that 42 million people in the world are living with HIV/AIDS which has killed the mother or both parents, making 13 million children orphans. Tuberculosis and malaria taking a toll of 2 million and 1 million people respectively a year are two other preventable diseases making havoc in many countries. But provision of bed-nets, affordable antibiotics, trained birth attendants, basic hygiene and health services, and education, all of them not very difficult to organize, could make a great headway in saving lives in millions. In general, the miserly habits of many national governments on health systems (the per capita health expenditure amounted to just $ 88 in low human development countries in 2000 against $ 1,061 per head in high human development countries), particularly in rural areas, with minimum focus on basic health and inefficiency in integrating programmes for specific diseases with general health systems are the main reasons for the spread of killer diseases and their unchecked destructive role. The links between health, education and income can play an effective role in subduing many of these preventable diseases since it is poor people lacking access to water, sanitation and economic ability to pay for the drugs who suffer most without having proper education about HIV prevention and family planning. In this respect, women are at greater risk than men, accounting for about half of adult HIV/AIDS cases but far more among young adults, the prevalence of HIV infections among 15 to 25 year-olds being up to six times higher for women than men in many African countries.

Use of condoms is essential for reducing HIV infections among people involved in risky sex, and especially for the sex workers (in India, for example, the 2 million sex workers having an average of two encounters a day, could be used as a key group to educate about HIV risks). The supply of condoms is considered to be the best available technology for this — it is estimated that when roughly 12 million condoms a year are required in the developing world and Eastern Europe, only 2.5 million are being distributed, leaving a deficit of 9.5 million condoms, the total cost of which would amount to only $ 2 billion (Brown, 2006: 135). Condoms seem to offer the most inexpensive as well as a sure preventive and therefore much better means than very costly treatment of HIV far less easily available. For example, the 4.7 million with symptoms of AIDS in Sub-Saharan Africa in 2005, only 500,000 got the drug treatment (anti-retroviral) while it is widely available in industrial countries

Very impressive health gains have been obtained from oral dehydration therapy introduced by the UNICEF in treating the symptoms of diarrhoeal disease, reducing deaths from 4.6 million in 1980 to 1.5 million in 1999, for example bringing them down by 82 per cent in Egypt between 1982 and 1989. Another very leading cause of premature death is smoking, taking away 4.9 million lives in 2000. Including cancer, there are some 25 diseases caused by tobacco use, and cigarette smoking kills more people each year than air pollutants taken together do, nearly 5 million against 3 million (Brown, 2006). (*See* Note *B*). Fortunately, smoking has started declining, a decline of 52 per cent in the United States between 1976 and 2003, 22 per cent in France since 1984, 20 per cent in Japan since 1991 and 5 per cent in China since 1989. Meanwhile, smoking has been banned nationwide in 2004 in Ireland, at all public places in India, Norway has banned it in bars and restaurants, Scotland has banned it in public buildings, in Bangladesh and Italy it has been banned since 2005 in public places, and in bars and restaurants in New Zealand. Polio is another menace still surviving by mid-2003 in pockets of countries like Nigeria, Niger, Egypt, India, Pakistan and Afghanistan with new infections appearing after that in Saudi Arabia, Yemen, Cote d'Ivoire, Burkina Faso, the Central African Republic, Chad, Mali, Sudan, Indonesia and Somalia.

According to WHO, providing the most basic healthcare services in developing countries, which can even be handled by village level clinics, would require $ 37 billion in 2007 and $ 38 billion in 2015 or an average of $ 33 billion per year which, apart from providing basic services would also cover funding for fighting deadly diseases like AIDS, tuberculosis, malaria, and providing universal childhood vaccinations (Brown, 2006: 136)

5. WATER AND SANITATION AS LIFE SUSTAINING BASICS

Access to safe, drinkable water and adequate improved sanitation are not only most crucial for survival, they are also essential for the environment, food security and sustainable development with a direct impact on eliminating poverty and removing hunger. In 2000, about 1.1 billion people had no access to safe water, while almost 2.4 billion people in the world lacked improved sanitation. In their geographical distribution 65 per cent of people without safe water were in Asia, and 28 per cent were in Africa. According to another estimate, 'about 1.7 billon people, a third of the developing world's people, live in countries facing water stress', which means more than 20 per cent of their renewable water supply is consumed, and the trend continuing, will bring 5.0 billion people by 2025 to this condition of living under water stress (UNDP, 2003). As pointed out earlier, of the two basic water needs, one involves satisfying the 'needs of everyone for drinking, cooking, sanitation, and other essential purposes', legally a 'non-negotiable allocation', while the other comprises water needed to support 'river ecosystem functions' so that biodiversity and valuable ecosystems supporting human life can be conserved or sustained. (Brown, 2006).

Balancing water demand and supply is the most essential step towards keeping an equilibrium between growing human demand for water and the supply in such a way that the water table does not fall, as is often happening now, making more rivers dry and more lakes disappear. The limited supply of water made more scarce and expensive for irrigation will ultimately reduce crop production and the resulting food shortage will further increase hunger.

Practices of water conservation are the only way to check further water loss through wastage and overuse in different areas, along with better management systems, including safeguards against natural disasters. For instance, floods account for over 65 per cent of people affected by natural disasters compared to 20 per cent affected by famine, the former making an average of about 66 million people per year its victims between 1973 and 1997. Of the flood disasters, 44 per cent took place in Asia in 1987 to 1996, claiming 228,000 lives (93 per cent of the world's total number of flood-caused deaths), accounting for a damage of US $ 136 billion to the Asian economy. Massive deforestation, overgrazing and relentless urban sprawl reducing water storage capacity 'amplify' flooding. However, integrated risk-based management gives an opportunity in deriving benefits from floods, such as nutrients carried by flood water and deposited on river basins are important for various aquatic species proving useful for deriving economic benefits.

Droughts, on the other hand, have a greater damaging effect than floods, taking a toll of over 280,000 lives between 1991 and 2000, affecting more people, such as 110 million in Sub-Saharan Africa during 1991-92. By 2025, between 1 and 2.4 billion people are expected to be living in water-scarce areas (United Nations, 2003: 274-275). Durations and extent of droughts vary greatly. "Examples of severe, persistent droughts over large geographical areas include the Sahel, covering 7.3 million km^2 from 1970 to 1988, continental Europe covering 9 million km^2 from 1988 to 1992, and India covering 3 million km^2 from 1965 to 1967. There are other examples of extreme droughts in North America and Australia" (United Nations, 2003: 275).

The key to stabilizing water tables is raising water productivity, application of more water-efficient irrigation practices and technologies aimed at cutting down wastages since according to Postel and Vickers "surface water irrigation efficiency ranges between 25 and 40 per cent in India, Mexico, Pakistan, the Philippines, and Thailand; between 40 and 45 per cent in Malaysia and Morocco; and between 50 and 60 per cent in Israel, Japan and Taiwan" (Brown, 2006:167). Shifting from flood or furrow system irrigation to overhead sprinklers or drip irrigation, considered as the 'gold standard of irrigation' is a definite way of raising irrigation water efficiency which, at the same time, increases yields from a steady supply of water. Postel believes that India has a potential of using drip irrigation profitably for 10 million hectares of her cropland, or nearly one-tenth of the total and she finds similar potential for China to save scarce water. Since, low water productivity is often the result of low water price, raising water price offers a double benefit, such as, it stops wastage and makes water available for other uses. Water productivity can also be increased by shifting to water-efficient crops, for example wheat for rice. Further, as Brown (2006) suggests, moving down the food chain, with a modest reduction of animal protein (meat, milk and eggs using grain as feed) 'could easily cut grain use per person by 100 kilograms' which indirectly means that for 300 million

Americans such a reduction would reduce use of grain by '30 million tonnes and irrigation water use by 30 billion tonnes'.

Such steps including rainwater harvest, particularly in developing countries like India and China, along with revival of age-old techniques of irrigation water storage and recycling urban waste water, are essential requirements for sustaining aquifers and rivers worldwide.

Adequate sanitation is another basic service lacked by nearly 37 per cent of South Asia's population. Not only is it a very delicate issue involving human dignity particularly for women as according to UNDP, 2003 report, 1.4 million people in South Asia defecate in open areas or use unsanitary bucket latrines (even in urban areas of small towns, while the actual number of people defecating in open areas is many times larger than mentioned) which is a highly unhygienic and harmful practice for human health as faeces from open areas washed into running water become a great health risk and to get out of the dangerous prospect of such a situation appropriate technologies matching consumer demands in both price and quality are needed which include 'pour-flush latrines, single-pit latrines, ventilated pit latrines and connections to septic tanks or covered public sewers. In rural areas waste disposal through composting' useful for agricultural production is sometimes appropriate. (UNDP, 2003: 104-05).

6. INTERNATIONAL TRADE FOR ECONOMIC GROWTH

With participation in export trade dealing with higher value-added goods remaining still confined to only countries of East Asia, the developing countries have a very small hold on international trade. The key to success in global markets depends largely on industrial capabilities but this should not remain confined to big producers alone in the developing countries. However, big producers quite often are found to derive benefit from the efforts made by small countries exporting high-value fruits and vegetables (e.g. Kenya and Zambia) where these are concentrated in the hands of large, capital-intensive farms with weak links to the rest of the country and as such all the financial benefit earned goes into the pockets of the big farms while the small farmers derive no benefit out of such exports. Similarly, in Brazil large commercial and agribusiness farms dominate the export trade. Besides, price fluctuations at the international market make industrial operations quite vulnerable, as in Bangladesh and Nepal, the workers losing their jobs at times of such crisis.

Over and above all these, the taxation system is very unfair to developing countries which have to pay much higher import taxes imposed by industrial countries than what is paid between the industrial countries. Imposition of higher taxes on imports from developing countries is quite often intentional, compelling producers in developing countries to sell unprocessed products (for example, cocoa beans from Cote d'Ivoire and Ghana, iron ore from India) to benefit processing in the industrial countries, which earn much more from selling value-added products.

So far not much headway has been made in making conditions of trade at international level easier and fairer for developing countries though some interesting signs are shown in reducing average tariff imposed by developed countries on exports from developing countries indicating a slight thaw in the highly rigid system. This applies to slight reduction in tariffs on the imports of agricultural products and textiles from the developing countries between 2000 and 2005.

Furthermore, the role of the TNCs in relocating their activities to developing countries is not, in general, beneficial to the developing countries as the relocation of their firms are principally guided by the major interest of reducing the cost of production utilizing local cheap labour and some other incentives (lower taxes, cheap land, lower infrastructural costs, etc.), which deprives the host country of benefit from value-added manufacturing taken hold of by the TNCs.

Finally, on weighing factors 'making poverty history', one cannot confuse between the priorities though it is, indeed, true that one cannot ignore any of the related factors enhancig or reducing deprivation. Nevertheless, one taking the responsibility of making a quick decision to overcome this hurdle, the author believes that universal education and healthcare along with making provision for basic services comprising safe drinking water and sanitation take a top priority, as was done in China, since they ultimately provide the necessary human capital for building up a robust economy which entailing unobstructed trade will bring unlimited benefits to the society. May be, this is the only way of salvation, capable of rescuing people from their untold miseries and a desperate state.

This study uses the latest data – mainly from UN agencies – available to the author. That some of this data is 4 to 5 years old viewed from the present (2012) does not invalidate it. The book attempts an historical as well as geographical perspective to what is analysed, But trends will be affected by recent changes

Problems in the developed world (especially the USA) led to a global financial crisis in 2008-9. The interdependence of the global economy meant serious effects on world trade and employment in the developing world. There was an effect on hunger, with 2009 being a year of increased hunger in the world according to the UN., Governments in the rich world bailed out their banks creating a problem of large scale public sector debt. The remedy of cutting public spending further dampened economic growth, notably in Europe, where 'eurozone' prolonged the financial crisis. Some developing countries, for example, China, India and Brazil, continued with fast economic growth, helping to support the world economy. But the dip in demand from richer countries in Europe and North America was affecting trade and growth throughout the world.

It will be sometime before the effect of the financial crisis on world poverty can be traced.

APPENDICES

APPENDIX A
The Top 10 Seeds Corporations of the World

Company	Seed Sales 2007 ($ billions)	% Global Proprietary (patented) seed market
1. Monsanto (US)	4,964	23
2. Du Pont (US)	3,300	15
3. Syngenta (Switzerland)	2,018	9
4. Group Limagrain (France)	1,226	6
5. Land O'Lakes (US)	917	4
6. KWS AG (Germany)	702	3
7. Bayer Crop Science (Germany)	524	2
8. Sakata (Japan)	396	2
9. DLF-Trifolium (Denmark)	391	2
10. Takii (Japan)	347	2
Total	14,785	67

Source: New Internationalist, September Issue 435, 2010.

APPENDIX B

Global Food Situation and Its Future

About 925 million people still remaining hungry in the world without access to 'sufficient major macronutrients' consisting of carbohydrates, fats while another billion are thought to suffer from 'hidden hunger' missing vitamins and minerals in contrast to a situation where a "billion people are substantially over-consuming", the mankind is in a great need of a global policy for a food production system to sustain the growing requirement of a population increasing from 7 billion today to 8 billion by 2030, which is likely to reach 9 billion by 2050, involving a greater demand for a "more varied, high-quality diet" in tune with higher prosperity of people as a result of economic growth in many countries.

Food production adequate to feed the growing world population has become a highly daunting challenge in the face of many of the prevailing systems of agricultural practices degrading the environment in the form of soil erosion, loss of soil fertility, salinization and other forms of land degradation processes further exacerbated by fast depletion of water resources through over as well as wasteful use, heavy reliance on 'fossil-fuel derived energy for the production of fertlizers and pesticides'.

This is clearly focused in the project report entitled, 'The Future of Food and Farming: Challenge and Choices for Global Sustainability', prepared by a team of about 400 experts on behalf of the Department of Innovation and Skills, UK, taking a strategic view of the challenges likely to be faced in this connection over the next four decades, till about 2050. The Most important issues determining any action towards developing a sustainable food production system to meet global requirement during this period taken into account are (i) a growing world population factoring in not only with its size but also with its qualifying characteristics, such as "GDP growth, educational attainment, access to contraception and gender equality", including the extent of female education as the most important factor. Besides, such "transformational changes, particularly in low-and middle-income countries as rising numbers of people" moving "from rural areas to cities that will need to be serviced with food, water and energy", (ii) The second important issue to be taken care of is the "changes in the size and nature of per capita demand" involving dietary changes" which "are "very significant for the future food system because, per calorie some food items (such as grain-fed meat) require considerably more resources to produce than others" though any prediction on the pattern of dietary changes, as governed by "cultural, social and religious influences", becomes very difficult. It is already predicted that the per capita consumption of meat will increase from "32 to 52 kg by the middle of the century", though with large variations and what is of major concern in this connection is that "grain-fed meat would have serious implications for competition for land, water and other inputs" with great impact on sustainability of food production.

Simultaneously, demand for fish is likely to increase substantially, "particularly in parts of East and South Asia", requiring further expansion of aquaculture with "significant consequences for the management of aquatic habitats and for the supply of feed resources".

According to this report, the 'future governance of the food system at both national and international levels' depends on a number of major decisions having "significant impact on the workings of the food system". Such as –

(a) 'globalization of markets';
(b) "emergence and continued growth of new food super powers", like Brazil, China, India and Russia, the latter "with a large supply of underutilized agricultural land;
(c) the role of large transnational companies "in agrobusiness, fisheries, food processing, distribution and retail sectors";
(d) how "the production subsidies, trade restrictions and other market interventions" operate in future, and
(e) to what extent the governments are going to "act collectively or individually to face future challenges, particularly in shared resources, trade and volaltility in agricultural markets";
(f) it is not clear how far adequate is the "current international institutional architecture (in its response) to future threats to global food system";
(g) at the same time "the control of increasing areas of land for food production (such as in Africa) will be influenced by both past and future land-purchase and leasing agreements".

The 'Report' further considers the situation in the perspective of climate change which is likely to interact with the global food production system in a very important way, such as, the growing demand for food has to meet with "rising global temperatures and changing patterns of precipitation "which will "affect crop growth and livestock performance, the availability of water, fisheries and aquaculture yields and the functioning of ecosystem services". Besides, "crop production will also be indirectly affected by changes in sea level and river flows".

It is observed that a number of "critical resources on which food production relies will come under more pressure in future which, "conversely" will be the result of growth in the food system. The first of them is land supply of it having increased very little in recent decades. As for example, land under agriculture standing at approximately 4,600 million hectares has increased by only 8 per cent between 1967 and 2007 when global crop yields grew by 115 per cent. The prospect of bringing more land under cultivation dims further in view of growing competition between agriculture and other uses of land lost to urbanization, desertification, salinization and sea level rise. Besides, although about "16 per cent of total land area including cropland, rangeland and forests is improving, the International Soil Reference and Information Centre has estimated (2009) that of the 11.5 billion hectares of vegetated land on earth, about 24 per cent has undergone human induced soil degradation, in particular through erosion". The expanding population with its greater demand on land needed for non-agricultural uses along with increase in biofuel production are likely to reduce land for crop production in future. In the backdrop of a situation like this conversion of rainforest to agricultural use "should be avoided as it will increase greenhouse gas emissions very significantly and accelerate the loss of biodiversity".

Production of crops worldwide depends largely on the use of energy, use of which inclusive of all purposes is "projected to increase by 45 per cent between 2006 and 2030 and could double between now and 2050. Increase in oil prices making five-fold increase in fertilizer price between 2005 and 2008" had a great effect on agriculture and experience had also been very depressing, affecting "financial viability of fishing (particularly capture fisheries").

Use of water has to be factored in as a deciding factor for food production and agriculture already consuming 70 per cent of total global 'blue water' with prospect increasing its demand by over 30 per cent by 2030, while the total global demand may rise by 35 to 60 per cent between 2000 and 2025 which may double by 2050 "owing to pressure from industry, domestic use and the need to maintain environmental flows". The overall situation becomes more critical when in some regions – such as — "in Punjab, Egypt, Libya and Australia" – the "major non-renewable fossil aquifers are increasingly depleted and cannot be replenished".

The recommendations made in the 'Report' on "balancing future demand and supply sustainably" are as follows:

The application of existing knowledge and technology, according to the body of the experts involved in this project may "increase average yields two- to threefold in many parts of Africa, and twofold in the Russian Federation. Similarly, global productivity in aquaculture could, with limited changes to inputs, be raised by around 40 per cent". However, the role of both "positive and negative externalities associated with different forms of food production" is needed to be considered as per situational demands

In pursuance of this policy, the "project provides support for four classes of intervention aimed at raising agricultural productivity" which relates "mostly to middle-and low-income countries" as they have greater prospect of gaining from such interventions for "increasing yields sustainably".

1. The first of such interventions includes "the revitalisation of extension services to increase the skills and knowledge base of food producers in both low-income and high-income countries" with "particular consideration" of the role of women "in view of their often significant role as food producers in low- income countries".
2. The second one is "improving the functioning of markets and providing market access, particularly in low-income countries" since in many of them the "food markets function poorly or only very locally" which demand facilitating entrepreneurship in food production sector for increasing "food production, household revenue, livelihood diversification (aquaculture, specialist crops" etc strengthening rural economies) offering greater access to capital.
3. The third most important factor is "strengthening the rights" of farmers "to land and natural resources, such as water, fisheries and forests (which) should be a high priority", effectiveness of such measures for development as provided by China and Ethiopia.
4. "Physical infrastructure", consisting of roads, ports, irrigation facilities, storage facilities, information and communication technology systems must be improved in middle-and low-income countries" providing "access to markets and investments in rural economies".

On the basis of the evidences reviewed by the Project Investigators, the following conclusions on research and development have been drawn:

(a) Taking the case of China as an example demonstrating the effectiveness of agricultural research in raising agricultural productivity, greater emphasis should be laid on "research on agriculture, fisheries and food system in most countries".

(b) "Recent scientific and technological advances offer significant new opportunities to address major environmental challenges such as climate change, water scarcity and soil degradation".

(c) "Research on climate change adaptation and mitigation in the food system should be a priority" so that agriculture and food production can better "adapt to a changing world with a higher likelihood of extreme weather events".

(d) "Investment in food production research needs to focus on raising yields in conjunction with improving sustainability and maintaining ecosystem services".

(e) "The magnitude of the challenges are so large that no single research avenue will address "the situation which requires a pluralistic research" approach.

(f) "New ways are required to incentivise research and development that meets the needs of low-income countries" where the returns may be low on investments to start with.

(g) "The contribution of funders to research from the public, private and third sector needs better coordination.

The priorities suggested for meeting the challenge in crop production are–

(i) "Development of new varieties or breeds of crops, livestock and aquatic organisms" are required to take advantage of "recent advances in biosciences".
(ii) "The preservation of multiple varieties, land races, rare breeds and closely related wild relatives of domesticated species" seems to be "very important in maintaining a genetic bank of variation that can be used in the selection of novel traits".
(iii) "Advances in nutrition and related sciences" offering "substantial prospects for improving the efficiency and sustainability of animal production (both livestock and aquaculture)" needs to be given top priority.
(iv) "Scientific and technological advances in soil science and related fields" offering "the best prospect for a better understanding of constraints to crop production and better management of soils to preserve their ecosystem functions, improve and stabilise output, reduce pollutant run-off and greenhouse gas emissions" need to be given serious consideration.
(v) "Targeted research in modern crop and animal science, agro-ecology, agricultural engineering and aquaculture management" have to be taken very seriously.

Among other more "revolutionary advances" considered by the 'Project' for attaining the desired object are "the development of perennial grain crops, the introduction of nitrogen fixation into non-legume crops, and re-engineering the photosynthetic pathways of different plants".

The 'Project' further looks into reducing food waste which amounts to as much as "30 per cent of all food grown worldwide" while some other estimates put it at "as high as 50 per cent ". Here, food waste is defined "as edible material intended for human consumption that is discarded, lost, degraded or consumed by pests as food travels from harvest to consumer"... and the definition also "includes food that is fit for human consumption but intentionally used as animal feed". In this connection it is pointed out in the 'Report' that if the total waste of 25 per cent could be halved, that "could reduce the food required by 2050 by an amount approximately equal to 25 per cent of today's production". As suggested in the 'Report', reduction of food waste depends on a number of conditions, such as— (a) improvement of storage and transport infrastructure using available technology, like "simple sealed storage drums for grain farmers in Afghanistan and elsewhere" provides one of the examples giving benefit to producers, (b) "Investment in new, appropriate technology to reduce post-harvest waste", such as "use of modern scientific advances to produce crops that are less susceptible to pest and spoilage, or better fish-smoking kilns that reduce losses and demand less fuel". Further measures in this respect includes "infrastructure, financial and market reforms" among which use of mobile phones is likely to "help improve market information and allow producers to make better decisions about timely supply to markets, avoiding or at least reducing seasonal gluts".

It is important to note that the 'Report' also emphasized "the extent of waste", chiefly in the high-income countries, "the development and use of cheap, mass-produced sensor technology that can detect spoilage in certain perishable foods "along with "productive recycling of surplus food deemed as non-premium quality" which includes "redistribution of good-quality surplus food to consumers ... or use of food no longer fit for human consumption as animal feed or as a source of energy"

In making all these work it has been asserted in the 'Report' that "Food security is best served by fair and fully functioning markets and not by policies to promote self-sufficiency". It maintains that "greater powers need to be given to international institutions to prevent trade restrictions at times of crisis." According to this report, "in high-income countries food production subsidies and related interventions act as a disincentive to efficient food production, raise consumer prices in protected countries, and are ultimately harmful to global food security. ...Agreement between low-, middle-and high-income countries on baseline standards of sustainability in food production and processing" need to be encouraged as an "important" step. There should exist a "more equitable trading system for poor agricultural producers" making it a "genuinely pro-development" approach.

The balance between supply and demand depends to a large extent on people's choice of food and this can be influenced by measures "aimed at influencing demand". It has been pointed out that a reduction in the amount of meat consumed in high-and middle-income countries would have multiple benefits, such as "a reduced demand for grain, leading to a lower greenhouse gas emissions, and a positive effect on health".

Another very intriguing issue threatening global food security is the future volatility in food prices which may be a major cause of increased hunger, specifically in the low-income countries. "The pattern of fluctuations in the price of five major commodities (wheat, rice, sugar, beef and palm) over the past 50 years shows that food prices can be strongly affected by shocks from outside the food system, such as the oil crisis of the early 1970s. It also shows that the last 20 years have been a period of relatively low volatility compared with the previous three decades – in particular, the spike in food prices of 2007-08" (caused by "conversion of agricultural land to the production of biofuels; a series of poor wheat harvest in 2006 and 2007 in agricultural regions such as Australia")

The ultimate aim of this Project was ending hunger with emphasis on chronic hunger which keeps 925 million people within its folds while another billion "suffer from 'hidden hunger' of not having enough vitamins and minerals". The Million Development Goal 1 has a target to halve the number of undernourished people from the "1990 level of 16 to 8 per cent in 2015". This has now come down to 13.5 per cent. "Although China met its target in the early 2000s, many countries in Africa and South Asia are unlikely to succeed by 2015; although there has been a slight fall (from 1,020 million to 925 million over the past 12 months), progress has been slow".

Constant hunger has serious physical and social implications on human life which has been very truthfully summed up in the Project 'Report'. "Hunger results in increased morbidity and mortality, through diseases caused by nutrient deficiency, and a greater susceptibility to disease more generally. It leads to distress behaviour that undermines development, including the sale of assets, the withdrawal of children from school (particularly girls) and into the labour force, the prompting of out migration and, at worst, permanent destitution, prostitution and child trafficking. It also contributes to the onset of armed conflict. The food-price rises of 2007-08 and other impacts, particularly on the poorest households, highlighted the inability of the current global food system to protect the most vulnerable from volatility in food prices"

The Project 'Report' has considered the steps to be taken by the "national and international actors" to "tackle hunger through the food system itself". They are summarised below:

> "In the countries where hunger is most chronic (South Asia and Sub-Saharan Africa) agriculture can make a major contribution to its eradication. For this to occur, agricultural development must be designed and incentivised" with its "triple role in ending hunger". Such as, "technologies, institutions, infrastructure and information that support increased, sustainable productivity and which are equitable (i.e. are desirable, available and practical for the farmers to adopt) can increase the supply of a diverse and locally desirable food at affordable prices". Second, "these technologies, institutions, infrastructure and information sources can improve economic access for all by raising farm income, generating employment on and off farm, and reduce food prices". Third, "production can address issues of social access by deliberately empowering women and other socially excluded groups"

It is observed that,"in the poorest countries, agriculture provides not only food for households, but also a very important means of broadly based income generation... (and) growth in agriculture generates welfare gains that are much stronger for the poorest parts of the population... The 2008 World Development Report shows that a 1 per cent gain in Gross Domestic Product (GDP) originating in agriculture generates a 6 per cent increase in overall expenditure of the poorest 10 per cent of the population, while the equivalent figure for GDP growth originating in non-agricultural sectors is zero growth".

It is suggested in the 'Report' that "agriculture needs to be repositioned within governments as a profession dedicated to multiple ends, of which hunger and poverty reduction are central". For this purpose, apart from seen the purpose of agriculture as production of food, it "requires flexibility to adapt to multiple agro-ecological niches" equipped for meeting the challenges of hunger and poverty. With this ultimate end in view the major thrust areas should be "innovation in how to involve producers in improving yields sustainably" which requires representatives of poor farmers to be involved in the programmes on technology development often taking place at 'greater distances from the farmer's plot' and taking greater care of smallholder farming long neglected.

It is true that "hunger cannot be ended by agriculture alone. Other policies and investments to increase food access, income, reduce differences in gender power and improve nutrition status are vital".

"A key argument of this 'Report' is that the global food supply will need to increase without the use of substantially more land and with diminishing impact on the environment: sustainable intensification is a necessity... The principle of sustainability implies the use of resources at rates that do not exceed the capacity of the earth to replace them. Thus water consumed in water basins at rates that can be replenished by inflows and rainfall, greenhouse gas emissions are balanced by carbon fixation and storage, soil degradation and biodiversity loss are halted, and pollutants do not accumulate in the environment. Capture fisheries and other renewable resources are not depleted beyond their capacity to recover. Sustainability also extends to financial and human capital: food production and economic growth must create sufficient wealth to maintain a viable and healthy workplace, and skills must be transmitted to future generations of producers".

In furtherance of the motive principle of ending poverty and hunger the 'Report' points out that "some of the most threatened and diverse habitats on earth exist in very low-income countries, and interventions to make farming more wildlife friendly, fishing less damaging, or set land aside as reserves may affect the livelihoods of the very poorest people. Whatever strategies are adopted, the human impacts need to be understood and quantified as there are strong ethical arguments against imposing the costs of protecting biodiversity on those least able to pay".

Source: *Foresight: The Future of Food and Farming* (2011), The Final Project Report The Government Office for Science, U.K., London

APPENDIX C
Africa Emerging

Africa presents a unique case with an immense potential for not only rescuing its people from chronic hunger and poverty, especially in Sub-Saharan Africa having one of the largest number of hungry people in the world where agricultural yields and poverty have remained stagnant over the last 30 years, but also becoming self-sufficient in its food requirements with properly guided development policies. This has very clearly been pointed our by Juma C in his new book, *The New Harvest* (2011).

In 1990, over 150 million Africans were hungry, increasing to 239 million by 2010 with undernourished people increasing in number since 2004. According to Juma, one-third of people in Sub-Saharan Africa are chronically hungry, "many of whom are small holders. High food prices in local markets price out the poorer consumers – forcing them to purchase less food and less nutritious food, as well as divert spending from education and health and sell their assets. This link of hunger and weak agricultural sector is self-perpetuating". At the same time, "half of African countries with the highest levels of hunger also have among the highest gender gaps. Agricultural productivity in Sub-Saharan Africa could increase significantly if such gaps were reduced in school and in the control of agricultural resources such as land". Juma points out the weakness of Africa's agriculture where "only 4 per cent of Africa's crop is irrigated, compared to 39 per cent in South Asia. Much of rural Africa is without passable roads, translating to high transportation costs and trade barriers. Over 40 per cent of the rural population lives in arid or semi-arid conditions, which have the least agricultural potential. Similarly, about 50 million people in Sub-Saharan Africa and 200 million people in North Africa and the Middle East live in areas with absolute water scarcity. Cropland per agricultural population has been decreasing for decades. Soil infertility has occurred due to degradation: nearly 75 per cent of the farmland is affected by excessive extraction of soil nutrients... Fertilizer use in Africa is less than 10 per cent of the world average of 100 kilograms per hectare...Part of the reason that fertilizer usage is so low is the high cost of imports and transportation; fertilizer in Africa is two to six times the average world price. This results in low usage of improved seeds; as of 2000, about 24 per cent of the cereal growing area used improved varieties, compared to 85 per cent in East Asia and the

Pacific. "As of 2005, 70 per cent of wheat crop and 40 per cent of maize crop area used improved seeds, a significant improvement". Besides, the level of mechanization is very low, with an average of only 13 tractors per 100 square kilometres of arable land against a global average of 200.

According to Juma, "Africa has abundant arable land (733 million hectares – 27.4% of world's total) compared with 570 million hectares for Latin America and 628 million hectares for Asia" while "only 3.8 per cent of Africa's surface and groundwater is harnessed and irrigation in Africa covers only 7 per cent of cropland and 3.6 per cent in Sub-Saharan Africa" and labour, which with sound policies could be translated into increased production, incomes and food security. This has not materialized because of lack of consistent policies and effective implementation strategies... "Thus, even though agriculture accounts for 64 per cent of the labour force, over 34 per cent of GDP, and over 20 per cent of businesses in most countries, it continues to be given low priority".

Nevertheless, there has been remarkable growth in agricultural production over the past 40 years "with per capita world food production" having grown by 17 per cent and "aggregate world food production growing by 145 per cent, In comparison, "in Africa food production is 10 per cent lower today than it was in 1960 because of low levels of investment in the sector" resulting in an increase of hungry people "by 20 per cent since 1990".

In his opinion, "strategies for transforming African agriculture have to address such challenges as low investment and productivity, poor infrastructure, lack of funding for agricultural research, inadequate use of yield-enhancing technologies, weak linkages between agriculture and other sectors, unfavourable policy and regulatory environments, and climate change".

But Sub-Saharan Africa makes these challenges far more daunting since it "ranks the lowest in the world in terms of yield-enhancing practices and techniques" which include "mechanization, use of agro-chemicals (fertilizers and pesticides), and increased use of irrigated land (where 96 per cent of agriculture is rain-fed making it highly vulnerable to weather shocks)... Land productivity in Africa is estimated at 42 and 50 per cent of that in Asia and Latin America respectively".

Despite all these drawbacks, Africa, as observed by Juma, "is among the most rapidly growing economies in the world today. Its real GDP grew approximately 4.9 per cent per year from 2000 to 2008" with a potential of "increasing agricultural output from US $ 280 billion to a projected US $ 880 billion by 2030.... (while Africa's collective GDP – at US $ 1.6 trillion in 2008 – is almost equal to that of Brazil or Russia – two emerging markets)".

"The picture is, therefore, promising though uncertain. Several recent successes demonstrate that the link between farm productivity and income growth for the poor indeed operates in Africa."

"Several countries have exhibited higher agricultural growth rates per capita over the last 10 years" resulting from a "better policy environment, increased usage of technology, and higher commodity prices".

Importantly, "about 36 per cent of all African labour potential is used in subsistence agriculture", but though "nearly 80 per cent of agricultural producers in Africa are women, only 69 per cent of female farmers receive visits from agricultural extension agents, compared to 97 per cent of male farmers". Increase in agricultural productivity depends largely on involving women in agriculture and innovation which can be done with women's training programmes focusing specifically on agriculture and prioritise female participation in extension work – "both as learners and extension agents". The Uganda Rural Development and Training Programme (URDT) establishing the African Rural University is the pioneer in developing agricultural innovation, the former using the proverb – "If you educate a man, you educate an individual. If you educate a woman, you educate a nation". Ghana also established a similar university giving courses on agricultural science, medicine and health sciences, applied sciences integrated development studies and interdisciplinary research and for all these it depended on the local resources available in the region. "The university believes that the most feasible and sustainable way of tackling underdevelopment is to start with what the people already know and understand".

The agricultural challenges have to be met in the context of climate change when "according to the World Bank, warming of 2 degrees C could result in a 4 to 5 per cent permanent reduction in annual income per capita in Africa and South Asia, as opposed to minimal losses in high-income countries and a global average GDP loss of about 1 per cent. These losses would be driven by impacts in agriculture, a sector important to the economies of both Africa and South Asia...The continent's economies are also highly dependent on natural resources. Nearly 80 per cent of Africa's energy comes from biomass and over 30 per cent of its GDP comes from rain-fed agriculture, which supports 70 per cent of the population. Stress is already being felt in critical resources such as water supply. Today, 20 African countries experience severe water scarcity and another 12 will be added in the net 25 years... The drying up of Lake Chad (shared by Nigeria, Chad, Cameroon, and Niger) is a grim reminder (the area of Lake Chad having decreased by 80 per cent over the last 30 years) that rapid ecological change can undermine the pursuit of prosperity... Technological innovation will be essential for enabling agriculture to adapt to a different climate" which will make it necessary to greater use of "diverse natural assets such as genetic resources for use in agriculture, forestry, aquaculture, and other productive activities. For example, the anticipated changes in the growing season of various crops will require intensified crop breeding", making necessary "the conservation of a wider pool of genetic resources of existing crops and breeds and their wild relatives to cope with shifts in agricultural production potential. This may be done through measures such as agricultural seed banks, zoos, and protected areas. Large parts of Africa may have to switch from crop production to livestock breeding...while others may have to change from cultivating cereals to growing fruits and vegetables". Enhancement of water supply becomes one of the most important among 'other measures' which can be done by "linking reservoirs, building new holding capacity in reservoirs, and injecting early snowmelt into groundwater reservoirs". Development and prosperity in Africa, according to Juma, has largely been the "consequences of recent neglect of sustainable agriculture and infrastructure as drivers of development ... the sustainable agriculture" having, through ages, serving as "the driving force behind national development. In fact, it has been a historical practice to use returns from investment in sustainable agriculture to stimulate industrial development. Restoring it to its right place", that is, "reshaping sustainable agriculture as a dynamic, innovative and rewarding sector in Africa will require world leaders to launch new initiatives. The value of sustainable agriculture in African economy needs to be given due recognition, to accept it as a "knowledge-intensive productive sector", requiring reforms "to be introduced in knowledge-based institutions to integrate research, university teaching, farmers' extension and professional training, and bring them into direct involvement with the production and commercialization of products".

Source: Calestous Juma 2011, *The New Harvest: Agricultural Innovation in Africa*, Oxford University Press

References and Notes

CHAPTER 1

Myrdal, Gunnar, 1963 (Reprinted), *Economic Theory and Underdeveloped Regions,* Universirty Paper Backs, Methuen, London: 8.

Brandt Report, 1990, North- South, A Programme for Survival, The Report of the Independent Commission on International Development Issues, Sydney: 49.

Sen, Amartya, 1976, Poverty: An Original Approach to Measurement, *Econometrica,* Vol 44, No 2, March 1976: 219.

— — , 1999, Development as Freedom, New York, Knoff: 87

— — , 1982, Poverty and Famines, Clarendon Press, Oxford: 21.

— — , 1999, Development as Freedom: 20

Fraaser, Bashabi (Ed), 2006, Bengal Partition Stories: An Unclosed Chapter, *Anthem South-East Asian Studies,* London: 15.

Special Note: Sen, Amartya, 1992: "Boats capable of carrying more than ten passengers were removed from a vast area of river-based Bengal to 'deny' them to the passively-arriving Japanese, and this interfered with river transport and fishing: 97, Footnote on the same page: "The boat denial policy was coupled with a 'rice denial' policy in May 1942, aimed also at the elusive Japanese...".

Watts, Harold, 1968, An Economic Definition of Poverty, in 'The Economies of Poverty and Inequality, An Elgar Reference Collection, ed. by Frank A Cowell, 2008, Vol II: 131.

M. Orshansky, 1965, Counting the Poor: Another Look at the Poverty Profile, in '*Economies of Poverty and Inequality, An Elgar Reference Collection,* Vol III: 4.

Pavallion Martin, 1996, Issues in Measuring and Modelling Poverty, in the Economies of Poverty and Inequality. An Elgar Reference Collection ed. by Frank A Cowell etc. 121.

Griffin, Keith, 1978, *International Inequality and National Poverty,* The MacMillan Press Ltd.,: 122.

Sachs, Jeffrey, 2005, *The End of Poverty: How Can We Make It Happen in Our Life Time,* Penguin Books: 20.

Daniels, Peter, Michael Bradshaw, Denis Shaw and James Sidaway, 2005, (Second Ed.), *An Introduction to Human Geography: Issues of the 21st Century,* Prentice Hall, England: 247.

Eastwood, Robert and Michael Lipton, 1999, The Impact of Changes in Human Fertility on Poverty, *Journal of Development Studies,* 36: 1: 4

CPAG (Child Poverty Action Group, The U.K.), 2004, Poverty, The Facts, London: 1.

Ibid: 2.

Mandle, J.R., 2003, *Globalization and the Poor*, Cambridge University Press: 21-23.

Wolf, Martin, 2005, *Why Civilization Works*, Yale University Press: 157-158.

Ibid: 160.

Daniels *et al.*, 2005: 89.

Sachs, 2005: 21.

United Nations 2007, Strengthening Efforts to Eradicate Poverty and Hunger: Dialogues at the Economic and Social Contact: 97.

Ibid: 108.

Note A.: *Definition of Poverty.*

"According to Prof. Tendulkar's findings, 2004-05, 37 per cent of the country's population was living below poverty line" (in India) (which) "is significantly higher than the figure released by the Planning Commission, according to which 27.5 per cent are below the poverty line": when defining poverty line it refers to per capita consumption of calorie norms of 2400 (rural) and 2100 (urban) per day to the percentage of people living in poverty, such as while it amounts to 28.3 per cent of the population in 2004-05 as estimated by the NSSO (National Sample Survey Organisation) the Arjun Sen Gupta Committee, formed by the Government, finds more than 77 per cent of people living on Rs. 20 (less than $ 1) or less per day which is insufficient even for meeting "the minimum requirements of a person's food, health, shelter and clothing". Reported by A. Mahajan, *The Statesman*, Kolkata Ed., 22. Jan 2010.

The World Bank Development Report 2000/2001, *Attacking Poverty*, Oxford University Press: 35.

Ibid: 36-37.

Ibid: 3.

The Government of India, Report on Conditions of Work and Promotion of Livelihoods in the Unorganised Sector: 6.

Lin, J Y, 1988, *The Household Responsibility System in China's Agricultural Reform: A Theoretical and Empirical Study*, Economic Development and Cultural Change, April,:199-224, and Johnson, DG, 1998, *Economic Reforms in the People's Republic of China*, Economic Development and Cultural Change, April: 225-245.

Yao, Sujie, 1999, Economic Growth, Income Inequality and Poverty in China under Economic Reforms, *Journal of Development Studies*, 35::106.

Ibid: 119.

Ibid: 123-124.

United Nations 2007, Strengthening Efforts to eradicate Poverty and Hunger etc.,: 247.

International Organisation for Migration (IOM), Migration, Development and Poverty Reduction in Asia, Academic Foundation, New Delhi (India): 33-34.

Census of India, 2001, Provisional Population Totals, Statement I

Castro, J De, 1955 (Third Impression), *Geography of Hunger*, Victor Gollancz Ltd., London: 5.

Ibid: 16.

Ibid: 22.

Ibid: 25.

Meadows, D H, Denis L Meadows, Jorgen Renders and William W Behrnes III, 1972, The Limits to Growth: A Report for the Club of Rome's Project on the Predicament of Mankind, A Signet Book from New American Library.

Note B Katie Allen, the Economic Correspondent of the Guardian, reporting on land grabbing mentions, "the trend of buying or taking out long-term leases on land first came to prominence during the 2008 World Food Crisis. As food riots rage from Mexico to Bangladesh, speculators and countries with their own food security make:. quietly sealed deals with African nations. Others call it 'neo-colonialism'. Expanding industrializing populations in need of plentiful food sources with cheap work-forces take over vast swaths of Africa".

However, very surprisingly even some at the World Bank "see these overseas investments in foreign farmland, particularly in Africa, as a way to modernize agriculture there, bringing more food and profits to the local population". But as Allen observes – which is very correct – "one of the difficulties in assessing the impact of the land grab is that many of the transactions are opaque and the extent of investment unclear... The UN's Food and Agricultural Organization (FAO) and other analysts estimated last year that nearly 20 million hectares (50 million acres) of farmland – an area roughly half the size of all arable land in Europe – was sold or had been negotiated for sale or lease in the space of just six months. Investors range from Gulf States dependent on food imports to Indian farming companies and European-base biofuel producers. Targets include Sudan, Tanzania, famine-prone Ethiopia and Madagascar.

To understand the real impact of land-grab policies, "the sheer numbers of Africans dependent on agriculture should raise alarm bells over what is at stake if opaque, bargain-based land deals are to continue. In west and east Africa, agriculture contributed about a third to GDP between 2003 and 2007, compared with just 1.6 per cent in developed economies. Agriculture accounted for more than half of total employment between 2002 and 2006. Those farmers stand to gain from new technology but they also risk being disenfranchised and left hungry. In reality, the dream of a win-win relationship is often a skewed flow of profits to the investors. Go back to the proposition: a 99 year lease on arable land at $ 1 a hectare. To put that in context, farmland in Britain is selling about 15,000 pounds per hectare, according to the Royal Institution of Chartered Surveyors... (Finally, the consequence of land grab under whatever plea it is done does not refute the fact that) food security for the rich does not result in more hunger for the poor". *The Guardian*, the U.K., 16 Aug 2010.

That the ultimate aim of land grab policy is more oriented to cultivate industrial rather than food crops comes out very clearly in the report of the 'Friends of Earth (FOE)'. The Friends of Earth want "European Union countries must drop their biofuel targets or risk plunging more Africans into hunger and raising carbon emissions. "The Charity", as reported by Katie Allen in the *Guardian* (16 Aug 2010), "accuses European companies of land grabbing in Africa to grow biofuel crops that directly competes with food crops" with an aim to produce 10 per cent of all transport fuels from biofuels by 2020. "A number of European companies have planted biofuel crops such as jatropha, sugarcane and palm oil in Africa and elsewhere to tap into rising demand... The trend has coincided with soaring global food prices and ignited a debate over the dangers of using agricultural land for fuel... Action Aid claimed this year that European biofuel targets could result in up to 100 million more hungry people, increased food prices and landlessness... Estimates of how much land in Africa is being farmed by foreign companies and governments, either for food or fuel crops, vary significantly. The FOE report focuses on 11 African countries in what it is a rush by foreign companies to farm there. In Tanzania, for example, it says that about 40 foreign-owned companies, including some from the UK, have invested in agro-fuel developments. It argues that such activities are actually raising carbon emissions in many cases because virgin forests are being cut down'. *The Guardian*, the U.K., 30

Aug 2010.

At the same time, there is an alarming report about the impact of industrial-scale production of crops which may be environmentally hazardous for a country. As reported by Felicity Lawrence in the Guardian, based on a study by 'The Development Charity Programme', made in Peru, it has been found that industrial production of asparagus in Peru's Ica Valley is depleting the area's water resources so fast that smaller farmers and local families are finding wells running dry. Water to the main city in the valley is also under threat, it says (the report), and warns that the export of the luxury vegetable, much of it to British super markets, is unsustainable in its current form... The UK is the world's sixth largest importer of 'virtual water', that is water needed to produce the goods it buys from other countries, according to the WWF... Now the UK is the third largest importer of fresh Peruvian asparagus, consuming 6.5 million kilograms a year. Peru meanwhile has become the largest exporter of asparagus in the world, earning more than $450 million a year from the trade. Around 95 per cent of that asparagus comes from the Ica Valley. The expansion of the agricultural frontier in the region – a desert area in the Andes and one of the driest places on earth – was made possible "with investments from the World Bank made in the late 1990s and within just 10 years "the asparagus cultivation has" expanded "to cover nearly 100 square kilometres of reclaimed desert... The trade has created around 10,000 new jobs in a very poor area... The water tragedy unfolding in this region of Peru should set alarm bells ringing for government, agribusiness and retailers involved in Ica's asparagus industry. "With gradual depletion of water resources there will be no water for the industry when people will be left to their destiny without any future,". *The Guardian*, the U.K., 15 Sept 1010.

The Guardian, The U.K., 29 October 2008, 22 November 2008, 28 November 2008, and 3 December 2008.

Brown, Lester R, Gary Gardner and Brian Halweil, 2000, Beyond Malthus: Nineteen Dimensions of the Population Change, *The Worldwatch Environmental Alert Series*, Earthscan Publications Ltd., London: 33-35.

Ibid: 35-36.

Smil, Vaclav, 1994, How Many People Can the Earth Feed ? *Population and Development Review*, Vol 20, June 1994: 264.

Ibid: 272

Note C: The history of making estimates of global carrying capacity in regard to 'how many people can the earth feed' goes back to about a hundred years or so, Revenstein (1891) perhaps being the first one to suggest this limit to 'just under 6 billion people' which is a little later Pfaundler (1902), "assuming no extension of the existing 2,174 million hectares (mha) of cropland and good grazing land, average density of 5 people per hectare supported by traditional farming methods, and recycling of organic matter, put at a minimum of 11 billion people. After that numerous estimates followed and among them more recently Brown (1954) did an estimate that increased yields from existing cultivated land supplemented by output from about 500 mha of new fields and 80 mha of newly irrigated lands would raise the 1950 food supply six-fold, enough to support 15 billion people". Estimates made later by others, such as Baade (1962), Zierhofter (1966), Meadows *et al.* (1972), Revelle (1976) etc. put the capacity of the earth in feeding people which varied widely between 11.5 billion to 30 and 40 billion people based on different assumptions. Such wide variations are the results of errors in data available on total cropland, output of continuous crops (many vegetables and fruits) and incomplete harvests etc, giving rise to as much as 20-25 per cent errors. Smil, Vaclov, 1994.

Tim Dyson, 1996, *Population and Food: Global Trends and Future Prospects*, Routledge: 204.

Wolf, Martin, 2005 '*Why Globalization Works*, Yale University Press: 165.

United Nations 2007, Strengthening efforts to Eradicate Poverty etc.,: 97.

Ibid: 214.

Willy, Brandt *et al.*, 1980: 90.

Griffin, Keith, 1978: 127.

Note D: The severity as well as the universality of poverty may further be gauged from the hard reality of the situation in an economically advanced country as revealed by a report of the government-sponsored Millennium Cohort study showing that "a fifth of British seven-year-olds live in severe poverty with both parents together earning less than half the average national income... The London University's Institute of Education researchers found that despite governments having spent billions to eliminate child poverty since 1999 almost one-fifth of seven-year-olds live in severe poverty-homes where the total income, including benefits, is less than 254 pounds a week. The UK average income for a family with one child is a 563 pounds a week... Almost three-quarters of children whose parents are of Pakistani or Bangladeshi origin live in poverty... largely because of high unemployment rates for mothers and fathers... Just over half (51%) of black seven-year-olds and just over a quarter of white seven-year-olds live in poverty".: Jessica Shepherd, Education Correspondent, The Guardian, The U.K., 15 Oct 2010.

CHAPTER 2

UNDP, Human Development Report 2003: Millennium Development Goals: A Compact Among Nations to End Poverty: 2-3.

Castro, J D, 1955, *Geography of Hunger*, Victor Gollanez Ltd., London: 235.

Dreze J and Amartya Sen, 1989, *Hunger and Public Action*, Clarendon Press, London: 23.

Note A; "While entitlement of a person is a set of alternative *commodity* bundles, the capability of a person is a set of alternative *functioning* bundles. Larger entitlements contribute to wide capabilities, but the relationship is not the same for different persons. For example, a pregnant woman has greater nutritional requirements and also a special need for medical attention, and hence having the same command over food and health care as another non-pregnant – person may not give her the same capability to be well-nourished and healthy" Dreze and Sen,1989: 4.

UNDP, 2003: 4.

Bhagwati, Jagadish, 2004, *In Defense of Globalization*, Oxford University Press: 54.

Ibid: 57.

Yunus, Muhammad, 1998, *Banker to the Poor*, The University Press Limited, London: 89.

Ibid: 116.

UNICEF, 2007: 15.

Note B: As reported by FAO, "the emerging biofuel market is a significant source of demand for some agricultural commodities, such as sugar, maize, cassava, oilseeds and palm oil. The stronger demand for these commodities having caused a surge in their prices has, in turn, raised food prices. While biofuel production and consumption is supported by government policies in a number of countries, rapid increase in crude oil prices have further contributed to growing demand for agricultural commodities for biofuel feedstock. Biofuel production will utilize an estimated 100 million tonnes of cereals (4.7% of global cereal production) in 2007-2008" (FAO, The State of Food Security in the World 2008: 10-11.

It may sound highly important to note that the push for increased ethanol-biofuel production was mandated by the United States Government, calling for 9 billion gallons of ethanol to be produced in 2008.

IFPRI (International Food Policy Research Institute), 2002, Reaching Sustainable Food Security for All by 2020: 1.

Note C: According to the Director of the International Charity Body 'Save the Children', 10 million more children have been left undernourished since the beginning of global meltdown or recession and an additional 400,000 children could die by the end of the year. Around 3.5 million children die every year due to malnutrition and the global recession may kill another 2.8 million children around the world while one million children in India is dying of malnutrition every year.

FAO, *The State of Food Insecurity in the World* 2008: 10.

Ibid: 11.

United Nations 2007, Strengthening Efforts to Eradicate Poverty etc.: 98

Sachs, J, 2005, *The End of Poverty: How Can We Make It Happen*, Penguin Books: 51.

Wolf, Martin, 2005, *Why Globalization Works*, Yale University Press, New Haven and London: 103.

Ibid: 104.

Ibid: 108.

Dicken, Peter, 2007, *Global Shift: Mappig the Changing Contours of the World Economy*, Sage Publications: 35.

Ibid: 38.

Ibid: 43.

Note D: China's Trade Network (% of China's Total)

Region/Country	*Exports*	*Imports*
North America	22.4	9.3
United States	21.1	8.2
Latin America	2.7	3.6
Western Europe	17.5	13.9
European Union 15	16.5	12.8
Central Europe, Baltic States and CIS	3.6	3.6
Africa	2.2	2.0
Middle East	3.0	3.5
Asia	48.4	58.0
Japan	13.6	18.0
Hong Kong	17.4	2.7
Korea	4.6	4.5
Taiwan	2.1	12.1

Source: Peter Dicken, 2007, Global Shift (Calculated from WTO International Trade Statistics 2004)

Government of India, Ministry of Finance, Economic Survey 2007-08, Oxford University Press: 3.

Ibid: 182

United Nations, World Economic Situation and Prospects 2007, New York: III

United Nations 2007, Strengthening Efforts etc., 123.

Sachs, 2005: 56-64.

Wolf, Martin, 2005: 157.

Op cit.

Ibid: 157-158

FAO 2006, The State of Food and Agriculture; 118, Table A2.

United Nations, World Economic Situation etc.,2007:108.

Ibid: III

Dicken, Peter 2003, (4th Edition), *Global Shift: Reshaping the Global Economic Map in the 21st Century*, Sage Publications: 556.

Ibid: 505.

United Nations, World Economic Situation etc., 2007: 203-204.

Ibid: 206.

Narayan, Deepa, Robert Chambers, Meera K Shah and Patti Petesch, 2000, *Voices of the Poor: Crying Out for Change*, Oxford University Press for the World Bank: 63-64.

Osmani, S. R., 1991, Social Security in South Asia, in '*Social Security in Developing Countries*' Edited by E Ahmad, Jean Dreze, John Hills and Amartya Sen, United Nations University (OUP paperback in India in 1999): 319.

Wolf, Martin, 2005: 145.

Ibid: 164-166.

Government of India, Planning Commission, 2008, Report on Conditions of Work and Promotion of Livelihood in the Unorganised sector: 100-101.

Brown, Lester B, 2006, Plan B 20: Rescuing a Planet under Stress and a Civilization in Trouble, Earth Policy Institute: 142.

World Commission on Environment and Development (WECD), 1987, Oxford University Press: 87.

Hamm, B and Pandurang K. Muttagi, 2001, *Sustainable Developmet and the Future of Cities*, ITDG publishing: 21.

Brown Lester B, 2006: 150.

Ibid: 149-152.

UNDP, Human Development Report 2006: Beyond Scaracity: Power, Poverty and the Global Water Crisis: 2.

Bryson, Bill, 2008, *Shakespeare*, Harper Perenial Edition: 22-23.

Ibid: 48.

UNDP, Human Development Report 2006: 6.

Op. cit

United Nations, 2003, Water for People, Water for Life, World Water Development Report: 10.

UNDP Human Development Report, 2006: 14-15.

Note E: Two very serious cases of water-stress are presented by Egypt as well as Iraq. The Nile Delta in Egypt is likely to be one of the three most vulnerable areas to a rise in sea level and "even the most optimistic predictions of global temperature increase will ... displace millions of Egyptians from one of the most densely populated regions of earth". The delta stretching out into 10,000 square miles of farmland, fed by Nile's branches is home to two-thirds of the country's rapidly growing population and is responsible for more than 60 per cent of its food supply. A 1 metre rise in sea level will drown 20 per cent of the delta and the 14 metre rise that is likely to result from the disappearance of Greenland and western Antarctica will take the entire delta under water. According to Dr. Rick Tutwiler, Director of the American University in Cairo's Desert Development Centre, "the delta is a kind of Bangladesh story. You have got a massive population, overcrowding, a threat to all natural resources from the pressure of all the people, production, pollution, cars and agricultural chemicals. And on top of all that there's the rising sea....The pressure on limited land is going to encourage the desertification process, yet slow but becoming unavoidable". There is already a major water deficit in Egypt, with only 700 cubic metres of fresh water available for a person, far short of 1000 cubic metres as stipulated by the United Nations, and is 'likely to go down to 450 cubic metres per person with population increase without taking into account the inescapable impact of climate change, as a result of increased evaporation and larger demand of water in upstream areas the amount of Nile water reaching the delta is likely to drop by about 70 per cent over the next few years with staggering ecological consequences, and as predicted by the experts at Egypt's Soil Water and Environment Research Institute, both wheat and maize production is likely to go down by 40 per cent and 50 per cent respectively in the next three decades making farmers losing around $ 1000 per hectare for each degree rise in the average temperature. Source: *The Guardian*, The U.K., 21.8.2009.

Almost a similarly grim future awaits Iraq, where water shortage has already reached the most critical stage in its entire history of a few millennia, threatening up to 2 million people in the south of the country without electricity and nearly as many without drinking water with Euphrates River gradually drying up as a result of dams built on the upper reaches of the river in Turkey (a series of dams built during the last five years on the Euphrates and its small tributaries shared by countries, such as Turkey, Syria and Iran, have worsened the situation further converting this life-sustaining river into a "slick black ooze" at many places today). Those dams controlling the level of the river in its lower reaches drastically reduces the supply of water essential for the generation of electricity as well as irrigation and raising the level of salinity makes it unusable already triggering evacuation of people from two towns on the northern edge of Basra. as said by Salah Aziz, Director of Planning in Iraq's agricultural ministry, land, which was "100 per cent in use" ..."this year (2009) less than 50 per cent of (it) is in use and most of the yields are marginal" so that the country cannot meet "even 40 per cent of Iraq's fruit and vegetable demand". *Source*: *The Guardian*, The U.K., 27.8.2009.

UNDP, 2006, Human Development Report: 16

Note F: As a recent UN report points out, there is no new land or water to develop for meeting the requirements of the extra 1.5 billion people expected to live in the continent of Asia alone by 2050 which will double the demand of food. The existing irrigation systems in countries like India, China and Pakistan are often 50 to 70 years old already under great pressure and water is evaporating from them. In the words of Colin Chartres, the Director of IWMI, "the agriculture of tomorrow will need a lot of more water. Given that one litre of water is used to produce one calories of food, the world will need up to 6,000 kilometres of additional water every year to feed another 2.5 billion people 2,500 calories per day" ..."almost twice what we use today and is not sustainable". The extra 2.5 billion mouths to be fed within 40 years from now will mostly belong to the developing countries where Africa's population "could double, Asia's could grow by nearly 30 per cent and Pakistan's by 85 per cent" Source: John Vidal, *The Guardian*, The U.K., 17 Aug 2009.

Brown, Lester B, 2006: 158-162.

Yao, Shujie, 1999, Economic Growth, Income Inequality and Poverty In China under Economic Reforms, *Journal of Development Studies*, 35: 6-107.and 109.

Note G: In a recent report by Olivier De Schutter, the UN Special Rapporteur on the right to food, questioning the long-term self-sufficiency in food in China, he is of opinion that "the shrinkage of arable land and the massive land degradation threatening the ability of the country to maintain current levels of agricultural production, while the widening gap between rural and urban is an important challenge to the right to food". The decline of soil quality because of fertilizers, pollution and drought has affected 37 per cent of China's territory with degradation and "8.2 million hectares (20.7 million acres) of arable land has been lost since 1997 to cities, industrial parks, natural disasters and forestry programmes". Further, an increasingly carnivorous diet, requiring more grain to feed livestock has aggravated the situation, "driving up" food inflation. 'In the last year, rice has gone up by 13 per cent, wheat by 9 per cent, chicken by 17 per cent, pork by 13 per cent and eggs by 30 per cent. Climate change is likely to 'increase price volatility and cut agricultural productivity by 5 per cent to 10 per cent by 2030. According to De Schutter, 'it was essential for China to wean itself off fossil fuel-intensive farming and adopt more sustainable agricultural techniques, including organic production, and to make even better use of its two great strengths ": a huge strategic grain reserve (accounting for 40% of the nation's 550 million tonne grain supply) and a large rural population.'. De Schutter cautions against 'a shift towards industrial-scale farming, which increases economic competitiveness at the cost of natural productivity'. He finds 'small-scale farming more efficient in its use of natural resources' though doubts how far it may by feasible in the face of more of China's 200 million farmers moving to the cities, giving rise to such contrasts that 'while some poor rural families in western China scrape by with two meals a day, wealthy urban households on the eastern seaboard eat so well that they are prone to the rich diseases of obesity and diabetes'. Source: Jonathan Watts, Beijing for *The Guardian*, The U.K., 14 December 1010,.

Brown, Lester, 2006:16.

Note H: John Vidal, the Environmental Editor of the *Gurardian* reports that "the world may be on the brink of a major new food crisis caused by environmental disasters and rampant market speculations... after a heatwave and wildfires in Russia led to a draconian wheat export ban and food riots broke out in Mozambique, killing 13 people... A near doubling of staple food prices in 2007 and 2008 led to riots in more than 30 countries and an estimated 150 million more people going hungry". The destruction of wheat harvest by wildfires in Russia and tte subsequent imposition of export ban automatically raised wheat prices. "In addition, other markets such as sugar and oilseeds are witnessing significant price increases". According to De Schutter, "the only long term way to resolve the crisis (food) would be to shift to ways of growing food that do not depend on pesticides or heavy machinery" *Source: The Guardian*, The U.K., 25 September 2010.

United Nations 2007, Strengthening Efforts etc., 124.

Ibid: 131.

Note I: Professor C. Juma of Harvard University claims in his new book, '*The New Harvest: Agricultural Innovation in Africa*' that "Africa can feed itself. And it can make the transition from hungry importer to self-sufficiency in a single generation".According to Juma, "Africa is the only continent with arable land readily available to expand agriculture, and that southern Sudan alone could feed all Africans if it was properly developed". In his words, again, "An African agricultural revolution is within reach, provided the continent can focus on supporting small-scale farmers to help meet national and regional demand for food". Juma says in this context that 'food self-sufficiency would require big shifts in policies that have led to dependence on food aid and imports' ($ 3 billion of food aid and $ 33 billion in food imports annually, according to the United Nations), with greater direct involvement in sectors like water, energy, transport, communication and education. His proposal includes the modernization of farms, with new machinery and storage and processing facilities, and the selective use of genetically modified crops. He calls for new roads, energy sources and irrigation projects. About 70 per cent of Africans are involved in agriculture, but almost 250 million people, or a quarter of the population, are undernourished. The number has risen by 100 million since 1990, while food production has grown globally by 145 per cent over the past 40 years. African food production has fallen by 10 per cent since 1960.Only 4 per cent of the continent's cropland is irrigated. Fertilizers, pesticides and high-quality seeds are prohibitively expensive and in short supply. 'Fears have been expressed that "land grabbing" – in which wealthy countries buy or lease land in Africa to farm it for their own populations – could hurt Africa's potential for self-sufficiency' ... which 'needs to be regulated'. Source: David Smith, Johannesburgh for *The Guardian*, The U.K., 3 December 2010.

CHAPTER 3

United Nations, World Population Prospects, *The 2006 Revision: Sex and Age Distribution* of the World, Vol II: 7.

Op cit.

Note: Median Age: 'the age at which half the population is older and half is younger than that age, are indicating of population ageing'.

Computed from data available from UN: World Population Prospects, *The 2006 Revision*, Table 3.

Note A: According to UN Report, the number and proportion of people, aged 80 years or more, is rising. In 2005, there were 88 million of them, representing 1.3 per cent of the world population. By 2050, the number is projected to reach 402 million, or 4.4 per cent of the world population. Besides, this is the fastest growing segment of the world population, the rapid increase of which is expected particularly in the less developed regions where, persons aged 80 or more are projected to increase from 43 million in 2005 to 284 million in 2050. They tend to be concentrated in the most populous countries, such as, in 2005, 15 million lived in China, 10.6 million lived in the United States, and 7.8 million in India. In 2050, those countries will still have the largest numbers of them, 103 million in China, 51 million in India and 31 million in the United States

In the opinion of Professor Kaare Christensen and colleagues at the Ageing Research Centre at the University of Southern Denmark, at least half the babies born in the UK in 2000 will reach their 100th birthday. "Life expectancy is increasing so that half the babies born in 2007 will live to be at least 103, while half the Japanese babies born in the same year will reach the age of 107'. According to a report in the Lancet, at least 50 per cent of babies born in 2007 are expected to be – 104 in Canada;101 years in Denmark; 104 years in France; 102 years in Germany; 104 years in Italy; 104 years in the United States. Source: Sarah Boseley, Health Editor, *The Guardian*, The U.K., 20 October 2009.

WHO, World Health Report, 2008: 7.

Ibid: 7-8.

Ibid: 8.

Murray, C J L and A D Lopez, 1997, Global Mortality, disability, and the contribution of risk factors: Global Burden of Disease Study, *in The Lancet*, Vol 349, May 17, 1997:1441.

Jamison, Dean T, 2007, *Solution for the World's Biggest Problems: Costs and Benefits,*Ed. by Bjorn Lomborg, Cambridge University Press: 296-297.

Demographic Year Book 2007: 52.

Coale, Ansley J, 1991, Excess Female Mortality and the Balance in the Population: An Estimate of the Number of "Missing Women", in *Population and Development Review*, Vol 17, No 3: 517.

Ibid: 518.

Krishnaraj, M, Ratna, M.S. and Abusaleh Shariff, 1998, *Gender, Population and Development*, Oxford University Press: 7.

UNDP: Human Development Report 2005: 61.

Coale, 1991: 521-522.

Klassen, Stephen, 1994, *"Missing Women"* Reconsidered, in World Development, Vol 22, No 7: 1061.

Klassen, Stephen and Claudia Wink, 2009, Gender Bias in mortality ? An Update on Missing Women, in *Gender and Discrimination: Health, Nutritional Status and Role of Women in India*, ed. by M. Pal, P. Bharati, B Ghosh and T S Vasula, Oxford University Press: 297.

Ibid: 305.

Note B: "While the number of missing women has risen in absolute numbers over the past 10 years, there appears to be a relative improvement in the share of missing women, suggesting that the phenomenon has stabilized at a high level... Regional analyses point to definite improvement in North Africa and West Asia, some improvements in South Asia which, however, appear overstated and are partly related to better enumeration of females, and deterioration in China related largely to its strict FP policies. Improvements in Bangladesh and Pakistan ensure that India now has the largest share of missing females in the world., and only a slightly lower absolute number of missing women than China". *Source*: Stephen Classen and Claudia Wink, 2009, Gender Bias in Mortality etc., 317-318

Clark, Helen and Elizabeth Carneigie, 2006, *"She was Aye Workin": Memoirs of Tenement Women in Edinburgh and Glasgow*, White Cockade Publishing, Oxford: 27-28 and 35, 42.

Note C: Sarah Bosley Health Editor, *The Guardian*, the U.K., reports unsafe abortions kill 70,000 women every year in the world and in Asia out of 29 abortions for each 1000 women aged 15-44, 11safe in 2003, the highest of them – 29 out of a total of 31 – happening in Latin America and the Caribbeans. *Soucre: The Guardian*, The U.K., 14 October 2009.

In a report of the United Nation, women in the world's least developed countries are 300 times more likely to die during childbirth or because of pregnancy. According to the UNICEF, the death toll is more than half a million women a year. *Source: The Guardian*, The U.K., 16 January 2009.

Note D: As reported by the UNICEF, October 2007, more than a third of the world's child brides are from India, leaving children at an increasing risk of exploitation despite Asia's modernity with rising literacy and economic wealth. According to this source, nearly 25 million young women in India were married in the year 2007 by the age 18, noting further that children in India, Nepal and Pakistan may be engaged or even married before the age of 10, prompted by tradition and religious practices despite existing ban on child marriage. *Source: The Guardian*, The U.K., 7 October 2007.

Gupta, Y P, 2009, The Disappearing Girl Child, *The Statesman*, Kolkata Issue, 2 February 2009.

Note E: "The TFR (Total Fertility Rate) gives us the average number of births per woman in a society. The higher the TFR, the greater is the vulnerability and burden faced by women. At the same time, too many births too close together, or at too young or old an age are a major cause of illness, disability, poor nutrition and premature death among both women and children. To that extent, fewer births can contribute dramatically to improve the lives of women and children in society". Ed. S Mukhopadhyay, 2001, *Women's Health, Public Policy and Community Action*, Manohar: 57.

Sayeed, A T, 2009, Victims of Earthquake and Patriarchy: The 2005 Pakistan Earthquake, in '*Women, Gender and Disaster*', Ed. by Elaine Enarson and P G Dhar Chakraborti, Sage Publications: 144.

Sen, Amartya, 1992, 'Missing Women', British Medical Journal, Vol 304, 586-587

——, 2001, Many Faces of Gender Inequality, *Frontline*, Vol 18, Issues 22, 27.

Miller, B, 2001, Gender Discrimination in Intrahousehold Food Allocation in South Asia, Paper presented at the meetings of the *American Anthropological Association*, 2-6 Nov, san Fransisco, US.

Smith, L.C. and Elizabeth N., Byron, 2009, Will Increasing Women's Power Reduce Discrimination Against Girl Children in South Asia ?, in '*Gender and Discrimination: Health, Nutritional Status and Role of Women in India*', Ed. by M Pal, P. Bharati, B. Ghosh and T.S. Vasulu, Oxford University Press: 116-117.

UNIFEM 2008/2008, Progress of the World's Women 2008/2009: Who Answers to Women ? Gender and Accountability: 127.

UNDP, Human Development Report 2002: 26.

UNICEF 2007, The State of the World's Children: Women and Children – The Double Dividend of Gender Equality: 5.

UNIFEM 2008/2009: 126-127.

UNDP, Human Development Report 2007/2008: 369.

UNIFEM 2008/2009: 127.

Note F: Admittedly, under-weight children are the results of undernourishment and malnutrition quite often leading to further consequences of producing stunted children. The data from a survey conducted in 33,000 rural households – encompassing 195,000 individuals – spread over 1,765 villages in 195 districts in 16 states of India, commissioned by the Indian Planning Commission yielded highly interesting results. According to the findings of this study, "the prevalence of undernourishment in India, even relative to other poor countries, is shockingly high. Thus in the 1990s, 36 per cent of children in India below the age of five, compared to 21 per cent in SSA (Sub-Saharan Africa) were severely stunted". As this study explored, "maternal literacy in particular, affects the likelihood of children being malnourished (and this may vary according to the literacy status of the mothers). As has been found in another study made of 53,207 children, between the ages of 0-12 years, "43 per cent were stunted and 9 per cent were underweight. Of the children that were underweight, 81 per cent were stunted and of the children that were stunted, 16 per cent were underweight... The incidence of stunting and of being underweight was lower for children with literate mothers (40 per cent and 7 per cent) than it was for children whose mothers were illiterate or proximate literate. In addition to benefiting from having a literate mother, children also benefited from living in households where at least one of the women was empowered (in decision-making on allocation of income for buying food, arranging for healthcare, education, etc.) and, in particular, they benefited from living in households headed by a female". *Source*: V K Barooah, *"Maternal Literacy and Child Nutrition in India'*, in (Ed) by M. pal *et al.*, 2009, *Gender and Discrimination* etc., Oxford University Press: 142-143, 150,

UNICEF 2007: 16.

UNICEF 2008, The State of the World's Children: Child Survival: 8.

Ibid: 4.

Ibid: 11

Ibid: 23.

Note G: Reports of widespread female foeticide pour in every now and then from various parts of India. Recently, "the West Bengal Commission for women has found more than 100 medical establishments in Kolkata (Calcutta) and an equal number of city-based medical practitioners, if not more, responsible for sex-selective abortions aided by Sex Discrimination (SD) of foetus, something held illegal in this country. And it seems that the affluent among the city's residents are opting more for the SD-influenced termination of pregnancy... According to the report, almost two-thirds of the women (64% among the women surveyed for this purpose) came from affluent households and only 14 per cent from poor households. The rest (about 22%) belonged to middle-class households. We could perhaps surmise that women from relatively affluent families would be in a position to have greater access to sex determination of fetuses and sex-selective practices". *Source*: Rajib Chatterjee, *The Statesman*, Kolkata Issue, 5 February 2010.

Sudha, S. and S. Irudaya Rajan, 2001, Female Demographic Disadvantage in India 1981-1991: Sex selective Abortions and Infanticide, in *'Gendered Poverty and Well-being'*, Ed. by Shahra Razvi, Blackwell, Oxford: 181

Krishnaraj *et al.*, 1998: 12.

UNICEF 2007: 23.

The Guardian, The U.K., 5 December 2008.

The Guardian, The U.K., 4 December 2008.

Note H: That women become the worst victims during war is demonstrated by one of the most glaring incidents reported from Congo where nearly 200 women and 4 boys - aged one month, six months, a year and 18 months - were repeatedly raped systematically by as many as 400 Rwandan and Congolese rebels near a UN Peace=Keepers' base between 30 July and 4 August 2010. Nobody came to their immediate rescue when husbands were forced to watch their wives and children were gang-raped ". *Source*: David Smith, *Africa Correspondent, The Guardian*, The U.K. 25 Aug 2010 and Lindsey Hilman, *The Guardian*, The U.K., International Edition, 13 October 2010.

UNICEF 2007: 5.

Sinha, Indrani and Sharmila Dasgupta, 2009, *Mothers For Sale: Women in Kolkata's Sex Trade*, Dasgupta-Alliance, Kolkata: 6.

Ibid: 32-33.

Note I: As reported in the 9th annual Trafficking in Persons (TIP) report released by the US Secretary of State, Mr. Abdul Latif Mandal, former Secretary to the Bangladesh Government, claimed its goal 'to stimulate action and create partnerships around the world to fight against the modern-day slavery'. Bangladesh has been identified in this report (TIP) 'as a source and transit point for men, women and children trafficked for labour and sexual exploitation'. Women and children are trafficked primarily to India and Pakistan 'with the promise of jobs and marriage', boys usually are under 10 and trafficked girls are between 11 and 16 years. Tourism in the developed western countries, as well as in a few developing countries in Asia, thrives to a considerable extent on the sex industry. Young and beautiful girls are trafficked from the developing countries, including Bangladesh for the booming sex industry worldwide. *Source: The Statesman*, Kolkata Issue, 26 June 2009

UNDP: human Development Report 1999: 89.

Sinha and Dasgupta: 181.

Ibid: 185.

Chatterjee, Shoma, A., 2009, Trafficking by Social Sanction, *The Sunday Magazine, The Statesman*, Kolkata Issue, 8 March 2009.

Pande, Rekha, 2008, Ritualized Prostitution: Devadasis and Jogins – A Few Case Studies, in '*Prostitution and Beyond: An Analysis of Sexwork in India*', Sage Publications: 102-103.

Sinha and Dasgupta, 2009: 21-22.

Pande, 2008: 106-107.

Chatterjee, Shoma A, 2009.

Elizabeth, V.S., 2009: *Violence Against Women in the Family* etc., 146 – 147.

Note J: Introducing the 'Illegal Traffic in women and Girls (Prevention) Bill in 1986 in Indian Parliament the then MP (Member of Parliament) Ms Margaret Alva said;"The exploitation of women and girls for the purpose of prostitution... has persisted since time immemorial, it has all through been considered an evil that wrecks the foundations of the family and the community, as basic units of human society... The most disturbing aspect today is the organized racketeering in the sexual abuse of innocent children and young persons in the trade ... A number of individuals, advocacy groups and women's voluntary organizations, have been urging upon the government to enlarge the scope of the Act". *Source*: Sahni, R, V Kalyan Shankar and Hemant Apte (Ed), 2008, *Prostitution and Beyond: An Analysis of Sex work in India*, Sage Publications: 29.

Elizabeth V.S., 2009: 153-154.

Note K: There were two news on the same day in *The Statesman*, Kolkata Issue on dowry-related violence. In one of them it was reported that a 23-year old woman was set ablaze allegedly by her husband after her father failed to pay dowry. She died. In the second news, a two-month old son was murdered by his father, alleged by his life to police as her father refused to pay (Rs. 20,000 or about US $ 445) demanded by son-in-law. *Source: The Statesman*, Kolkata Issue, 12 May

2009.
UNDP, Human Development Report: 40.
Kumari, Ranjana, 2009, Globalizing India and Domestic Violence Against Women, in "*Family Violence in India* etc., By Swati Shirwaskar: 211.
Ibid: 212-213.
Ibid: 218-219.

Note L: Domestic violence costs British economy 5.8 billion pounds a year. It significantly depresses woman's ability to contribute to economic growth, including the costs of business alone – through absence, loss of productivity and rapid turn over of employees – which is estimated at 2.7 billion pounds while the cost to the public sector, in terms of medical and social services is another 3.1 billion pounds *Source*: Hirsch Afna, *The Guardian*, The U.K., 25 November 2008.

In Britain domestic violence accounts for one in six of all reported attacks (many going unreported) according to the National Domestic Violence Delivery Plan, and repeat victimization is an enormous factor in domestic violence involving more than one in four victims – 27 per cent – having been victimized at least three or four times. Audrey Gillan, *The Guardian*, The U.K., 1 November 2008.

Note M: Perhaps, the first mention of '*Sati*' (Sutti), that is the practice of burning of wife still alive on dead husband's pyre was made by Greek writers that occurred in the fourth century B.C., and the practice seems to have continued among the Hindus through centuries later during the Hindu as well as Muslim rules in India. Majumdar et al. (1970) describing the social conditions in India in mid-seventh century A.D. mention that the custom of burning widows on the funeral pyre of their husbands was coming into general practice. In the early fourteenth century the 'horrible rite', the '*jauhar*' used to be practiced for immolating the females so that they were saved from 'pollution or captivity'. "The rite of Sati, or woman burning themselves on the funeral piers of their husbands, was very common in Vijaynagar (in the fourteenth century), and the Brahmins freely sanctioned it" and this was widely prevalent among certain classes. Its popularity may be assumed from the fact that even "some Muslims of aristocratic Hindu origin, or living in a Hindu environment, assimilated the Hindu custom of *Sati* or *Jauhar*". However, Aurangzeb, the Mughal Emperor of India "forbade Sati in December 1663" but evidences show that the "royal prohibition was seldom observed." One of his predecessors, Emperor Akbar, also tried to control social practices of *Sati*, child-marriage and dowry-system but "his attempts do not seem to have been effective in practice. "More successful drive against these practices came much later, starting sometime in mid-nineteenth century during British rule in India, beginning with the promulgation of 'Bengal Regulation XXI of 1795' and 'Regulation VI of 1802 dealing with widespread prevalence of infanticide of girl-child, especially among the Rajputs, Jats and Mewars in Central and Westen India, followed subsequently by passing the famous 'Regulation XVII' on 4th December 1829 which "declared Sati illegal and punishable by Courts, the major drive for which came from Raja Rammohan Roy, the leading reformist of his time. This was followed by another Revolutionary Act, enactment of the Hindu Widows Remarrigae Act of 1856 primarily initiated by the famous educationist and reformist, Pandit Iswar

Chandra Vidyasagar (Bandopadhyay). *Source*: Majumdar, R.C., H.C. Roy Chaudhuri and Kalikinkar Datta, 1970 (Reprinted), *An Advanced History of India*, Macmillan St. Martin's Press: 190, 294, 370, 393, 395, 489-90, 561, 766, 807, 816-819.

Mathur, Kanchan, 2009, Witchcraft, Witches and Social Sanction, in '*Family Violence in India*, Ed. by Swati Shirwadkar, Rawat Publications, India:286.

Sakaria, Ajay, 1997, Women, Witchcraft and Gratuitous Violence in Colonial Western India, in '*Past and Present*', May Issue.

Mathur, Kanchan, 2009: 287-288.

Ibid: 293-299.

UNICEF 2007: 4.

Ibid: 23.

Ibid: 27.

Psacharopoulos, George and Zafiris Tzannatos, 1991, Female Labour Force Participation and Education in '*Essays on Poverty and Growth*', Ed. by George Psacharpoulos, Pregmanon Press for World Bank: 275.

Flegg, A.T., 1982, Inequality, Illiteracy and Medical Care as Determinants of Infant Mortality in Underdeveloped Countries, in *Population Studies*, Vol 36, No. 3 (ov 1982): 443.

Ibid: 454.

Caldwell, John C, 1986, Routes to Low Mortality in Poor Countries in *Populatioin and Development Review*, Vol 12, No 2 (June 1986): 171-200,

UNICEF 2007: 37-38.

UNIFEM, 2008/2009: 37-38.

Note N: According to a report of the Press Trust of India, researchers in Britain found through their study that women spend an average of 74 hours a week on household chores and childcare against only 53 hours a week spent by men on domestic tasks. *Source: The Statesman*, Kolkata Issue, 17 March 2009.

UNICEF 2007: 39.

Ibid: 40

Ibid: 41.

Note: O: In a report, Jenny Percival claims that men are paid 17.1 per cent more than women for full-time work, according to government figures and in part-time, the gender gap is 36.6 per cent according to the Office of National Statistics. "Over a lifetime, that disparity means that women in full-time employment will be paid 369,000 pounds less their male counterparts, the EHRC estimated" for Britain. *Source: The Guardian*, The U.K., 6 January 2009.

Further, as repoted by Denis Campbell, Health correspondent of the Guardian, "female doctors working in the NHS (National Health Service) are paid thousands of pounds a year less than their male colleagues as a result of widespread discrimination... There is an average salary gap of 15,245 pounds between men and women ... (and) female consultants typically earn 5,500 pounds less than their male peers and female junior doctor's pay is around 2,000 pounds below that of their male counterparts... Among Britain's 40,521 consultants, men are on average earn 13,729 pounds more than similarly successful female colleagues. The pay gap worsens the longer a female consultant has worked". The Guardian, The U.K., 10 November 2009.

Chatterjee, Shoma A, 2009.
UNICEF: 43-44.
Sen, Amartya, 2000 (2008: 15th Impression), *Development as Freedom*. (Indian Paperbacks): 190-191.
Ibid: 191.
Ibid: 197-198.
Hulme, David and Paul Mosley, 1986, *Finance Against Poverty*, Routledge, London and New York,: 125.
UNICEF 2007: 21.
Op cit.
UNICEF 2007: 20.
Ibid: 22.
UNIFEM 2008/2009: 26.
UNICEF 2007: 56-57.
UNIFEM 2008/2009: 21.
UNDP: Human Development Report 2007/2008: Table 33.
Butt, Riazat, 2008, *The Guardian*, The U.K., 1 November 2008.
UNIFEM 2008/2009: 78.
Ibid: 79.
Ibid: 105:
Fig. 6.9. UN Millennium Project: Task Force on Education and Gender Equality, 2005, Earthscan: 90.
UNIFEM 2008/20098: 58.
UNDP: Human Development Report 2000: Human Rights and Human Development: 41.
Sen A, 2008: 30.
Ibid: 114-115.
Government of India, 2002, Planning Commission, National Human Development Report, Oxford University Press: 94-95.
Bajpai, Asha, 2009 (Third Impression), Child Rights in India: Law, Policy and Practice, Oxford University Press: 151-153.
Sekar, Helen, 2001, Ensuring their Childhood, V V Giri National Labour Institute, New Delhi: 107
Government of India, 2002: 97.

Note P: Official estimate put the figure of child laboureres in India at 12.6 million. A recent survey shows that 'in 40 per cent of Kolkata's slum families, one child, and in 59.6 per cent families, more than one sibling joins the labour force: 55.8 per cent of these children do not attend school; 45.3 per cent discontinue school due to poverty (69.9%), father's apathy (8.4%), children's own attitude (10.4%) and frequent absence and physical assaults by school teachers (5.0%) …wages for child labourers are exploitative and low: as little as Rs. 50 (a little over one US Dollar) per month in 20.8 per cent, for work of an average 10-12 hours per day. *Source: The Statesman*, Kolkata Issue, 12 June 2008.

According to the Labour Commissioner, Government of West Bengal, 'there are around 8 lakh (800,000) child labourers in the state, between six and fourteen, who are engaged in domestic work, tea stalls, sericulture, brick-building, tannery industry and agriculture."

Source: The Statesman, Kolkata Issue, 13 June 2008.
Bajpai, Asha, 2009: 428.
Burra, Neera, 2006, Born to Work, in '*Born Unfree; Child labour, education and the State in India*' Ed. by Myron Weiner, Neera Burra and Asha Bajpai, Oxford University Press: 27-28.
Ibid: 191-192.
Bajpai, Asha, 2009: 201.
Weiner, 2006: XXXIII.
Weiner, Myron, 2006, The Child and the State in India, in "Born Unfree etc., 27-29.
Burra Neera, 2006: 193 and 205.
Ibid: 56-59, 79, 81, 110-111, and 163.
Bajpai, Asha, 2009: 212-214.
Ibid: 214.
Randerson, James, Science Correspondent, *The Guardian*, The U.K., 8 December 2008.
UNDP, 2005, International Cooperation at a crossroads: Aid, Trade and Security in an Unequal World: 61.
Ibid: 62.
UNICEF 2007: 28-29.
Ibid: 30-31.
Flegg, A.T., 1982: 442-443.
Bosley, Sarah, Health Editor, *The Guardian*, the U.K., 3 December 2008.
— —, *The Guardian*, The U.K., 10 December 2008.
UNDP, Human Development Report 2003: 45
Phansalkar-Joshi, S.S., 2009, Child Sexual Abuse: A Violation of Human Rights ? in '*Family Violence in India: Human Rights*, Issues, actions etc., Ed. by Swati Shirwaskar: 162.
Ibid: 162-163.
Sinha and Dasgupta, 2009,: 27-28.
Bajpai, Asha, 2009: 420.

CHAPTER 4

Deepa, Narayan, Robert Chambers, Meera K Shah and Patti Petesch, 2000, *Voices of the Poor: Crying out for Change*, Oxford University Press: 21.
Sen, Amartya, 1995 (5th Impression), Inequality Re-examined, Oxford University Press: 72.
UNDP, 2002: Human Development Report: Deepening Democracy in a Fragmented World: 11.
Ibid. 22.
Ibid 22-23.
UNICEF, The State of the World's Children: 71.
UNIFEM: Progress of the World's Women etc. 2008/2009: 37.
Ibid: 38-39.
UN: Millennium Project, 2005: 47.
Sachs, Jeffrey, 2005, *The End of Poverty* etc., 254-255.
Hawthorn, Geoffrey, Ed., 1998, *The Standard of Living*, Cambridge University

Press: 1.

UNDP: Human Development Report 2007/2008: Fighting Climate Change: Human Solidarity in a Divided World: 249-250.

Collier, Paul, 2008, The Bottom Billion: Why the Poorest Countries are Failing and What Can be Done About It ?, Oxford University Press: 93-94.

Government of India, Planning Commission, 2002, National Human Development Report, Oxford Univesity Press: 49.

Viswanathan, Brinda, 2007, Living Conditions of Women, in '*Solutions For the World's Biggest Problems, Costs and Benefits*", Ed. by Bjorn Lomnborg, Cambridge University Press: 377.

Op cit.

Government of India, Planning Commission, 2002 etc.: 51.

— —, 2003, Tenth FiveYear Plan 2002-2007, Vol I: 17.

— —, Planning Commission, 2008, Eleventh Five Year Plan, 2007-2012, Oxford University Press: 2, 13.

Note A: "The Social Composition of out-of-school children indicates that 9.97 per cent of Muslim children, 9.54 per cent of Scheduled Tribes (STs), 8.17 per cent of Schduled Castes (SCs), and 6.97 per cent of Other Backward Class (OBCs) children were out of school and an overwhelming majority (68.7%) was concentrated in five states, viz. Bihar (23.6%), UP (22.2%), West Bengal (9%), MP (8%), and Rajasthan (5.9%)". *Source*: Government of India, Planning Commission, 2008, Eleventh Five Year Plan (2007-2012), OUP: 3.

Sen, Amartya, 2008 (15th Impression),: 41-42.

Sachs, Jeffreys, 2005: 317,

Note B: The 'Right to Education Act' – making free and compulsory education up to class eight a fundamental right for every child in the country – came into force on 1 April 2010 in India where, at present, 4.6 per cent of the nearly 22 crore (220 million) children in the 6-14 year age group (or nearly 92 lakhs, or 9.2 million) are out of school, according to Human Resource Developmnent (HRD) ministry. The child had to wait 60 years to get this right though it is not very clear of becoming a total success as "uncertainty deepens with the states reluctant to bear more than ten per cent of the expenditure" (Editorial, *The Statesman*, Kolkata Issue, 2 April, 2010), more so as the State Education Minister of West Bengal stating that, 'the state did not have proper infrastructure to implement the Right of Children to free and compulsory education. *Source: The Statesman*, Kolkata Issue, 2 April, 2010.

Sen, Amartya, 2008: 197-198.

Mukhopadhyay and Garimella, 2001: 99-101.

Sen, Amartya, 2008: 221.

Orazem, Peter F, 2007, Lack of Education, in "*Solutions for the World's Biggest Problems* etc. 249.

Op cit. Sen, Amartya, 2008: 292-294.

UNDP: Human Development Report, 2003: 85.

WHO, The World Health Report, 2008: 2-4.

Ibid: 7.

Ibid: 8.

Murray Christopher J.L. and Alkan D. Lopez, 1997: 1436.

Ibid: 1438.

Ibid: 1440.

Gatrell, Anthony C, 2002, Geographies of Health, an Introduction, Blackwell: 5.

Bhopal, R, 2002, *Concepts of Epidemiology: An Integrated Approach to the Ideas, Theories and Methods of Epidemiology*, Oxford University Press: 3.

Sachs, Jeffrey, 2005: 213.

Ibid: 204.

WHO, 2005, Preventing Chronic Diseases: A Vital Investment: 2-3.

Ibid: 6-7.

Ibid: 8-9.

Ibid: 5.

United Nations, 1999: Health and Mortality: Issues of Global Concern: 447.

Ibid: 449.

United Nations, 1997: Fig 133.

Ibid: 448.

Misra Rajib, Rachel Chatterjee and Sujatha Rao, 2003, India Health Report, Oxford University Press: 87-88.

United Nations, 1997: 462.

Sachs, Jeffrey, 2005: 322-323.

Note C: Sub-Saharan Africa accounts for almost 90 per cent of Paediatric HIV infections worldwide. An estimated 530,000 children were newly infected in 2006, mostly through mother- to-child transmission. Without treatment, half of the "infants born with the virus will die before age two. Girls are at particular risk of contacting HIV, both because of their physiology and because of social and cultural power imbalances in their relationships with men and boys" (UNICEF, 2008: 52). "Among all adults living with HIV/AIDS, the share of women living with AIDS has increased from 45 per cent in 1990 to 50 per cent in 2007. In developed countries, it can reach 30 per cent. But in regions where the problem is most serious, the pandemic has become feminized. In Sub-Saharan Africa the share of women among adults living with HIV/AIDS has increased from 54 per cent in 1990 to over 60 per cent in 2007. In the Caribbean, this has grown from 24 to 43 per cent. Of particular concern is the elevated level of HIV prevalence among young women, who are two to three times more likely to be infected with HIV than men in the same age-group in high prevalence environments. One reason for this is the lower proportion of young women than men with access to comprehensive and correct knowledge about HIV/AIDS... According to WHO, violence is both cause and consequence of HIV infection. In some countries, the percentage of women reporting that their first sexual experience was forced – one of the reasons for increasing infection rates among young women – is as high as 30 per cent. Infected women sometimes experience further violence from their partners and communities, due to stigma and discrimination". *Source*: UNIFEM: Progress of the World's Women 2008/2009.: 128.

UNDP: Human Development Report 2002: 27.

UNDP: Human Development Report 2003: 43.

Taylor, V., 1998, HIV/AIDS and Human Development, Pretoris, Amabhuka Publications

Myeza, Nokujabula and Dennis Francis, 2009, 'Openly Positive': Living and working with HIV/AIDS, in *'Working Women: Stories of Strife, Struggle and Survival,* Ed. by Kogi Naidoo and Fay Patel, Sage Publications, New Delhi: III

Gopal, N and Reshma Sookrajh, 2009, Children Left Behind: Voices (Ukhuhebeza) of HIV + Mothers, in 'Working Women etc., Ed.by Kogi Naidoo and Fay Patel:123-124.

UNDP: Human Developmet Report 2001: 106.

Ibid: 106-107.

UNDP: Human Development Report 2003: 43.

United Nations 2007, Strengthening Efforts etc., 252.

Sachs, Jeffrey, 2005: 196-197.

Sachs, Jeffrey and Pia Malaney, 2002, The Economic and Social Burden of Malaria, *Nature* 415, 7 February 2002: 680-685.

Sachs, Jeffrey, 2005: 198-199.

Greenwood, Brian and Theonest Mutabingwa, 2002, Malaria in 2002, *Nature*, 415, & February 2002: 670-672.

Misra *et al.*, 2003: 58.

Op cit.

UNDP: Human Development Report 2003: 103.

Sachs and Malaney, 2002: 680-685.

UNDP: Human Development Report, 2003: 100.

Sachs, Jeffrey, 2005: 215.

United Nations, 1999, Health and Mortality: Issues of Global Concern: 54-55.

Murray, Christopher, J.L. and Alan D Lopez: *The Lancet*, Vol 349, May 17: 1440.

United Nations, 1999: 27.

Peto, R., A.D. Lopez, J. Boreham, M Thun and C Heath, 1994, *Mortality from Smoking in Developed Countries*, 1050-2000, New York, Oxford University Press.

United Nations, 1999: 400-401.

Ibid: 397.

Murray and Lopez, 1997: 1440.

United Nations, 1999: 403.

Note D: Thirty Leading Causes of Death Worldwide in 1990

Rank	*Causes of Death*	*Number of Deaths x 1000*	*Rank*	*Causes of Death*	*Number x 1000*
1.	Ischaemic Heart disease	6,250	16	Diabetes mellitus	571
2.	Cerebrovascular disease	4,381	17	Violence	563
3.	Lower Respiratory Infections	4,299	18	Tetanus	542
4.	Diarrhoeal Diseses	2,946	19	Nephritis and Mephrosis	536
5.	Perinatal Disorders	2,443	20	Drowning	504
6.	Chronic Obstructive Disease	2,211	21	War Injuries	502
7.	Tuberculosis (not HIV)	1,960	22	Liver Cancer	501
8.	Measles	1,058	23	Inflamatory Heart Disease	495

9.	Road Traffic Accident (RTAs)	999	24	Colon and Rectum Cancers	472
10.	Trachea, Bronchus and Lung Cancers	945	25	Protein-energy Malnutrition	372
11.	Malaria	856	26	Oesophagus Cancer	358
12.	Self-Inflicted Injuries	786	27	Pertussis	347
13.	Cirrhosis of Liver	779	28	Rheumatic Heart Disease	340
14.	Stomach Cancer	752	29	Breast Disease	322
15.	Congenital Abnormalities	589	30	HIV	312

Source: UN 1999, Table 76 (Taken from Murray and Lopez, 1997a).

Note E: Sarah Bosley, the Health Editor of *the Guardian*, The U.K., quoting from a recent report published in '*Lancet*' Medical Journal, says that alcohol has been found to be the "most dangerous drug in the UK by a considerable margin beating heroin and crack cocaine into second place and third place. The medical report (in Lancet) examined nine categories of harm that drugs can do to the individual – "from death to damage to mental functioning and loss of relationships" – and seven types of harms to others. The maximum possible harm scores was 100 and the minimum zero. Overall, alcohol scored 72 against 55 for heroin and 54 for crack. *Source: The Guardian*, The U.K., 1 November 2010.

Another report, presented by Denis Campbell as Health Correspondent of the Guardian, shows that the "number of underage drinkers admitted to hospitals in England rose by 32 per cent between 2003 and 2007, from around 11,000 in 2003 to more than 14,000 in 2007. A total of 92,220 children and young people under 18 were admitted to hospital between 2002 and 2009, or 36 under 18 a day. Girls are 1.3 times more likely than boys need to be admitted due to alcohol."

Source: The Guardian, the U.K., 23 October 2010.

United Nations, 1999 : 419.

Note F: According to a report by Randeep Ramesh in the Guardian, road accidents take 3,500 people everyday on the world's roads, about 3,000 of whom are from developing countries and this is likely to rise to "more than 5,700 a day in 10 years' time unless the governments act. In the opinion of Kevin Watkins, Oxford University, "more lives among those aged five to fourteen were lost on the roads than to malaria, diarrhoea and HIV/AIDS ... While rich countries have brought down death tolls on the roads, the poor have a rising body count ... On the Sub-Saharan highways the projected deaths are set to double, while in Europe they will fall by 36 per cent in the next 10 years. In Ethiopia figures show 100 people were killed for every 10,000 cars. The comparable for Japan is one death for every 10,000 cars... (making) the fatality risk in excess of 100 times (for Ethiopeans) than faced by a Japanese road user...... Watkins said that road deaths and injuries were not accidents but caused by the criminal neglect of road safety". Source: *The Guardian*, The U.K., 13 September 2010.

United Nations, 1999: 436.

Note G: As James Meikle reports in *The Guardian*, "the worlwide cost of dementia this year (2010) will be 388 billion pounds in social care by relatives and medical bills", equating to "more than 1 per cent of Global GDP". The world Alzheimer Report 2010, prepared by the Acadmics at the King's College, London, and the Karolinska Institute in Stockholm said "governments worldwide should follow the example of countries such as England, France and Australia in developing national plans to tackle the problem". A similar report in 2009 "suggesting that there were 35.6 million people with dementia worldwide, and that this figure would grow to 65.7 million by 2030 and 115.4 million by 2050", making it one of the greatest health problems to be cared for taking further into account the fact that the actual dimension of the problem is much larger than the figures shown in these reports when (one knows that) many (similar) cases occurring in the developing countries go unrecorded in the absence of proper records as well as right diagnosis. *Source: The Guardian*, The U.K., 22 September 2010.

Who, TheWorld Health Report 2008: Now More than Ever: 105.

Op cit.

UNDP, 2003: Millennium Development Goals etc. 2-3.

UNDP: Human Development Report, 2006: Beyond Scarcity: Power, pcverty and the Global Water Crisis: 3.

L'vovich, Mark I and Gilbert F. White, 1990, Use and Transformation of Terrestrial Water Systems, in '*The Earth as Transformed by Human Action: Global and Regional Changes in the Biosphere over the Past 300 years*', Ed. by B L Turner II, William C Clark, Robert W Kates, John F. Richards, Jessica T Mathews and William B Meyer, Cambridge University Press: 245.

Ibid: 246.

UNDP, 1998: Human Development Report: 68.

Brown Lester, R., 2006, Plan B 20 etc., 42.

Brown Lester, R., Gary Gardner and Brian Halweil, 2000, Beyond Malthus etc., 37.

Brown Lester, R., 2006: 42-43.

UNDP, 2006: Human Development Report: 5-6.

Op cit.

UNIFEM, 2008/2009: 130.

UNDP, 2006: 7.

Brown *et al.*, 2000,: 38.

Brown, Lester R., 2006: 44.

Ibid: 44-48.

Ibid: 49.

McCully, Patrick, 1998, *Silenced Rivers: The Ecology and Politics of Large Dams*, Orient Longman: 3-4.

Brown, Lester R., 2006,: 48-52.

Ibid: 52-53.

United Nations, 2003, World Water Development Report: 13.

Brown *et al.*, 2000: 39.

UNDP, 2006, Human Development Report: 14.

Ibid: 15.

Ibid: 27.

Ibid: 34-35.

United Nations, 2003: World Water Development Report: 91.

Ibid: 90-92.

UNDP, 2006: Human Development Report: 49.

Ibid: 52.

Ibid: 57.

UNDP, 2006: 36.

Government of India, 2002: 40.

Note H: A source quoting municipal authorities in Kolkara city says that 75 per cent of the city population has access to clean water which has not increased during the last five years. The latest area adjoining the Eastern Metropolitan By Pass containing about 16 per cent of the city population is deprived of this facility. South Kolkata, comprising Behala, Jadavpur and Tollugunge areas, suffers from a scarcity of water despite receiving clean water from the supply system because of its fast growing population, having doubled in the last 25 years.

Source: Bartaman, Kolkata Issue (a Bengali Newspaper), 29 March 2010

UNDP, 2006: 58-59.

UNDP, 2006: 59.

Ibid: 135-138.

Ibid: 157.

Ibid: 169-170.

Ibid: 176.

Ibid: 61.

Ibid: 62.

Government of India, Planning Commission, Eleventh Five Year Plan, 2007-2012, Vols I and Ii.

Notes I A and I B: Achievement of Millennium Development Goals

Prior to the meeting of the world leaders in New York in the fourth week of September 2010, for an overall assessment of the scope of accomplishing the Millennium Development Goals set for 2015, '*The Guardian*" (The U.K.) has brought out a special report (14th September 2010) on the success so far achieved in this regard, giving a correct insight into the existing situation, as summarized in the following paragraphs "

Target 1 was to halve the proportion of people suffering from hunger by 2015. The number of hungry people having reduced by 98 million stood at 925 million in 2008. While China has halved its poverty, in India the poverty level is stable, "but UN is confident to hit the target by 2015 though the recent food crisis has pushed the hunger target off track". The progress has been quite patchy in Africa. Egypt, Ethiopia and Angola have halved but "Niger's record is poorer." There has been progress on hunger in more than half of countries but in 28 countries the levels have remained the same while in 24 it has increased. The UN is confident the poverty levels will be halved by 2015.

Target 2: on achieving universal primary education nearly 90 per cent of countries have made progress with the "greatest strides"made in Sub-Saharan Africa, Ethiopia recording the largest absolute increase. However, the rates of dropout are higher among poor households. The pace of progress is too slow to fulfil the target.

Target 3: To eliminate gender disparity in primary and secondary education, preferably by 2025, and in all levels of education by 2015.

There has been significant progress in parity in primary education, "with Ethiopia and Guinea having been the topper-performers. China has achieved parity and India is moving closer" (interestingly, the number of female parliamentarians in India was 11 per cent in 1995 and it was 19 per cent in 2010). However, "girls are still out of school and the gender gap is wide in secondary education. Share of women employed outside agriculture is increasing, but very slowly", and "unless the pace of progress increases dramatically, the MDG is unlikely to be achieved ".

Target 4: To reduce under five mortality by two-thirds between 1995 and 2015.

Progress is steady, the mortality rate having fallen from 100 deaths per 1000 live births in 1990 to 72 in 2008, which means that the number of under fives dying globally fell from 12.5 million to 8.8 million during this", and "unless the pace of progress increases dramatically, the MDG is unlikely to be achieved ".

"Immunisation has increased globally, but children from poorer rural households are less likely to be immunized. Immunisation rates in China and VietNam has declined". Hence, there is much doubt about meeting MDG unless further fund is committed.

Target 5: To reduce maternal mortality ratio by three-quarters between 1990 and 2015. With wide variation in progress, "in about one-third of developing countries, skilled health workers now attend 95 per cent of all births, and nearly 20 per cent have almost universal access". But,"birth attendance is still low in countries in Sub-Saharan Africa and South Asia. Access and use of Contraceptives is still lowest among poor countries". As a result of progress being "too low", the MDG cannot be met.

Target 6: To reverse the spread of HIV/AIDS, malaria and other major diseases, and to ensure universal success. HIV infection rates have so far dropped in 26 per cent countries, and 41 per cent have recorded "no change. Sub-Saharan Africa has the highest HIV infection rates in the world, but 21 of the 27 countries" have reduced rates. Progress has been made on malaria. Interestingly, "HIV infection rates have increased in Russia and the former Soviet Block. In Latin America, only Honduras has reduced the number of infection". Prospect of reaching MDG in HIV is nil though"treatment targets could be reached by 2020 Malaria target may be reached with more funding".

Target 7: To reverse the loss of environmental resources and biodiversity, halve the number with no access to clean water and sanitation, and improve the life of slum dwellers. There has been huge improvements on access to clean water. "India and China have already met the target. Deforestation is declining but is still high. CO^2 emissions dipped but are projected to rise again. The proportion of slum dwellers is lower but absolute numbers have risen". There is no likelihood of meeting the MDG though "access to clean water is the only target expected to be met".

Target 9: To develop global partnership for development and address the needs of the least developed countries. "Debt levels have dropped and 40 countries are now eligible for debt relief under the heavily indebted poor countries initiatve. Developing countries have had greater access to global markets since tariffs were reduced but the financial crisis hit the least developed countries the hardest". Pledges made by the wealthiest countries in 2005 for providing aids are yet to be fulfilled. It is very difficult to say if the MDG could be met by the stipulated time. *Source: The Guardian,* The U.K., 14 September 2010.

Note I B: According to another report made by Randeep Ramesh in the Guardian, "not enough is being done to cut the number of child deaths across the globe". Though"only 3.5 million more children survive past their fifth birthday today than 20 years ago, the scale of task is such that at least 1 million more children every year will need to live past five to achieve the UN's 2015 target. Nearly three-quarters of the countries with the highest child mortality burden will not reach the goal on current trends". *Source: The Guardian,* The U.K., 6 September 2010.

UNDP, 2006: 59.

UNDP, 2006: 135-138.

Ibid: 157.

Ibid: 169-170.

Ibid: 176.

Ibid: 61.

Ibid: 62.

Ibid: 120.

CHAPTER 5

Note A: Amartya Sen, in defining poverty, notes that "starvation, clearly, is the telling aspect of poverty" though according to him the nutritional requirements are difficult to define as dependent on 'physical features, climatic conditions and work habits' of people and in his opinion malnutrition takes a central place in the conception of poverty. Deprivation and poverty are closely associated and in this regard the "measurement of poverty must be seen as an exercise of description assessing the predicament of people in terms of the prevailing standards of necessities" *Source*: Amartya Sen, 2008, *Poverty and Famines,* OUP: 12-21.

Dreze, Jean, Amartya Sen, 1999, Public Action for Social Security: Foundations and Strategy, in "*Social Security in Developing Countries*', Ed. by E. Ahmad, J. Dreze, J Hills and A Sen, Published by United Nations University in 1991: First published in India by OUP in 1999 as Paperback; 8.

Ibid: 9.

Ibid: 22.

Hulme, David and Paul Mosley, 1996, *Finance Against Poverty,* Vol I, Routledge: 105.

Ibid: 105-106.

Eastwood, Robert and Michael Lipton, 1999, The Impact of Changes in Human Fertility on Poverty, in '*Journal of Development Studies*', 36:1:4.

Khan, Azizur Rahman, 2006, Employment Policies for Poverty Reduction, in '*Fighting Poverty: The Development-Employment* Link, ed. by Rizwanul Islam, Viva Books Private Ltd.,: 63.

Vandemoortele, Jan, 2004 (Reprint), Are We Really Reducing Poverty ?, in '*World Poverty, New Policies to Defeat an Old Enemy*', Ed. by Peter Townsend and David

Gordon, The Policy Press: 378.

Note B: "According to the preliminary data released by the Planning Commission (of India) last week, India's poverty is estimated to have declined to 32 per cent in 2009-10 from 37.2 per cent five years ago", *Source: The Statesman,* Kolkata Issue, 27 April 2011.

Gordon, David, 2004 (Reprint), The International Measurementof Poverty and Anti-Poverty Policies, in *'World Poverty, New Policies to Defeat an Old Enemy* etc.,: 61-62.

Collins, Daryl, Jonathan Morduch, Stuart Rutherford and Orlanda Ruthven, (2010), Portfolios of the Poor: How the World's Poor Live on $ 2 A Day, Permanent Black, Ranikhet: 5.

Note C: As reported by Ewen Macskill from Washington, "one in seven Americans now live on or below the poverty line, according to figures published yesterday (16 Sep. 2010) by the Census Bureau. It is the sharpest annual rise for three decades and analysts predicted next years figures will be even worse. According to the Bureau, 43.6 million people or 14.3 per cent of the population were in poverty in 2009 up from 39.8 million in 2008. The hardest hit are African-Americans and Hispanics". The same source reports that 'apalling levels of poverty still existed in the 1950s, the numbers on or below poverty line being 22.4 per cent in 1959 which, though, dropped to 19 per cent by 1964 in the "Great Society" where, according to the Census Bureau Report, the number of people without health insurance benefit rose from 46.3 million in 2008 to 50.7 million in 2009.' The recent introduction of a healthcare reform package by President Obama is likely to improve the situation greatly but full implementation is not expected before 2014. *Source: The Guardian,* The U.K., 17 September 2010.

Islam, Rizwanul, 2006, The Nexus of Economic Growth, Employment and Poverty Reduction, An Empirical Analysis, in *'Fighting Poverty: The Development-Employment Link*", Ed. by Rizwanul Islam etc.,: 34.

Ibid: 43.

Ibid: 59.

Khan, Azizur Rahman, 2006, Employment Policies for Poverty Reduction etc.,: 64.

Ibid: 64-65.

Ibid: 97-98.

Dreze and Sen, 1999: 24.

UNDP, Human Development Report, 2003: 18.

Ibid: 19.

Ibid: 76-78.

United Nations, 2007: 250.

UNDP, 2003: 79.

Osmani, S.R., 1999, Social Security in South Asia, in *'Social security in Developing Countries,* ed by E. Ahmad *et al.*, etc.,: 306-307

Ibid: 318-319.

Rahman, R. Islam and K.M. Nabiul Islam, 2006, Bangladesh: Linkages Among Economic Growth, Employment and Poverty, in *'Fighting Poverty: The Developmeny-Employment Link* etc.,: 118.

Ibid: 130.

Ibid: 155-156.

Collins, Daryl, Jonathan Morduch, Stuart Rutherford and Orlanda Ruthven, 2009 etc.,: 155.

Ibid: 171-173.

Braun, Joachim von, 1999, Social security in Sub-Saharan Africa: Reflections on Policy Challenges, in '*Social Security in Developing Countries*', Ed by E. Ahmad *et al.*,: 395.

Ibid: 402-403.

United Nations, 2007, Strengthening Efforts to Eradicate Poverty etc.,: 125.

Kanji, Nazneen, 2004, Social Funds in Sub-Saharan Africa: How Effective for Poverty Reduction ?, in '*World Poverty* etc., Ed by Peter Townsend and David Gordon: 246.

United Nations, 2007: 126-127.

Islam, Iyanatul, 2006, Indonesia: Poverty, Employment and Wages, in '*Fighting Poverty* etc., Ed. by Rizwanul Islam *et al.*: 314-315,

Ibid: 324.

Ibid: 355.

Ibid: 356.

Ahmad, Ehtisham and Athar Hussain, 1999, Social Security in China: A Historical Perspective, in '*Social Security in Developing Countries*', Ed by E. Ahmad *et al.*,:250

Ibid: 267.

Ibid: 268.

Ibid: 273-274.

Ibid: 276-278.

Hussain, Athar, 2002, Urban Poverty in China: Incidence and Policy Responses in '*World Poverty etc.*' Ed by Peter Townsend *et al.*,: 315.

Mesa-Lago, Carmelo, 1999, Social Security in Latin America and the Caribbean: A Comparative Assessment in '*Social Security in Developing Countries* etc., 359.

Ibid: 368-369.

Note D: "Surveys carried out in seven countries – Argentina, Costa Rica, Cuba, Chile, Mexico, Peru and Uruguay – show that the historical appearance of coverage of various occupational groups was largely determined by the power of pressure groups, with a gap of almost 200 years between the first and the last groups covered by pensions. Those in the armed forces and civil service began to be covered between the beginning of the nineteenth century and the beginning of the twentieth; teachers between 1880 and 1930; the public between 1890 and the 1940s; the labour aristocracy (public utilities, banks, merchant marine) between 1910 and 1940s; the great mass of the uban labout force (white- and blue-collar workers) between 1920 and the 1940s; agricultural workers between 1930 and 1950s; domestic servants between 1930 and the 1970s; and the self-employed between 1930 ands the 1970s. One has to take account of the fact that in the majority of these countries all groups are covered, although there exists substantial differences in the degree of coverage, in spite of the measures taken in most of them to make it more universal, unified and uniform... the states or departments which are most highly developed ... (with) the highest per capita income are covered to a substantially higher degree than the states... which are least developed... (with) lowest per capita incomes. Coverage in a particular geographical region varies between 6 and 100 per cent in Argentina, 3 and 25 in Colombia, 54 and 100 in Costa Rica etc. "*Source: Mesa-Lago*, 1999: 371-372.

Government of India, Planning Commission, Eleventh Five Year Plan, 2007-2012, Vol II: 80.

Op cit.

Note E: Arjun Sengupta Committee's categorization of the poor in India into four groups is more logical. The four categories – as mentioned earlier – are "extremely Poor", "poor". "marginally Poor", and "vurnerable" and the poor and vulnerable are defined as those whose per capita expenditure falls below Rs. 20 (less than half-a-dollar) a day. The significance of the Arjun Sengupta (Economic Advisor to the Government of India and a member of the Planning Commission) Committee's cut off line of Rs. 20 per day expenditure is very relevant in understanding the plight of people living Below the Poverty Line (BPL) since according to the very recently issued guidelines of the central government for the Mahatma Gandhi NREGA, the minimum wage to be paid to a labour is Rs. 100 (a little over US $ 2) but as only one member of any given family can become an MGNREGA employee on any given day, the daily wage comes down to Rs. 20 per head (per day of a family of five members) a very modest number in the context of an Indian rural family). That means about 90 million Indians are living on less than the minimum recommended wage under the MGNREGA.

The World Bank, 2000, India: Reducing Poverty, Accelerating Development, Oxford University Press: 13.

Government of India, Planning Commission, 2007-2012: 85.

Ibid: 94-98.

Note F: Based on the 61st report of the NSSO (National Sample Survey Organization), the Planning Commission's latest estimates for 2004-05 put poverty incidence at 27.5 per cent, implying that 301.7 million people in India live Below the Poverty Line (BPL), with 28.3 per cent in the rural and 25.7 per cent in the urban areas. Contrary to the Arjun Sengupta Committee's report it was claimed that 77 per cent of the Indian population cannot spend even Rs 20 per day, as per world normal earning of US $ 1.25 a day set as the global poverty line. According to the World Bank's criterion, 456 million Indians or 42 per cent of the total population are poor while the Saxena Committee, appointed by the Ministry of Rural Development, Government of India, supervising the social sector schemes during the Eleventh FYP period, estimated that 50 per cent of the Indians are BPL, but the Tendulkar Committee, appointed by the Government of India, finds 37.2 per cent of people living below the poverty line in the country. This is corroborated by the Global Hunger Index, 2008, prepared by the International Food Policy Research Institute, showing nearly 350 million Indians are food insecure. *Source*: Biswadb Catterjee, *Perspective, in the Statesman*. Kolkata Issue,27 April 2010.

Planning Commission, Eleventh FYP Vol III: 130.

Ibid: 133.

Note G: "As per the Sarva Shiksha Abhiyan, Twelfth Joint Review Mission Report, all states (in India) have managed to reduce OUSC (out of school children) rate and increase enrolment. It has been revealed that the number of OUSC is 8.1 million in 2009, which is a reduction of 5.3 million children from the number reported in 2005. Uttar Pradesh, Bihar, Rajasthan, West Bengal and Orissa account for more than three-fourths of all the OUSCs in the country, the report states. In fact, Bengal ranks fourth, along with Uttar Pradesh, registering the highest OUSC, followed by Bihar and Rajasthan". This is due to the "improper implementation of mid-day meal scheme (MDMS)". In West Bengal, the scheme has not been implemented in a majority of schools of Kolkata while it was withdrawn from several schools in the districts where mid-day meals were being distributed earlier. *Source: The Statesman*, Kolkata Issue, 13 May 2011.

Planning Commission, Eleventh Five Year Plan, Vol III: 136.

Op cit.

Kumar, Sunil, 2004, Round Pegs and Square Holes: Mismatches between Poverty and Housing Policy in Urban India, in '*Worked Poverty: New Policies to Defeat an Old Enemy* etc., Ed by Peter Townsend etc.,: 276.

Planning Commission, Eleventh Five Year Plan, Vil III: 406.

Ibid: 410.

Basu, Kaushik, (Ed), 2006, (3rd Impression), India's Emerging Economy: Performances and Prospects in 1990s and Beyond, Oxford University Press: 154.

Ibid: 157.

Agarwal, Bina, 1995, Gender and Legal Rights in Agricultural Land in India, in '*Economic and Political Weekly*', March 24, 1995: 52.

Basu, Kaushik (Ed), 2006: 169-170.

Datt, Gaurav and Martin Ravallion, 2002, Is India's Economic Growth Leaving the Poor Behind ? in '*Journal of Economic Perspectives*, Vol 16, No 2, 2002: 97-98.

Government of India, Planning Commission, Eleventh Five Year Plan, Vol III: 79-81.

Datt and Ravallion, 2002,: 106.

Dealton, Angus and Jean Dreze, 2002, Poverty and Inequality in India: A Re-examination, in '*Economic and Political Weekly*, September 7, 2002: 3734-3735.

Brandt, Willy *et al.*, 1980, North-South: A Programme for Survival, Pan Books: 90-91.

International Food Policy Research Institute, 2002.

Brown, Lester R, 2006, Plan B 2.0 etc.: 163.

Op cit.

Note H: There is some confusion about the number of people living below the poverty line in India. While the Prime Minister of India claimed sometime in 2009 that poverty was declining in India, the government later accepted the Tendulkar Committee Report estimating people below the poverty line (BPL) at 37.2 per cent, much above 27.5 per cent taken by the Planning Commission for poverty line as recommended earlier by the N C Saxena Committee, the Wardha Committee and other such exercises. *Source*: N k Singh, *The Statesman*, Kolkata Issue, 30 April 2010.

Meadows, D.H, D.L. Meadows, J Randers and W.W. Behrens III, 1972, the Limits to Growth etc.,: 41 and 44.

World Population Prospects, The 2006 Revision, Vol III: Table 1.

Demographic Year Book, 2007.

Keilman, Nico, 2001, Uncertain Population Forecasts, *in Nature*, Vol 412, 2 August 2001, The U. K.

Daniels, Peter *et al.*, 2005: 89.

Op cit.

Ibid: 92.

Note I: "According to Weeks (1988), 'Islamic nations are the world's fastest growing population group' ...The majority is due to the high levels of natural increase, resulting from high birth rate and youthful age structure. "Data for 1988 put the fertility of Islamic nations at 6.0 children per woman, a third as much again as the 4.5 figure for other developing nations, while the proportion of people aged under 15 is put at 43 per cent. Analysis of the recent UN population data indicates the scale of growth over the past half century and how things are expected to change over the next five decade. These datas are for the 40 countries where Muslims made up at least half the population in 1988... The population of these 40 countries in 1950 was 302 million, but had more than trebled to 952 million in 1998, and is expected to reach 1,869 million by 2050... By 2050, these 40 countries would account for 21 per cent of world population, up from 12 per cent in 1950 and 16 per cent in 1998. These projections, however, assume that the Islamic world moves to replacement-rate (around 2.1 births per woman) before 2050". *Source*: Daniels *et al.*, 2005, 93.

UN, 2007, Strenghtening Efforts to Eradicate Poverty etc., 96-97.

Fao, The State of Food Insecurity in the World, 2008: 12.

Ibid: 14.

Ibid: 15-16.

Ibid: 8.

Note J: John Vidal, Environment Editor for the Guardian, reporting on 26 October 2010, stated that "rising food prices and shortages could cause instability in many countries as the cost of staple foods and vegetables reached their highest levels in two years., with scientists predicting further wide spread droughts and floods... Global wheat and maize prices recently jumped 30 per cent in a few weeks while meat prices are at 20-year high.... There is growing concern among countries about continuing volatility and uncertainty in food markets (as said by the President of the World Bank)...Mounting anger has greeted food price inflation of 21 per cent in Egypt in the last year, along with 17 per cent rise in India and similar amounts in many countries. Prices in the UK have risen 22 per cent in three years... The Govts of Kenya, Uganda, Nigeria, Indonesia, Brazil and the Philippines have all warned of possible food shortages next year, citing floods and droughts in 2010, expected extreme weather next year, and speculation by traders who are buying up food stocks for release when price rises". *Source: The Guardian*, The U.K., 26 October 2010.

Smil, Vaclav, 1994: 263.

Ibid: 262.

Ibid: 264.

Medows *et al.*,1972: 60.

Dyson, Tim, 1996: 117.

Smil, Vaclav, 1994,: 275.

Op cit.
Brown, Lester R *et al.*, 2000: 62-64.
Government of India, *Economic Survey* 2008-09:194.
Brown, Lester R, 2004: 148.
Ibid: 146.
Brown, Lester R, 2006: 164-165.
Dyson, Tim, 1996: 117.
Ibid: 117-118 and 122.
Ibid: 124-125.
Brown, Lester R, 2004: 154.
Brown, Lester R, 2006: 167.
Note K: In dry North China Plain, with 550 million people, two-thirds of the country's cropland producing 25 per cent of grain harvest and one-fifth of water, is experiencing fast fall of its water table where 2.6 million wells were dug by the end of 1997, the bulk of them meant for irrigation. During that year, 99,900 wells were abandoned as they ran dry due to fall of water table while 221,900 new wells were dug, where water table was falling more than 5 feet (1.5 metres) a year. According to World Bank Report (2001),'deep wells around Beijing have to reach 1,000 metres (more than half a mile) to tap fresh water'. Currently, the annual water use in North China Plain amounts to 55 billion cubic metres but the sustainable supply is 35 billion cubic metres, 'leaving an annual deficit of 21 billion cubic metres' to be satisfied from groundwater sources. The depletion of the aquifer providing water now will cut down the basin's water supply by 40 per cent. As the World Bank estimates, with growing population, China's requirement of urban water will increase from present 50 billion cubic metres to 80 billion cubic metres by 2020, while industrial water use is likely to increase from 127 billion cubic metres to 206 billion cubic metres, or 62 per cent which has to be met through more efficient use of water or at the cost of water for irrigation. *Source*: Brown, Lester R, 2004: 43-44.
Government of India, Economic Survey 2008-09: 177.
Brown, Lester R, 2004: 157.
Ibid: 157-158
Ibid: 158.
Note L: "Raising water productivity – the output or benefit derived per liter of water extracted from the natural environment – is critical to remaining within any sustainability boundary established to protect river health. Higher water productivity in the global food system is particularly important. Nearly 70 per cent of all the water withdrawn from freshwater ecosystems goes to irrigated agriculture, yet drip irrigation – which often doubles yield per liter of water compared with conventional irrigation methods – accounts for only 1 per cent of world irrigated area. From California to Central Asia, water-thirsty crops are grown in dry climates, often with inefficient irrigation methods. At the individual level, dietary choices make a large difference on per capita human impact on rivers and aquifers. A typical American diet requires about 1,970 cubic meters of water per year (in the form of evapotranspiration) to produce, while an equally nutritious vegetarian diet takes only 950 cubic meters. A large-scale shift from animal to vegetable protein could save large quantities of water. Indeed, for the

same volume of water consumption, two people could be fed instead of one". *Source*: Sandra Postel and Brian Richter, 2003, *Managing Water for People and Nature*, Island Press, USA.

Agarwal, A and Sunita Narain (Ed), 1999, *Dying Wisdom: Rise, Fall And Potentialsof India's Traditional Harvesting System*, Centre for Science and Development: 314-315.

Ibid: 331 and 335.

Note ML: The earliest evidence of irrigation practices goes back to about 3000 B.C. or there about providing evidences of earliest dams to retain water in large quantities constructed in Jawa (Jordon) and in Wadi Garawi (Egypt) sometime in 2630 B.C., though one cannot be sure if the technology for dam construction emereged first there or elsewhere. According to a water management expert in Germany, ancient dams were not built on the basis of mathematical calculations, but 'they were apparently built ... by using empirical experience and rules, technical skills and an intimate comprehension of hydraulic forces'. The great earliest river valley civilizations of the Nile, the Tigris-Euphrates, the Indus and the Hwang Ho in China flourished on the basis of water management systems. Both Indus Valley civilizations (2600-1500 B.C.) and Mesopotamian civilizations, developed (circa 3500 B.C.) on the joint valleys of the Tigris and the Euphrates, had excellent water supply and sewage disposal systems as prominent features of those cultures. "The standard of their hydro-technical infrastructure was not equalled even 2000 years later by the Romans... As in Mesopotamia, protection against annual flooding of the Indus river, irrigation to secure and increase crop yield and drainage of alluvial areas, were prerequisites to the very existence of kingdoms in the Indus Valley" (Agarwal and Narain, Ed., 1999: 13)). According to Sjoberg (1960), "the large size of these (Indus Valley) cities and the abundance of towns and villages point to a highly productive system of agriculture. The remains of embankments to control riverine floods suggest irrigation of a sort" (Sjoberg, Gideon, 1960: *The Pre-Industrial City: Past and Present*: 41). Further, available evidence indicate that wells as well have come into use for irrigation in the Indus Valley, starting about 2600 B.C.

The oldest known specimens of masonry cisterns for storage of water collecting rainwater from roofs, paved squares and sometimes also from water-bearing subsoil strata were found in Palestine and Greece. Covered cisterns with storage volumes of 75,000 cubic metres were being built by the middle of first century A.D.

Vedic hymns written by the Aryans around 800-600 B.C., appearing in *Rigveda*, make profuse references to 'irrigated agriculture, river courses, dykes, water reservoirs, wells and water lifting structures'. Kautilya, writing his '*Arthashastra*' sometime in 321-297 B.C. refers to 'water harvesting system' in his days consisting of 'setu' for embankment or dam for storing water, parvaha channel, '*Nadinibandhayatana*' for a structure dependent on a river, such as a dam, '*nibadhayatana*' for canals from a river dam, '*khata*' for a well. He refers to two types of '*setu*', such as '*sahodaka*', where there is a natural spring or flow of water, and the '*aharyodaka*', which is a sort of storge tank with water brought into it through channels (Agarwal and Narain, Ed, 1999: 11 and 14). The authors of 'Dying Wisdom, Agarwal and Narain, further refer to the fact that there are numerous inscriptions of different periods giving 'wide range of information about dams, tanks and embankments, their upkeep and management' found 'almost in every part' of India.

Smil, Vaclav, 1994: 273.

Dyson, Tim, 1996,: 119.

Brown, Lester R, 2004: 151.

Brown, Lester R, 2006: 166.

Ibid: 170.

Smil, Vaclav, 1994,: 277.

Ibid: 278.

Brown, Lester R, 2006: 171.

Note N: "The spark of this explosive growth came in 1965 when an enterprising young Indian, Dr. Verghese Kurien, organized the National Dairy Board, an umbrella organization of dairy cooperatives. The dairy coop's principal purpose was to market the milk from tiny herds that typically averaged two to three cows each, providing the link between the growing appetite for dairy products and the millions of village families who had only a small marketable surplus. Catering the market for milk spurred the four-fold growth in output. In a country where protein shortages stunt the growth of so many children, expanding the milk supply from less than half a cup per person a day 30 years ago to more than a cup represents a major advance. What is more remarkable is that India has built the world's largest dairy industry almost entirely on roughage – wheat straw, rice straw, corn stalks, and grass collected from the road side. Even so the value of the milk produced each year now exceeds that of the rice harvest". *Source*: Brown, Lester R., 2006, 174-175.

Brown, 2000: 50.

Brown, 2004: 160.

Smil, Vaclav, 1994: 272.

Ibid: 279-280.

Tim, Dyson 1996: 152-153.

Brown, Lester R:2006: 167-168.

Dyson, Tim, 1996: 158.

Ibid: 159.

Brown, Lester R 2006: 173.

Ibid: 173-174.

Ransom, David, 2010, Seed Savers, *New Internationalist*, September: 5

New Internationalist, September, 2010: 5.

Ransom, 2010: 10.

Ibid: 6.

Ibid: 7.

Op cit.

Branford, Sue, 2010, Merchants of Death, *New Internationalist*, September: 19.

Op cit.

Hardicar, Joydeep, 2010, Crops of Truth, *New Internationalist*, September: 9

Op cit.

New Internationalist, September, 2010: 11'

Op cit.

CHAPTER 6

Richards, John F, 1993, Land Transformation, in '*The Earth as Transformed by Human Action: Global and Regional Changes in the Biosphere over the Past 300 Years*', Ed. by Turner II, B L, William C Clark, Robert W Kates, John F Richards, Jessica T Mathews, and William B Meyer, Cambridge University Press: 163.

Ehrlick, P R, Anne H Ehrlick and G C Daily, 1993, Food Security, Population and Environment, in *Population and Development Review*, 19, Vol I (March 1993): 7.

Note A: UN Special Rapporteur on the right to food, Olivier de Schutter, says a combination of environmental degradation, urbanization and large-scale land acquisition by foreign investors for biofuels is squeezing land suitable for agriculture. "Worldwide, 5 m to 10 m hectares of agricultural land are being lost annually due to severe degradation and another 19.5 m are lost for industrial uses", he says in a new report.

"But the pressure on land resulting from these factors has been boosted in recent years by policies favouring large-scale industrial plantations. According to the World Bank, more than one-third of large-scale land acquisitions are intended to produce biofuels" *Source*: John Vidal, Environment Editor, *The Guardian*, The U.K., 26 October 2010.

Ehrlick, P. R. *et al.*, 1993: 8.

Brown, Lester R, 2006: 150.

Ibid: 151.

L'Vovich, Mark I, G.F. White *et al.*, 1993, Use and Transformation of Water Systems, in 'The Earth as Transformed by Human Action etc, Ed by Turner II B L *et al.*,: 237.

Note B: Water runoff from melting ice and snow in Greenland amounts to 240 km^3/ years, contributions from calving ice are considered to be 237 km^3/year, making total input of water and ice from glacial cover of the arctic region about 477 km^3/year. As for the Antarctic area, nearly all water reaching the sea is in the form of ice and comprises 2,400 km^3/year. Liquid run off in the Antarctic, as calculated by Krenke, accounts for only 15 km^3/year.

Direct underground runoff into oceans by paths other than rivers also is a second imrportant source of fresh water resource... and represents great potentialities. Studies of these elements of the water cycle give the following figures, in km^3/ years: Europe – 167; Asia-328; Africa – 236; North America – 344: South America and Oceania – 385; Australia – 938: World Total - 2,398. *Source*: L'Vovich *et al.*, 1993.

United Nations: Water Development Report, 2003: 12-13.

Brown, Lester R, 2006: 152.

United Nations: World Water Report, 2003: 110-112.

Dyson, Tim, 1996: 150.

Ibid: 151.

Ibid: 154.

Brown, Lester R, 2006: 153.

Ehrlick *et al.*, 1993: 13.

Smil, 1994: 268 and 270.

UNDP: Human Development Report, 2007/2008: 95

Note C: As reported recently by Adam Vaughan in *the Guardian,* quoting a report published in Nature, "the world's rivers are so badly affected by human activity that the water security of almost 5 billion people, and the survival of thousands of aquatic species, are threatened, scientists warned yesterday. The study conducted by institutions across the globe, is the first to simultaneously look at all types of human intervention on fresh water – from dams and reservoirs to irrigation and pollution... Even the world's great rivers, such as the Yangtse, the Nile and the Ganges, are suffering serious biodiversity and water security stress. Despite their size, more than 30 of the largest rivers showed at least moderate threats to water security, due to a range of human impacts such as pollution and irrigation... Globally between 10,000 and 20,000 aquatic wild-life species are at risk or face extinction because of the human degradation of global rivers... The world's least affected rivers, the authors found, were those furthest from populated areas, such as remote parts of the tropics, Siberia and elsewhere in the polar regions". *Source: The Guardian,* The U.K., 30 September 2010.

UNDP 2007/2008: 91.

Smil, 1994: 276.

Note D: In the high valleys of the Andes, hunter-gatherers may have unearthed and consumed one of the plants later domesticated, "the potato, by 8000 B.C., but millennia passed before Andeans first cultivated that tuber. Early farmers domesticated many other wild species round the world, selecting and sowing the seeds of plants with such desirable characteristics as high yield and resistance to pests, climate stress, and diseases. Growers cultivated thousands of different strains, each with its own hereditary material or germ plasm. These traditional varieties are today known as 'land races'. Some growers continue to raise many of the old varieties... in the regions where they originated. These pockets of traditional agriculture thus serve as natural repositories for the diverse genes of land races and their wild forbears. Since, plant breeders now manipulate germ plasm to produce a relatively few 'improved' varieties it has become essential to rediscover and protect the old strains. Their vigour and genetic diversity help to provide insurance for the future of our food supply... When growers (today) abandon old land races to plant modern varieties, often in single-variety monocultures, the potential for widespread disaster is greatly increased. Clear-cutting across virgin lands destroys valuable natural vegetation, and in tropical rain forest – home to half of earth's plant and animal species – some 60 acres a minute are denuded.A quarter of the total falls to the saw-toothed bite of commercial operations.

"The resulting loss of naturally diverse germ plasm – much of it not yet studied – is called genetic erosion. It is estimated that by the middle of the next century (21st), one-quarter of the world's 250,000 plant species may vanish, victims of deforestation, the shift to monocultures, overgrazing, water-control projects and urbanization. (In the Gene Bank at the International Rice Research Institute, Los Banos, the Philippines, 86,000 varieties of rice from all over the world are stored)... Ironically, the loss of genetic diversity accelerated with the green revolution in the 1960s... The miracle seeds (rice and wheat by carefully crossbreeding plants to increase food production... resistant to insects and diseases) were not perfect, however, (against) opportunistic insects and viruses and unlocked genetic resistance of the new seeds... In Sri Lanka, where farmers grew some 2000 traditional varieties of rice as recently as 1959, only five principal varieties are grown today. In India, which once had 30,000 varieties of rice, more than 75 per cent of the total production comes from fewer than ten varieties. The trend toward single-veriety monoculture, the planting of one strain instead of many varieties, leaves modern plant breeders little margin for error, says Garrison Wilkes, a professor of biology at the University of Massachusetts and a leading authority on genetic erosion. The extinction of local land races by the introduction of improved varieties is analogous to removing stones from the foundation to repair the roof. At present rates of extinction, as many as 60,000 plant species – one-fourth of the world´s total – may be lost or engendered within the next 50 years. Meanwhile, there are more mouths than ever to be fed.

"When farmers began harvesting the first domesticated plants aboiut 8000 B.C., the earth's population was around four million. Today that many people are born every ten days. If the trend continues beyond the year 2000 we will have to grow as much food in the first two decades of the new century as was produced over the past 10,000 years.

"Human domesticated wheat about 10,000 years ago from plants whose relatives still survive. Today land races and modern varieties number about 22,000 and belong to two major species: wheat used to make bread and wheat used to make a pasta". *Source*: Robert E Rhoades, 1991, The World's Food Supply at Risk, in *National Geographic*, Vol 179, No 4, April 1991: 74-105.

Ehrlick *et al.*, 1993: 15-16.

Brown, Lester R, 2004: 240.

Brown, Lester R, 2006: 137.

Ibid: 138.

Op cit.

Brown, Lester R, 2004: 243

William, Michael, 1993, Forests, in '*Earth as Transformed by Human Action* etc., Ed by Turner II B L *et al.*: 181.

Ibid: 195-196.

New Scientist, the U.K., 12 May 2007: 44.

Brown, Lster R, 2004: 56.

Ibid: 57.

Brown, Lester R, 2006: 58.59.

Ibid: 143

Ibid: 144.

Brown, Lester R, 2004: 66-67.

Ehrlick *et al.*, 1993: 18

Dyson, Tim, 1994: 137-138.

Intergovernmental Panel of Climate, Fourth Assessment Report, Climate Change 2007: The Physical Science Basis: 2.

Ibid: 3-5.

Ibid: 7.

UNDP: Human Development Report 2007/2008: 100-101.
Ibid: 102.
Op cit.
Op cit.
Brown, lEster R, 2006: 19.
Brown, Lester R, 2004: 131.
Ibid: 133.
Ehrlick *et al.*, 1993: 20.
Balmford, A, A. Bruner, P. Cooper, R. Constanza, S. Farber, R.E. Green, M. Jenkins, P. Jefferiss, V. Jessamy, J. Madden, K. Munro, N. Myers, S. Naeem, J. Paavola, M. Rayment, S. Rosendo, J. Roughgarden, K. Trumper and R.K. Turner, 2002, Economic Reasons for Conserving Wild Nature, in 'Science Magazine, Americal Association for the Advancement of Science', 5583, 2002: 950.
Ibid: 951.
Brown, Lester R, 2004: 213.
Daniels, Peter *et al.*, 2005 (Second Ed.).: 88-89.
Ibid: 89.
Ibid: 91.
Ibid: 98-99.
Brown, Lester R, 2006: 125.
Op cit.
Ibid: 126-127.
Note E: (a) Japan feels highly alarmed at experiencing the biggest fall in population because of low birth rates, with the average number of children born to a woman aged between 15 and 49 with record low of 1.36 in 2005. *Source: The Guardian*, The U.K., 15 March 2007 and 2 January 2009.
(b) In China, the Prime Minister reaffirmed the policy of one child per couples being continued for over three decades as "lower birth rate equals faster prosperity", pledging the provision of "gynecological examination and subsidise child birth for rural women" and "will continue to reward and support eligible rural families that comply with family planning regulations". *Source "The Statesman*, Kolkata Issue, 6 March 2010.
(c) Britain having experienced a sharp rise in infertility, the TFR (Total Fertility Rate) rising from 1.683 children per woman in 2001 to 1.96 in 2006 (the highest being 2.8 in Boston, Lincolnshire, called nation's fertility capital) has started seriously thinking about setting the limit of two children per family. At the current rate of fertility (1.96), the population in Britain is growing by 0.7 per cent every year, more than double the rate in the 1990s and three times the level of 1980s... *Source: The Guardian*, The U.K., 11 July 2007 and 28 August 2009.
Brown: 2006: 130.
Ibid : 129.
Brown, 2004: 224.
Daniels *et al.*, 2005: 103.
Puri, Lakshmi, 2008, Assuring Development Gains and Poverty Reduction from Trade: The Labour Mobility and Skills Trade Dimension, United Nations: 9.
Ibid: 12.
Ibid: 25-26.

Ibid: 32-33.
Ibid: 44.
Ibid: 47-48.
Ibid: 49.
UNDP: Human Development Report, 2005: 116.
Op cit
Ibid: 123.
Ibid: 126-127.
Kapstein, E B, 2006, *Economic Justice in an Unfair World: Toward A Level Playing Field,* Princeton University Press: 177.
Ibid: 183.
Ibid: 187.
Daniels *et al.*, 2005: 302.
Ibid: 307.
Note F: John Vidal, the Environment Editor of *the Guardian*, basing on papers written by scientists from many disciplines and published by the Royal Society (August 2010), reports that according to the academic assessment of future global food supplies, even new technologies such as genetic modification and nanotechnology may not be adequate in feeding the world population, leaving hundreds of millions hungry. However, it is pointed out that though the scope of further expansion of agricultural land is limited, an increase of as much as 70 per cent in food supply in next 40 years is not "insurmountable", and this may be derived from the presence of extra carbon in the air from global warming in combination with better fertilizers and chemicals to protect crops with concurrent reduction in water consumption. Some scientists, nevertheless, have warned about greater need of water, twice as much by 2050, for producing food to feed extra 3 billion people by then, meaning an "18 per cent reduction in water availability across the world creating serious problems in countries like northern Africa, India, China and parts of Europe and the western US". Reduction of the 30 to 40 per cent food-waste taking place in both rich and poor countries, better storage facilities and general consumption kept at level what is necessary are likely to largely improve the prospect of food availability. At the same time, much of the anticipated doubling of demand for dairy and meat products are expected to be met through conventional animal breeding and, as one scientist believes, artificial meat made in giant vats and nanotechnology delivering medication to livestock meeting food-challenge in future. *Source: The Guardian*, The U.K., 16 August 2010.

CHAPTER 7

UNDP: Human Development Report, 2003: 1.
Ibid: 5.
Meadows *et al.*, 1972: 58.
Brown, Lester R., *et al.*, 2000: 64.
FAO, The State of Food and Agriculture 2006: 96.
FAO, 2008: 9.

Note A: According to a very recent report of the United Nations,"world food prices have hit record levels and are now in danger territory", as "unpredictable weather, changing fast due to global warming as experienced by unprecedented rainfall in Argentina and floods in north-east Australia having record rainfall for 150 years, equally heavy floods forcing more than 300,000 people out of their homes in Sri Lanka, torrential rains with accompanying mudslides claiming at least 350 lives in Brazil making it "one of the worst disasters in Brazilian history", causing similar flood havoc in the Philippines, along with cold weather in the northern hemisphere when Britain had its chilliest December in 100 years with an average temperature of -10 degrees could cause further price rise this year". With 50 centimetres of snow recorded in New York's Central Park, making it one of the largest snowfall in the city's history, 2010 has become the 'year of floods, freak snowstorms and heatwaves, the warmest on record, competing with 2005 as the warmest year on record since it began in 1880, with world temperature 0.62 C (1.12 degrees F) above the 20th century average of 13.9 C (57. degrees F). The previous high in food prices in June 2008 (that sparked food riots leaving four dead in Haiti and costing 40 lives in Cameroon) was surpassed in December (2010), "to reach its highest since records began in 1990... Last year European wheat prices doubled, US corn prices rose more than 50 per cent, and US soybean prices jumped more than 30 per cent... Another rise in the oil price – currently $95 a barrel – could exacerbate the problem". *Source*: Jill Treanor, *The Guardian*, The U.K., 6 January 2011.

In another report made by the Reuters, "fresh riots broke out in Algiers yesterday (7.1.2011) ... after protests over food prices and unemployment... The cost of flour and salad oil has doubled in recent months, reaching record highs. A kilogram of sugar, which a few months ago cost 70 dinars, is now 150 dinars. (1.28 pounds) ". *Source: Guardian*, The U.K., 8 January 2011.

Brown, Lester R,. 2006: 171.

Brown, Lester R, 2004: 52-53.

Ibid: 55.

Brown, Lester R, 2006: 173.

International Food Policy Research Institute (IFPRI) 2002, Reaching Sustainabile Food Security for All by 2020: 4.

UNDP; Human Development Report, 2003: 87.

FAO: The State of Food Insecurity in the World 2008: 6-8.

Ibid: 10.

UNDP: Human Development Report, 2003: 87.

IFPRI 2002: 3.

UNDP: Human Development Report, 2003: 97.

Brown, Lester, 2006: 135.

Note B: According to a study by WHO (2010), the number of girls and boys smoking is about equal in half in the 151 countries surveyed though the number of girls smoking is higher than boys in some countries, such as in Bulgaria, Chile, Colombia, Cook Islands, Croatia, Czech Republic, Mexico, New Zealand, Nigeria and Uruguay. According to this report,'fewer than 1 in 10 women smoke, making it 200 million women around the world. It further noted that 15 million women die of smoking-related causes across the world every year and the toll could rise to 25 million women by 2030. *Source*: Anjali Sharma, *The Statesman*, Kolkata Issue, 8 june 2010.

Brown, Lester R, 2006, 136.

United Nations, World Water Development Report, 2003: 274-275.

Ibid: 275.

Brown, Lester R, 2006: 167.

UNDP: Human Development Report 2003: 104-105.

Bibliography

Agarwal, A and Sunita Narain (Ed), 1999, Dying Wisdom, Rise, Fall and Potentials of India's Traditional Harvesting System, Centre for Science and Development, New Delhi.

Agarwal, Bina, 1995, Gender and Legal Rights in Agricultural Land in India, in *'Economic and Political Biweekly'*, March 24, 1995.

Ahmad, Ehtishan and Arthar Hussain, 1999, Social Security in China: A Historical Perspective, in *'Social Security in Developing Countries'*, Ed. by E. Ahmad, J. Dreze, J.

John Hills and Amanya Sen, Published by United Nations University in 1991: first published as Paperback in India by Oxford University Press.

Bajpai, Asha, 2009 (Third Imp), *Child Rights in India: Law, Policy and Practice*, Oxford University Press.

Balmford, A., A. Bruner, P. Cooper, R. Constanza, S. Farber, R.E. Green, M. Jenkins, P. Jefferies, V. Jessamy, J. Madden, K. Munro, N. Myers, S. Naeem, J. Paavola, M. Rayment, S. Rosendo, J. Roughgarden, K. Trumper and R.K. Turner, 2002, *Economic Reasons for Conserving Wild Nature, in 'Science Magazine*, American Association for the Advancement of Science.

Basu, Kaushik (Ed), 2006, (Third Imp), *India's Emerging Economy: Performances and Prospects in 1990s and Beyond*, Oxford University Press.

Bhagwati, Jagadish, 2004, *In Defense of Globalization*, Oxford University Press

Bhopal, R., 2002, *Concept of Epidemiology: An Integrated Approach to the Ideas, Theories and Method of Epidemiology*, Oxford University Press.

Bosley, Sarah, Health Editor, *The Guradian*, the U.K,3 and 10 December 2008.

Brandt, W., 1990, Brandt Report: North-South, A Programme for Survival, The Report of the Independent Commission on International Development Issues, Sydney

Braun, Joachim Von, 1999, Social Security in Sub-Saharan Africa: Reflections on Policy Challenges, in *'Social Security in Developing Countries'*, Ed by E. Ahmad *et al.*

Brown, Lester R, Gary Gardner and Brian Halweil, 2000, *Beyond Malthus: Nineteen Dimensions of the Population Challenge*, The Worldwatch Environmental Alert Series, Earthscan Publications Ltd., London.

Brown, Lester R, 2004 (Reprint), *Eco-Economy: Building an Economy for the Earth*, Orient Longman Private Limited.

— —, 2006, Plan B 20: Rescuing A Planet Under Stress and a Civilization in Trouble, Earth Policy Institute.

Bryson, Bill, 2008, *Shakespeare*, Harper Perennial Edition.

Burra, Neera, 2006, Born to Work, in '*Born Unfree: Child Labour, Education and the State in India*', Ed. by Myron Weiner, Neera Burra and Asha Bajpai, Oxford University Press.

Butt, Riazat, 2008, *The Guardian*, The U.K., 1 November 2008

Caldwell, John C, 1986, Routes to Low Mortality in Poor Countries, in '*Population and Development Review*', Vol 12, No 2 (June 1986)

Castro, J De, 1955 (Third Imp), *Geography of Hunger*, Victor Gollanez Ltd., London.

Chatterjee, Shoma A, 2009, Trafficking by Social Sanction, *The Sunday Magazine, The Statesman*, Kolkata Issue, 8 March 2009.

Census of India, 2001, Provisional Population Totals, Statement I.

Clark, Helen and Elizabeth Carnegie, 2006, *"She Was Aye Workin": Memoirs of Tenement Women in Edinburgh and Glasgow*, White Cockade Publishing, Oxford.

Coale, Ansley J, 1991, Excess Female Mortality and the Balance in the Population: An Estimate of the Number of "Missing Women", in '*Population and Development Review*', Vol 17, No 3.

Collier, Paul, 2008, *The Bottom Billion: Why the Poorest Countries are Failing and What Can be Done About* It ?, Oxford University Press.

Collins, Daryl, Jonathan Morduch, Stuart Rutherford and Orlanda Ruthven, 2009, *Portfolios of the Poor*, Priceton University Press (first published in India in 2010 by Permanent Black, Ranikhet).

CPAG (Child Poverty Action Group, The U.K.), 2004, *Poverty*, The Facts, London.

Daniels, Peter, Michael Bradshaw, Denis Shaw and James Sidaway, 2005 (Second Ed), *An Introduction to Human Geography: Issues of the 21st Century, Prentice* Hall, England.

Datta, Gaurav and Martin Ravallion, 2002, Is India's Economic Growth Leaving the Poor Behind ?, in '*Journal of Economic Perspectives*', Vol 16, No 2, 2002.

Deaton, Angus and Jean Dreze, 2002, Poverty and Inequality in India: A Re-Examination, in '*Economic and Political Weekly*', September 7, 2002.

Deepa, Narayan, Robert Chambers, Meera K Shah and Patti Petesch, 2000, *Voices of the Poor: Crying Out For Change*, Oxford University Press.

Demograhic Year Book, 2007.

Dicken, Peter, 2003 (Fourth Ed), *Global Shift: Reshaping the Global Economic Map in the 21st Century*, Sage Publications.

——, 2007, *Global Shift: Mapping the Changing Contours of the World Economy*, Sage Publications.

Dreze, J. and Amartya Sen, 1989, *Hunger and Public Action*, Clarendon Press, London.

——, 1990, Public Action for Social Security: Foundations and Strategy, in '*Social Security in Developing Countries*, Ed by E. Ahmad, J. Dreze, J. Hills and A Sen, Published by United Nations University in.1991; first published as Paperback in India by Oxford University Press.

Dyson, Tim, 1996, *Population and Food: Global Trends and Future Prospects*, Routledge.

Eastwood, Robert and Michael Lipton, 1999, The Impact of Changes in Human Fertility on Poverty, in '*Journal of Development Studies*, 36, 1:4.

Ehrlick, P R, Anne H Ehrlick and G C Daily, 1993, Food Security, Population and Environment, in '*Population and Development Review*', Vol I (March 1993)

Elizabeth, V.S., 2009, 'Violence Against Women in the Family: A Human Rights Violation ? in *Family Violence in India: Human Rights, Issues, Actions and International Comparisons*', Ed. by Swati Shirwadkar, Rawat Publications., Jaipur, India.

FAO, 2006, *The State of Food Insecurity in the World*, Food and Agriculture Organization of the United Nations, Rome.

—— *The State of Food Insecurity in the World*, 2008.

Flegg, A.T., 1982, Inequality, Illiteracy and Medical Care as Determinants of Infant Mortality in Underdeveloped Countries, in '*Population Studies*', Vol 36, No.3.

Gatrell, Anthony C, 2002, *Geographies of Health, An Introduction*, Blackwell.

Government of India, 2002, Planning Commission, National Human Development Report, Oxford University Press.

——, 2003, Tenth Five Year Plan 2002-2007, Vol I

——, Ministry of Finance, Economic Survey 2007-08, Oxford University Press.

——, 2008, Planning Commission, Report on Conditions of Work and Promotion of Livelihood in the Unorganised Sector.

——, Planning Commission, 2008, Eleventh Five Year Plan, 2007-2012, Vols I, II and III.

Greenwood, Brian and Theonset Mutabingwa, 2002, Malaria, in '*Nature*', 415 and February 2002.

Griffin, Keith, 1978, *International Inequality and National Poverty*, The Macmillan Press Ltd.

Hamm, B and Pandurang K Muttagi. 2001, *Suatainable Development and the Future of Cities*, ITDG Publishing.

Hawthorn, Geoffrey (Ed), 1998, *The Standard of Living*, Cambridge University Press.

Hulme, David and Paul Mosley, 1996, *Finance Against Poverty*, Routledge, London and New York.

Hussain, Arthar, 2002, Urban Poverty in China: Incidence of Policy Responses, in '*World Poverty: New Policies to Defeat an Old Enemy*, Ed by Peter Townsend and David Gordon, The Policy Press, Bristol, U.K.

Intergovernmental Panel of Climate, Fourth Assessment Report, Climate Change.

International Food Policy Research Institute (IFPRI), 2002, Reaching Sustainable Food Security for All by 2020.

International Organization For Migration (IOM), Migration, Development and Poverty Reduction in Asia, Academic Foundation, New Delhi (India).

Islam, Iyanatul, 2006, Indonesia: Poverty, Employment and Wages, in '*Fighting Poverty: The Development – Employment Link*', Ed by Rizwanul Islam, Viva Books Private Lrd.

Islam, Rizwanul, 2006, The Nexus of Economic Growth, Employment and Poverty Reduction: An Empirical Analysis, in '*Fighting Poverty: The Development – Employment Link*', Ed by Rizwanul Islam, Viva Books Private Ltd.

Johnson, D.G., 1998, Economic Reforms in the People's Republic of China, in '*Economic Development and Cultural Chnge*', April Issue.

Kanjee, Nazneen, 2004, Social Funds in Sub-Saharan Africa: How Effective for Poverty Reduction?, in '*World Poverty* etc', Ed by Peter Townsend and David

Gordon.

Kapstein, E.B., 2006, *Economic Justice in an Unfair World: Toward A Level Playing Field,* Princeton University Press.

Keliman, Nico, 2001, Uncertain Population Forecasts, in '*Nature*', Vol 412, 2 August 2001.

Khan, Azizur Rahman, 2006, Employment Policies for Poverty Reduction, in '*Fighting Poverty: The Development – Employment Link*', Ed by Rizwanul Islam, Viva Books Private Ltd.

Klassen, Stephen, 1994, "Missing Women" Reconsidered, in '*World Development*', Vol 22, No 7.

Klassen, Stephen and Claudia Wink, 2009, Gender Bias in Mortality? An Update on Missing Women, in '*Gender Discrimination: Health, Nutritional Status and Role of Women in India*', Ed by M. Pal, P. Bharati, B. Ghosh and T Vasula, Oxford University Press.

Krishnaraj, M, Ratna M S and Abusaleh Shariff, 1998, *Gender, Population and Development,* Oxford Univewrsity Press.

Kumar, Sunil, 2004, Round Pegs and Square Holes: Mismatches between Poverty and Housing Policy in Urban India, in '*World Poverty: New Policies to Defeat an Old Enemy* etc', Ed by Peter Townsend and David Gorden, The Policy Press, University of Bristol.

Kumari, Ranjana, 2009, Globalizing India and Domestic Violence Against Women, in '*Family Violence in India* etc', Ed by Swati Shirwaskar.

Lin, J Y, 1988, The Household Responsibility System in China's Agricultural Reform: A Theoretical and Empirical Study, in '*Economic Development and Cultural Change*', April, 1988.

L'Vovich, Mark I and Gilbert F White, 1990, Use and Transformation of Terrestrial Water Systems, in '*The Earth as Transformed by Human Action: Global and Regional Changes in the Biosphere over the Past 300 Years*', Ed by B L Turner II, William C Clark, Robert W Kates, John F Richards, Jessica T Mathews and William B Meyer, Cambridge University Press.

Majumdar, R C, H C RoyChaudhuri and Kalikinkar Datta, 1970 (Reprinted), An *Advanced History of India,* MacMillan St. Martin's Press.

Mandle, J.R., 2003, *Globalization and the Poor,* Cambridge University Press.

Mathur, Kanchan, 2009, Witchcraft, Witches and Social Sanction, in '*Family Violence in India*', Ed by Swati Shirwadkar, Rawat Publications, India.

McCully, Patrick, 1998, *Silenced Rivers: The Ecology of Politics of Large Dams,* Orient Longman.

Meadows, D.H., Denis L Meadows, Jorgen Renders and William W Behrnes III, 1972, *The Limits to Growth: A Report for the Club of Rome's Project on the Predicament of Mankind,* A Signet Book from New American Library.

Mesa-Lago, Carmelo, 1999, Social Security in Latin America and the Caribbean: A Comparative Assessment, in '*Social Security in Developing Countries*', Ed By E Ahmad *et al.*

Millar, B, 2001, *Gender Discrimination in Intrahousehold Food Allocation in South Asia,* Paper presented at the meetings of the American Anthropological Association, San Fransisco.

Misra, Rajib, Rachel Chatterjee and Sujatha Rao, 2003, *India Health Report*, Oxford University Press.

Mukhopadhyay S. (Ed), 2001, *Women's Health, Public Policy and Community Action*, Manohar (India).

Mukhopadhyay and Garimella, 2001, The Contour of Reproductive Choice for Women: Findings from Micro-Surveys in '*Women's Health, Public Policy and Community Action*', Ed by S. Mukhopadhyay, Manohar, New Delhi.

Myeza, Nokujabula and Dennis, Francis, 2009, 'Openly Positive': Living and Working with HIV/AIDS, in '*Working Women: Stories of Strife, Struggle and Survival*', Ed by Kogji Naidoo and Fay Patel, Sage Publications.

Myrdal, Gunnar, 1963 (Reprinted), *Economic Theory and Underdeveloped Regions*, University Paper Backs, Methuen.

New Scientist, The United Kingdom, 12 May 2007.

Orshansky, M. 1965, Counting the Poor: Another Look at the Poverty Profile, in '*Economies of Poverty and Inequality*', An Elgar Reference Collection, Vol III,2008.

Orzem, Peter F, 2007, Lack of Education, in '*Solutions for the World's Biggest Problems: Costs and Benefits*', Ed by Bjorn Lomborg, Cambridge University Press.

Osmani, S.R., 1999, Social Security in South Asia, in '*Social Security in Developing Countries*', Ed by E Ahmad et al.

Pande, Rekha, 2008, Ritualized Prostitution: Devadasis and Jogins – A Few Case Studies, in '*Prostitution and Beyond: An Analysis of Sex Work in India*', Ed by Rohini Sahni, V. Kalyan Shankar and Hemant Apte, Sage Publications.

Peto, R., A.D. Lopez, J. Boreham, M. Thun and C Health, 1994, *Mortality from Smoking in Developed Countries* 1950-2000, New York, Oxford University Press.

Phansalkar – Joshi, S.S., 2009, Child Sexual Abuse: A Violation of Human Rights?, in '*Family Violence in India: Human Rights, Issues, Actions* etc.', Ed by Swati Shirwadkar etc.

Psacharopoulas, George and Zafiris Tzannatos, 1991, Female Labour Force Participation and Education, in '*Essays on Poverty and Growth*', Ed by George Psacharpoulas, Pregmanon Press for the World Bank.

Puri, Lakshmi, 2008, *Assuring Development Gains and Poverty Reduction from Trade: The Labour Mobility and Skills Trade Dimension*, United Nations.

Rahaman, R. Islam and K.M. Naibul Islam, 2006, Bangladesh: Linkages among Economic Growth Employment and Poverty, in '*Fighting Poverty: The Development-Employment Link* etc', Ed by Rizwanul Islam, Viva Books etc.

Ravallion, Mastain 1996, Issues in Measuring and Modelling Poverty, in 'The Economics of Poverty and Inequality, Ed. by Frank A Cowell etc.

Renderson, James, Science Correspondence, *The Guardian*, The U.K., 8 December 2008.

Richards, John F, 1993, Land Transformation, in '*The Earth as Transformed by Human Action: Global and Regional Changes in the Biosphere over the Past 300 years*', Ed by Turner II, B L William, C Clark etc, Cambridge University Press.

Sachs, Jeffrey, 2005, *The End of Poverty: How Can We Make It Happen in Our Life Time*, Penguin Books.

Sachs, Jeffrey and Pia Malaney, 2002, The Economic and Social Burden of Malaria, in '*Nature*', 415., February 2002.

Sarkaria, Ajay, 1997, Women, Witchcraft and Gratuitous Violence in Colonial

Western India, in '*Past and Present*'. May Issue.

Sayeed, A.T., 2009, Victims of Earthquake and Patriarchy: The 2005 Pakistan Earthquake, in '*Women, Gender and Disaster*', by Elaine Emerson and P.G. Dhar Chakraborti, Sage Publications.

Sekar, Helen, 2001, *Ensuring Their Childhood,* V.V. Giri National Labour Institute, New Delhi.

Sen, Amartya, 1976, Poverty: An Original Approach to Mesurement, in '*Econometrica*', Vol 44, No 2, March 1976.

— —, 1982, *Poverty and Famines,* Clarendon Press, Oxford

— —, 1992, '*Missing Women*', in 'British Medical Journal, Vol 304.

— —, 1995 (5th Imp), *Inequality Reexamined,* Oxford University Press.

— —, 2008 (15th Imp), *Development as Freedom,* New York, Knoff.

— —, 2001, Many Faces of Gender Inequality, in '*Frontline*', Vol 18, Issues 22,27

Sinha, Indrani and Sharmila Dasgupta, 2009, *Mothers For Sale: Women in Kolkata's Sex Trade,* Dasgupta-Alliance, Kolkata.

Smil, Vaclav, 1994, How Many People Can the Earth Feed? in '*Population and Development Review*', Vol 20, June 1994.

Smith, L.C. and Elizabeth N. Byron, 2009, Will Increasing Women's Power Reduce Discrimination Against Girl Children in Asia? In '*Gender Discrimination: Health, Nutritional Status and Role of Women in India*', Ed by M Pal, P Bharati, B Ghosh and T S Vasulu, Oxford University Press.

Sudha, S. and S. Irudaya Rajan, 2001, Female Demographic Disadvanatage in India 1981-1991: Sex Selective Abortions and Infanticide, in '*Gendered Poverty and Well-being*', Ed by Shahra Razvi, Blackwell, Oxford.

Taylor, V, 1998, *HIV/AIDS and Human Development,* Pretoris, Amabhuka Publication.

The World Bank Development Report 2000/2001, Attacking Poverty, Oxford University Press

UNDP, Human Development Report, 1998,.

UNDP, Human Development Report, 1999,

— —, Human Development Report, 2000, Human Rights and Human Development.

— —, Human Development Report, 2001

— —, Human Development Report, 2002, Deepening Democracy in a Fragmented World.

— —, Human Development Report, 2003, Millennium Development Goals: A Compact Among Nations to End Poverty.

— —, Human Development Report, 2005, International Cooperation at a Crossroads: Aid, Trade and Security in an Unequal World.

— —, Human Development Report, 2006, Beyond Scarcity: Power, Poverty and the Global Water Crisis.

— —, Human Development Report, 2007/2008, Fighting Climate Change: Human Solidarity in a Divided World

UNICEF, 2007, The State of the world's Children.

— —, 2008, The State of the World's Children: Child Survival.

UNIFEM,2008/2009, Progress of the World's Women: Who Answers to Women? Gender and Accountability.

United Nations, 1997,

— —, 1999, Health and Mortality: Issues of Global Concern.

——, 2003, Water for People, Water for Life, World Water Development Report.

——, 2006, World Population Prospects, The 2006 Revision, Sex and Age Distribution of the World, Vol II.

——, 2007, World Economic Situation and Prospects.

——, 2007, Strengthening Efforts to Eradicate Poverty and Hunger: Dialogues at the Economic and Social Contact.

United Nations Millennium Project: Task Force on Education and Gender Equality, 2005, Earthscan

Vandermoortele, Jan, 2004 (Reprint), Are We Really Reducing Poverty ?, in '*World Poverty, New Policies to Defeat an Old Enemy*', Ed by Peter Townsend and David Gordon, The Policy Press.

Viswanathan, Brinda, 2007, Living Conditions of Women, in '*Solutions For the World's Biggest Problems, Costs and Benefits*', Ed by Bjorn Lomborg, Cambridge University Press.

Watts, Harold, 1968, An Economic Definition of Poverty, in '*The Economies of Poverty and Inequality, An Elgar Reference Collection*, Ed by Frank A Cowell, 2008, Vol II.

Weiner, Myron, 2006, The Child and the State in India, in '*Born Unfree: Child Labour, Education and the State in India*', Ed by Myron Weiner, Neera Burra and Asha Bajpai, Oxford University Press.

WHO, 2005, Preventing Chronic Diseases: A Vital Investment 2008, The World Health Report.

William, Michael, 1993, Forests, in '*The Earth as Transformed by Human Action* etc', Ed by B L Turner II et al., Cambridge University Press.

Wolf, Martin, 2005, *Why Civilization Works*, Yale University Press.,

World Commission on Environment and Development, 1987, Oxford University Press.

Yao, Sujie, 1999, Economic Growth, Income Inequality and Poverty in China under Economic Reforms, '*Journal of Development Studies*', 35.

Yunus, Muhammad, 1998, *Banker to the Poor*, The University Press Limited, London.

Index